XX Winter School on Continuous Media Mechanics 2017

IOP Conference Series: Materials Science and Engineering
Volume 208

Perm, Russia
13 – 16 February 2017

ISBN: 978-1-5108-4223-6
ISSN: 1757-8981

Printed from e-media with permission by:

Curran Associates, Inc.
57 Morehouse Lane
Red Hook, NY 12571

Printed by Curran Associates, Inc. (2017)

For permission requests, please contact the Institute of Physics at the address below.

Institute of Physics
Dirac House, Temple Back
Bristol BS1 6BE UK

Phone: 44 1 17 929 7481
Fax: 44 1 17 920 0979

techtracking@iop.org

Additional copies of this publication are available from:

Curran Associates, Inc.
57 Morehouse Lane
Red Hook, NY 12571 USA
Phone: 845-758-0400
Fax: 845-758-2633
Email: curran@proceedings.com
Web: www.proceedings.com

Table of contents

Volume 208

Winter School on Continuous Media Mechanics13–16 February 2017, Perm, Russian Federation

Accepted papers received: 1 June 2017 Published online: 8 June 2017

Preface

Papers

IOP Conf. Series: Materials Science and Engineering **208** (2017) 011001 doi:10.1088/1757-899X/208/1/011001

PREFACE

Winter Schools on Continuum Mechanics are held biennially by the Institute of Continuous Media Mechanics of the Ural Branch of the Russian Academy of Sciences beginning with 2017. The papers in this issue of the journal were selected on the basis of their bearing on the materials of the XX Winter School held in 2017.

About 350 people from 30 cities of 7 countries (Russia, Austria, Germany, Poland, Israel, India, Japan) took part in the work of the XX-th Winter School on Continuum Mechanics. Russian participants represented 47 organizations in 23 cities (Perm, St.-Petersburg, Moscow, Yekaterinburg, Krasnoyarsk, Novosibirsk, Kazan, Chelyabinsk, Samara Rostov-on-Don, etc.)

The scope of the problems addressed at the Conference is extremely wide and includes the following topics:

- computational continuum mechanics
- coupled problems of solid mechanics
- physics and mechanics of meso- and nano-structured systems
- convection, hydrodynamic stability and turbulence
- hydrodynamics of multiphase media
- hydrodynamics of non-Newtonian liquids and liquids with special characteristics
- interdisciplinary studies

All these topics are among the central themes of modern continuum mechanics, as evidenced by the increasing number of publications in the leading Russian and foreign journals

During the panel sessions a great number of papers on the fundamental and applied problems of continuum mechanics were presented. Of particular interest were the invited papers of the well-known specialists in different areas of continuum mechanics (full members of RAS, corresponding members of RAS, doctors of sciences), and the presentations of representatives of industries manufacturing hi-tech products.

Chairman of the
XX Winter School
On Continuum Mechanics academician of RAS V.P. Matveenko

Peer review statement

All papers published in this volume of *IOP Conference Series: Materials Science and Engineering* have been peer reviewed through processes administered by the proceedings Editors. Reviews were conducted by expert referees to the professional and scientific standards expected of a proceedings journal published by IOP Publishing.

IOP Conf. Series: Materials Science and Engineering **208** (2017) 012001 doi:10.1088/1757-899X/208/1/012001

Jet formation at interaction of a vibrating plate with liquid

V A Aleksandrov, S P Kopysov, L Y Tonkov

Institute of Mechanics, Ural Branch of the Russian Academy of Sciences
34 ul. T. Baramzinoy, Izhevsk, Russia 426067

E-mail: ava@udman.ru

Abstract. In this work, we experimentally investigate the mechanism of jet formation on liquid surface near the edge of a vibrating plate that is partially immersed in the liquid. Under the conditions of resonant bending vibrations, the vibrating plate excites capillary waves in the form of Faraday ripples on the surface of the liquid layer wetting the plate surface. Two-dimensional capillary waves also appear on the curved surface of the liquid near the edges of the vibrating plate. The vibrations of the plate area generate hydrodynamic pressure on the liquid surface, which initiates surface eddy flows. At a certain position of the vibrating plate in the liquid, capillary oscillations in the form of standing waves appear along the boundary of the wetting layer of the liquid, directly under the free edge of the plate. The vibrations of the plate edge modulate the standing waves in the transverse direction, which results in the periodic variation of the surface curvature of the wetting liquid layer, from negative to positive values. The inertia forces, periodically varying with the frequency of the plate vibrations, combined with the Laplace pressure, in the excited standing capillary waves on the surface flow under the plate edge, initiates the periodic ejection of particles of the liquid forming a jet.

1. Introduction

In our previous work, we experimented with a bended rod with a wedge-shaped end surface and a bended plate with free edges, which are partially immersed in liquid at an angle. We showed that the high-frequency vibrations excited by a piezoelectric transducer can atomize the liquid at the wetting layer in the areas with the antinode of vibrations and therefore generate a liquid jet from under the end surface of the rod and the plate edge [1]. In the case of a rectangular plate fixed to the source of vibrations by one edge, a jet can be generated by the areas with the antinode of bending vibrations on the lateral edges of the plate. The conditions to generate a steady jet are the following: the large amplitude of vibrations of a rod or plate (i), and the retention of a certain position of the rod end and the plate edge relative to the free surface of the liquid (ii). In [1], we suggested the mechanism of generation of a liquid jet by a vibrating rod. It can be a high-frequency cumulative effect associated with periodic entry of the flat area of the rod end surface into the liquid at an angle. In this case, the amplitude of the velocity of the liquid wetting layer free surface should be larger than the amplitude of the vibrational speed of the rod surface.

The investigation of plate vibrations in liquid demonstrated that, when a jet is generated by the plate and a liquid wetting layer is present on other areas of the plate surface, with the antinode of bending vibrations, there will be standing capillary waves on the wetting layer

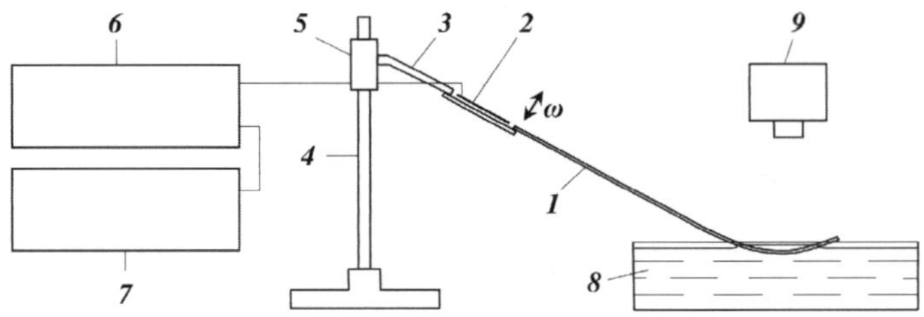

Figure 1. The schematic view of the experiment: 1 — plate; 2 — piezoelectric transducer; 3 — holder; 4 — guide; 5 — shaft; 6 — signal amplifier; 7 — audio-frequency generator; 8 — liquid; 9 — digital camera.

surface, in the form of Faraday ripples [2]. A jet has not been generated when the free edge of the plate is completely immersed in the liquid or there is no liquid wetting layer on the plate edge. The behavior of the liquid near the free edge of the partially immersed vibrating plate was out of the scope of that work.

Bending vibrations of a plate are normal vibrations with distributed amplitude. The vibrations of the plate surface are maximal in the antinodes and they are absent in the nodes of vibrations. The excitation of the bending vibrations of the plate partially immersed in liquid leads to the vibrations of the liquid layer on the interphase boundary. Depending on the geometry of the interaction of the plate and liquid, the thickness and area of the free surface of the layer wetting the plate surface can vary. The liquid layer on the area of the vibrating plate surface is the volume of liquid bounded by the free surface and the surface of the contact with the plate. The appearance of Faraday ripples on the surface of the liquid layer is due to the parametric excitation of capillary waves, which is described in [3] in more detail.

The production of sprays from a vibrating liquid film is used in ultrasonic atomization [4, 5, 6]. In related works, the liquid contact has a solid support. The goal of our work is to investigate the mechanism of jet formation at the interaction of liquid and a vibrating deformable plate excited by a piezoelectric transducer.

2. Experimental devices and method

In the experimental facility, the schematic view of which is shown in Figure 1, rectangular plate 1 was made from transparent polyethylene terephthalate (PET). One edge of the plate is glued to one side of piezoelectric disk transducer 2 which is the source of bending vibrations. The other side of the piezoelectric transducer is soldered to metal holder 3 fixed to the element containing vertically positioned shaft 4 and cylindrical guide 5 with a slide, which allows the guide position adjustment with a screw.

The movement of the plate with the piezoelectric transducer together with the guide was monitored by a micrometer. The plate was bent nearby the free edge. For the excitation of bending vibrations of the plate, on the piezoelectric transducer electrodes the alternating voltage with amplitude about 20 V was supplied by amplifier 6 of signals generated by audio-frequency generator 7 GZ-35. The plate was brought close to the surface of liquid 8 so that the concave area was wetted from above and from below, and at the plate partial immersion in the liquid the free edge of the plate was above the liquid surface. Hydrodynamic processes on the liquid surface nearby the plate and on the surface of the liquid layer wetting the plate were recorded by digital camera 9 Canon EOS 650D or microscope camera OMAX A35140U3. After the excitation of the plate bending vibrations, the frequency of the signals of the generator was adjust to the magnitude at which the formation of a jet outgoing from under the plate edge

IOP Conf. Series: Materials Science and Engineering **208** (2017) 012001 doi:10.1088/1757-899X/208/1/012001

Figure 2. The generation of an alcohol jet by the vibrating PET plate excited by the piezoelectric transducer.

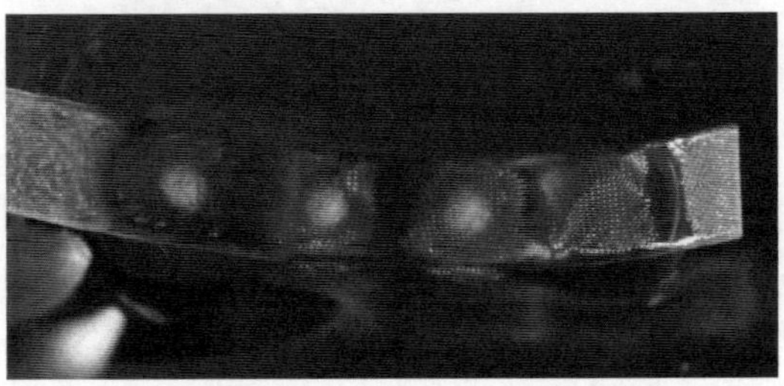

Figure 3. Faraday ripples, cavities with air bubbles under the plate.

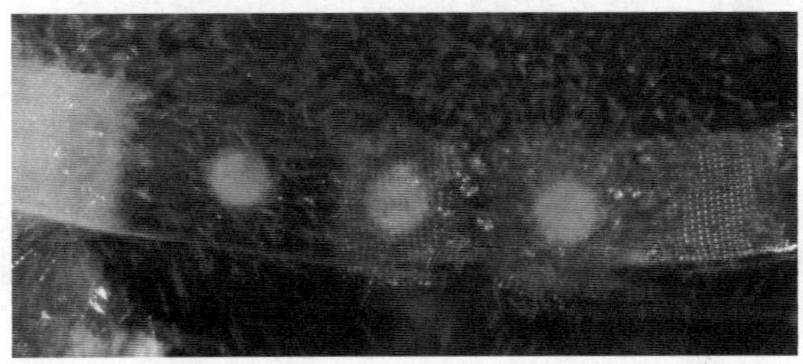

Figure 4. Liquid atomization in the areas of the antinodes of bending vibrations.

was observed. The voltage amplitude and the signal frequency were monitored with the use of an oscillograph C1-55 and frequency meter ChZ-34. When it was necessary, the trimming of the plate length was conducted for obtaining a jet at the minimal value of the amplitude of the vibrations. A piezo buzzer FML-20T-4.5A1-100 was used as a piezoelectric transducer. The plate had the size of $48.80 \times 4.60 \times 0.36$ mm. The length of curved part which immersed in the liquid is 18 mm, the radius of curvature of the plate is 15 mm, the depth of the plate dipping is 2 mm. The frequency of the resonance vibrations of the facility, when the interaction of the plate with the liquid led to the jet generation, was 4.7 kHz. The vibrations of the plate were excited at the 10th mode of bending vibrations. The amplitude of the vibrations of the plate free edge in the air reached 60 μm. Distilled water and ethyl alcohol were used as liquid.

3. Results and discussion

The jet formation by the vibrating polymer plate at the interaction with liquid is shown in Figure 2. Here the amplitude of the vibrations of the plate edge is $30 \div 40$ μm; at the frequency of 4.7 kHz the maximal value of the instantaneous velocity of the plate edge is $0.9 \div 1.2$ m/s. The velocity of droplets of water and alcohol in the jet is $1.0 \div 1.6$ m/s and $1.4 \div 1.7$ m/s respectively, which is slightly higher than the vibrational speed of the plate edge.

IOP Conf. Series: Materials Science and Engineering **208** (2017) 012001 doi:10.1088/1757-899X/208/1/012001

Simultaneously with the jet generation, the liquid atomization from the layer is observed in the antinode of bending vibrations near the interphase boundary on the side of piezoelectric transducer (see Figure3, 4). On the liquid layer wetting the upper side of the plate and on the areas of the plate surface with antinodes of vibrations, two-dimensional capillary waves in the form of Faraday ripples also appear. On the underside of the plate in the centre of the antinodes of vibrations, cavities with microscopic air bubbles are formed as the result of artificial hydrodynamic cavitation.

The partial immersion of the plate bent at an angle in the liquid and the wetting of the plate surface with the liquid leads to a liquid capillary rise over the plate surface on the phase boundary (Figure 5). The height of the liquid capillary rise increases at the interphase boundary when the plate bending vibrations are excited by the piezoelectric transducer (Figure 6). The additional capillary rise of the liquid layer over the vibrating plate surface indicates that in this liquid layer the pressure is smaller than that in the liquid layer wetting the surface of the nonexcited plate. This is due to the fact that the liquid layer close to the vibrating plate surface obtains the vibrational speed and energy which results in the appearance of the dynamic pressure equal to the liquid specific kinetic energy $p = \rho v^2/2$, where ρ is the liquid density, v is the velocity of the plate surface area.

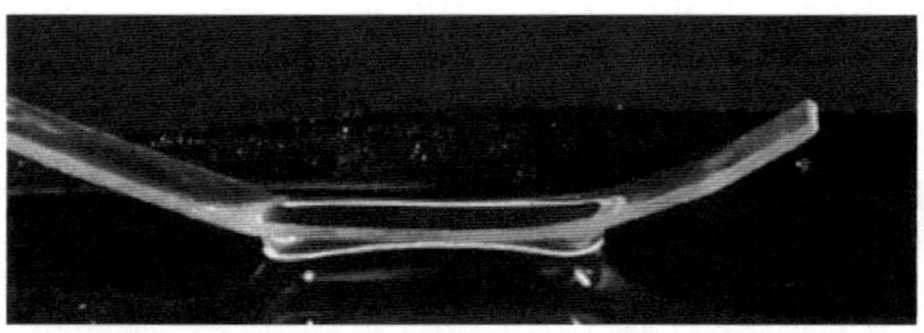

Figure 5. The capillary rise of the liquid over the polymer plate surface in the absence of vibrations.

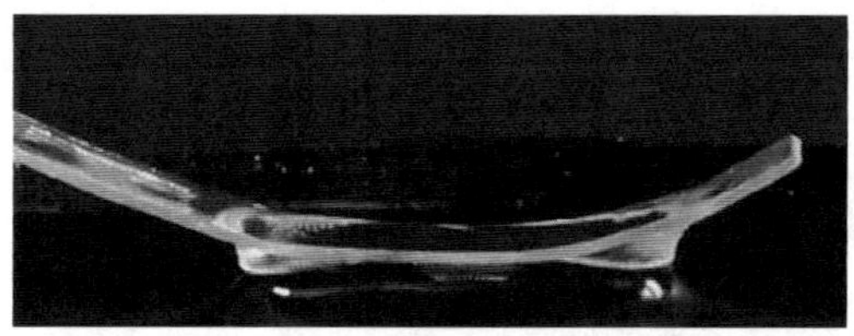

Figure 6. The capillary rise of the liquid over the polymer plate surface at the vibrations with a frequency of 4.7 kHz.

A sufficiently long narrow plate with a free edge presents by itself a quarter-wavelength resonator of bending vibrations with several antinodes, one of which is on the free edge. The distributed bending plate vibrations and capillary waves are similar in shape. In this case, the displacement of the plate surface can be presented as follows

$$s = s_0 \cos(k_{bx}x)\cos(\omega t), \tag{1}$$

where s_0 is the amplitude of the vibrations in the antinodes of bending vibrations, k_{bx} is the wave number of the bending waves depending on the plate parameters, ω is the circular frequency.

The velocity of the vibrating plate surface is

$$v = \omega s_0 \cos(k_{bx}x)\sin(\omega t). \tag{2}$$

The particles of the liquid wetting the plate surface have the same velocity and therefore the specific kinetic energy W_l of the liquid is

$$W_l = \frac{\rho}{2}(\omega s_0 \cos(k_{bx}x)\sin(\omega t))^2, \tag{3}$$

or

$$W_l = \frac{\rho}{4}\omega^2 (s_0)^2 \cos^2(k_{bx}x) - \frac{\rho}{4}\omega^2 (s_0)^2 \cos^2(k_{bx}x)\cos(2\omega t). \tag{4}$$

The second summand on the right side of equation (4) changes with the doubled frequency of the plate vibrations; therefore, the specific kinetic energy of the liquid layer on the vibrating plate surface and the hydrodynamic pressure in the layer pulsate. The value of the above summand for the period of vibrations is zero, and the specific energy of the liquid layer on the vibrating plate surface is on average

$$W_l = \frac{\rho}{4}\omega^2(s_0)^2\cos^2(k_{bx}x). \tag{5}$$

The value of the specific energy of the liquid layer vibrating together with the area of the plate surface depends on the coordinate and is maximal at the areas of the plate with the antinode of bending vibrations. The phenomenon of the shift of the liquid layer over the vibrating plate surface towards the antinodes of vibrations indicates the presence of the hydrodynamic pressure gradient

$$\nabla p = -\frac{\rho}{4}\omega^2(s_0)^2\sin^2(k_{bx}x). \tag{6}$$

The maximal value of the pressure gradient is on the layer wetting the areas of the plate surface in the center between the antinodes and nodes of vibrations. The increase of the amplitude of the plate vibrations leads to the appearance of two-dimensional capillary waves in the form of Faraday ripples on the surface of the liquid layer wetting the plate surface in the antinodes of vibrations. When the concave part of the plate is immersed in the liquid, two boundaries of the wetting liquid layer are formed, which envelope the plate surface from above and from below; one of the boundaries is formed near the free edge of the plate. The vertical shift of the plate makes it possible to change the distance from the boundary of the wetting liquid layer to the border of the plate free edge and to select the plate position so that the boundary of the liquid layer will be near the plate edge. When the plate bending vibrations are excited, the amplitude on the area of the free edge surface is maximal, and the same vibrations are passed to the liquid layer on the wetting boundary. It has experimentally been established that for the plate vibrations in liquid, on the phase boundary the parametric excitation of capillary waves takes place, the ridges of which are directed perpendicularly to the boundary of the wetting liquid layer and the plate edge (Figure 7). The width of the steel plate in Figure 7 is 8.8 mm, and that of the PET plate in Figure 8 is 4.6 mm. The calculations made for different frequencies of the excitation of capillary waves both by the PET plate and by the steel plate show that the distance between the ridges of waves is half of the length of a capillary wave, excited by the vibrations with a frequency two times lower than the frequency of the plate vibrations. In addition, an increase in the amplitude of the plate vibrations leads to the increase of the frequency range of the vibrations at which the standing capillary wave remains, which is characteristic of the excitation of parametric vibrations [7]. It has been noticed that the length of the two-dimensional standing capillary waves on the surface of the thin liquid layer on the plate surface on the area with the antinodes of vibrations is a little shorter than the length of the waves on the curved surface of the liquid nearby the plate edge.

The dependence of the length λ of the capillary waves on the frequency f of the excited vibrations can be obtained from the expression for the dependence of the circular frequency ω on the wave vector k for capillary-gravity waves on the liquid surface [8]

$$\omega^2 = (gk + \sigma k^3/\rho)\tanh(kH), \tag{7}$$

where g is the acceleration due to gravity, σ is the surface tension, ρ is the density and H is the liquid depth; the multiplier $\tanh(kH)$ is the hyperbolic tangent of the product of the wave vector and the depth.

For purely capillary waves excited by the vibrating plate, the first term on the right-hand side of the equation can be ignored and we have

$$\omega^2 = \frac{\sigma k^3}{\rho}\tanh(kH). \tag{8}$$

IOP Conf. Series: Materials Science and Engineering **208** (2017) 012001 doi:10.1088/1757-899X/208/1/012001

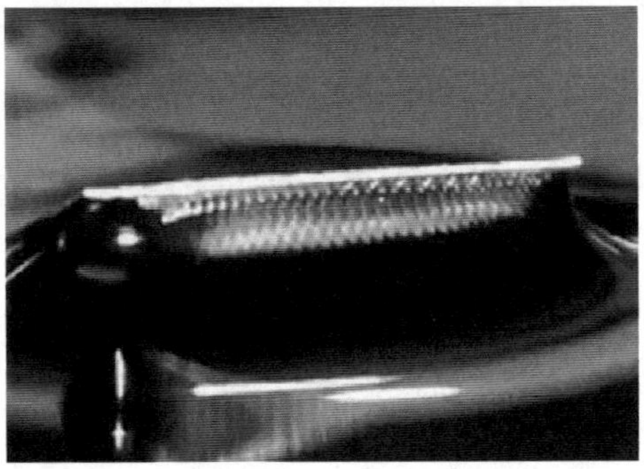

Figure 7. Capillary waves on the water surface under the free edge of the steel plate at the vibrations with a frequency of 5.4 kHz.

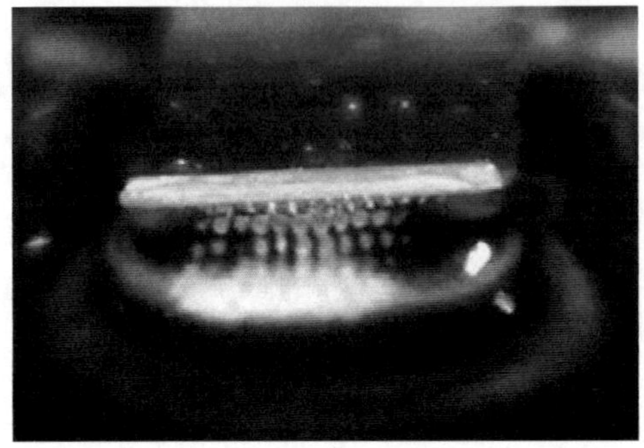

Figure 8. Capillary waves on the water surface under the PET plate edge at the vibrations with a frequency of 4.7 kHz.

Taking into account that $\omega = 2\pi f$ and $k = 2\pi\lambda$, we obtain the following dependence of the capillary wave length on the frequency

$$\lambda^3 = \frac{2\pi\sigma}{\rho f^2}\tanh(2\pi H/\lambda). \tag{9}$$

Equation 9 can be used for a thin liquid layer with the thickness $H < \lambda/2$, when the value $\tanh(2\pi H/\lambda) < 1$. In this case there is the necessity of the experimental determination of the thickness of the liquid layer, on the surface of which capillary waves are excited.

For capillary waves in deep liquid and in the layer of liquid with the thickness $H \geqslant \lambda/2$, the value $\tanh(2\pi H/\lambda) = 1$ and the relation

$$\lambda^3 = \frac{2\pi\sigma}{\rho f^2} \tag{10}$$

is fulfilled. This equation is applicable to capillary waves on the liquid surface near the edges of the vibrating plate. The surface tension of water $\sigma = 72.310^{-3}$ N/m and its density $\rho = 1000$ kg/m^3, and alcohol has $\sigma = 22.310^{-3}$ N/m and $\rho = 790$kg/m^3. Using these values we can obtain convenient analysis expressions for the length of a capillary wave in millimeters on the surface of water and alcohol and, respectively, $\lambda = 0.77f^{-2/3}$, $\lambda = 0.56f^{-2/3}$ where the frequency is given in kilohertz.

The parametric excitation of capillary waves in the liquid layer which is in contact with the vibrating plate surface is due to that at the vibrations of the plate the particles in the liquid get both impulse and acceleration

$$\lambda^3 = \frac{2\pi\sigma}{\rho f^2}. \tag{11}$$

The volume of the liquid layer with the thickness H and the surface area ΔS has the mass $m = rho\Delta SH$. At the accelerated movement this mass of the liquid layer acts on the plate with the force equal to the force of inertia

$$F = m\frac{d^2s}{dt^2} = \rho\Delta SH\omega^2 s_0\cos(k_{bx}x)\cos(\omega t). \tag{12}$$

Correspondingly, in the liquid layer at vibrations together with the plate, the pressure p_i of the forces of inertia appears, which in the antinode of the plate vibrations is

$$p_i = F/\Delta S = \rho H\omega^2 s_0\cos(\omega t). \tag{13}$$

The inertial force pressure in the liquid layer depends on the thickness in the direction of the vibrations and it periodically changes with the frequency of the plate area vibrations. For each period of vibrations with the beginning of the plate motion towards the liquid layer the pressure has a maximal positive value and at the moment, when the direction of the motion is from the liquid layer, the pressure has a maximal negative value. The inertial force pressure also produces an effect on the curved surface areas on the phase boundary; at the periodic variation of the inertial force pressure, the vibrations of the wetting layer boundaries are excited. Such process can be considered as a periodic variation of the area of the liquid free surface as the result of the action of additional periodically changing surface forces due to the inertial forces. This leads to the parametric excitation of capillary vibrations in the form of standing waves on the boundary of the liquid layer wetting the plate surface.

The increase of the amplitude of the plate vibrations causes the space modulation of capillary waves in the crosswise direction with the result that the capillary waves become two-dimensional (Figure 8). Moreover, if in the process of vibrations, the liquid layer boundary reaches the border of the plate edge, the curvature of the free surface of the liquid layer along the plate edge periodically changes and, as a result, the ridges of two-dimensional capillary waves are stretched up to the detachment of particles of the liquid. In general, this process leads to the atomization of the liquid in the layer wetting the surface of the vibrating plate. It is established that on the curved free surface under the edge of the vibrating plate, surface eddy flows appear on either side of the plate edge. The liquid surface in these flows moves in the opposite directions so that between them a rising stream appears moving under the edge of the plate in the direction perpendicular to this edge (Figure 9). The observed motion of the liquid can be explained by that the pressure of the liquid under its free surface near the vibrating plate is smaller than that under the undisturbed surface far from the plate. As has been mentioned above, the pressure difference is determined by the liquid specific energy got from the vibrating plate area.

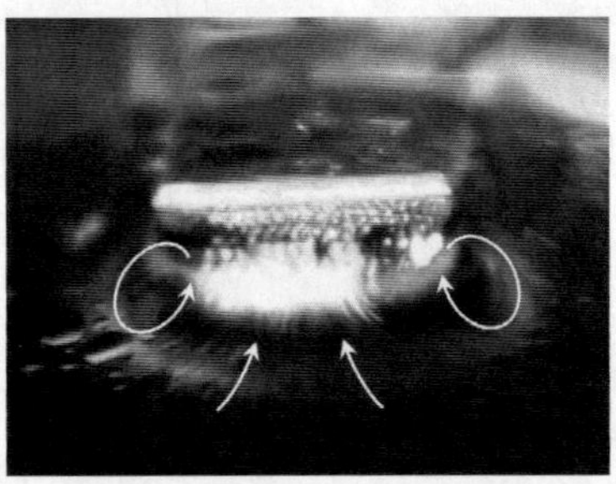

Figure 9. The surface flow and capillary vibrations in the liquid under the vibrating plate edge.

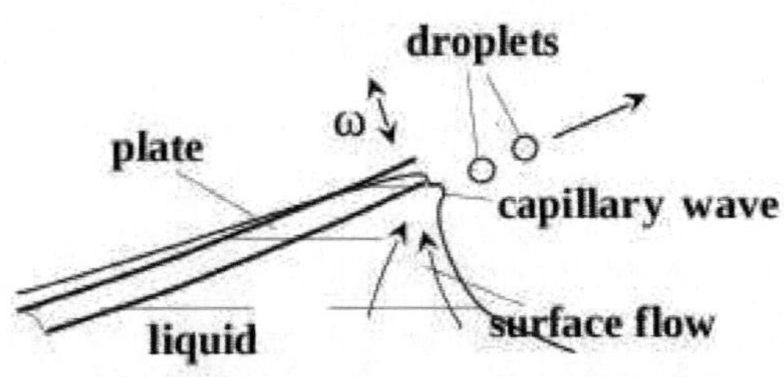

Figure 10. The mechanism of the jet formation under the free edge of the vibrating plate.

The further increase of the amplitude of the plate bending vibrations leads to the periodic inertial detachment of relatively large droplets of the liquid from under the vibrating plate edge creating a jet. The formation of the liquid jet by the vibrating plate takes place as the result of the appearance of the surface flow on the curved surface of the liquid and the excitation of two-dimensional capillary waves on the moving liquid surface (Figure 10) at the same time.

In general, the discussed interactions of liquid and a vibrating plate excited with piezoelectric transducer are threshold. If the excitation of capillary waves on the piezoelectric transducer plate electrodes enough to apply a voltage of about 1 V, the generation of the jet occurs at an amplitude voltage of at least $15 \div 20V$. Figure 11 shows the regions of amplitude and frequency of voltage on the electrodes of the piezoelectric transducer, within which vibrating plate sizes specified excites capillary waves, produces spraying and forms a jet.

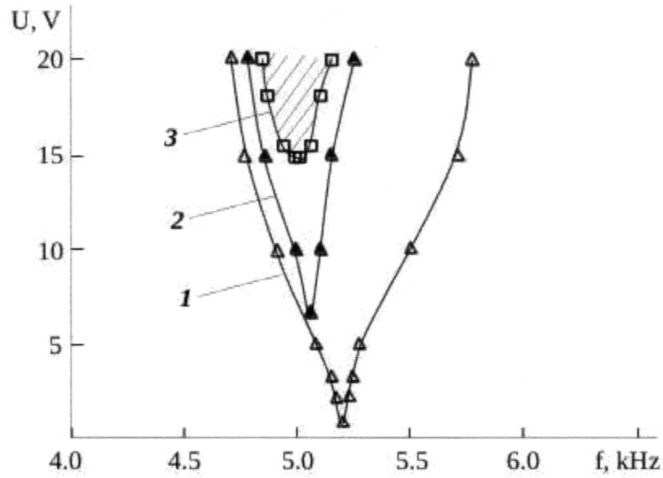

Figure 11. Regions of amplitudes and frequencies of electrode voltage of the piezoelectric transducer FML-20T-6.0A1-100 in case of excitation in water of capillary waves — 1, liquid atomization — 2 and generations of a jet — 3 by steel curved plate $40.0 \times 8.6 \times 0.1$ mm.

4. Conclusion

The vibrations of a plate partially immersed in liquid excite capillary waves not only on the surface of the liquid layer wetting the plate surface, but also on the curved surface of the liquid near the plate edges. Since the capillary waves formed on the boundary of the liquid layer that wets the surface of the plate near its free edge are modulated by the plate vibrations, two-dimensional capillary waves appear. The capillary oscillations with the frequency of the plate vibrations periodically change the surface curvature of the wetting liquid layer near the plate edge from negative to positive. In addition, the plate vibrations cause the appearance of the hydrodynamic pressure in the vibrating liquid. As a result, the surface eddy flows appear in the liquid near the plate. The liquid flow under the free edge of the vibrating plate is directed up to the free edge. The overlapping of the pressure of the inertial forces and the Laplace pressure of the two-dimensional capillary waves on the surface flow generated by the plate vibrations results in periodic inertial detachment of the liquid particles from under the plate edge, and thus forms the jet.

4.1. Acknowledgments

The work is supported by the Russian Foundation for Basic Research (grant 16-41-180276_r_ural_a).

References

[1] Aleksandrov V 2008 *Technical Physics Letters* **34** 100
[2] Faraday M 1831 *Philosophical Transactions of the Royal Society of London* **121** 299
[3] Miles J 1993 *J. of Fluid Mechanics* **248** 671
[4] James A J, Vukasinovic B, Smith M K and Glezer A 2003 *Journal of Fluid Mechanics* **476** 1–28
[5] Yule A J and Al – Suleimani Y 2000 *Proceedings of the Royal Society of London A: Mathematical, Physical and Engineering Sciences* **456** 1069
[6] Ashgriz N (ed) 2011 *Handbook of Atomization and Sprays* (Springer US)
[7] Bolotin V (ed) 1984 *Random vibrations of elastic systems* vol 8 (Amsterdam: Springer Netherlands)
[8] Landau L and Lifshitz E 1987 {Chapter} {XV} - {Relativistic} {Fluid} {Dynamics} *Fluid Mechanics* ed Landau L and Lifshitz E (Pergamon) p 505

IOP Conf. Series: Materials Science and Engineering **208** (2017) 012002 doi:10.1088/1757-899X/208/1/012002

Mathematical Simulation of the Problem of the Pre-Critical Sandwich Plate Bending in Geometrically Nonlinear One Dimensional Formulation

I B Badriev [1, a], V V Banderov [1] and M V Makarov [1,2]

[1] Kazan Federal University, 18 Kremlyovskaya Street, 420008, Kazan, Russian Federation

[2] Kazan National Research Technical University, 10 Karl Marx Street, 420111, Kazan, Russian Federation

E-mail: [a] ildar.badriev1@mail.ru

Abstract. In this paper we consider the geometrically nonlinear problem of determining the equilibrium position of a sandwich plate consisting of two external carrier layers and located between transversely soft core, connected with carrier layer by means of adhesive joint. We investigate the generalized statement of the problem. For its numerical implementation we offer a two-layer iterative process and investigate the convergence of the method. Numerical experiments are carried out for the model problem.

1. Introduction

Multilayer structures, in particular plates, are widely used in various fields of modern technology: aerospace, aviation, shipbuilding; industrial, civil and transport construction, chemical and power engineering [1–6]. The interest in layered plates associated primarily with the fact that they have a set of properties and features that are qualitatively different from conventional constructions. In this paper the construction and study of a generalized statement of geometrically nonlinear problem of the bending of sandwich plate with transversal-soft core are carried out. To solving the problem an iterative method is proposed and its convergence is investigated. In the study correctness of a generalized statement and the convergence of method we use the pseudomonotone operators apparatus. Note that the generalized statement of physically non-linear and geometrically linear problem in the form of saddle problems, as well as a method for its solution were considered in [7–10]. The study of nonlinear problems of the shells theory, including the approximate methods for their solution in [11–18] is carried out.

2. Problem statement

In this paper we consider the problem of determining the stress-strain state of an infinitely long sandwich plate with transversal-soft core. The width of the plate is a, the core thickness is $2h$, the thickness of the carrier layers are equal $2h_{(k)}$, where k is the number of the layer. To describe the stress-strain state in the carrier layers are used the equations of Kirchhoff-Love model, in the core the equations of elasticity theory, simplified by the accepted model of transversely soft layer and integrated over its thickness with the satisfaction of the of conjugation conditions of the layers on the displacements [19–21]. According to [19–21], we introduce the following notations (here and in what follows we assume that $k = 1, 2$), $X^1_{(k)}$, $X^3_{(k)}$ are surface load components reduced to the middle surface of the k-th layer, $w^{(k)}$ and $u^{(k)}$ are bending and axial displacements of the middle the surface points of k-th layer, respectively, $T^{11}_{(k)}$, $M^{11}_{(k)}$ are the inner membrane forces and bending moments in the k-th layer respectively. The plate edges are assumed fixed, so that the conditions $u^{(k)}(x)=0$, $w^{(k)}(x)=0$, $d\,w^{(k)}/dx=0$ are satisfied at $x=0$, $x=a$. Let us consider the problem in geometrically nonlinear statement, i.e., $T^{11}_{(k)} = B_{(k)}\left(d\,u^{(k)}/dx+\frac{1}{2}(d\,w^{(k)}/dx)^2\right)$, $M^{11}_{(k)}=-D_{(k)}\,d^2 w^{(k)}/dx^2$, where $B_{(k)}=2h_{(k)}E^{(k)}/(1-\nu^{(k)}_{12}\nu^{(k)}_{21})$ is the stiffness of k-th layer on the tension-compression, $E^{(k)}$ and $\nu^{(k)}_{12}$, $\nu^{(k)}_{21}$ are the first-order elastic modulus and Poisson's coefficients of the material of the carrier k-th layer, $D_{(k)}=B_{(k)}h^2_{(k)}/3$ is the bending stiffness of k-th layer. Let $U=(w^{(1)},w^{(2)},u^{(1)},u^{(2)})$ be a displacements vector of points of the middle surface of the k-th layer, q^1 be a tangential stresses in the core. For q^1 we assume the boundary conditions

$$q^1(0)=q^1(a)=0 \tag{1}$$

are satisfied. In [19–21] to describe the stress-strain state of sandwich plate following strain energy functional was constructed $L(U,q^1)=P(U,q^1)-A_e(U,q^1)-A_q(U,q^1)$, where

$$\begin{aligned}P(U,q^1)=\frac{1}{2}\int_0^a\Big\{\sum_{k=1}^{2}\Big[B_{(k)}\left(d\,u^{(k)}/dx+\frac{1}{2}(d\,w^{(k)}/dx)^2\right)^2+D_{(k)}(d^2w^{(k)}/dx^2)^2\Big]+\\+c_1(q^1)^2+c_2(d\,q^1/dx)^2+c_3(w^{(2)}-w^{(1)})^2\Big\}dx\end{aligned} \tag{2}$$

is the strain potential, $c_1=2h/G_{13}$, $c_2=2h^3/3E_3$, $c_3=E_3/(2h)$, G_{13}, E_3 are the transverse shear and compression modules of a core,

$$A_e(U,q^1)=\int_0^a\sum_{k=1}^{2}[X^1_{(k)}u^{(k)}+M^1_{(k)}\,d\,w^{(k)}/dx+X^3_{(k)}w^{(k)}]dx \tag{3}$$

is the work of given external forces and moments, $M^1_{(k)}$ is the surface moment of external forces, reduced to the middle surface of the k-th layer, and

$$A_q(U,q^1)=\int_0^a\Big[(u^{(1)}-u^{(2)})-\sum_{k=1}^{2}H_{(k)}d\,w^{(k)}/dx+c_1q^1-c_2d^2q^1/dx^2\Big]q^1dx \tag{4}$$

is the work of unknown tangential stresses on corresponding displacements. It has been established [22] that the solution of the problem of the sandwich plate equilibrium is a stationary point of the functional L.

IOP Conf. Series: Materials Science and Engineering **208** (2017) 012002 doi:10.1088/1757-899X/208/1/012002

3. Generalized statement of the problem

Let $V_k = \overset{o}{W}{}_2^{(k)}(0,a)$ be the Sobolev spaces [23, 24] with inner products $(u,\eta)_k = \int_0^a d^k u / dx^k \, d^k \eta / dx^k dx$, V_q be the Sobolev space of functions with compact support on (0, a) having the first generalized derivative, integrable with the square, with the inner product $(y,z)_+ = \int_0^a [c_2 y(x) z(x) + c_3 d\, y / dx \; d\, z / dx] dx$. We denote the inner product in V through $(\cdot,\cdot)_V$. Using the Sobolev embedding theorem and the Rellich-Kondrashov theorem [23, 24], it is easy to verify that the functional L is well defined on $V \times V_q$. We obtain the equations for the stationary points of the functional L. For this purpose we find the Gateaux its derivatives [24, 25]. Let us rewrite the functional L in the form $L(U, q^1) = \Phi_0(U) + \Phi_1(U, q^1) - \Phi_2(q^1)$, where

$$\Phi_0(U) = \frac{1}{2}\int_0^a \{ \sum_{k=1}^{2} \Big[B_{(k)} \big(d\,u^{(k)} / dx + \frac{1}{2} (d\,w^{(k)} / dx)^2 \big)^2 + D_{(k)} (d^2 w^{(k)} / dx^2)^2 \Big] + $$
$$+ c_3 (w^{(2)} - w^{(1)})^2 \} dx - \int_0^a \sum_{k=1}^{2} [X^1_{(k)} u^{(k)} + M^1_{(k)} d\,w^{(k)} / dx + X^3_{(k)} w^{(k)}] dx, \tag{5}$$

$$\Phi_1(U, q^1) = \int_0^a \left[\sum_{k=1}^{2} H_{(k)} \frac{d\,w^{(k)}}{dx} + (u^{(2)} - u^{(1)}) \right] q^1 dx, \quad \Phi_2(q^1) = \frac{1}{2}\int_0^a \left[c_1 (q^1)^2 + c_2 \left(\frac{d\,q^1}{dx} \right)^2 \right] dx \tag{6}$$

Defining the Gateaux derivatives we have that the stationary points of the functional L are the solutions of variational equations (integral identities)

$$\int_0^a \sum_{k=1}^{2} B_{(k)} \Big(\frac{du^{(k)}}{dx} + \frac{1}{2}\left(\frac{dw^{(k)}}{dx} \right)^2 \Big) \frac{d\eta^{(k)}}{dx} dx + \int_0^a \sum_{k=1}^{2} B_{(k)} \Big(\frac{du^{(k)}}{dx} + \frac{1}{2}\left(\frac{dw^{(k)}}{dx} \right)^2 \Big) \frac{dw^{(k)}}{dx} \frac{dz^{(k)}}{dx} dx +$$
$$+ \int_0^a \sum_{k=1}^{2} D_{(k)} \frac{d^2 w^{(k)}}{dx^2} \frac{d^2 z^{(k)}}{dx^2} dx - \int_0^a \sum_{k=1}^{2} [X^1_{(k)} \eta^{(k)} + M^1_{(k)} \frac{dz^{(k)}}{dx} + X^3_{(k)} z^{(k)}] dx + \tag{7}$$
$$+ c_3 \int_0^a (w^{(2)} - w^{(1)})(z^{(2)} - z^{(1)}) dx + \int_0^a \{ \sum_{k=1}^{2} H_{(k)} \frac{d\,z^{(k)}}{dx} + (\eta^{(2)} - \eta^{(1)}) \} q^1 dx = 0 \qquad \forall Z \in V,$$

$$\int_0^a \{ [\sum_{k=1}^{2} H_{(k)} \frac{d\,w^{(k)}}{dx} + (u^{(2)} - u^{(1)}) + c_1 q^1] y + c_2 d q^1 / dx \, d y / dx \} dx = 0 \quad \forall y \in V_1 \tag{8}$$

Summing the integral identities, we obtain the variational equation

$$b((U, q^1), (Z, y)) = ((f_1, 0), (Z, y))_{V \times Vq} \qquad \forall (Z, y) \in W = V \times V_q. \tag{9}$$

Again using the Sobolev embedding theorem and the Rellich-Kondrashov theorem [23, 24] it is easy to verify that the form $b(\cdot,\cdot)$ defined on $W \times W$ is linear and bounded to the second argument, and therefore, generates an operator $A: W \to W$, the right side (9) generates an element $F \in W$. Thus (3) can be written as operator equation

$$A(U, q^1) = F. \tag{10}$$

Recall that the operator $A_0 : V \to V^*$ is called pseudomonotone [26, 27] if it is bounded and the weak convergence of the sequence $\{u_k\}_{k=1}^{+\infty}$ in V to u^* and the inequality $\limsup_{k\to+\infty} \langle A_0 u_k, u_k - u^* \rangle \le 0$ imply that the following relationship holds $\liminf_{k\to+\infty} \langle A_0 u_k, u_k - \eta \rangle \ge \langle A_0 u^*, u^* - \eta \rangle$ for all η. The following result holds.

Theorem 1. Operator A is continuous, coercive and pseudomonotone.

From Theorem 1 and general results of the theory of monotone operators [25, 26] it implies that the following theorem is true.

Theorem 2. The problem (10) has at least one solution.

4. Iterative method

To solve the problem (10) in analogy to [28–32] we propose the following iteration process has been. Let (U_0, q_0^1) be any element from W. For $n = 0,1,2,\ldots$ we find (U_n, q_n^1) as a solution of the problem

$$J((U_{n+1}, q_{n+1}^1) - (U_n, q_n^1)) = \tau(F - A(U_n, q_n^1)), \tag{11}$$

where $J : W \to W$ is a duality operator [26], $\tau > 0$ is an iterative parameter.

Following [33–36], we can prove the following theorem on the convergence of the iterative method.

Theorem 3. There is a constant τ_0 such that for $0 < \tau < \tau_0$ the sequence (U_n, q_n^1) weakly converges to some solution of the problem (10) as $n \to +\infty$.

The proposed method (11) for solving the problem have been implemented numerically. Software package in Matlab environment was developed and carried out calculations for the model problem. For the model problem, numerical experiments were carried out. Iteration parameter was chosen empirically. The calculations were performed for the following characteristics: $a = 1$ cm, $h_{(1)} = h_{(2)} = 0.005$ cm, $h = 0.05$ cm, $G_{13} = 15$ MPa, $E_3 = 25$ MPa, $X_{(1)}^3 = 0.0319$ MPa, $X_{(2)}^3 = 0$, $E^{(k)} = 7\cdot 10^4$ MPa, $\nu_{12}^{(k)} = \nu_{21}^{(k)} = 0.3$, $X_{(k)}^1 = 0$, $M_{(k)}^1 = 0$, $k = 1,2$. The results of numerical experiments are shown in figures 1–3.

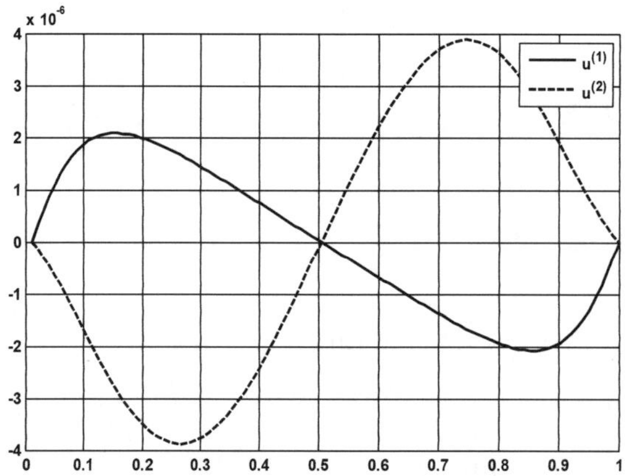

Figure 1. Axial displacements in carrier layers.

It should be noted that formulated for q^1 boundary conditions (1) correspond to the presence of, at the edges $x = 0$, $x = a$ the diaphragms, this leads to the formation of the maximum transverse tangential stresses in sections the filler a distance of about its thickness $2h$, as observed in figure 2. The limiting the free end sections $x=0$, $x=a$ displacements in axial direction leads to the formation to the formation of a significant in magnitude membrane forces $T_{(1)}^{11}$, $T_{(2)}^{11}$ in the carrying layers $T_{(1)}^{11}$, $T_{(2)}^{11}$, the force $T_{(1)}^{11}$ of which is a contraction in cross-section $x=a/2$. For this reason, in the neighborhood of this section should be expected buckling carrier layers in a mixed form. It is easy to verify that the first two equations of (1) implies that $T_{(1)}^{11}+T_{(2)}^{11}=\text{const}$, in the implementation of which can be verified on the basis of the results shown in figure 3.

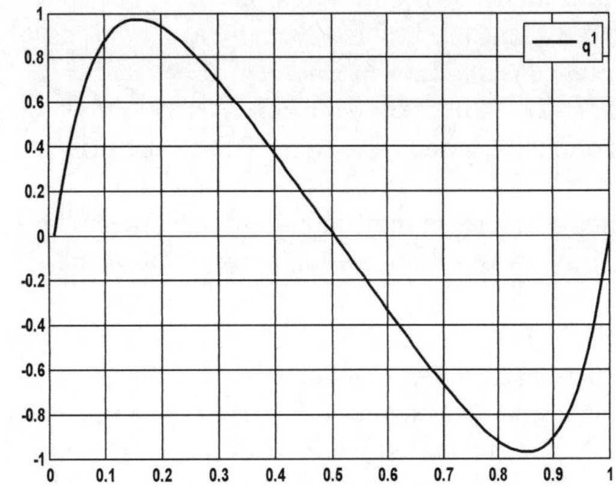

Figure 2. Tangential stresses in core.

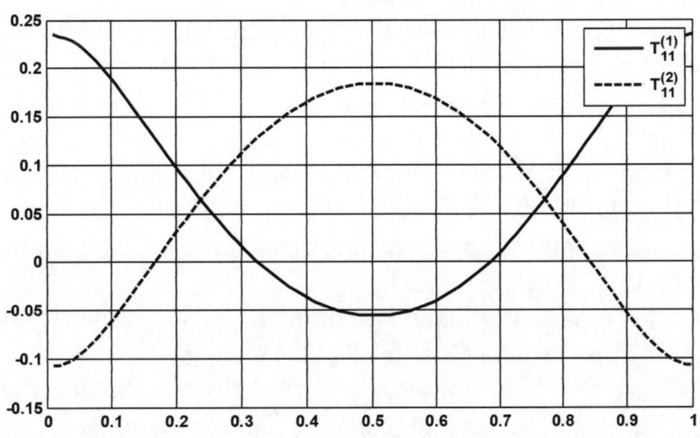

Figure 3. Membrane forces in carrier layers.

5. Conclusion

In this paper, we construct a generalized statement for the geometrically nonlinear problem of the bending of a sandwich plate in a one-dimensional formulation in the form of an operator equation in the Sobolev space. The solvability of the equation is proved on the basis of general results of the monotone operators theory. To solve the problem, a two-layer iterative process is proposed and its convergence is investigated. The results of numerical experiments, their correspondence to the physical picture of the problem, showed both the adequacy of the mathematical model and the effectiveness of the proposed iterative method. The results of the work can be used in the design of the sandwich structures.

IOP Conf. Series: Materials Science and Engineering **208** (2017) 012002 doi:10.1088/1757-899X/208/1/012002

Acknowledgements
This work was supported by the Russian Science Foundation (project 16-11-10299).

References

[1] Berezhnoi D V and Paimushin V N 2011 Two formulations of elastoplastic problems and the theoretical determination of the location of neck formation in samples under tension *Journal of Applied Mathematics and Mechanics* **75 (4)** 447-62

[2] Akishev N I, Zakirov I I, Paimushin V N and Shishov M A Theoretical-experimental method for determining the averaged elastic and strength characteristics of a honeycomb core of sandwich designs *Mechanics of Composite Materials* **47 (4)** (2011) 377-386

[3] Abdrakhmanova A I and Sultanov L U 2016 Numerical modelling of deformation of hyperelastic incompressible solids *Materials Physics and Mechanics* **26 (1)** 30-2

[4] Rahmani O and Lashkari M J 2015 Bending analysis of sandwich plates with composite face sheets and compliance functionally graded syntactic foam core *Journal of Mechanical Engineering Science* **1** 1-24

[5] Paimushin V N 2008 Static and dynamic beam forms of the loss of stability of a long orthotropic cylindrical shell under external pressure *Journal of Applied Mathematics and Mechanics* **72 (6)** 738–47

[6] Liang Y and Izzuddin B A 2016 Large displacement analysis of sandwich plates and shells with symmetric/asymmetric lamination *Computers & Structures*. **1** 11–32

[7] Badriev I B, Makarov M V and Paimushin V N 2015 Solvability of physically and geometrically nonlinear problem of the theory of sandwich plates with transversally-soft core *Russian Mathematics*. **59 (10)** 57-60

[8] Badriev I B, Garipova G Z, Makarov M V and Paimushin V N 2015 Numerical Solution of the Issue about Geometrically Nonlinear Behavior of Sandwich Plate with Transversal Soft Filler *Research Journal of Applied Sciences* **10 (8)** 428-35

[9] Badriev I B, Garipova G Z, Makarov M V, Paimushin V N and Khabibullin R F 2015 Solving Physically Nonlinear Equilibrium Problems for Sandwich Plates with a Transversally Soft Core *Lobachevskii Journal of Mathematics* **36 (4)** 474-81

[10] Badriev I B, Makarov M V and Paimushin V N 2016 Numerical Investigation of Physically Nonlinear Problem of Sandwich Plate Bending *Procedia Engineering* **150** 1050-5

[11] Paimushin V N and Polyakova N V 2009 The consistent equations of the theory of plane curvilinear rods for finite displacements and linearized problems of stability *Journal of Applied Mathematics and Mechanics* **73 (2)** 220-36

[12] Berezhnoi D V, Sachenkov A A and Sagdatullin M K 2014 Geometrically nonlinear deformation elastoplastic soil *Applied Mathematical Sciences* **8 (127)** 6341-8

[13] Badriev I B, Banderov V V and Zadvornov O A 2013 On the equilibrium problem of a soft network shell in the presence of several point loads *Applied Mechanics and Materials* **392** 188-90

[14] Abdrakhmanova A I, Gariffulin I R, Davydov R L, Sultanov L U and Fakhrutdinov L R 2015 Investigation of Strain of Solids for Incompressible Materials *Applied Mathematical Sciences* **9 (118)** 5907-14

[15] Davydov R L and Sultanov L U 2015 Numerical Algorithm for Investigating Large Elasto-Plastic Deformations *Journal of Engineering Physics and Thermophysics* **88 (5)** 1280-8

[16] Badriev I B, Makarov M V and Paimushin V N 2016 Mathematical Simulation of Nonlinear Problem of Three-point Composite Sample Bending Test *Procedia Engineering* **150** 1056-62.

[17] Sultanov L U 2016 Analysis of finite elasto-plastic strains. Medium kinematics and constitutive equations *Lobachevskii Journal of Mathematics* **37 (6)** 787-93

[18] Berezhnoi D V, Paimushin V N and Shalashilin V I 2010 Studies of quality of geometrically nonlinear elasticity theory for small strains and arbitrary displacements *Mechanics of Solids* **44 (6)** 837-51

[19] Paimushin V N and Bobrov S N 2000 Refined geometric nonlinear theory of sandwich shells with a transversely soft core of medium thickness for investigation of mixed buckling forms *Mechanics of composite materials* **36(1)** 59-66
[20] Paimushin V N 1983 Variational-methods in the non-linear 3-dimensional problems of conjugation of deformable-bodies *Doklady Akademii Nauk* **273 (5)** 1083-86
[21] Paimushin V N 1987 Nonlinear theory of the central bending of three-layer shells with defects in the form of sections of bonding failure *Soviet Applied Mechanics* **23 (11)** 1038-1043
[22] Paimushin V N 1987 Generalized Reissner variational principle in nonlinear mechanics of three-dimensional composite solids, with applications to the theory of multilayer shells *Mechanics of solids* **22 (2)** 166-174
[23] Adams R A 1975 *Sobolev Spaces* (New York, San Francisco, London, Academic Press)
[24] Gajewskii H, Gröger K and Zacharias K 1974 *Nichtlineare Operatorgleichungen und Operatordifferentialgleichungen* (Berlin, Akademie-Verlag)
[25] Ekeland I and Temam R 1976 *Convex Analysis and Variational Problems*, (Amsterdam, North-Holland)
[26] Lions J-L 1969 *Quelque problèmes méthodes de résolution des problèmes aux limites nonlinéaires* (Paris, Dunod)
[27] Brézis H 1968 *Équations et inéquations non linéaires dans les espaces vectoriels en dualité Annales de l'institut Fourier* **18 (1)** 115-75
[28] Dubrovin V T, Chebakova V Ju and Zheltukhin V S 2016 Radio-Frequency Discharge at Low Pressure: a Non-Local Problem Statement Approach *Procedia Engineering* **150** 1041–5
[29] Badriev I B 2013 On the Solving of Variational Inequalities of Stationary Problems of Two-Phase Flow in Porous Media *Applied Mechanics and Materials* **392** 183-7
[30] Badriev I B and Karchevskii M M 1989 Convergence of the iterative Uzawa method for the solution of the stationary problem of seepage theory with a limit gradient *Journal of Soviet Mathematics* **45 (4)** 1302-9
[31] Badriev I B and Nechaeva L A 2013 Mathematical simulation of steady filtration with multivalued law *PNRPU Mechanics Bulletin* **(3)** 37-65
[32] Badriev I B 1989 Application of duality methods to the analysis of stationary seepage problems with a discontinuous seepage law *Journal of Soviet Mathematics* **45 (4)** 1310-4
[33] Chebakova V J, Gerasimov A V and Kirpichnikov A P 2016 On the solving of one type of problems of mathematical physics, IOP Conference Series: Materials Science and Engineering **158 (1)** 012023
[34] Badriev I B, Zadvornov O A and Saddek A M 2001 Convergence Analysis of Iterative Methods for Some Variational Inequalities with Pseudomonotone Operators *Differential Equations* **37 (7)** 934-942
[35] Badriev I B, Zadvornov O A, Ismagilov L N and Skvortsov E V 2009 Solution of plane seepage problems for a multivalued seepage law when there is a point source *Journal of Applied Mathematics and Mechanics* **73 (4)** 434-42

IOP Conf. Series: Materials Science and Engineering **208** (2017) 012003 doi:10.1088/1757-899X/208/1/012003

Conversion of elastic waves as a result of diffraction in anisotropic layer

Y N Belyayev

Syktyvkar State University, 55, Oktyabrskii ave., Syktyvkar, 167001, Russia

E-mail: ybelyayev@mail.ru

Abstract. The transfer matrix method for the theory of elastic wave scattering in anisotropic layered medium is developed. The cases of the six-beam diffraction are considered. The method of the transfer matrix $\exp(Wz)$ computation is based on the use of polynomials of principal minors in combination with scaling of matrix W. This method, in contrast to the Lagrange-Sylvester and others known polynomial approaches does not require solution of eigenvalue problem. Evaluations of the scaling parameter and relative truncation error of the method are made. Some features of the conversions of shear waves and longitudinal wave are demonstrated by calculations of the diffraction on a crystalline silicon layer. It is shown that the conversion SH wave into SV waves and SV wave into SH waves are equivalent.

1. Introduction

The anisotropy of the elastic properties in a layered medium leads to the possibility of interactions between elastic waves of all possible types. As a result of these, each type of the incident elastic wave, namely, longitudinal (P), shear horizontal (SH) and shear vertical (SV) waves, can produce six elastic waves in an anisotropic layer (Fig. 1a).

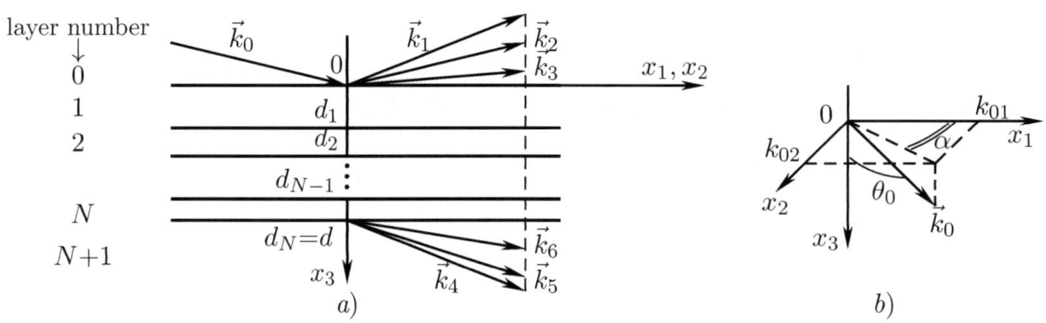

Figure 1. The geometry of scattering.

The most effective way of wave field calculation in a multilayer medium is the transfer matrix method. According to this method, the stress-strain state of an elastic medium is characterized by a vector function $\Psi(x_3) = \|\psi_i(x_3)\|$, where ψ_j, $i = 1, \ldots, 6$ are components of displacement vector and stress tensor. So-called transfer matrix T_i relates the values of vector function $\Psi(x_3)$ at thicknesses d_i and d_{i+1}: $\Psi(d_{i+1}) = T_i\Psi(d_i)$. Thus $\Psi(d) = T\Psi(0)$, where $T = \prod_{i=N}^{1} T_i$ is a global transfer matrix and $d = d_1 + \cdots + d_N$ is the thickness of N-layer plate.

In the usual approaches [1, 2, 3] transfer matrices T_i are represented as a functions of six exponentials of wave numbers. In the general case of six-beam diffraction in N-layer structure the result of transfer matrix computation depends on accuracy of $6N$ wave numbers definitions and reliability of corresponding exponentials computations. That's why one of the substantial values that affect the accuracy of the transfer matrix calculation is propagation parameter $\kappa = \nu d$, where ν is a frequency. The computation of transfer matrix for inhomogeneous (or evanescent) waves at large values of κ can lead to unsatisfactory results [4].

This paper uses calculations of transfer matrix $T = \exp(Wd)$ by means of method of polynomials of principal minors (MPPM) of the matrix W. MPPM, unlike the Lagrange-Sylvester method and other polynomial approaches [5], does not require finding the eigenvalues of the matrix W. An elastic wave scattering by an anisotropic structure bounded by isotropic solid media (layers 0 and $N+1$ in Fig. 1a) are considered.

2. Method of computations

2.1. Main equations

The system of equations of motion and Hooke's law can be written in the following form

$$\begin{aligned}
&\rho\frac{\partial^2 u_1}{\partial t^2} = \frac{\partial P_1}{\partial x_1} + \frac{\partial P_6}{\partial x_2} + \frac{\partial P_5}{\partial x_3}, &&\frac{\partial u_1}{\partial x_1} = \sum_{g=1}^{6} S_{1g}P_g, &&\frac{\partial u_2}{\partial x_3} + \frac{\partial u_3}{\partial x_2} = \sum_{g=1}^{6} S_{4g}P_g,\\
&\rho\frac{\partial^2 u_2}{\partial t^2} = \frac{\partial P_6}{\partial x_1} + \frac{\partial P_2}{\partial x_2} + \frac{\partial P_4}{\partial x_3}, &&\frac{\partial u_2}{\partial x_2} = \sum_{g=1}^{6} S_{2g}P_g, &&\frac{\partial u_3}{\partial x_1} + \frac{\partial u_1}{\partial x_3} = \sum_{g=1}^{6} S_{5g}P_g,\\
&\rho\frac{\partial^2 u_3}{\partial t^2} = \frac{\partial P_5}{\partial x_1} + \frac{\partial P_4}{\partial x_2} + \frac{\partial P_3}{\partial x_3}, &&\frac{\partial u_3}{\partial x_3} = \sum_{g=1}^{6} S_{3g}P_g, &&\frac{\partial u_1}{\partial x_2} + \frac{\partial u_2}{\partial x_1} = \sum_{g=1}^{6} S_{6g}P_g,
\end{aligned} \tag{1}$$

where ρ is a density; t is a time; x_1, x_2, x_3 are Cartesian coordinates; u_1, u_2, u_3 are components of displacement vector; P_g and S_{gl}, $g, l = 1, \ldots, 6$ are accordingly components of stress tensor ($P_1 = p_{11}$, $P_2 = p_{22}$, $P_3 = p_{33}$, $P_4 = p_{23} = p_{32}$, $P_5 = p_{13} = p_{31}$, $P_6 = p_{12} = p_{21}$) and compliance tensor in the matrix form [6, § 52].

The density $\rho = \rho(x_3)$ and elastic parameters of medium $S_{gl} = S_{gl}(x_3)$ depend only on one coordinate x_3 along an axis perpendicular to the surface of anisotropic layer. The deformations in an anisotropic layer $0 \le x_3 \le d$ are generated by a plane wave $\vec{u}_0 = \vec{\mathscr{A}}_0 \exp[i(\vec{k}_0 \cdot \vec{r} - \omega t)]$ incident from the region $x_3 < 0$. Here $\vec{u}_0$, $\vec{\mathscr{A}}_0$ $\vec{k}_0$ and ω are displacement vector, amplitude, wave vector and circular frequency of incident wave respectively and i is imaginary unit. These assumptions make it possible to reduce the system (1) to the equation

$$\frac{\partial}{\partial x_3}\Psi = W\Psi \equiv \begin{Vmatrix} w_{11} & w_{12} & w_{13} & w_{14} & w_{15} & w_{16}\\ w_{21} & w_{11} & w_{23} & 0 & w_{25} & w_{26}\\ w_{25} & w_{15} & w_{33} & w_{34} & w_{35} & w_{36}\\ w_{26} & w_{16} & w_{43} & 0 & w_{36} & w_{46}\\ w_{23} & w_{13} & w_{53} & 0 & w_{33} & w_{43}\\ 0 & w_{14} & 0 & w_{64} & w_{34} & 0 \end{Vmatrix}\Psi, \tag{2}$$

IOP Conf. Series: Materials Science and Engineering **208** (2017) 012003 doi:10.1088/1757-899X/208/1/012003

where column vector $\Psi = \left\| u_2 \quad p_{23} \quad u_1 \quad u_3 \quad p_{13} \quad p_{33} \right\|^T$,

$$\begin{aligned}
&w_{11} = -i(k_{01}a_7 + k_{02}a_4), && w_{12} = S_{14}a_1 + S_{24}a_4 + S_{46}a_7 + S_{44},\\
&w_{13} = -i(k_{01}a_1 + k_{02}a_7), && w_{14} = -ik_{02},\\
&w_{15} = S_{14}a_2 + S_{24}a_5 + S_{46}a_8 + S_{45}, && w_{16} = S_{14}a_3 + S_{24}a_6 + S_{46}a_9 + S_{34},\\
&w_{21} = k_{01}^2\gamma_6 + k_{02}^2\gamma_4 - 2k_{01}k_{02}\gamma_5 - \rho\omega^2, && w_{23} = k_{01}^2\gamma_3 + k_{01}k_{02}(\gamma_6 - \gamma_1) - k_{02}^2\gamma_5,\\
&w_{25} = -i(k_{01}a_8 + k_{02}a_5), && w_{26} = -i(k_{01}a_9 + k_{02}a_6),\\
&w_{33} = -i(k_{01}a_2 + k_{02}a_8), && w_{34} = -ik_{01},\\
&w_{35} = S_{15}a_2 + S_{25}a_5 + S_{56}a_8 + S_{55}, && w_{36} = S_{15}a_3 + S_{25}a_6 + S_{56}a_9 + S_{35},\\
&w_{43} = -i(k_{01}a_3 + k_{02}a_9), && w_{46} = S_{13}a_3 + S_{23}a_6 + S_{36}a_9 + S_{33},\\
&w_{53} = k_{01}^2\gamma_1 + 2k_{01}k_{02}\gamma_3 + k_{02}^2\gamma_6 - \rho\omega^2, && w_{64} = -\rho\omega^2,
\end{aligned}$$

$$\begin{aligned}
&a_1 = S_{24}\gamma_2 - S_{14}\gamma_1 - S_{46}\gamma_3, && a_2 = S_{25}\gamma_2 - S_{15}\gamma_1 - S_{56}\gamma_3, && a_3 = S_{23}\gamma_2 - S_{13}\gamma_1 - S_{36}\gamma_3,\\
&a_4 = S_{14}\gamma_2 - S_{24}\gamma_4 + S_{46}\gamma_5, && a_5 = S_{15}\gamma_2 - S_{25}\gamma_4 + S_{56}\gamma_5, && a_6 = S_{13}\gamma_2 - S_{23}\gamma_4 + S_{36}\gamma_5,\\
&a_7 = S_{24}\gamma_5 - S_{14}\gamma_3 - S_{46}\gamma_6, && a_8 = S_{25}\gamma_5 - S_{15}\gamma_3 - S_{56}\gamma_6, && a_9 = S_{23}\gamma_5 - S_{13}\gamma_3 - S_{36}\gamma_6,
\end{aligned}$$

$k_{01} = k_0 \sin\theta_0 \cos\alpha$, $k_{02} = k_0 \sin\theta_0 \sin\alpha$ are projections of the wave vector $\vec{k}_0$ on the axes x_1 and x_2 (see Figure 1*b*) and $\gamma_j = \Delta_j/(S_{11}\Delta_1 - S_{12}\Delta_2 + S_{16}\Delta_3)$, $j = 1, \ldots, 6$, $\Delta_1 = S_{22}S_{66} - S_{26}^2$, $\Delta_2 = S_{12}S_{66} - S_{16}S_{26}$, $\Delta_3 = S_{12}S_{26} - S_{16}S_{22}$, $\Delta_4 = S_{11}S_{66} - S_{16}^2$, $\Delta_5 = S_{11}S_{26} - S_{16}S_{12}$, $\Delta_6 = S_{11}S_{22} - S_{12}^2$.

2.2. Transfer matrix

The integration of the equation (2) for a uniform layer $0 \le x_3 \le d$ gives solution in the form

$$\Psi(d) = T\Psi(0), \tag{3}$$

where matrix

$$T = \sum_{j=0}^{\infty} \frac{W^j d^j}{j!} \equiv \exp(Wd) \tag{4}$$

is called *transfer matrix*. If a structure consists from a set of N layers $0 \le x_3 < d_1$, $d_1 \le x_3 < d_2$, …, $d_{N-1} \le x_3 < d_N = d$ with transfer matrices T_1, T_2, $T_3, \ldots, T_N$, then $\Psi(d) = T_N T_{N-1} \ldots T_1 \Psi(0)$. In particular, $\Psi(d) = T^N \Psi(0)$, if $T_i = T$, $i = 1, \ldots, N$.

Transfer matrix T can be decomposed in the basis of the first six powers of the matrix Wd:

$$T = \exp(Wd) = \sum_{h=0}^{5} \mathscr{T}_h (Wd)^h. \tag{5}$$

We use representation of coordinates $\mathscr{T}_h$ by means of method [8]:

$$\mathscr{T}_h = \frac{1}{h!} + \sum_{j=6}^{\infty} \frac{1}{j!} \sum_{g=0}^{h} p_{6-h+g} \mathscr{B}_{j-1-g}(6), \quad h = 0, 1, \ldots, 5, \quad j = 0, 1, \ldots; \tag{6}$$

p_m $(m = 1, \ldots, 6)$ are coefficients of characteristic equation of matrix $Wd = \|w_{gh}d\|$, and polynomials $\mathscr{B}_g(6)$ are defined by recurrence relations

$$\mathscr{B}_0(6) = \ldots = \mathscr{B}_4(6) = 0; \quad \mathscr{B}_5(6) = 1; \quad \mathscr{B}_g(6) = \sum_{h=1}^{6} p_h \mathscr{B}_{g-h}(6), \; g \ge 6. \tag{7}$$

It should be noted that the calculation according to the formulas (5) – (7) does not use the eigenvalues of the matrix Wd, in contrast to the method of Lagrange-Sylvester, which was applied in propagator matrix method[3]. We use here a well known representation of coefficients p_m as a sums of principle minors of the matrix:

$$p_1 = \sum_{g=1}^{6} w_{gg}d,\ p_2 = -\sum_{j>i} \begin{vmatrix} w_{ii}d & w_{ij}d \\ w_{ji}d & w_{jj}d \end{vmatrix},\ \ldots,\quad p_6 = -\det(Wd). \tag{8}$$

Therefore, the functions $\mathscr{B}_g(6)$, which are determined by formulas (7), are called in this paper polynomials of principal minors.

Calculation of coefficients p_m conveniently carried out by the recurrence formulas (see method of Le Verrier[7]):

$$mp_m = s_m - p_1 s_{m-1} - \ldots - p_{m-1}s_1,\quad m = 1, 2, \ldots, 6,$$

where s_m are traces of the matrices $(Wd)^m$, $m = 1, \ldots, 6$.

For numerical calculations of transfer matrix T by the formula (5) one must approximate the series $\sum_{j=6}^{\infty}$ in the formulas (6) by sum of finite number of terms, let's say J:

$$\mathscr{T}_h \approx \frac{1}{h!} + \sum_{j=6}^{J-1} \frac{1}{j!} \sum_{g=0}^{h} p_{6-h+g}\mathscr{B}_{j-1-g}(6),\quad J = 6 + N,\ N = 0, 1, \ldots. \tag{9}$$

2.2.1. The relative truncation error. Obviously, the values (9) depend on J: $\mathscr{T}_h = \mathscr{T}_h(J)$. Equality in the formula (9) becomes exact, when $J \to \infty$, i.e. $\mathscr{T}_h(\infty)$, $h = 0, \ldots, 5$ conform the exact values of coordinates of the matrix $T = \exp(Wd)$ in the basis $I, Wd, \ldots, (Wd)^5$.

The value

$$\epsilon = \max \left| \frac{\mathscr{T}_h(\infty) - \mathscr{T}_h(J)}{\mathscr{T}_h(\infty)} \right| \tag{10}$$

is called the relative truncation error in calculating the coordinates $\mathscr{T}_h$ by the formulas (9).

It can be shown that if

$$\eta = 8.5\frac{\omega d}{\min v} < 1, \tag{11}$$

then

$$\epsilon < \frac{\eta^{N_1+1}}{(6 + N_1)\prod_{h=1}^{N_1}(6 + h)}. \tag{12}$$

The following notation is used in formulas (11) and (12): $\min v$ is minimum propagation velocity of elastic waves in the layer, $N_1 = J - 7$ is a difference between the limits of summation in $\sum_{j=6}^{J-1} \mathcal{C}_{jl}/j!$. The values of parameters η and N_1 can be regulated by scaling.

2.2.2. Scaling. If the condition (11) for the matrix Wd is not satisfied, we can always choose an integer m, for which

$$\eta = 8.5\frac{\omega}{\min v}\frac{d}{m} < 1.$$

In other words, we first calculate the matrix $\exp(Wd/m)$ with a given accuracy. And then we use the remarkable property of the exponential function:

$$T = \exp(Wd) = \left[\exp\left(\frac{Wd}{m}\right)\right]^m, \tag{13}$$

where the integer m is called *scaling parameter.*

Example. Let us consider the scaling procedure on the example of crystalline silicon. The density of Si $\rho = 2.329 \cdot 10^3$ kg/m^3. Non-zero elements of compliance tensor have values: $S_{11} = S_{22} = S_{33} = 7.69$, $S_{12} = S_{13} = S_{23} = -2.14$, $S_{44} = S_{55} = S_{66} = 12.58$ (in units 10^{-12}m^2/N). $d = 10^{-3}$m, $\nu = 10^7$Hz. The value $\min v$ can be estimated by the formula: $\min v \approx 1/(\sqrt{\rho \max S_{ij}})$. So, from (2.2.2) we find $m > 17\pi\nu d\sqrt{\rho \max S_{ij}} > 91$. The smallest number of matrix multiplications (namely seven) and, consequently, the smallest round-off errors will be when $m = 2^7 = 128$ and the last operation in (13) is performed by repeated squaring.

3. Six-wave scattering

If the regions $x_3 < 0$ and $x_3 > d$ bounding the layered structure are solid isotropic, then six scattered plane waves can propagate in them: SH wave with amplitude $\mathscr{A}_1$ and wave number $k_1 = 2\pi\nu\sqrt{\rho_0/\mu_0}$; SV wave with amplitude $\mathscr{A}_2$ and wave vector $\vec{k}_2 = \vec{k}_1$; P wave with amplitude $\mathscr{A}_3$ and wave number $k_3 = 2\pi\nu\sqrt{\rho_0/(2\mu_0 + \lambda_0)}$; SH wave with amplitude $\mathscr{A}_4$ and wave number $k_4 = 2\pi\nu\sqrt{\rho_d/\mu_d}$; SV wave with amplitude $\mathscr{A}_5$ and wave vector $\vec{k}_5 = \vec{k}_4$; P wave with amplitude $\mathscr{A}_6$ and wave number $k_6 = 2\pi\nu\sqrt{\rho_d/(2\mu_d + \lambda_d)}$. Here ρ_0, λ_0, μ_0 and ρ_d, λ_d, μ_d are density and Lamé parameters for regions $x_3 < 0$ and $x_3 > d$.

Possible directions of wave propagation in layered medium are defined by the relations $k_j \sin\theta_j = k_0 \sin\theta_0$, where θ_j is angle between wave vector $\vec{k}_j$ and axis x_3. All wave vectors $\vec{k}_j$ lie in the plane of incidence and

$$k_{j1} = k_{01} = k_0 \sin\theta_0 \cos\alpha, \quad k_{j2} = k_{02} = k_0 \sin\theta_0 \sin\alpha, \quad j = 1, \ldots, 6,$$
$$k_{13} = k_{23} = -k_1 \cos\theta_1, \quad k_{33} = -k_3 \cos\theta_3, \quad k_{43} = k_{53} = k_4 \cos\theta_4, \quad k_{63} = k_6 \cos\theta_6.$$

3.1. Components of vector functions $\Psi(0)$ and $\Psi(d)$

From the solutions of the equations (1) for an isotropic layer it follows that vector functions $\Psi(0)$ and $\Psi(d)$ are expressed in terms of the amplitudes $\mathscr{A}_h$ by formulas:

$$\psi_g(0) = \sum_{h=0}^{3} \beta_{gh}\mathscr{A}_h, \quad \psi_g(d) = \sum_{h=4}^{6} \beta_{gh}\mathscr{A}_h, \quad g = 1, \ldots, 6, \tag{14}$$

where $\beta_{1h} = c_{h2}$, $\beta_{2h} = i\mu(c_{h3}k_{h2} + c_{h2}k_{h3})$, $\beta_{3h} = c_{h1}$, $\beta_{4h} = c_{h3}$, $\beta_{5h} = i\mu(c_{h1}k_{h3} + c_{h3}k_{h1})$, $\beta_{6h} = i\Big[2\mu c_{h3}k_{h3} + \lambda\sum_{j=1}^{3} c_{hj}k_{hj}\Big]$,

$$\lambda,\ \mu = \begin{cases} \lambda_0,\ \mu_0, & \text{if } h = 0,1,2,3 \text{ (for region } x_3 < 0), \\ \lambda_d,\ \mu_d, & \text{if } h = 4,5,6 \quad \text{(for region } x_3 > d) \end{cases}$$

and direction cosines c_{h1}, c_{h2}, c_{h3} of vectors $\vec{\mathscr{A}}_0, \ldots, \vec{\mathscr{A}}_6$ are defined by Figures 2a, Figures 2b and Figures 3a, respectively.

3.2. Conversion coefficients

The equality (3) expresses the vector function $\Psi(d)$ through the $\Psi(0)$ with the help of transfer matrix $T = \|t_{ij}\|$. Substitution of expressions (14) in the formula (3) gives a system of algebraic equations for the amplitudes $\mathscr{A}_h$, $h = 1, \ldots, 6$:

$$d_{i1}\mathscr{A}_1 + d_{i2}\mathscr{A}_2 + \ldots + d_{i6}\mathscr{A}_6 = f_i\mathscr{A}_0, \quad i = 1, \ldots, 6, \tag{15}$$

where $f_i = -\sum_{j=1}^{6} t_{ij}\beta_{j0}$; $d_{il} = \sum_{j=1}^{6} t_{ij}\beta_{jl}$, if $l = 1, 2, 3$ and $d_{il} = -\beta_{il}$, if $l = 4, 5, 6$.

The solution of the system of equations (15) allows us to find in an obvious way the *conversion coefficients*

$$C_{ij} \equiv \left| \frac{\mathscr{A}_j}{\mathscr{A}_{0i}} \right|^2, \quad j = 1, \ldots, 6,$$

where $i = 1, 2, 3$ correspond to incident wave SH-type, SV-type and P-type, respectively.

Dependencies of the coefficients C_{ij} on the angles of incidence θ_0, α and on the propagation parameter κ were investigated for some crystals, the elastic constants of which are available in the monograph [9]. Examples of spectra for a silicon crystal are shown in Figures 2 and 3. Numbers $h = 1, \ldots, 6$ indicates curves $C_{gh}(\theta_0)$. They were calculated for the following conditions: $\rho_0 = \rho_d = 2.65{\cdot}10^3\mathrm{kg/m^3}$, $\lambda_0 = \lambda_d = 1.67{\cdot}10^{10}\mathrm{N/m^2}$, $\mu_0 = \mu_d = 3.27{\cdot}10^{10}\mathrm{N/m^2}$.

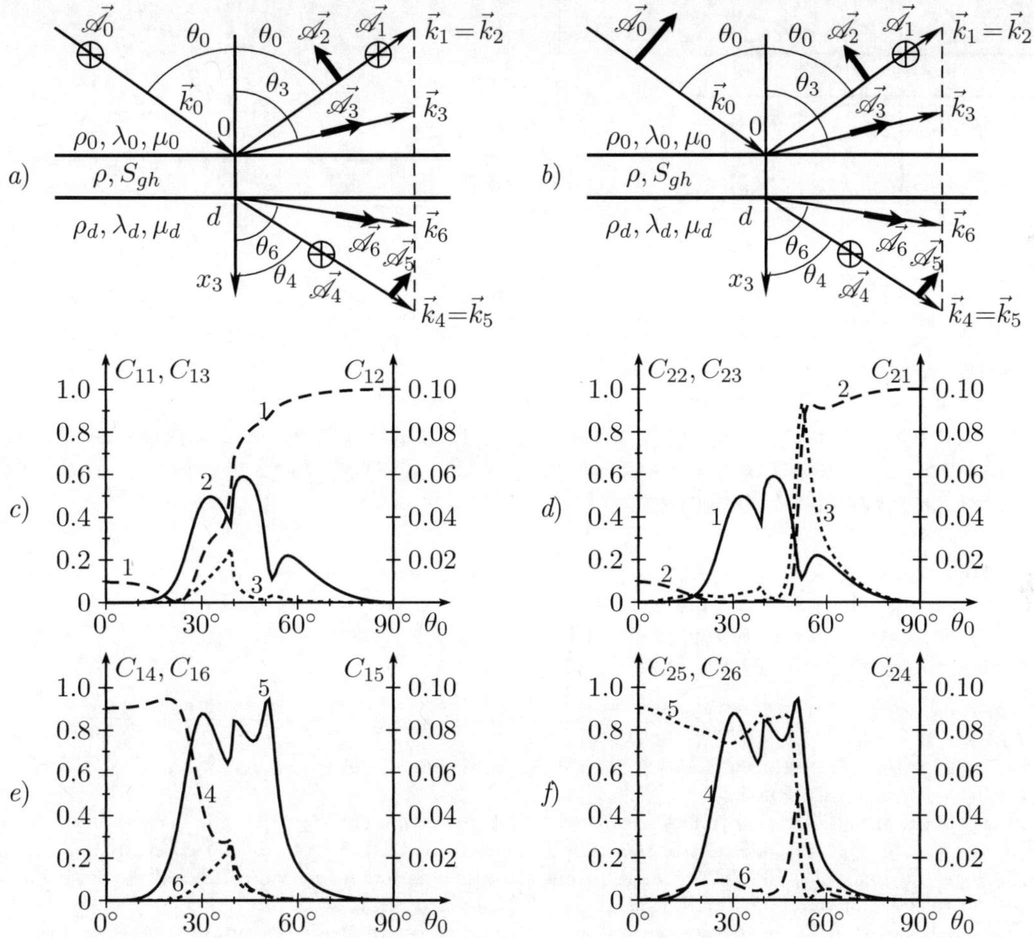

Figure 2. Conversion coefficients of the shear waves, $\alpha = 23°$, $\kappa = 2{\cdot}10^3\mathrm{m/s}$.

3.3. Conversions of shear waves

The results of modeling allow us to state that

$$C_{12}(\theta_0, \alpha) = C_{21}(\theta_0, \alpha), \quad C_{15}(\theta_0, \alpha) = C_{24}(\theta_0, \alpha). \tag{16}$$

Figure 2 illustrates this relations. Here functions $C_{12}(\theta_0), C_{21}(\theta_0), C_{15}(\theta_0), C_{24}(\theta_0)$ are shown by solid lines. It should be noted that the corresponding coefficients of SH and SV waves coversions into longitudinal waves differ substantially (see Figure 2*c* and Figure 2*d*; Figure 2*e* and Figure 2*f*).

Another important feature of the spectra $C_{1j}(\theta_0)$ and $C_{2j}(\theta_0)$ is the resonance character of the shear waves transformation into longitudinal waves. For the spectra shown in Figure 2, this can be observed in the regions near the angles $\theta_0 = 39°$ and $\theta_0 = 50°$. It is possible completely transformation of incident SH wave into SV-type in reflected wave. For example $C_{11} = 0$, $C_{12} = 0.375$, when $\theta_0 = 39.96°$, $\alpha = 14.95°$ and $\kappa = 9290$m/s (for Si).

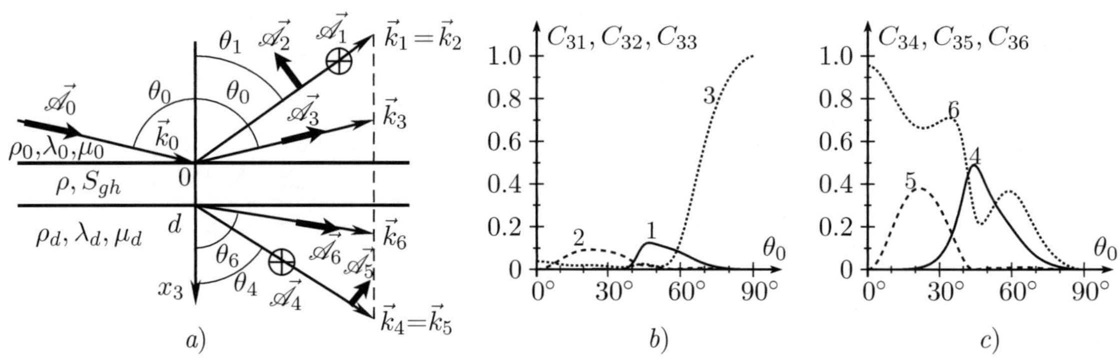

Figure 3. Scattering of P wave: *b*) for $\kappa = 5230$m/s, $\alpha = 35°$; for $\kappa = 5350$m/s, $\alpha = 24°$.

3.4. *P wave scattering*

Conversion of a longitudinal wave into the *SH* wave can be as significant as transformation $P \to SV$ and even stronger. Figure 3*b* shows, that for $\theta_0 = 52°$ *SH* wave prevails among the reflected waves. An analogous situation among the transmitted waves is observed in Figure 2*c* for the angle of incidence $\theta_0 = 35°$.

4. Conclusion

When performing calculations, the implementation of the law of conservation of energy flow was monitored. High accuracy of realizations of this law and relations (16) indicate the reliability of the transfer matrix calculations by means of MPPM.

References

[1] Molotkov L A 1984 *Matrichnyiy metod v teorii rasprostraneniya voln v sloistych uprugich i zhidkih sredah* (Leningrad: Nauka) (in Russian)
[2] Brekhovskikh L M and Godin O A 1990 *Acoustics of layered media* (Berlin: Springer-Verlag)
[3] Aki K, Richards P G 2002 *Quantitative seismology* (Sausalito CA: University Science Books)
[4] Potel C and de Belleval J-F 1993 Acoustic propagation in anisotropic periodically multilayered media: A method to solve numerical instabilities *J. Appl. Phys.* **74** 2208–2215
[5] Moler C and Van Loan C 2003 Nineteen dubious ways to compute the exponential of matrix, Twenty-five years later *SIAM Review* **45** 3–49
[6] Shaskol'skaya M P 1982 *Crystallography* (Moskow: Vysshaya Shkola Publishing House) (in Russian)
[7] Gantmacher F R 1959 *The thory of matrices*, Vol. 1 (New York: Chelsea)
[8] Belyayev Yu N 2013 On the calculation of functions of matrices *Mathematical Notes* **94** 177–184
[9] Dieulesaint E and Royer D 1974 *Ondes élastiques dans les solides. Application au traitment du signal* (Paris: Masson)

IOP Conf. Series: Materials Science and Engineering **208** (2017) 012004 doi:10.1088/1757-899X/208/1/012004

On static contact of belt and different pulleys

A K Belyaev[1], V V Eliseev[2], H Irschik[3] and E A Oborin[3]

[1]Institute for Problems in Mechanical Engineering, Russian Academy of Sciences, 61, Bolshoy ave., V.O., St. Petersburg, 199178, Russia

[2] Peter the Great Saint-Petersburg Polytechnic University, 29, Polytechnicheskaya St., St. Petersburg, 195251, Russia

[3]Institute of Technical Mechanics, Johannes Kepler University, Linz, Austria

E-mail: evgenii.oborin@jku.at

Abstract. The fitting of a looped belt on two pulleys with different radii is considered. A geometrically nonlinear model with account for tension and transverse shear is applied for modeling the belt. The pulleys are considered rigid bodies, and the belt-pulley contact is assumed frictionless. The problem has an axis of symmetry, therefore the boundary value problem is formulated and solved for a half of the belt. The considered part consists of three segments, two contact segments and a free span segment between them. The introduction of a dimensionless material coordinate at all segments leads to a system of ordinary differential equations of fifteenth order. The nonlinear boundary value problem for this system and boundary conditions is solved numerically with the shooting method and the finite difference method. As a result, the belt shape including the rotation angle, the forces, moments and contact pressure are determined. The contact pressure increases near the end point of contact areas, however no concentrated contact forces occur.

1. Introduction

The first systematic study of belt drives (and elastic creep in it) was reported by Reynolds in [1], where he used the string model. Until recently, one-dimensional models of elastic strings were widely used, cf. [2, 3, 4]. The exact solution for nonlinear steady state equations of extensible string is obtained in [3] by assuming zones of perfect and sliding friction contact between the belt and the pulley. A spatial description of the belt contour motion is suggested in [2]. An idealized point friction model allows extending this result to the transient dynamics, see [4].

However it turned out that the model of extensible string describes just a part of important effects in belt mechanics. Friction forces transmit power between the belt and the pulleys. They are applied on the belt from one side and result not only in tangent forces, but also in distributed moments. The model of extensible string without bending stiffness cannot resist the moment loading. Therefore, we use the rod model (i.e., the one with at least bending stiffness).

The calculation of belt-pulley interaction can be found in works [3, 5, 6, 7, 8, 9]. The study of interaction should be made with account for the transverse shear, because the importance of shear in contact problems of the rod theory is well-known, cf. [10, 11, 12, 13]. A contrary point of view is presented in [14]. The introduction of shear deformation causes the absence of lumped contact forces and promotes better understanding of the contact force distribution. The shear is also required to describe the effect of elastic microslip, see, e.g. [15, 16] for general friction modelling. Particularly the shear is of importance in friction belt drive operation [17, 18, 19].

IOP Conf. Series: Materials Science and Engineering **208** (2017) 012004 doi:10.1088/1757-899X/208/1/012004

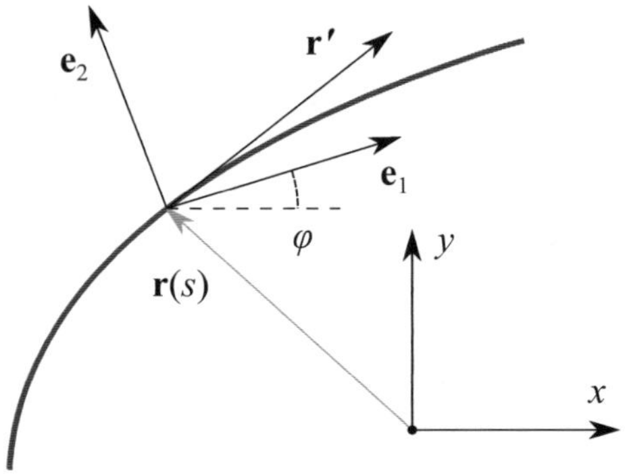

Figure 1. Nonlinear elastic rod with account for shear and tension. Deformed configuration, unit vectors $\mathbf{e}_1$, $\mathbf{e}_2$, tangent vector $\mathbf{r}'$, and angle φ.

However, shear is known to be taken into account only together with tension (compression) if the nonlinear theory of rods is applied. The goal of this work is to model the belt as a rod with bending and shear stiffness in the static problem where the belt is set on the pulleys. The system of equations of the corresponding rod theory can be found in [20, 21, 22, 23] (without direct application to belt drive mechanics, though).

The penalty formulation is in common use in numerical modeling of contact problems [24] and includes the choice of penalty coefficient or its distribution (see e.g. [25] for application to contact between rods). This approach is applied for belt drives in [9, 26, 27]. In the present paper we obtain the contact pressure and the stress-strain state in a different manner by describing the form of the belt lying on the rigid pulleys (the constraint formulation or the Hertz–Signorini–Moreau conditions for frictionless contact according to [24]).

We assume that the contact is full in a classical sense, i.e. it is continuous in an interval of the belt which length is to be determined. This assumption will be verified by obtaining the positive contact pressure. In the present study we use computer mathematics to solve difficult boundary value problems for the systems of ordinary differential equations (ODE). The application of standard methods is possible due to introduction of transformed material coordinate.

The presented study is an extension of the work [28]. Here we introduce significant complexity by rejecting the assumption of equal pulleys accepted there.

2. Equations of nonlinear contact problem

Figure 1 displays a general calculation scheme of a rod with transverse shear and tension in the xy plane. The rod is considered a Cosserat line [20, 21, 22, 23]. The position vector in the deformed configuration $\mathbf{r}(s)$ is a function of material coordinate s, and $\partial(...)/\partial s = (...)'$. The angular orientation of each particle of the rod is given by the orthonormal unit vectors $\mathbf{e}_1$, $\mathbf{e}_2$. The angle φ denotes the angle between unit vector $\mathbf{e}_1$ and Cartesian axis x, the latter having the unit vector $\mathbf{i}$. For plane deformations the system of equations of nonlinear theory of rods is as follows [20, 21, 22, 23]:

$$\begin{aligned} &\mathbf{Q}' + \mathbf{q} = 0, \;\; M' + \mathbf{k} \;\cdot \mathbf{r}' \times \mathbf{Q} + m = 0, \\ &\varphi' - \varphi_0' = AM, \;\; \mathbf{r}' = \mathbf{P} \cdot \mathbf{r}_0{}' + \mathbf{B} \cdot \mathbf{Q} = \mathbf{e}_1(1 + B_1 Q_1) + \mathbf{e}_2 B_2 Q_2. \end{aligned} \quad (1)$$

Here we denote: $\mathbf{Q}$ is the internal force vector, M is the bending moment, and $\mathbf{q}$, m stand for the external distributed forces and moments, respectively; $\mathbf{k}$ is the unit vector of Cartesian axis z perpendicular to the plane of drawing. The index zero indicates the values in an initial state, and $\mathbf{P} = \mathbf{e}_i \mathbf{e}_{i0}$ is the rotation tensor ($\mathbf{P} \cdot \mathbf{r}_0{}' = \mathbf{e}_1$). Elastic characteristics of the rod in plane are determined by three scalar compliances: the bending compliance A, the tension compliance B_1 and the shear compliance B_2. The unit vectors $\mathbf{e}_1$, $\mathbf{e}_2$ are directed along the principal axes of the compliance tensor $\mathbf{B}$.

Algorithms and results for the contact of belt and pulleys without shear are presented in [5, 6]. In the contact segment we introduce the spatial coordinate σ which is the arc coordinate in the actual configuration. The peculiarity of the considered problem is due to the fact that the function $\mathbf{r}(s) = \mathbf{R}(\sigma)$ is unknown because the arc coordinate σ is not yet determined.

We use the obvious geometric relations and the expression of $\mathbf{r}'$ from (1):

$$\begin{aligned} &\mathbf{e}_1 = \mathbf{i}\cos\varphi + \mathbf{j}\sin\varphi,\ \mathbf{e}_2 = -\mathbf{i}\sin\varphi + \mathbf{j}\cos\varphi; \\ &\mathbf{r}'(s)ds = \dot{\mathbf{R}}(\sigma)d\sigma,\ \left|\dot{\mathbf{R}}\right| = 1, \\ &\sigma' = |\mathbf{r}'| \equiv D = \sqrt{(1+B_1Q_1)^2 + (B_2Q_2)^2},\ \dot{s} = D^{-1}. \end{aligned} \tag{2}$$

The last equations determine the relation of material coordinate s with arc coordinate σ (a dot indicates the derivative with respect to σ).

Consider the full contact with the pulley. We write the tangent unit vector and integrate it

$$\begin{aligned} &\dot{\mathbf{R}} = \mathbf{i}\sin\kappa\sigma + \mathbf{j}\cos\kappa\sigma,\ \mathbf{r}' = D\dot{\mathbf{R}}, \\ &\mathbf{R} = \kappa^{-1}\left[\mathbf{i}\left(1-\cos\kappa\sigma\right) + \mathbf{j}\sin\kappa\sigma\right]. \end{aligned} \tag{3}$$

Here κ is the curvature of the pulley. We also assume $\mathbf{R}(0) = 0$ and note that

$$\begin{aligned} &\mathbf{r}'\cdot\mathbf{e}_1 = 1 + B_1Q_1 = D\sin(\varphi+\kappa\sigma), \\ &\mathbf{r}'\cdot\mathbf{e}_2 = B_2Q_2 = D\cos(\varphi+\kappa\sigma). \end{aligned} \tag{4}$$

Hence it is possible to express the transverse force Q_2 as a function of arc coordinate, angle and axial force in the following form:

$$Q_2(\sigma,\varphi,Q_1) = B_2^{-1}(1+B_1Q_1)\cot\alpha,\quad \alpha(\sigma,\ \varphi) \equiv \varphi + \kappa\sigma. \tag{5}$$

As a result we have four unknown functions of s which are $\sigma,\ \varphi,\ M,\ Q_1$.

We assume no tangential contact between the belt and the pulley. Therefore the vectors $\mathbf{q},\ \mathbf{r}'$ are orthogonal:

$$\begin{aligned} &\mathbf{Q}'\cdot\mathbf{r}' = 0;\ \mathbf{e}_1{}' = \varphi'\mathbf{e}_2,\ \mathbf{e}_2{}' = -\varphi'\mathbf{e}_1; \\ &(Q_1{}' - \varphi'Q_2)(1+B_1Q_1) + (Q_2{}' + \varphi'Q_1)B_2Q_2 = 0. \end{aligned} \tag{6}$$

In the last equation we express the derivatives using (1), (2), and $Q_2 = Q_2(\sigma,\varphi,Q_1)$ as follows:

$$\begin{aligned} &\varphi' = \varphi_0{}' + AM, \\ &Q_2{}' = \frac{\partial Q_2}{\partial Q_1}Q_1{}' + \frac{\partial Q_2}{\partial \sigma}D + \frac{\partial Q_2}{\partial \varphi}\varphi' = \frac{B_1}{B_2}Q_1{}'\cot\alpha - \frac{1+B_1Q_1}{B_2\sin^2\alpha}(\varphi' + \kappa\sigma'). \end{aligned} \tag{7}$$

The combination of (6) and (7) provides the differential equation for Q_1 not stated here because it is cumbersome. Then we substitute $\mathbf{r}'$ from (1) into the balance of moment equation and obtain

$$M' = \left((B_2 - B_1)Q_1 - 1\right)Q_2. \tag{8}$$

The expressions $\sigma',\ \varphi',\ M',\ Q_1'$ form a fourth order ODE system:

$$Y' = F(s,Y),\ Y \equiv \begin{pmatrix} \sigma & \varphi & M & Q_1 \end{pmatrix}^T. \tag{9}$$

Finally we determine the contact pressure p:

$$Dp = \mathbf{q}\cdot\mathbf{k}\times\mathbf{r}' = \mathbf{k}\cdot\mathbf{Q}'\times\mathbf{r}' = (Q_1{}' - \varphi'Q_2)B_2Q_2 - (Q_2{}' + \varphi'Q_1)(1+B_1Q_1). \tag{10}$$

The contact pressure must be non-negative. Unlike the case without shear, in the considered problem we expect no lumped contact reactions. The similar case of the classical contact between an initially straight nonlinear shearable rod and a rigid straight obstacle is discussed e.g. in [11]. Also the distributed contact moment which plays the key role in the elastic microslip (see [18]) in loaded drive is zero because of the frictionless contact.

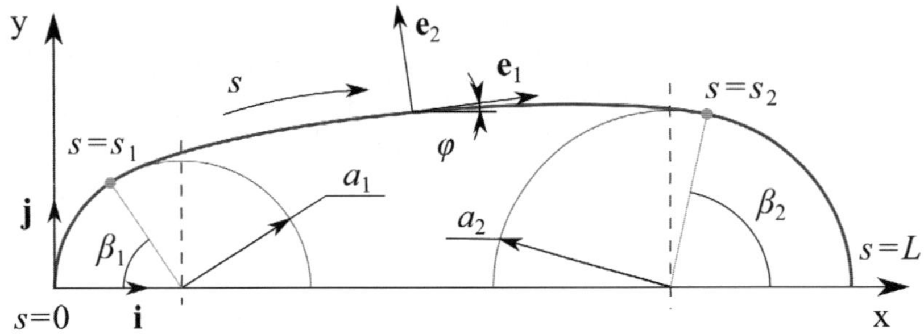

Figure 2. Contact of belt and pulleys. Thick line is deformed belt, thin half-circles are pulleys.

3. Contact of belt with pulleys

The scheme of fitting the belt on the pulleys is shown in Fig. 2. In the initial state the belt is the circle of radius a_0. Hence the initial angle and its derivative are $\varphi_0 = \pi/2 - a_0^{-1}s$, $\varphi_0' = -a_0^{-1}$. The pulleys have radii a_1 and a_2. The centre distance (i.e. the distance between the pulley centres) is $2a_0 - a_1 - a_2 + \delta$, where the pulley displacement δ increases from zero under force P that takes the pulleys apart.

Because of the problem symmetry it is sufficient to consider only the upper half of the belt $0 \le s \le L = \pi a_0$. In the segment $0 \le s \le s_1$ the full contact is assumed, however the coordinate s_1 is unknown. The formulation for this segment is presented above. For the segment of contact with another pulley we have $s_2 \le s \le L$, s_2 is also unknown. We will use upper indices: $(...)^{(1)}$ for the values in the left contact segment and $(...)^{(3)}$ for the values in the right contact segment. We assign the spatial coordinate origin by the equation $\sigma^{(3)}(L) = 0$, inside the segment $\sigma^{(3)}$ is negative. Similarly to (3) we write

$$\begin{aligned} \dot{\mathbf{R}} &= -\mathbf{i}\sin\kappa_2\sigma^{(3)} + \mathbf{j}\cos\kappa_2\sigma^{(3)}, \quad \mathbf{r}' = D^{(3)}\dot{\mathbf{R}}; \\ \mathbf{R} &= \mathbf{i}\left[2R + \delta - \kappa_2^{-1}\left(1 - \cos\kappa_2\sigma^{(3)}\right)\right] - \mathbf{j}\kappa_2^{-1}\sin\kappa_2\sigma^{(3)}. \end{aligned} \tag{11}$$

We make use of the formula (11) as the boundary condition below. Also we transform the equation (4) to the form:

$$\begin{aligned} \mathbf{r}' \cdot \mathbf{e}_1 &= 1 + B_1Q_1^{(3)} = -D^{(3)}\sin\left(\varphi^{(3)} + \kappa_2\sigma^{(3)}\right) \\ \mathbf{r}' \cdot \mathbf{e}_2 &= B_2Q_2^{(3)} = -D^{(3)}\cos\left(\varphi^{(3)} + \kappa_2\sigma^{(3)}\right). \end{aligned} \tag{12}$$

This gives four equations for the segment similar to (9).

Let us turn to the free segment $s_1 < s < s_2$. As follows from the force balance equation in (1), the force $\mathbf{Q}$ is constant. Using the symmetry we can determine this force:

$$\mathbf{Q} = \frac{P}{2}\mathbf{i} + Q_y\mathbf{j}. \tag{13}$$

The value P is prescribed, however the dependence $\delta(P)$ should be determined. Q_y is the unknown constant. Due to the obvious absence of lumped contact reaction at point s_1 (and at s_2 also) we have the continuity of $\mathbf{Q}$:

$$\begin{aligned} Q_1 &= \frac{P}{2}\cos\varphi + Q_y\sin\varphi, \\ Q_2 &= -\frac{P}{2}\sin\varphi + Q_y\cos\varphi \end{aligned} \tag{14}$$

(it also holds everywhere in the free segment). Because the force is constant, we are able to integrate the equation of moments (1):

$$M = \frac{P}{2}y - Q_y x + M_*. \tag{15}$$

We use the continuity of moment at points s_1, s_2 in order to determine the constants M_*, Q_y. Then we derive the following equations from (1):

$$\varphi' = -a_0^{-1} + AM,\ x' = (1+B_1Q_1)\cos\varphi - B_2Q_2\sin\varphi,\ y' = (1+B_1Q_1)\sin\varphi + B_2Q_2\cos\varphi. \tag{16}$$

The functions φ, x, y are continuous at points s_1, s_2.

To solve the formulated problem as a single boundary value problem, we introduce the new nondimensional coordinate $0 \le \xi \le 1$:

$$\begin{aligned} &s \le s_1:\ s = \xi s_1 \Rightarrow \frac{d}{d\xi} = s_1\frac{d}{ds};\\ &s_1 \le s \le s_2:\ s = s_1 + \xi(s_2 - s_1) \Rightarrow \frac{d}{d\xi} = (s_2 - s_1)\frac{d}{ds};\\ &s_2 \le s \le L:\ s = L - \xi(L - s_2) \Rightarrow \frac{d}{d\xi} = (s_2 - L)\frac{d}{ds}. \end{aligned} \tag{17}$$

We combine the equations at both segments into a single system and distinguish the values by the indices $(...)^{(1)}$, $(...)^{(2)}$, $(...)^{(3)}$. We note that this transformation extends the formulation for the problem of laying process of underwater pipeline [11] where contact of an initially straight beam with a flat rigid surface is considered. Thus we obtain the boundary value problem for the fifteenth order ODE system:

$$Z' = G(\xi, Z),\ Z(\xi) \equiv \left(\sigma^{(1)}\ \varphi^{(1)}\ M^{(1)}\ Q_1^{(1)}\ \sigma^{(3)}\ \varphi^{(3)}\ M^{(3)}\ Q_1^{(3)}\ \varphi^{(2)}\ x\ y\ M_*\ s_1\ s_2\ Q_y\right)^T. \tag{18}$$

Here a prime indicates the derivative with respect to the new coordinate ξ (and the functions written in (18) are the functions of ξ). We must not forget the relations between the derivatives (17). For the first eight unknown variables we have already derived the equations of type (9). The remaining seven equations are equations (16) and conditions for the unknown constants: ${M_*}' = 0$, ${s_1}' = 0$, $s_2' = 0$, $Q_y' = 0$.

Now we consider the boundary conditions for the system (18). We note that the right end $\xi = 1$ corresponds to the unknown boundary of the contact segment $s = s_1$ for the first segment and $s = s_2$ for the second and third segments whereas the left end $\xi = 0$ corresponds to the start point $s = 0$, to the unknown boundary $s = s_1$, and to the end point $s = L$ for three segments sequentially. For the points with material coordinate $s = 0$ and $s = s_2$ the conditions are

$$\begin{aligned} &\xi = 0:\ \sigma^{(1)} = 0,\ \sigma^{(3)} = 0,\ \varphi^{(1)} = \frac{\pi}{2},\ \varphi^{(3)} = -\frac{\pi}{2},\\ &\xi = 1:\ \varphi^{(2)} = \varphi^{(3)},\ Q_1^{(3)} = \frac{P}{2}\cos\varphi^{(2)} + Q_y\sin\varphi^{(2)},\ -a_2\sin\kappa\sigma^{(3)} = y;\\ &Q_2\left(\sigma^{(3)}, \varphi^{(3)}, Q_1^{(3)}\right) = -\frac{P}{2}\sin\varphi^{(2)} + Q_y\cos\varphi^{(2)},\ M^{(3)} = \frac{P}{2}y - Q_y x + M_*. \end{aligned} \tag{19}$$

And for $s = s_1$ the conditions are

$$s = s_1:\ \varphi^{(1)}\Big|_{\xi=1} = \varphi^{(2)}\Big|_{\xi=0},\ M^{(1)}\Big|_{\xi=1} = \left(\frac{P}{2}y - Q_y x + M_*\right)\Big|_{\xi=0},$$

IOP Conf. Series: Materials Science and Engineering **208** (2017) 012004 doi:10.1088/1757-899X/208/1/012004

$$a_1\left(1-\cos\kappa_1\sigma^{(1)}\right)\Big|_{\xi=1}=x|_{\xi=0}\,,\ a_1\sin\kappa_1\sigma^{(1)}\Big|_{\xi=1}=y|_{\xi=0}\,,$$
$$Q_1^{(1)}\Big|_{\xi=1}=\left(\frac{P}{2}\cos\varphi^{(2)}+Q_y\sin\varphi^{(2)}\right)\Big|_{\xi=0},$$
$$Q_2\left(\sigma^{(1)},\varphi^{(1)},Q_1^{(1)}\right)\Big|_{\xi=1}=\left(-\frac{P}{2}\sin\varphi^{(2)}+Q_y\cos\varphi^{(2)}\right)\Big|_{\xi=0}. \tag{20}$$

4. Numerical results

The formulated nonlinear boundary value problem of fifteenth order (18),(19),(20) is solved by the shooting method in Wolfram Mathematica using the standard built-in function `NDSolve`. Also we use the finite difference method for this system following the scheme described in [11]. The results of both methods agree.

In the finite difference method we divide the segments into N uniform intervals with the step $\varepsilon = 1/N$ (with respect to the transformed coordinate ξ). To approximate the system of equations we use implicit symmetric one-step difference scheme that has second order of accuracy [29, 30, 31]: $(Z_{i+1}-Z_i)/\varepsilon = (G_i + G_{i+1})/2$, $(i = 0, ..., N-1)$. As a result of transformations we obtain the system of $11N$ nonlinear algebraic equations. Total number of unknowns consists of the values of eleven desired functions at the nodes $11(N+1)$ and four additional constants M_*, s_1, s_2, Q_y. The system of equations is complemented with the fifteen boundary conditions (in our case the Dirichlet boundary conditions) projected into mesh. Furthermore we apply the Newton method using built-in function `FindRoot` (Mathematica). We use our solution of the problem with equal pulleys [28] as an initial guess in order to begin the computations performed with the numerical continuation method.

The form of the belt before and after the deformation is shown in Fig. 3. The parameters are: the Young modulus is $E = 10^9$, the Poisson coefficient is $\nu = 0.5$, the sides of the square cross-section are $h = 10^{-2}$, the initial radius of the belt is $a_0 = 0.25$, the pulley radii are $a_1 = 0.15$, $a_2 = 0.1$, the current force is $P = 200$ (all values are in SI units). The results of computations are: the end coordinate of the first contact segment $s_1 = 0.163$, the second contact segment has the length $L - s_2 = 0.0585$, the displacement of the pulley $\delta = 0.132$. The central angles of the contact zones are introduced in Fig. 2, their values are $\beta_1 = 1.088$, $\beta_2 = 0.586$ in the example considered.

Then we determine the contact pressure using the formula (10). We take into account equation (6) and rewrite it as follows:

$$p = D^{-1}\left({Q_1}' - \varphi' Q_2\right)\left[B_2Q_2 + \frac{\left(1+B_1Q_1\right)^2}{B_2Q_2}\right]. \tag{21}$$

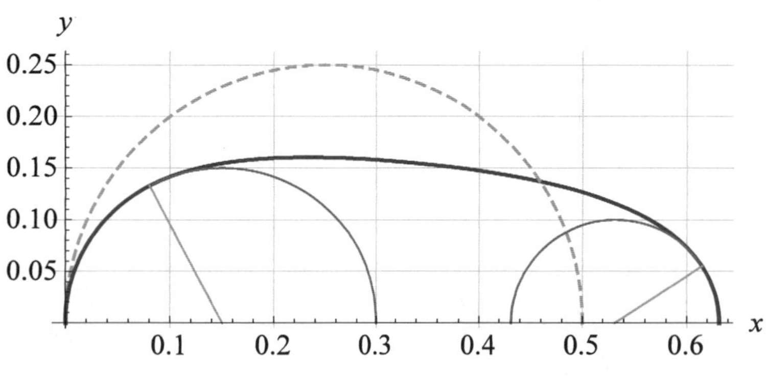

Figure 3. Belt and pulleys. Thick line is deformed belt, thin half-circle is pulley, dashed half-circle is undeformed belt, thin line is contact area boundary.

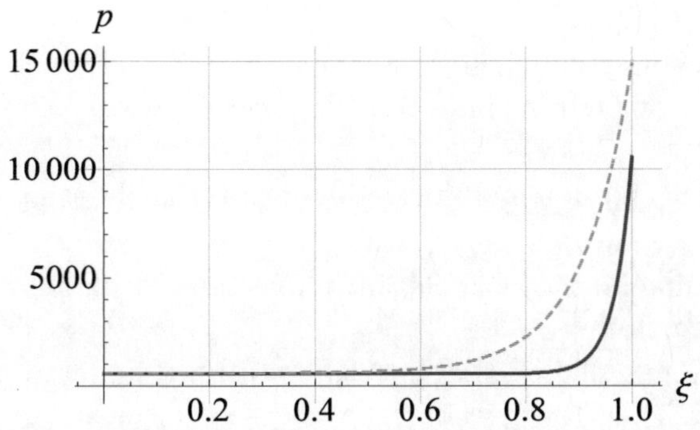

Figure 4. Contact pressure over transformed coordinate.
Full line is first pulley, dashed line is second pulley.

The result of calculation is shown in Fig. 4. The pressure is positive. We see the concentration of pressure at the boundary of the contact segment which is characteristic for the problems with shear [11]. For the unshearable models we have the constant pressure distribution with the lumped contact reaction at the contact boundary [5, 6]. In the present case there is no lumped reaction force at the boundary. The right contact segment is smaller, therefore the pressure distribution is smoother in it. The maximal values of pressure are $p^{(1)} = 10574$, $p^{(2)} = 14879$.

Finally we demonstrate the loading diagrams. The load P increases from zero while other parameters are the same as above. We solve a number of static problems with the numerical continuation method to show the dependences. The calculations were performed until the contact segments reduce itself to a point.

The dependence of force $P(\delta)$ on the pulley displacement is depicted in Fig. 5. It is nonlinear from the beginning, its behaviour is similar to the one obtained for problem with equal pulleys.

The dependence $\beta(\delta)$ of the angle of contact zone on displacement of pulley is shown in Fig. 6. The angle of the left contact zone is wider than the angle of the right contact zone as expected.

5. Conclusion

The main results of the present work are listed below:

- a system of ODE for the belt as plane nonlinear elastic rod with account for tension and

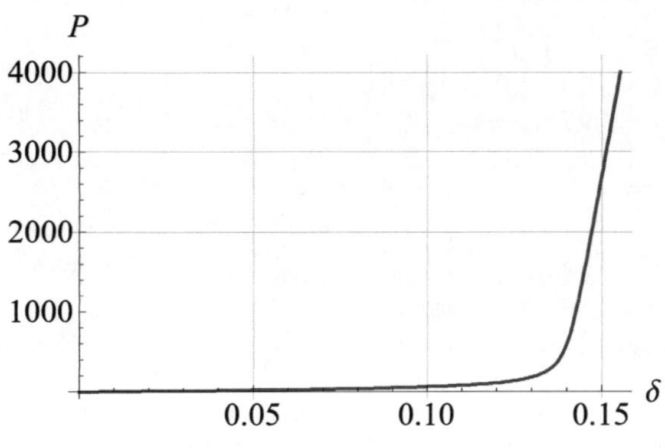

Figure 5. Force taking pulleys apart over displacement of pulley.

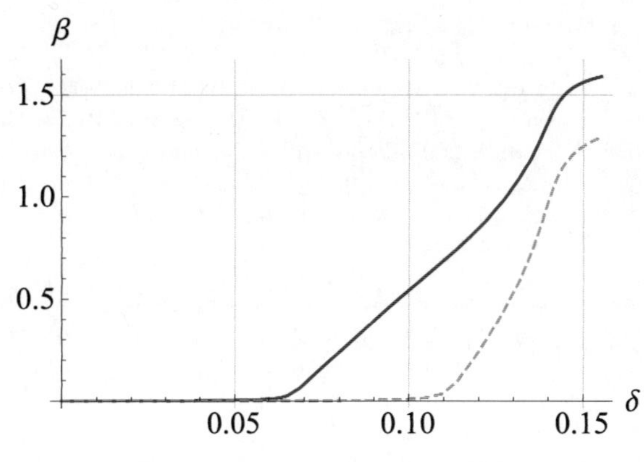

Figure 6. Contact angle over displacement of pulley. Full line is first pulley; dashed line is second pulley.

transverse shear is derived;

- the equations for the free and contact segments which have qualitatively different formulation are combined in a single system of ODE. This transformation is done in a novel way by introducing the new coordinate;
- the formulated boundary value problem is suitable for numerical solving by the shooting method and finite difference method;
- computer mathematics is utilized for determining all the variables including the coordinates of boundaries of contact areas and contact pressure;
- the effect of different pulley radii on the belt-pulley interaction problem is studied in detail.

These results will be used in further work including the study of belt-pulley interaction with friction and the dynamical modeling, both stationary and transient.

Acknowledgments

This research is carried out in the framework of the joint project of the Russian Foundation for Basic Research (grant No. 14-51-15001) and the Austrian Science Fund (grant No. I 2093 International Project).

References

[1] Reynolds O 1874 *The Engineer* **38** 396
[2] Eliseev V and Vetyukov Y 2012 *Acta Mechanica* **223** 1657–1667
[3] Rubin M B 2000 *J Mech. Des.* **122** 311–316
[4] Vetyukov Y, Oborin E, Krommer M and Eliseev V 2017 *Mathematical and Computer Modelling of Dynamical Systems* **23** 40–54
[5] Belyaev A, Eliseev V, Irschik H and Oborin E 2016 *PAMM* **16** 11–14
[6] Belyaev A, Eliseev V, Irschik H and Oborin E 2017 *Dynamics & Control of Advanced Structures and Machines* ed Irschik H, Belyaev A and Krommer M (Springer International Publishing) pp 195–203
[7] Kong L and Parker R G 2005 *J Appl. Mech.* **72** 25–34
[8] Morimoto T and Iizuka H 2012 *Int J Mech. Sci.* **54** 234–240
[9] Pechstein A and Gerstmayr J 2013 *Multibody System Dynamics* **30** 343–358
[10] Chen J S 2011 *Acta Mechanica* **222** 225–232
[11] Eliseev V and Zinovieva T 2012 *Comp. Cont. Mech.* **5** 70–78
[12] Gasmi A, Joseph P F, Rhyne T B and Cron S M 2011 *Int. J Solids & Structures* **48** 843 – 853
[13] Gasmi A, Joseph P F, Rhyne T B and Cron S M 2012 *Int. J Solids & Structures* **49** 2604–2616
[14] Osipenko M, Nyashin Y I and Kasatkin A 2015 *PRNPU Mechanics Bulletin* 121–129
[15] Berger E 2002 *Applied Mech. Reviews* **55** 535–577
[16] Menq C H, Bielak J and Griffin J 1986 *J Sound & Vibration* **107** 279–293
[17] Alciatore D and Traver A 1995 *ASME J Mech. Des.* **117** 506–511
[18] Belyaev A K, Eliseev V V and Oborin E A 2016 *Int. Review Mech. Eng.* **10** 333–338
[19] Firbank T 1970 *Int J Mech. Sci.* **12** 1053–1063
[20] Antman S 2005 *Nonlinear problems of elasticity* (New York: Springer)
[21] Eliseev V 1988 *J Appl. Math. & Mech.* **52** 493–498
[22] Eliseev V 2006 *Mechanics of deformable solid bodies* (St. Petersburg: St.-Peterbg. State Polytech. Univ. Publ. House)
[23] Villaggio P 2005 *Mathematical models for elastic structures* (Cambridge University Press)
[24] Wriggers P and Laursen T A 2006 *Computational contact mechanics* (Berlin: Springer)
[25] Durville D 2012 *Computational Mechanics* **49** 687–707
[26] Dufva K, Kerkkänen K, Maqueda L and Shabana A 2007 *Nonlinear Dyn.* **48** 449–466
[27] Čepon G and Boltežar M 2009 *J Sound & Vib.* **319** 1019–1035
[28] Belyaev A, Eliseev V, Irschik H and Oborin E 2017 *Acta Mechanica* to appear in 2017
[29] Bahvalov N S, Zhidkov N P and Kobelkov G G 2003 *Numerical methods* (Moscow: Binom. Laboratoria znanii)
[30] Pozrikidis C 2008 *Numerical computation in science and engineering* (Oxford: Oxford University Press)
[31] Samarskii A A and Vabishchevich P N 2008 *Numerical methods for solving inverse problems of mathematical physics* (*Inverse and Ill-Posed Problems* no 52) (Berlin, Boston: De Gruyter)

Investigation of deformation of elements of three-dimensional reinforced concrete structures located in the soil, interacting with each other through rubber gaskets

D V Berezhnoi, I S Balafendieva, A A Sachenkov and L R Sekaeva

Kazan Federal University, 18 Kremlyovskaya Street, Kazan, 420008, Russian Federation

e_xo@mail.ru

Abstract. In work the technique of calculation of elements of three-dimensional reinforced concrete substructures located in a soil, interacting with each other through rubber linings is realized. To describe the interaction of deformable structures with the ground, special "semi-infinite" finite elements are used. A technique has been implemented that allows one to describe the contact interaction of three-dimensional structures by means of a special contact finite element with specific properties. The obtained numerical results are compared with the experimental data, their good agreement is noted.

1. Introduction
In modern mechanics of a deformable solid, the problems of non-linear deformation of elements of three-dimensional structures with each other and with the surrounding soil are still relevant and in demand in construction, road construction, foundation engineering, etc. Many authors have dealt with the nonlinear problems of mechanics, in particular, papers [1-11] can be noted in which theoretical features of the construction of geometrically and physically nonlinear deformation of structures are noted. Practical problems in various areas of mechanics of a deformed solid are usually solved numerically [12-19]. It is also possible to note realized methods of contact interaction problems [20-24] and hyperelastic behavior of three-dimensional structures [25-29]. This paper develops a mathematical model and realizes a numerical algorithm for studying the stress-strain state of three-dimensional reinforced structures interacting with each other and with the water-saturated soil in which they are located.

2. Formulation of the problem
The purpose of this paper is to construct a technique for calculating the stress-strain state of three-dimensional structures located in the soil, the elements of which interact with each other. The object of calculation is the lining of the subway tunnel from prefabricated reinforced concrete, geometrically representing a cylindrical shell located in the soil mass, under the influence of the weight of the above rocks and the longitudinal compressive force that occurs when this structure is created. The main feature of the design under consideration is the fact that a similar lining of segment blocks is being constructed, which interact with each other through special lining that does not completely cover the interface surfaces, which entirely transmit only compressive forces and partially tangential forces

IOP Conf. Series: Materials Science and Engineering **208** (2017) 012005 doi:10.1088/1757-899X/208/1/012005

within the frictional forces. Fig. 1 partially shows three rings of lining, in real proportions, with conditionally remote blocks to illustrate their relative location.

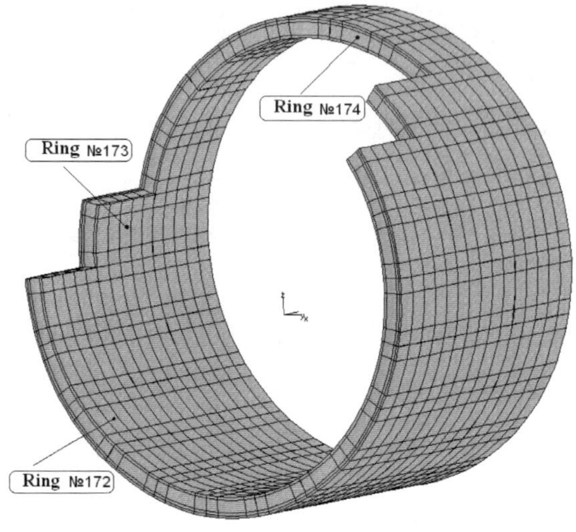

Figure 1. Finite-element partitioning of three rings of the lining of the subway tunnel (some blocks have been removed for clarity).

The calculation was carried out by FEM on the basis of twisted 20-node isoparametric finite elements of the three-dimensional theory of elasticity with quadratic approximation of geometry and displacements. The mesh of similar elements used to calculate the described construction is also shown on Fig.1.

3. Modeling of mechanical contact through gaskets

The mechanism of interaction of blocks through the lining can be illustrated on Fig. 2, where various variants of deformation of the overlays are depicted, depending on the forces exerted by the blocks on each other.

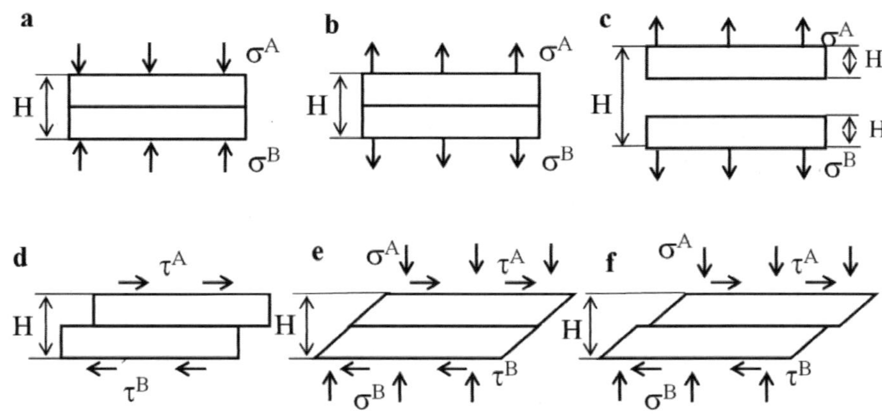

Figure 2. The mechanism of interaction of blocks through gaskets.

For the situation on Fig. 2a, we have that in the overlays there is a stress of compression $\sigma_H = \sigma^A = \sigma^B$ and deformation $\varepsilon_H = \sigma_H / E_H$, where E_H is the modulus of elasticity of the lining material. The geometrical condition for the presence of this situation is $H \le \left(H^A + H^B\right)$, where H^A, H^B is the original thickness of the linings, and H is the distance between the surfaces on which they are fixed.

The situation on Fig. 2b arises in the presence of preliminary reduction, i.e. $H<\left(H^A+H^B\right)$ and, in this case, too, $\sigma_H=\sigma^A=\sigma^B$, $\varepsilon_H=\sigma_H/E_H$ is true.

On Fig. 2c there is no force action and the patches move freely. In this case $H\geq\left(H^A+H^B\right)$, $\sigma_H=0$.

Fig. 2d shows free slip, in which tangential stresses do not arise, which is realized when $H\geq\left(H^A+H^B\right)$, and in this case $\tau_H=0$.

Fig. 2e illustrates the elastic interaction with compression and shear without slipping. A similar situation is possible with $H\leq\left(H^A+H^B\right)$ and for stresses and deformations in the overlays can be written: $\sigma_H=\sigma^A=\sigma^B$, $\tau_H=\tau^A=\tau^B$, $\varepsilon_H=\sigma_H/E_H$, $\gamma_H=\tau_H/G_H$. An additional condition here should be the condition $\tau_H\leq f\left|\sigma_H\right|$, where f is the linear coefficient of friction.

If these conditions are not fulfilled, the situation appears on Fig. 2f. In this case $\sigma_H=\sigma^A=\sigma^B$, $\tau_H=f\left|\sigma_H\right|$, $\varepsilon_H=\sigma_H/E_H$ and slip is observed.

All these situations are modeled within the framework of continuum mechanics, i.e. when two overlays are presented in the form of a single material with specific properties. The problem obtained is non-linear and requires the use of special techniques for solving it. A characteristic feature of this nonlinearity is that for normal stresses there are limitations on deformation ($H\leq H^A+H^B$, i.e. the mutual deformation of the linings can't be greater than their total thickness), and for tangential stresses, by their limiting values, that determine the possibility of slippage.

To solve the formulated physically nonlinear problem on the basis of the virtual work equation, an iterative process is constructed, which is a combination of the initial stress method and the additional deformation method. The basic for the determination of the *k*-th iteration is the following variational equation

$$\sum_m \iiint_{\Omega_m} \{\sigma^k\}^T \{\delta\varepsilon\} d\Omega + \sum_k \iiint_{\Omega_k} \left(\varepsilon_H^k E_H \delta\varepsilon_H + \gamma_H^k G_H \delta\gamma_H\right) d\Omega_k =$$
$$= \sum_m \left(\iiint_{\Omega m} \rho \{g\}^T (\delta V) d\Omega + \iint_{S_m^\sigma} \{p\}^T \{\delta V\} ds \right) + \sum_k \iiint_{\Omega_k} \left(\hat{\sigma}_H^k \delta\varepsilon_H + \hat{\tau}_H^k \delta\gamma_H\right) d\Omega_k, \tag{1}$$

where $\hat{\sigma}_H^k$ and $\hat{\tau}_H^k$ are determined as the differences between the corresponding values obtained from linear relationships of the theory of elasticity and their real values, which depend on the interaction mechanism (Fig. 2).

4. Contact finite element

To implement the previously described mathematical model of the interaction of overlays within the framework of the FEM, the so-called contact element is determined. Geometrically it is a shell quadrangular element with 16 nodes. As initial information for it, the radii-vectors of the points defining the lower (odd numbers) and the upper (even numbers) of the surface are given, and the initial thickness $H=H^A+H^B$, which can be constant on the element, may vary (in this case, their nodal values are specified).

Approximations of the face surfaces are introduced

$$\vec{r}^{(-)}(\xi,\eta)=\sum_{i=1}^{8}\vec{r}_{2i-1}N_i(\xi,\eta),\quad \vec{r}^{(+)}(\xi,\eta)=\sum_{i=1}^{8}\vec{r}_{2i}N_i(\xi,\eta,\zeta), \tag{2}$$

where $N_i(\xi,\eta)$ - are the known form functions for the two-dimensional quadratic approximation of the Sirendip family. To approximate the displacement vector, we use a similar representation

$$\vec{V}^{(-)}(\xi,\eta)=\sum_{i=1}^{8}\vec{V}_{2i-1}N_i(\xi,\eta),\quad \vec{V}^{(+)}(\xi,\eta)=\sum_{i=1}^{8}\vec{V}_{2i}N_i(\xi,\eta). \tag{3}$$

In the process of deformation, initially parallel facial surfaces $\vec{r}^{(-)}$ and $\vec{r}^{(+)}$ cease to be so. An illustration of this situation is given on Fig. 3, where it is clear that the unit vectors of the normals $\vec{P}_3^{(-)}$, $\vec{P}_3^{(+)}$ are essentially different. Therefore, we will determine all geometric, kinematic and power characteristics on both face surfaces independently. In other words, the stress-strain state will be determined independently in each overlay (adjacent, respectively, to the surfaces $\vec{r}^{(-)}$ and $\vec{r}^{(+)}$), which will allow more correctly simulate their state when slipping relative to each other.

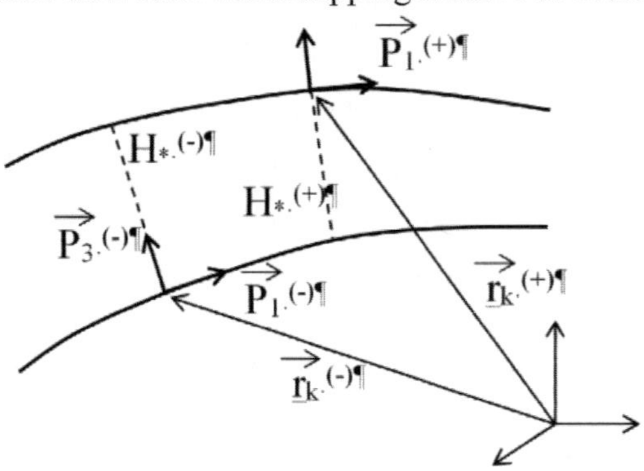

Figure 3. Schematic representation of the face surfaces of the contact elements.

The distances $H_*^{(-)}$ and $H_*^{(+)}$ from one face to the other, along the normal to it, are determined in the form

$$H_*^{(-)}=\vec{P}_3^{(-)}\cdot\left[\vec{r}^{(+)}+\vec{V}^{(+)}-\vec{r}^{(-)}-\vec{V}^{(-)}\right],\quad H_*^{(+)}=\vec{P}_3^{(+)}\cdot\left[\vec{r}^{(+)}+\vec{V}^{(+)}-\vec{r}^{(-)}-\vec{V}^{(-)}\right]. \tag{4}$$

These values are basic in judging the nature of the interaction between the overlays (the classification is shown in Fig. 2). The compression deformation is defined as

$$\varepsilon_H^{(+)}=\frac{H_*^{(+)}-H}{H},\varepsilon_H^{(-)}=\frac{H_*^{(-)}-H}{H}. \tag{5}$$

The definition of tangential deformations and stresses (for the estimation of frictional forces) requires the introduction of local coordinate systems x',y',z' oriented along the unit vectors $\vec{P}_1,\vec{P}_2,\vec{P}_3$ for each of the face surfaces. The calculation of the derivatives along these directions is performed in a known manner and the relations for the shear deformation are taken in vector form

$$\gamma_{z'y'}=\vec{P}_1\cdot\frac{\partial\vec{V}}{\partial z'}+\vec{P}_3\cdot\frac{\partial\vec{V}}{\partial x'},\quad \gamma_{y'z'}=\vec{P}_2\cdot\frac{\partial\vec{V}}{\partial z'}+\vec{P}_3\cdot\frac{\partial\vec{V}}{\partial y'}.. \tag{6}$$

5. Modeling the interaction of the lining with the soil

To reproduce the interaction of the lining blocks with the soil, without increasing the dimension of the problem, so-called "semi-infinite finite elements" or "finite elements with remote nodes" were used. They are 16-node crooked parallelepipeds, in which the inner curved surface mates with the upper surface of the lining blocks, and the outer surface is removed from the inner surface for considerable distances.

The main task of this element is to reproduce the elastic repulsion of the soil massif when deforming the rings of the lining. Therefore, the movements of remote nodes are considered to be zero. An additional argument in the validity of this assumption may be the following statement: the construction of the tunnel is conducted in a steady mass of soil and must not be accompanied by its deformations at a considerable distance from the rings of the lining of the tunnel.

IOP Conf. Series: Materials Science and Engineering **208** (2017) 012005 doi:10.1088/1757-899X/208/1/012005

The technology of constructing the stiffness matrix of such an element practically coincides with that described in [30-32] for a 20-node EM. The only difference is that in the mailing the component of the rigidity matrixes of these EMs only the stiffnesses that relate to the adjacent nodes are involved, and the nodes with zero displacements are not assigned global numbers. This allows us to limit ourselves only to the degrees of freedom defined on the outer surface of the lining blocks, which have already been introduced in modeling the corresponding concrete arrays.

6. Comparison of the results of calculations with experimental studies

To assess the reliability of the calculations and evaluate the developed methodology, their results were compared with the results of monitoring studies of the lining of the tunnel being built. In particular, special devices were installed on the ring blocks, measuring their deformations (strain gauges) before placing these blocks in the lining of the tunnel. Then, after sufficient time necessary to stabilize the stress state, both in the lining and in the soil massif surrounding it, the readings of circumferential and longitudinal deformations were taken. The time of this stabilization was determined by systematic observation of the instrument readings and evaluation of their variation over time.

A comparison of the results of experimental and numerical studies of these rings on the example of circumferential stresses for 172-th and 173-th lining rings is given on Fig. 4-5. The dashed line corresponds to experimental data, solid line - calculation without taking into account the water saturation of the soil, dash-dotted line - calculation taking into account the hydrostatic pressure of groundwater.

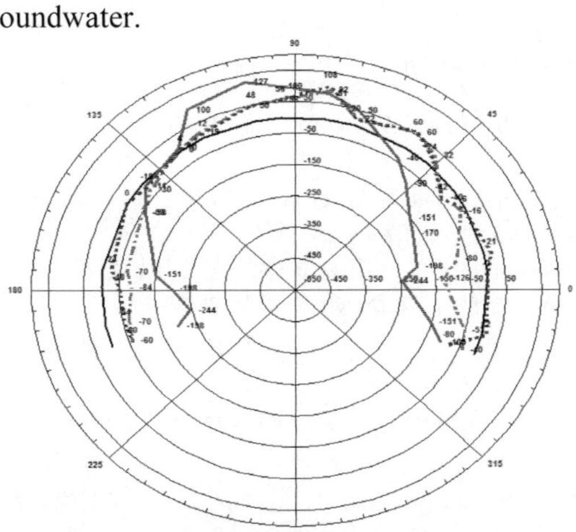

Figure 4. Circumferential stresses in one of the rings of the lining of the subway tunnel (numerical calculation and experimental data).

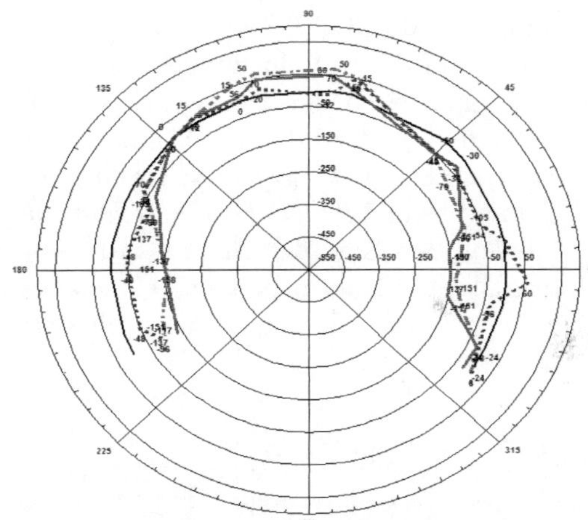

Figure 5. Longitudinal stresses in one of the rings of the lining of the subway tunnel (numerical calculation and experimental data).

7. Conclusion

Proceeding from the obtained results, it should be noted that the developed numerical method for studying the stressed-deformed state of the lining rings of the subway tunnel in a three-dimensional setting, with modeling of the contact interaction of the blocks with each other and with the introduction of "semi-infinite" finite elements to reproduce the elastic backing of the soil, gives results that are in good agreement with Data of full-scale tests. Consequently, it is possible to calculate such designs on its basis and obtain reliable results.

Acknowledgments

The reported study was supported by Government of the Republic of Tatarstan and RFBR research projects No. 15-41-02555 and 15-41-02557.

IOP Conf. Series: Materials Science and Engineering **208** (2017) 012005 doi:10.1088/1757-899X/208/1/012005

References

[1] Paimushin V N 2008 *Journal of Applied Mathematics and Mechanics* **72 (5)** 597–610
[2] Badriev I B, Banderov V V, Garipova G Z, Makarov M V and Shagidullin R R 2015 *Applied Mathematical Sciences* **9 (82)** 4095–4102
[3] Badriev I B, Banderov V V, Makarov M V and Paimushin V N 2015 *Applied Mathematical Sciences* **9 (78)** 3887–3895
[4] Badriev I B, Banderov V V and Zadvornov O A 2013 *Applied Mechanics and Materials* **392** 188–190
[5] Badriev I B, Garipova G Z, Makarov M V and Paimushin V N 2015 *Research Journal of Applied Sciences* **10 (8)** 428–435
[6] Badriev I B and Shagidullin R R 1995 *Journal of Mathematical Sciences* **73 (5)** 519–525
[7] Berezhnoi D V and Paimushin V N 2011 *Journal of Applied Mathematics and Mechanics* **75 (4)** 447-462
[8] Berezhnoi D V, Paimushin V N and Shalashilin V I 2009 *Mechanics of Solids* **44 (6)** 837-851
[9] Zienkiewicz O 1977 *The finite element method* (London Mc Graw-Hill) 787
[10] Wilkins M L 1967 *Vychislitelnye metody v gidrodinamice* (Mir) 212-263
[11] Evans M W and Harlow F H 1957 *Los Alamos Scientific Laboratory Report No. LA-2139* (Los Alamos)
[12] Sultanov L U 2015 *Procedia Earth and Planetary Science* **15** 119–124
[13] Davydov R L and Sultanov L U 2013 *PNRPU Mechanics Bulletin* **1** 81–93
[14] Golovanov A I and Sultanov L U 2005 *International Applied Mechanics* **41 (6)** 614–620
[15] Sultanov L U 2014 *Applied Mathematical Sciences* **8 (143)** 7117–7124
[16] Sultanov L U and Davydov R L and 2014 *Applied Mathematical Sciences* **8 (60)** 2991–2996
[17] Sultanov L U 2016 *Lobachevskii Journal of Mathematics* **37 (6)** 784–790
[18] Sagdatullin M and Berezhnoi D 2014 *Applied Mathematical Sciences* **8 (35)** 1731–1738
[19] Sagdatullin M and Berezhnoi D 2014 *Applied Mathematical Sciences* **8 (60)** 2965–2972
[20] Sachenkov O, Kharislamova L, Shamsutdinova N, Kirillova E and Konoplev Yu 2015 *IOP Conference Series: Materials Science and Engineering* **98** 012079
[21] Shigapova F A, Mustakimova R F, Saleeva G T and Sachenkov 2015 O A *International Journal of Applied Engineering Research* **10 (24)** 44711–44714
[22] Sachenkov O A, Mitryaikin V I, Zaitseva T A and Konoplev Yu G 2014 *Applied Mathematical Sciences* **8 (159)** 7889–7897
[23] Berezhnoi D V, Sachenkov A A and Sagdatullin M K 2014 *Applied Mathematical Sciences* **8 (127)** 6341–6348
[24] Berezhnoi D V, Sachenkov A A and Sagdatullin M K 2014 *Applied Mathematical Sciences* **8 (143)** 7107–7115
[25] Abdrakhmanova A I and Sultanov L U 2016 *Materials Physics and Mechanics* **26 (1)** 30–32
[26] Davydov R and Sultanov L 2013 *Sixth International Conference on Nonlinear Mechanics (ICNM-VI)* 64–67
[27] Davydov R L and Sultanov L U 2015 *Journal of Engineering Physics and Thermophysics* **88 (5)** 1280–1288
[28] Sultanov L U and Fakhrutdinov L R 2013 *Magazine of Civil Engineering* **44 (9)** 69–74
[29] Sultanov L U and Davydov R L 2013 *Magazine of Civil Engineering* **44 (9)** 64–68
[30] Berezhnoi D V, Balafendieva I S., Sachenkov A A and Sekaeva L R 2016 *IOP Conference Series: Materials Science and Engineering* **158** 012018
[31] Shamim M R and Berezhnoi D V 2016 *IOP Conference Series: Materials Science and Engineering* **158** 012083
[32] Berezhnoi D V and Sagdatullin M K 2015 *Contemporary Engineering Sciences* **8 (23)** 1091-1098

IOP Conf. Series: Materials Science and Engineering **208** (2017) 012006 doi:10.1088/1757-899X/208/1/012006

Investigation of the processes of deformation of super flywheels made by packing its layers on each other

D V Berezhnoi and L R Gajnulina[a]

Kazan Federal University, 18, Kremlyovskaya St., Kazan, 420008, Russia

E-mail: [a] leysan5684@yandex.ru

Abstract. In the work for different cases of nozzles of layers of the super-magnetic pendulum, exact solutions in the stresses are obtained, and estimates of the increase in its specific energy are given. Dependences of tensile stresses in a tape are obtained when winding it into a super-flywheel, an increase in the specific kinetic energy in the super flywheel is observed with a decrease in the relative radius of the inner hole.

1. Introduction

In modern realities, the problem of accumulating energy becomes more important. Simple enough is the technology of energy storage by means of kinetic storage. The main limitation in implementing such a technology is the rotational speed of the rotor part of the kinetic storage, which can lead to an increase in the stress level and its possible further destruction. In the literature, we give some estimates of the specific kinetic energy of rotation of the flywheels of the canonical form [1-9]. For flywheels of arbitrary shape, numerical methods for investigating the specific kinetic energy are proposed in the literature. In a number of works kinetic energy storage devices made of rubber-like materials are proposed, for the calculation of which special methods are proposed [10-14]. Calculation of super-flywheels, made by winding threads or tapes, presupposes methods for their contact interaction [15-21]. In this paper, we obtain some exact solutions in the stresses for super flywheels, made by winding and subsequent gluing of tapes.

2. Kinematics

A tape super flywheel with an acceptable degree of accuracy can be represented in the form of a system of concentric cylinders (Fig. 1), clad against each other with some interference. In this case, even before the start of loading in the super-flywheel, there will be some stress-strain state, which for two cylinders can be determined by calculating the so-called composite cylinders [22].

When calculating a compound cylinder (for definiteness consisting of two cylinders), the main thing is to establish the amount of pressure on the contact surface for a given interference, which is the difference between the outer diameter of the inner cylinder and the inner diameter of the outer cylinder (Fig. 2). It is obvious that the decrease in the outer radius of the inner cylinder and the increase in the inner radius of the outer cylinder are equal to half the interference

$$\left|u_r^{in}\right| + \left|u_r^{out}\right| = \delta/2 \qquad (1)$$

IOP Conf. Series: Materials Science and Engineering **208** (2017) 012006 doi:10.1088/1757-899X/208/1/012006

Given that the interference is very small compared to the contact surface radius, it can be assumed that the outer radius of the inner cylinder $r_{in\_cyl}^{out}$ and the inner radius of the outer cylinder $r_{out\_cyl}^{in}$ are equal to the radius of the contact surface

$$r_{in\_cyl}^{out} = r_{out\_cyl}^{in} = r_c \, . \tag{2}$$

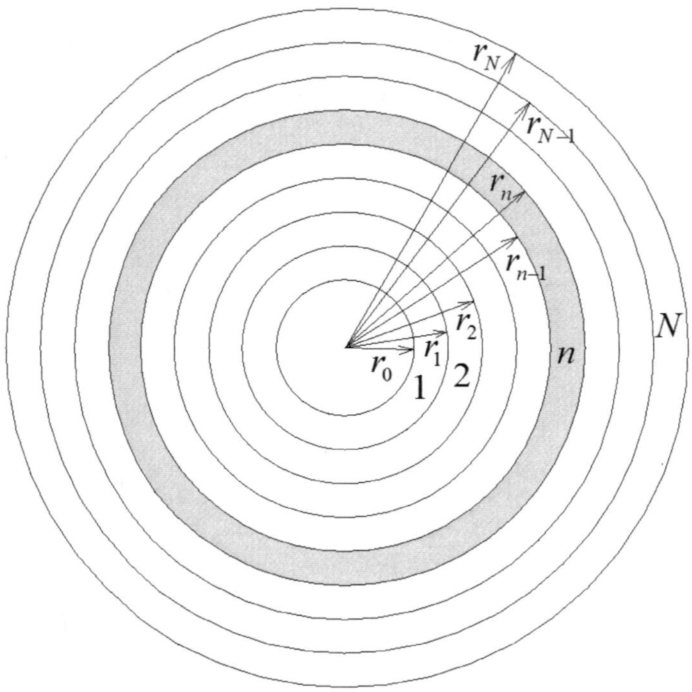

Figure 1. The system of concentric cylinders.

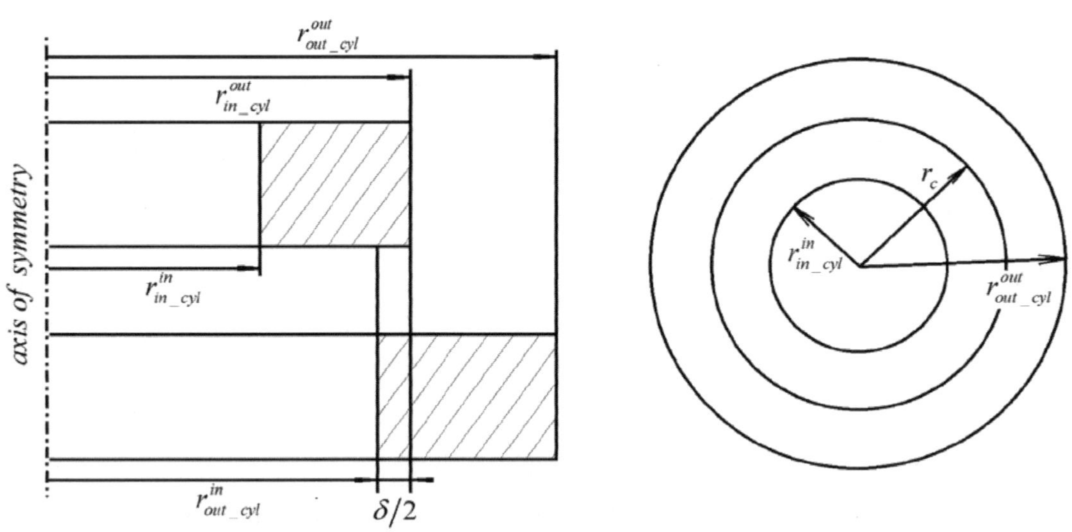

Figure 2. Scheme of attachment of two cylinders with interference.

The contact pressure will be external for the inner cylinder and internal for the outer cylinder. Thus, to determine the stress-strain state, we need a solution of the Lame problem on the deformation of cylinders under the influence of internal and external pressure. Formulas for determining radial displacements and circumferential and radial stresses in a cylinder with internal r_1 and external radius r_2 under internal p_1 and external p_2 pressure in the case of a plane stress state have the form

$$u_r = \frac{p_1 r_1^2\left[(1-\mu)r+(1+\mu)r_2^2 r^{-1}\right]}{E\left(r_2^2-r_1^2\right)} - \frac{p_2 r_2^2\left[(1-\mu)r+(1+\mu)r_1^2 r^{-1}\right]}{E\left(r_2^2-r_1^2\right)}, \tag{3}$$

$$\sigma_{rr} = \frac{p_1 r_1^2\left(1-r_2^2 r^{-2}\right)}{\left(r_2^2-r_1^2\right)} - \frac{p_2 r_2^2\left(1-r_1^2 r^{-2}\right)}{\left(r_2^2-r_1^2\right)}, \tag{4}$$

$$\sigma_{\varphi\varphi} = \frac{p_1 r_1^2\left(1+r_2^2 r^{-2}\right)}{\left(r_2^2-r_1^2\right)} - \frac{p_2 r_2^2\left(1+r_1^2 r^{-2}\right)}{\left(r_2^2-r_1^2\right)}, \tag{5}$$

where E and μ - is the Young's modulus and Poisson's ratio of the material.

Then the radial displacement of the contact surface of the inner cylinder is determined by the formula

$$u_r^{in}(r_c) = -\frac{p_c r_c^2\left[(1-\mu)r_c+(1+\mu)\left(r_{n\_cyl}^{in}\right)^2 r_c^{-1}\right]}{E\left(r_c^2-\left(r_{n\_cyl}^{in}\right)^2\right)}, \tag{6}$$

and the radial displacement of the contact surface of the inner cylinder is determined by the formula

$$u_r^{out}(r_c) = \frac{p_c r_c^2\left[(1-\mu)r_c+(1+\mu)\left(r_{n\_cyl}^{out}\right)^2 r_c^{-1}\right]}{E\left(\left(r_{n\_cyl}^{out}\right)^2-r_c^2\right)}. \tag{7}$$

If the inner and outer cylinders are made of the same material, then the relationship between the contact pressure and interference is obtained in the form

$$p_c = \frac{\delta E}{2r_c}\frac{\left(1-\left(r_{in\_cyl}^{in}/r_c\right)^2\right)\left(1-\left(r_c/r_{out\_cyl}^{out}\right)^2\right)}{\left(1+\left(r_{in\_cyl}^{in}/r_c\right)^2\right)\left(1-\left(r_c/r_{out\_cyl}^{out}\right)^2\right)+\left(1-\left(r_{in\_cyl}^{in}/r_c\right)^2\right)\left(1+\left(r_c/r_{out\_cyl}^{out}\right)^2\right)}. \tag{8}$$

If the thickness of each cylinder t is the same, then the inner and outer radii of the i-th cylinder are determined by the formulas

$$r_i^{in} = r_{i-1} = r_0 + (i-1)t,\ r_i^{out} = r_i = r_0 + it. \tag{9}$$

Then for the n-th cylinder the stresses will be determined by the formulas

$$\sigma_{rr}^{n} = p_n^{in}\frac{r_{n-1}^2}{r^2}\frac{r^2-r_n^2}{r_n^2-r_{n-1}^2} - \sum_{i=n}^{N-1} p_i^{out}\frac{r_i^2}{r^2}\frac{r^2-r_0^2}{r_i^2-r_0^2}, \tag{10}$$

$$\sigma_{\varphi\varphi}^{n} = p_n^{in}\frac{r_{n-1}^2}{r^2}\frac{r^2+r_n^2}{r_n^2-r_{n-1}^2} - \sum_{i=n}^{N-1} p_i^{out}\frac{r_i^2}{r^2}\frac{r^2+r_0^2}{r_i^2-r_0^2}, \tag{11}$$

where

$$p_i^{in} = \frac{\delta_{i-1}E}{2r_{i-1}}\frac{\left(r_{i-1}^2-r_0^2\right)\left(r_i^2-r_{i-1}^2\right)}{\left(r_{i-1}^2+r_0^2\right)\left(r_i^2-r_{i-1}^2\right)+\left(r_{i-1}^2-r_0^2\right)\left(r_i^2+r_{i-1}^2\right)}, \tag{12}$$

$$p_i^{out} = \frac{\delta_i E}{2r_i}\frac{\left(r_i^2-r_0^2\right)\left(r_{i+1}^2-r_i^2\right)}{\left(r_i^2+r_0^2\right)\left(r_{i+1}^2-r_i^2\right)+\left(r_i^2-r_0^2\right)\left(r_{i+1}^2+r_i^2\right)}, \tag{13}$$

and δ_i is the interference on the outer surface of the n-th cylinder ($\delta_0 = 0$), then

$$\sigma_{rr}^{n}=\frac{\delta_{n-1}E}{2}\frac{\left(r_{n-1}^{2}-r_{0}^{2}\right)r_{n-1}}{\left(r_{n-1}^{2}+r_{0}^{2}\right)\left(r_{n}^{2}-r_{n-1}^{2}\right)+\left(r_{n-1}^{2}-r_{0}^{2}\right)\left(r_{n}^{2}+r_{n-1}^{2}\right)}\frac{r^{2}-r_{n}^{2}}{r^{2}}-$$
$$-\sum_{i=n}^{N-1}\frac{\delta_{i}E}{2}\frac{\left(r_{i+1}^{2}-r_{i}^{2}\right)r_{i}}{\left(r_{i}^{2}+r_{0}^{2}\right)\left(r_{i+1}^{2}-r_{i}^{2}\right)+\left(r_{i}^{2}-r_{0}^{2}\right)\left(r_{i+1}^{2}+r_{i}^{2}\right)}\frac{r^{2}-r_{0}^{2}}{r^{2}}$$
$$\sigma_{\varphi\varphi}^{n}=\frac{\delta_{n-1}E}{2}\frac{\left(r_{n-1}^{2}-r_{0}^{2}\right)r_{n-1}}{\left(r_{n-1}^{2}+r_{0}^{2}\right)\left(r_{n}^{2}-r_{n-1}^{2}\right)+\left(r_{n-1}^{2}-r_{0}^{2}\right)\left(r_{n}^{2}+r_{n-1}^{2}\right)}\frac{r^{2}+r_{n}^{2}}{r^{2}}-$$
$$-\sum_{i=n}^{N-1}\frac{\delta_{i}E}{2}\frac{\left(r_{i+1}^{2}-r_{i}^{2}\right)r_{i}}{\left(r_{i}^{2}+r_{0}^{2}\right)\left(r_{i+1}^{2}-r_{i}^{2}\right)+\left(r_{i}^{2}-r_{0}^{2}\right)\left(r_{i+1}^{2}+r_{i}^{2}\right)}\frac{r^{2}+r_{0}^{2}}{r^{2}}. \quad (14)$$

The circumferential stresses in winding in the i-th turn are determined in the form can be defined as the circumferential stresses in the i-th cylinder when it is pressed with interference onto the already assembled part of the super flywheel

$$\sigma_{\varphi}^{i}=\frac{p_{i}^{in}r_{i-1}^{2}\left(1+r_{i}^{2}r^{-2}\right)}{\left(r_{i}^{2}-r_{i-1}^{2}\right)}=\frac{\delta_{i-1}E}{2r_{i-1}\left(r_{i}^{2}-r_{i-1}^{2}\right)}\frac{\left(r_{i-1}^{2}-r_{0}^{2}\right)\left(r_{i}^{2}-r_{i-1}^{2}\right)r_{i-1}^{2}\left(1+r_{i}^{2}r^{-2}\right)}{\left(r_{i-1}^{2}+r_{0}^{2}\right)\left(r_{i}^{2}-r_{i-1}^{2}\right)+\left(r_{i-1}^{2}-r_{0}^{2}\right)\left(r_{i}^{2}+r_{i-1}^{2}\right)}=$$
$$=\frac{\delta_{i-1}E}{2}\frac{r_{i-1}\left(r_{i-1}^{2}-r_{0}^{2}\right)\left(1+r_{i}^{2}r^{-2}\right)}{\left(r_{i-1}^{2}+r_{0}^{2}\right)\left(r_{i}^{2}-r_{i-1}^{2}\right)+\left(r_{i-1}^{2}-r_{0}^{2}\right)\left(r_{i}^{2}+r_{i-1}^{2}\right)}. \quad (15)$$

If the tightness of all the same, the formula for the radial and circumferential stress to the n-th cylinder will take the form

$$\sigma_{rr}^{n}=\frac{\delta E}{2}\left(\frac{\left(r_{n-1}^{2}-r_{0}^{2}\right)r_{n-1}}{\left(r_{n-1}^{2}+r_{0}^{2}\right)\left(r_{n}^{2}-r_{n-1}^{2}\right)+\left(r_{n-1}^{2}-r_{0}^{2}\right)\left(r_{n}^{2}+r_{n-1}^{2}\right)}\frac{r^{2}-r_{n}^{2}}{r^{2}}-\right.$$
$$\left.-\sum_{i=n}^{N-1}\frac{\left(r_{i+1}^{2}-r_{i}^{2}\right)r_{i}}{\left(r_{i}^{2}+r_{0}^{2}\right)\left(r_{i+1}^{2}-r_{i}^{2}\right)+\left(r_{i}^{2}-r_{0}^{2}\right)\left(r_{i+1}^{2}+r_{i}^{2}\right)}\frac{r^{2}-r_{0}^{2}}{r^{2}}\right), \quad (16)$$

$$\sigma_{\varphi\varphi}^{n}=\frac{\delta E}{2}\left(\frac{\left(r_{n-1}^{2}-r_{0}^{2}\right)r_{n-1}}{\left(r_{n-1}^{2}+r_{0}^{2}\right)\left(r_{n}^{2}-r_{n-1}^{2}\right)+\left(r_{n-1}^{2}-r_{0}^{2}\right)\left(r_{n}^{2}+r_{n-1}^{2}\right)}\frac{r^{2}+r_{n}^{2}}{r^{2}}-\right.$$
$$\left.-\sum_{i=n}^{N-1}\frac{\left(r_{i+1}^{2}-r_{i}^{2}\right)r_{i}}{\left(r_{i}^{2}+r_{0}^{2}\right)\left(r_{i+1}^{2}-r_{i}^{2}\right)+\left(r_{i}^{2}-r_{0}^{2}\right)\left(r_{i+1}^{2}+r_{i}^{2}\right)}\frac{r^{2}+r_{0}^{2}}{r^{2}}\right). \quad (17)$$

The circumferential stresses in winding in the i-th turn are determined in the form

$$\sigma_{\varphi}^{i}=\frac{\delta E}{2}\frac{r_{i-1}\left(r_{i-1}^{2}-r_{0}^{2}\right)\left(1+r_{i}^{2}r^{-2}\right)}{\left(r_{i-1}^{2}+r_{0}^{2}\right)\left(r_{i}^{2}-r_{i-1}^{2}\right)+\left(r_{i-1}^{2}-r_{0}^{2}\right)\left(r_{i}^{2}+r_{i-1}^{2}\right)}\approx$$
$$\approx\frac{\delta E}{2}\frac{(i-1)\left(2r_{0}+(i-1)t\right)}{2r_{0}^{2}+2r_{0}(i-1)t+(i-1)^{2}t^{2}+(i-1)\left(2r_{0}+(i-1)t\right)\left(r_{0}+(i-1)t\right)}. \quad (18)$$

If we assume that the contact pressure between all the cylinders are the same and are equal to p_c, then in this case the interference between the n-th cylinder and the $(n+1)$-th will be determined by the formula

$$\delta_{i}=\frac{2p_{c}r_{i}}{E}\frac{\left(r_{i}^{2}+r_{0}^{2}\right)\left(r_{i+1}^{2}-r_{i}^{2}\right)+\left(r_{i}^{2}-r_{0}^{2}\right)\left(r_{i+1}^{2}+r_{i}^{2}\right)}{\left(r_{i}^{2}-r_{0}^{2}\right)\left(r_{i+1}^{2}-r_{i}^{2}\right)}, \quad (19)$$

the ratio of stress take the form of

$$\sigma_{rr}^{n}=p_c\left[\frac{r^2-r_n^2}{r^2}\frac{r_{n-1}^2}{\left(r_n^2-r_{n-1}^2\right)}-\frac{r^2-r_0^2}{r^2}\sum_{i=n}^{N-1}\frac{r_i^2}{\left(r_i^2-r_0^2\right)}\right], \tag{20}$$

$$\sigma_{\varphi\varphi}^{n}=p_c\left[\frac{r^2+r_n^2}{r^2}\frac{r_{n-1}^2}{\left(r_n^2-r_{n-1}^2\right)}-\frac{r^2+r_0^2}{r^2}\sum_{i=n}^{N-1}\frac{r_i^2}{\left(r_i^2-r_0^2\right)}\right]. \tag{21}$$

The circumferential stresses in winding in the i-th turn are determined in the form

$$\sigma_{\varphi}^{i}=\frac{p_c r_{i-1}^2\left(1+r_i^2r^{-2}\right)}{\left(r_i^2-r_{i-1}^2\right)}\approx\frac{p_c r_{i-1}}{t}=\frac{p_c\left(r_0+(i-1)t\right)}{t}. \tag{22}$$

If we take the ratio

$$\delta_i=\frac{2p}{Er_i}\frac{\left(r_i^2+r_0^2\right)\left(r_{i+1}^2-r_i^2\right)+\left(r_i^2-r_0^2\right)\left(r_{i+1}^2+r_i^2\right)}{\left(r_{i+1}^2-r_i^2\right)}, \tag{23}$$

the ratio of stress take the form

$$\sigma_{rr}^{n}=p\left[\frac{\left(r_{n-1}^2-r_0^2\right)}{\left(r_n^2-r_{n-1}^2\right)}\frac{r^2-r_n^2}{r^2}-\frac{\left(r^2-r_0^2\right)(N-n)}{r^2}\right], \tag{24}$$

$$\sigma_{\varphi\varphi}^{n}=p\left[\frac{\left(r_{n-1}^2-r_0^2\right)}{\left(r_n^2-r_{n-1}^2\right)}\frac{r^2+r_n^2}{r^2}-\frac{\left(r^2+r_0^2\right)(N-n)}{r^2}\right]. \tag{25}$$

The circumferential stresses in winding in the i-th turn are determined in the form

$$\sigma_{\varphi}^{i}=\frac{p\left(r_{i-1}^2-r_0^2\right)\left(1+r_i^2r^{-2}\right)}{\left(r_i^2-r_{i-1}^2\right)}\approx\frac{p(i-1)\left(2r_0+(i-1)t\right)}{r_0+(i-1)t}. \tag{26}$$

3. An estimate of the increase in the energy capacity of a super flywheel under a monotonically varying interference

The circumferential stresses in the n-th cylinder of the untwisted flywheel are determined by the formula

$$\sigma_{\varphi\varphi}^{n}=p\left[\frac{(n-1)\left(2r_0+(n-1)t\right)}{\left(2r_0+2nt-t\right)}\frac{r^2+\left(r_0+nt\right)^2}{r^2}-\frac{r^2+r_0^2}{r^2}(N-n)\right]+$$
$$+\frac{3+\mu}{8}\rho\omega^2r_0^2\left(1+k^2\left(1+r_0^2r^{-2}\right)-\frac{1+3\mu}{3+\mu}r^2r_0^{-2}\right). \tag{27}$$

To estimate the specific energy intensity of the super flywheel, whose layers are stacked with interference, we equate the kinematic energy with the circumferential stresses along the inner and outer edges of the super flywheel

$$\sigma_{\varphi\varphi}\left(r_0\right)=\frac{\rho\omega^2r_0^2}{4}\left(\left(3+\mu\right)k^2+1-\mu\right)-2p\left(N-1\right)=.$$
$$=\frac{\rho\omega^2r_0^2}{4}\left(3+\mu+k^2\left(1-\mu\right)\right)+\frac{2p\left(N-1\right)\left(2r_0+\left(N-1\right)t\right)}{\left(2r_0+2Nt-t\right)}=\sigma_{\varphi\varphi}\left(r_N\right). \tag{28}$$

From here

$$2p\left(N-1\right)=\frac{\rho\omega^2r_0^2}{2}\frac{2r_0+2Nt-t}{4r_0+3Nt-2t}\left(1+\mu\right)\left(k^2-1\right) \tag{29}$$

and

$$\sigma_{\varphi\varphi}^{0}=\sigma_{\varphi\varphi}^{0}(r_0)=\frac{\rho\omega^2 r_0^2}{4}\left[\left((3+\mu)k^2+1-\mu\right)-\frac{2k-t/r_0}{1+3k-2t/r_0}2(1+\mu)(k^2-1)\right]. \tag{30}$$

If the number of winding layers is large, then

$$\sigma_{\varphi\varphi}^{0}\approx\frac{\rho\omega^2 r_0^2}{4}\left[\left((3+\mu)k^2+1-\mu\right)-\frac{2k}{1+3k}2(1+\mu)(k^2-1)\right]=$$
$$=\frac{\rho\omega^2 r_0^2}{4}\left[\frac{1-\mu+k\left(7+\mu+k\left(3-k(-5+\mu)+\mu\right)\right)}{1+3k}\right]. \tag{31}$$

If the flywheel is solid, then the stresses on the inner surface of the flywheel will coincide with those for the super flywheel at angular velocity ω_s

$$\frac{\rho\omega_s^2 r_0^2}{4}\left((3+\mu)k^2+1-\mu\right)=\frac{\rho\omega^2 r_0^2}{4}\left[\frac{1-\mu+k\left(7+\mu+k\left(3-k(-5+\mu)+\mu\right)\right)}{1+3k}\right]. \tag{32}$$

Then the ratio of the squares of the angular velocities (as well as the kinetic energies) will be equal to

$$\frac{\omega^2}{\omega_s^2}=\frac{(1+3k)\left((3+\mu)k^2+1-\mu\right)}{1-\mu+k\left(7+\mu+k\left(3-k(-5+\mu)+\mu\right)\right)}. \tag{33}$$

For a small hole $(k\to\infty)$, the ratio of the squares of the angular velocities will be

$$\frac{\omega^2}{\omega_s^2}=\frac{3(3+\mu)}{5-\mu}. \tag{34}$$

4. Numerical example

The stresses in the layers were calculated for a super flywheel with the following geometric and mechanical characteristics [23, 24]: the Young's modulus $\AA=200000\,MPa$, Poisson's ratio $\mu=0.3$, internal radius $r_0=1\,m$, external radius $r_N=2\,m\,(k=2)$, tensile strength $\sigma_{ts}=1100\,MPa$, material density $\rho=7860\,kg/m^3$. In the continuous flywheel for these geometric and mechanical parameters, the angular velocity $\omega=200\ rad/s$, while the maximum stress in the flywheel (hoop stresses on the inner surface of the flywheel) are equal to $\sigma_{\varphi\varphi}^{\max}=1092.54\,MPa$.

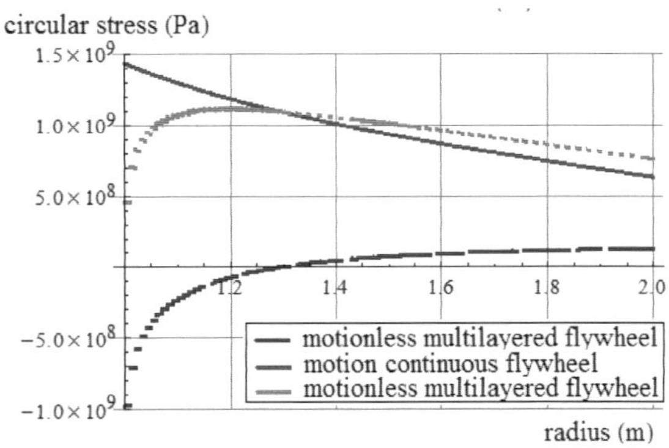

Figure 3. Circular stresses for the case of constant interference.

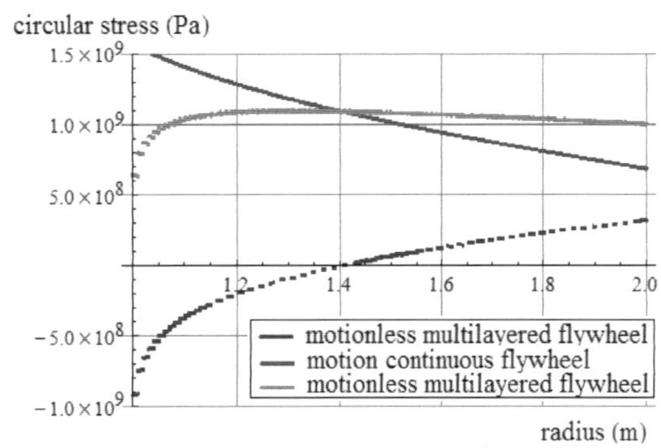

Figure 4. Circular stresses for the case of equal contact.

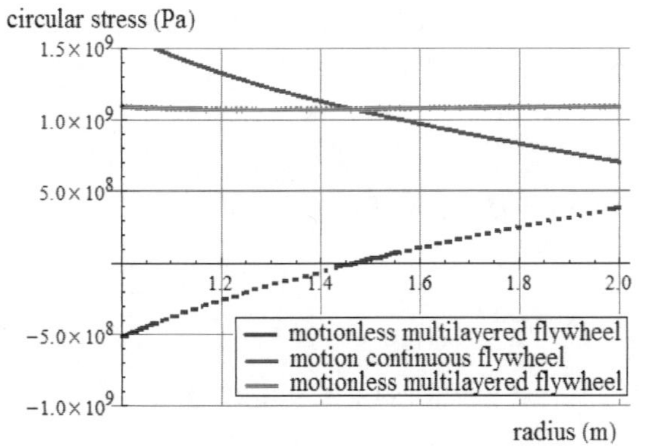

Figure 5. Circular stresses for case of a monotonically changing interference.

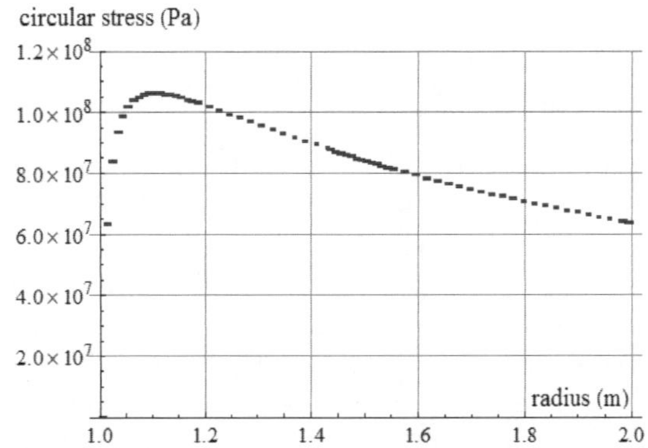

Figure 6. Circumferential stresses in the layers of the super flywheel when winding for constant interference.

In Figure 3, for the case of the same interference, the diagrams of the circumferential stresses in a 100-layer super flywheel are given for the maximum possible angular velocity and the diagram of the circumferential stresses in a continuous flywheel for the same angular velocity. Figures 4 and 5 show the corresponding diagrams of the circumferential stresses for the case of equal contact pressures and the case of a monotonically varying interference.

The circumferential stresses in the layers of the super flywheel when winding for constant interference δ, constant contact pressure p_c and a constant monotonic interference parameter p for the case of 100 turns are shown in Figures 6-8, respectively.

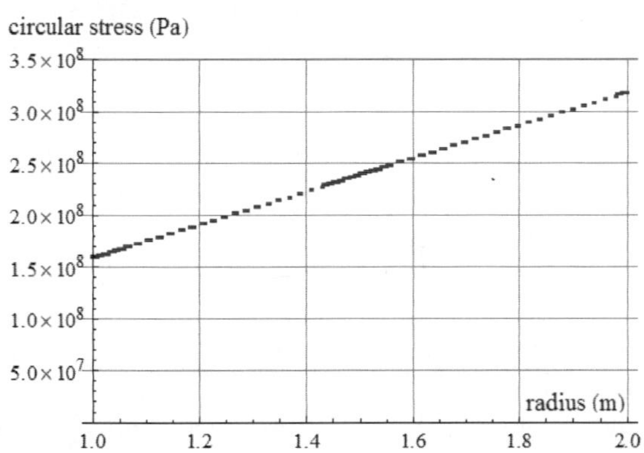

Figure 7. Circumferential stresses in the layers of the super flywheel when winding for constant contact pressure.

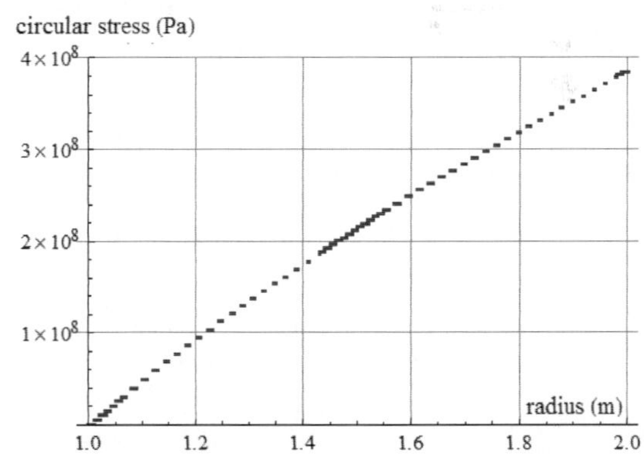

Figure 8. Circumferential stresses in the layers of the super flywheel when winding for case of a monotonically changing interference.

5. Conclusion

Analysis of the results shows that the manufacture of super-flywheels with concentric packing on each other with the tightness of thin-walled cylinders (with their further gluing or welding) with an appropriate choice of interference can lead to an increase in the specific energy capacity of the super-flywheel design. The relative decrease in the radius of the inner opening of the super flywheel, the increase in the number of layers (thin-walled cylinders planted on each other) and the increase in the

Poisson's ratio of the material from which the super flywheel is made also leads to an increase in the specific energy intensity. Precise solutions for the distribution of stresses in the layers of the super flywheel are obtained.

Acknowledgments
The reported study was supported by Government of the Republic of Tatarstan and RFBR research projects No. 15-41-02555 and 15-01-05686.

References

[1] Drum T O 2011 *Energy storage-flywheel* http://www.resilience.org/stories/2011-10-05/energy-storage-flywheel
[2] Gulia N V 1976 *Machovichnye dvigately* (Moscow: Mashinostroenie) p 163
[3] Gulia N V 1980 *Nakopitely Energii* (Moscow: Nauka) p 152
[4] Matveev V V, Pisarenko G V and Yakovlev A P 1988 *Spravochnik po soprotivleniyu materialov* (Kiev: Naukova Dumka) p 736
[5] Seong-yeol Y, Wook-ryun L, Yong-chae B and Myounggyu N 2010 *Journal of Mechanical Science and Technology* **24** 231-235
[6] Berezhnoi D V, Chickrin D E and Galimov A F 2014 *Applied Mathematical Sciences* **8 (124)** 6181–6190
[7] Berezhnoi D V, Chickrin D E, Kurchatov E Yu and Galimov A F 2014 *Applied Mathematical Sciences* **8 (163)** 8125–8135
[8] Berezhnoi D V, Chickrin D E, Gajnulina L R, Kurchatov E Yu, Kokunin P A, Sozutov I G and Shigapov M I 2015 *Contemporary Engineering Sciences* **8 (36)** 1703–1712
[9] Berezhnoy D V, Gajnulina L R and Sekaeva L R 2015 *International Journal of Applied Engineering Research* **10(24)** 44672–44676
[10] Abdrakhmanova A I and Sultanov L U 2016 *Materials Physics and Mechanics* **26 (1)** 30–32
[11] Golovanov A I and Sultanov L U 2005 *International Applied Mechanics* **41 (6)** 614–620
[12] Davydov R L and Sultanov L U 2015 *Journal of Engineering Physics and Thermophysics* **88 (5)** 1280–1288
[13] Sultanov L U and Fakhrutdinov L R 2013 *Magazine of Civil Engineering* **44 (9)** 69–74
[14] Sultanov L U and Davydov R L 2013 *Magazine of Civil Engineering* **44 (9)** 64–68
[15] Sachenkov O, Kharislamova L, Shamsutdinova N, Kirillova E and Konoplev Yu 2015 *IOP Conference Series: Materials Science and Engineering* **98** 012015
[16] Sachenkov O, Hasanov R, Andreev P and Konoplev Y 2016 *IOP Conference Series: Materials Science and Engineering* **158 (1)** 012079
[17] Shigapova F A, Mustakimova R F, Saleeva G T and Sachenkov O A *International Journal of Applied Engineering Research* **10 (24)** 44711–44714
[18] Sachenkov O A, Mitryaikin V I, Zaitseva T A and Konoplev Yu G 2014 *Applied Mathematical Sciences* **8 (159)** 7889–7897
[19] Badriev I B, Banderov V V, Makarov M V and Paimushin V N 2015 *Applied Mathematical Sciences* **9 (78)** 3887–3895
[20] Badriev I B, Banderov V V and Zadvornov O A 2013 *Applied Mechanics and Materials* **392** 188–190
[21] Badriev I B, Garipova G Z, Makarov M V and Paimushin V N 2015 *Research Journal of Applied Sciences* **10 (8)** 428–435
[22] Rabotnov Y N 1962 *Soprotivlenye materialov* (Moscow: Fizmatgiz) p 456
[23] Berezhnoi D V and Gajnulina L R 2016 *IOP Conference Series: Materials Science and Engineering* **158** 012017
[24] Berezhnoi D V, Chickrin D E, Gajnulina L R, Kurchatov E Yu, Kokunin P A, Sozutov I G and Shigapov M I 2016 *Contemporary Engineering Sciences* **9 (32)** 1591–1598

IOP Conf. Series: Materials Science and Engineering **208** (2017) 012007 doi:10.1088/1757-899X/208/1/012007

Elastic properties of magnetorheological elastomer: description with the two-particle mesoscopic model

A M Biller, O V Stolbov and Yu L Raikher

Institute of Continuous Media Mechanics UB RAS, 1, Acad. Korolev St., Perm, 614013, Russia

E-mail: kam@icmm.ru

Abstract. A pair of magnetizable solid particles embedded in a cylinder made of high-elasticity material is considered as a model of a mesoscopic structure element of a magnetorheological elastomer. An applied magnetic field induces ponderomotive interaction of the particles making them to move relative to one another so as to balance the counteracting magnetic and elastic forces. In a certain parameter range, the system exhibits bistability due to which under the increase / decrease of the field, the interparticle distance changes in a hysteretic manner. This behavior has a significant effect on the ability of the mesoscopic element to resist external load. Using the developed two-particle model prone to the magnetomechanical hysteresis, we extend it to the case of a virtually macroscopic sample presenting the latter as a superposition of such elements with distributed interparticle distances. In spite of its simplicity, this scheme in a generally correct way describes the field-induced changes of the internal structure and elastic modulus of the magnetorheological composites.

1. Introduction

Magnetorheological elastomers (MREs) are a special type of smart composites distinguished by their ability for significant shape and elastic properties changes in response to applied magnetic field. Under magnetization, the microparticles of the ferromagnet filler are got coupled by ponderomotive forces that entails a number of interesting effects: magnetically induced deformation and stiffening of MREs, the magnetic shape memory, and others, all of which possessing a substantial practical potential. At the qualitative level, the picture is simple. The applied external field magnetizes the particles and imparts magnetic moments to them. The arising ponderomotive interaction strives to arrange the particles into a spatial structure that corresponds to the minimum of magnetostatic energy. For the particles embedded in a polymer, this tendency is opposed by the elastic restoring forces, which turn up in the MRE matrix as soon as any particle shifts from its initial position. Both experiment [1, 2, 3] and theory [4, 5, 6] show that such changes can strongly modify the properties of simulated MRE samples, thus entailing their qualitatively different behavior. The key issue underlying these peculiarities is the magneto-elastic interaction of the particles in a magnetorheological elastomer at the mesoscopic level, i.e., at the scale of the particle size and the interparticle distance.

2. Mesoscopic element

To clarify the above-mentioned points, in our works [7, 8] we considered a system consisting of a pair of spherical magnetizable particles embedded in an elastomer matrix. Such a mesoscopic system is the smallest representative entity that displays basic magnetomechanical effects inherent to MREs. The scheme is illustrated in figure 1. Two particles of identical radius a are

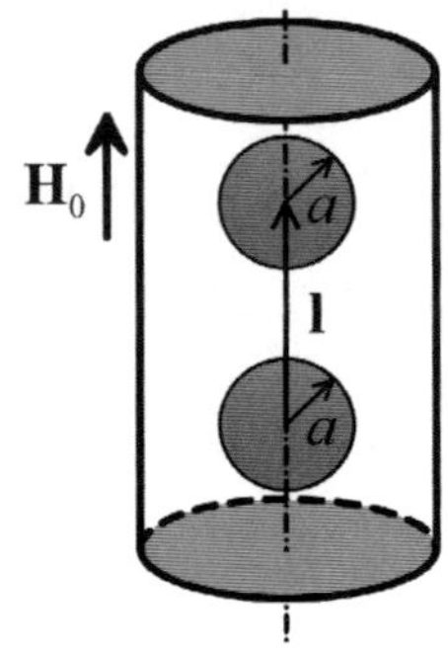

Figure 1. Schematic view of the mesoscopic element.

positioned insde an elastomer cylinder in such a way that the center-to-center line points along its axis. The outer dimensions of this two-particle (2P) element: length L_0 and cross section area S, are chosen so that the particles occupy 30% of its geometric volume; this reference value of the volume content of ferromagnetic phase is most interesting for applications.

An external magnetic field $\mathbf{H}_0$ is applied along the axis of the cylinder. Under the action of the field, the particles magnetize and get involved in the magnetostatic (ponderomotive) interaction that in the configuration of figure 1 results in mutual attraction. The magnetic energy U_{mag} of the pair depends on the center-to-center distance $l = |\mathbf{l}|$ and the magnitude of the applied field $H_0 = |\mathbf{H}_0|$ assuming minimum when the particles come into tight contact.

As full adhesion at the particle-elastomer interface is assumed, this implies that any particle displacement causes deformation of the matrix. As the polymer is non-magnetic, the elastic energy U_{el} accumulated due to that, is independent of the magnetic field being a function of only the initial l_0 and actual l distances between the particles; apparently, U_{el} is minimal in the absence of deformations ($l = l_0$).

Given that, the total energy U of the considered 2P cylindrical element of a MRE could be presented as a sum of the magnetic U_{mag} and elastic U_{el} contributions. Accordingly, under the action of external magnetic field two types of forces, competing with each other, arise inside the system: the magnetic forces tending to bring the particles closer, and the elastic ones resisting any deformation of the matrix. The algorithm that enables one to evaluate the total energy of the considered mesoscopic element, is similar to that described in our paper [8]. In below we adduce it briefly focusing on the modification made to the model and on the characteristic properties of the considered system.

The magnetic part U_{mag} of the total energy is considered, as in Ref. [8] using the numerical solution of the magnetostatic problem for two spherical particles, whose magnetization $\mathbf{M}$ is related to the internal field by the empirical Fröhlich-Kennelly formula [9], so that the magnetic response of the particles is described by two material parameters: the initial magnetic susceptibility χ_0 and the saturation magnetization M_s. For carbonyl iron particles, most commonly used as MRE fillers, the corresponding reference values are $\chi_0 \sim 10^4$ and $M_s \sim$ 1500 kA/m. However, in below we use for U_{mag} dimensionless units denoting: the field intensity as $h_0 = H_0/M_s$, the distance between the particle centers as $q = l/a$ and the magnetic energy as $\tilde{U}_{mag}(q, h_0) = U_{mag}(l, H_0)/(\mu_0 M_s^2 a^3)$ with μ_0 being the magnetic permeability of vacuum.

When treating the elastic energy U_{el} induced by the particle displacements in the matrix, the approach of Ref. [8] was modified by assuming that the sample has finite dimensions. Since the elastomer can undergo significant deformations, we describe it as an incompressible Mooney-

Rivlin body [10]. The emerging nonlinear elasticity boundary problem was formulated as an axisymmetrical one and solved by the finite element method. The constants of the model, viz. c_1 and c_2, determine the elastic properties of the polymer matrix and are directly connected with its shear modulus G. To pass to dimensionless units, we introduce parameter $\tilde{c}_2 = c_2/c_1$ and elastic energy $\tilde{U}_{el}(q, q_0) = U_{el}(l, l_0)/(c_1 a^3)$. The MREs, which display the highest magnetically induced deformation and stiffening, have small Young moduli ($E \simeq 10 - 30$ kPa), see for example [11]. Taking that as a reference value, one finds that for a MRE with Young modulus $E \sim 25$ kPa the elastic constant c_1 of our model equals to 1.1 kPa for a fixed $\tilde{c}_2 = 0.2$.

To enable analytic calculations, the numerical data for the magnetic and elastic components of the energy were interpolated and their sum arranged in the functional form:

$$U(q, q_0, h_0)/(c_1 a^3) = \beta \tilde{U}_{mag}(q, h_0) + \tilde{U}_{el}(q, q_0), \tag{1}$$

which proved to be a good approximation for the true magnetoelastic energy of the mesoscopic element. As seen from (1), the ratio of the magnetic and elastic contributions is determined by parameter $\beta = \mu_0 M_s^2/c_1$.

Analysis of function (1) shows that on the increase of the field the system passes a threshold above which two different equilibrium configurations of the particles exist simultaneously. One state is "remote", where the interparticle distance is slightly reduced in comparison with its initial value ($l \simeq l_0$). Another possible equilibrium is a "collapsed" one, where the particles converge to one another ($l \simeq 2a$). This bistabilty regime exists within a finite field-strength interval and ends up on further enhancement of the field, where only the collapsed state is stable. The consequence of this bistability is a hysteresis of the interparticle distance if the applied magnetic field change in cycle, see figure 2. In the systems where particle magnetization saturates, the possibility of hysteresis for a given initial distance between the particles $q_0 = l_0/a$ is controlled by parameter β that has the meaning of effective compliance of the system. For the reference magnetic characteristics of a real MRE and elasticity of the matrix $c_1 = 1.1$ kPa, parameter β of our model is about 2500.

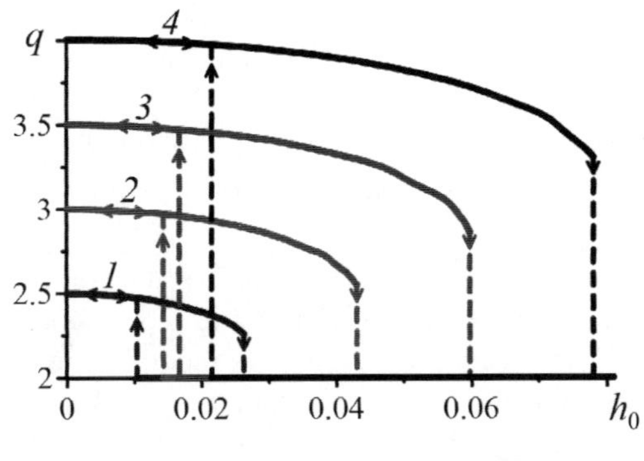

Figure 2. Magnetomechanical hysteresis for the mesoscopic element with $\beta = 2500$ and initial center-to-center distances q_0 equal 2.5 *(1)*, 3 *(2)*, 3.5 *(3)*, 4 *(4)*.

3. Elastic properties of the mesoscopic element

In the presence of external magnetic field h_0, the interparticle distance in the 2P element changes from the initial value l_0 to the new equilibrium one l. Simultaneously, the total length of the element decreases from L_0 to L. To account for that, we introduce the interparticle $\varepsilon_p = 1 - l/l_0$ and total $\varepsilon = 1 - L/L_0$ strains, and denote the equilibrium state of the element at h_0 as (ε_p, ε).

An applied small (probing) force f causes an additional small deformation u of the element. Then the internal magnetic and elastic forces must change in order to balance this extra load, i.e., the sum of the sample energy and the work of the applied force must assume minimum:

$$U(\varepsilon_p, \varepsilon) - fu = \min . \tag{2}$$

This problem can be solved by expanding the energy in the vicinity of the equilibrium at $f = 0$. From that solution one obtains the parameter $K = f/\varepsilon$, that we term *effective rigidity*, characterizing the ability of the considered model to resist deformation:

$$\frac{f}{\varepsilon} = \left(A_{22} - \frac{A_{12}^2}{A_{11}} \right) \frac{1}{L_0}; \tag{3}$$

here A_{ij} are the second partial derivatives of the energy. Then the ratio $\tilde{K} = K(h_0)/K_0$, where $K(h_0)$ is the rigidity of the mesoscopic sample under a magnetic field and K_0 that calculated in its absence renders the field effect on the sample rigidity.

Since the interparticle distance in the element undergoes hysteresis, the rigidity should behave similarly. This conclusion is illustrated by figures 2 and 3, where, respectively, the field strength dependencies of the interparticle distances q and the reduced rigidity $\tilde{K}$ are presented.

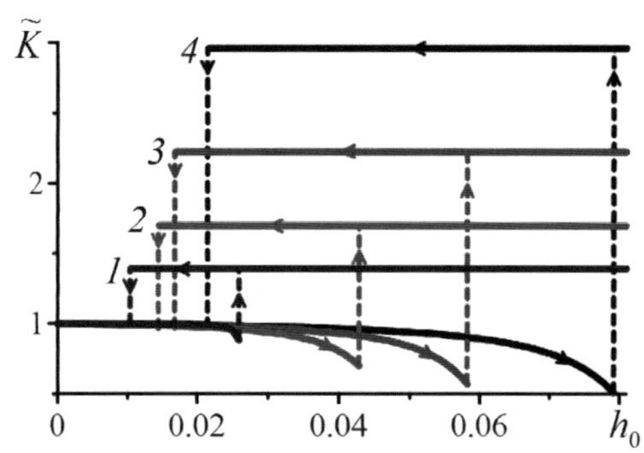

Figure 3. Rigidity of 2P mesoscopic elements with $\beta = 2500$ as a function of applied magnetic field; the initial interparticle distances q_0 are: 2.5 *(1)*, 3 *(2)*, 3.5 *(3)*, 4 *(4)* .

As figures 2 and 3 show, under the increase of the external field, first, the particles approach each other gradually that very slightly affects the rigidity of the system. Above a certain field strength (it depends on the interparticle distance), a second equilibrium configuration for the particle pair emerges, and this entails the second possible value of rigidity. As it is inherent to hysteretic systems, the actual rigidity of the system depends on the strain history. With further growth of the field, the bistability regime ends up, the collapsed state of the particles remains the only possible and, accordingly, the mesoscopic element acquires the enhanced rigidity.

4. Mesoscopic element under the action of external forces

The stress-strain dependence is a fundamental mechanical property of a material. To get a notion of how the external field affects this characteristic of our 2P structure element, let us consider a system with interparticle distance $q_0 = 3.25$ and a relative compliance $\beta = 2500$. Let a uniform magnetic field of strength $h_0 = 0.015$ be applied to the element along its axis. The set of parameters is chosen in such a way that in the force-free state the sample is in the bistability regime. Then on the end walls of the element a force of magnitude f, i.e., either compressing (negative) or stretching (positive), is exerted. Using our model, we obtain the stress-strain diagram shown in figure 4a; here the dimensionless stress is defined as $p/c_1 = f/(c_1 S)$, where S is the cross section of the cylinder and c_1 the Mooney-Rivlin constant. Two possible rigidities of the system reflecting the mechanical responses of two different configurations of particles within the element cause the "doubling" of the diagram curves. In particular, they have different slopes in the absence of any mechanical load (see points *1* and *3*). An important circumstance revealed

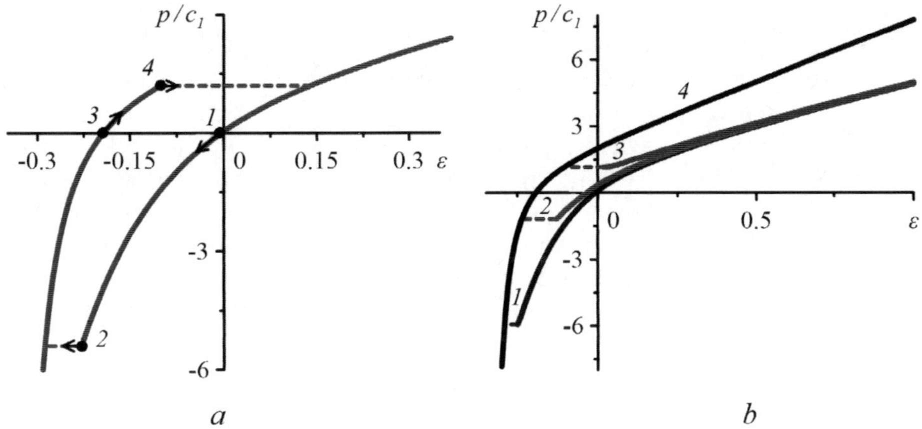

Figure 4. a) The stress-strain diagram for the mesoscopic element with compliance $\beta = 2500$ and initial interparticle distance $q_0 = 3.25$ under applied field $h_0 = 0.015$. b) The stress-strain diagram for the same mesoscopic element under variation of the applied field: $h_0 = 0.02$ *(1)*, 0.04 *(2)*, 0.06 *(3)*, 0.08 *(4)*; note that the latter diagram does not have a loop in the considered range of deformation and coincides with the upper curve of the other diagrams.

by figure 4 is that the increase of rigidity (caused by the particle collapse) can be provoked by an external mechanical load. If the 2P element is magnetized, it is already slightly shrunk even in the absence of an external mechanical force (see point *1*). This "load-free" compression is due to the well known *magnetostriction* effect inherent to MREs. A compressive force f causes further reduction of the element length, and at weak f the shrinking is gradual. However, when the force reaches a certain finite value, the particles collapse, and the strain changes abruptly (see point *2*). In terms of diagram 4a this means that the representing point jumps from the lower branch of the curve to the upper one and after that moves along the latter. It is noteworthy that, when the load is removed, the system does not return to its initial position (see point *3*) but remains in a compressed state, thus justifying the concept of a "magnetic staple" [12]. Indeed, one needs some effort in the opposite direction, a tensile force (see point *4*), to destroy this structure. Because of that, the graphic image of the loading / unloading cycle in figure 4 has a loop shape, whose boundaries projected on the vertical axis mark the boundaries of the bistability interval in terms of stress.

The above-presented considerations show that as soon as the mesoscopic element is prone to magnetomechanical hysteresis, then its stress-strain curve would be loop-shaped under the magnetic field of any intensity. This is illustrated in figure 4b that presents deformation cycles for the same mesoscopic element with $q_0 = 3.25$ and $\beta = 2500$. Note that a "magnetic staple" that has formed inside an element with high compliance is hard to destroy. For that, the sample should be stretched several times. Thus, in the diagram 4b for the element in the field 0.02 (cf. the diagram 4a) there is no reverse transition, see point *4*. In other words, as soon as "the magnetic staple" has once formed, it can break only under a substantial reduction or switching off of the field. The figure shows also that the position of the loop depends on the magnitude of the applied field. Therefore, one sees that the mechanical properties of the 2P mesoscopic element could be field-tuned, thus increasing or reducing its ability to resist deformation.

5. Homogenization

The above-presented 2P model allows one to understand the basics of the MRE internal structure rearrangement and the subsequent change in the properties of those composites on the qualitative level. In order to predict the properties of macroscopic samples, it is necessary to subject our discrete model to some "homogenization". Several approaches of that kind were developed recently [13, 14, 15]. However, as our model takes into account only a single geometry—axially symmetrical layout of the particles with respect to the external magnetic field—in here we restrict our consideration by a very simple averaging procedure.

We present a macroscopic MRE sample to be an assembly of cylindrical samples—our 2P mesoscopic elements—whose initial interparticle distances q_0 are distributed according to a certain law. All the cylinder axes are pointing the same direction along which the external load is applied. Each element responds to it independently, as if it was cut from the whole sample. Thus, any interaction at the boundaries between the individual elements is neglected. Under these conditions, to calculate the elasticity modulus or the stretching / compression stress, one just needs to average the corresponding mesoscopic characteristics over q_0. This way of material modelling is known as the Voigt hypothesis.

As is easy to understand, the resulting continuum model of a MRE has as its prototype a set of 2P mesoscopic elements connected in parallel. Assuming that the number of structure elements in a macroscopic sample is very large, the averaging is done with a continuous distribution of the interparticle distances. Thus, the "homogenized" rigidity of the model MRE is defined as

$$\langle K(H_0)\rangle = \left[\int_0^{q_0^{max}} K(q_0, h_0)\, f(q_0)\, dq_0\right] \Big/ \left[\int_0^{q_0^{max}} F(q_0)\, dq_0\right], \tag{4}$$

where function $F(q_0)$ renders distribution of the pairs with respect to q_0. As an example, we assume quasi-normal distribution of the distances. That means that when discretizing the standard formula for $F(q_0)$ we, first, limit ourselves by only non-negative values of q_0 and, second, truncate the distribution at $q_0 = 4.25$ since this is the largest center-to-center distance that our calculational scheme of the 2P element allows us to deal with. These restrictions are taken into account in order to have $F(q_0)$ normalized to unity. Keeping the form of the function, we set in it the "mean value" $\langle q_0\rangle = 3.0$ and "standard deviation" $\sigma = 0.3$. The resulting curve is shown in figure 5. As seen, with the field growth the rigidity of the model macroscopic sample, first, slightly decreases, and then increases. The rate of increase slows down gradually and finally tends to zero when saturation regime is attained. The second part of the cycle (field diminution) reveals the hysteresis loop of $\tilde{K}(h_0)$ of a considerable width.

The behavior of $\tilde{K}(h_0)$ can be easily explained with the help of figure 3 that renders the rigidity changes of 2P mesoscopic elements with different q_0's. When the field is turned on, the particles move towards each other and the stiffness decreases in all the elements. At the field strength about 0.02, two different rigidities become available, see figure 5. As the field is increased, the elements with smallest q_0's are first to collapse. However, the increase of rigidity induced by such pairs is not enough to counterbalance the decrease induced by the elements with larger q_0, where the particles are yet at the first stage of convergence. In the considered assembly, the majority of the pairs has the initial interparticle distances close $q_0 = 3$. Because of that, in the field $h_0 = 0.045$ that provokes the collapse of those pairs, the rigidity undergoes a fast increases. In the field $h_0 = 0.06$ even particles that were at the farthest initial distance are forced to collapse, and the sample rigidity saturates.

The stress–strain diagrams could be averaged similarly. In figure 6a the dashed line shows the diagram for the MRE sample loaded in the absence of a magnetic field. Solid lines show

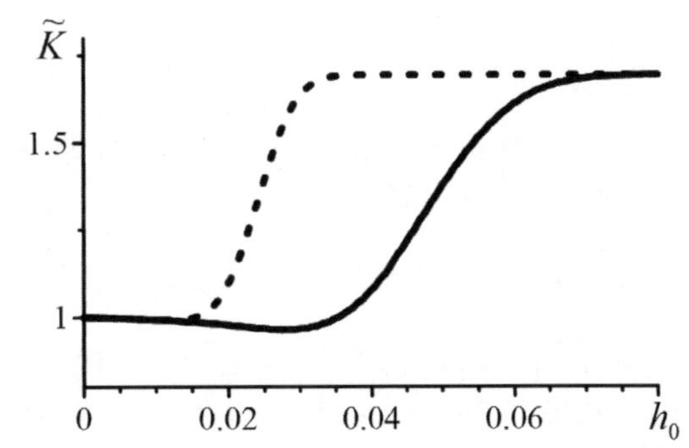

Figure 5. Relative rigidity of the model MRE sample corresponding to quasi-normal distribution of the 2P elements with $\langle q_0 \rangle = 3.0$ and $\sigma = 0.3$ as a function of external magnetic field.

the diagram obtained in the field $h_0 = 0.04$. The point marks the initial state of the sample in the field, that is, the equilibrium magnetically induced deformation under zero external stress. According to figure 5, the rigidity of this MRE in the field $h_0 = 0.04$ increases by about 10%, so the slopes of the lower curve of the diagram in the initial position and the dashed curve differ just a little.

So, one sees that, in general, the considered macroscopic MRE sample inherits the behavior of its mesoscopic elements, with exception that the transitions between the the lower and upper branches occurs smoothly. The conclusion that the stronger applied field the higher rigidity of the composite in the initial state is also confirmed by the stress-strain diagram 6b, where the slopes of the curves increase from point *1* to point *3* for successively increasing field strengths.

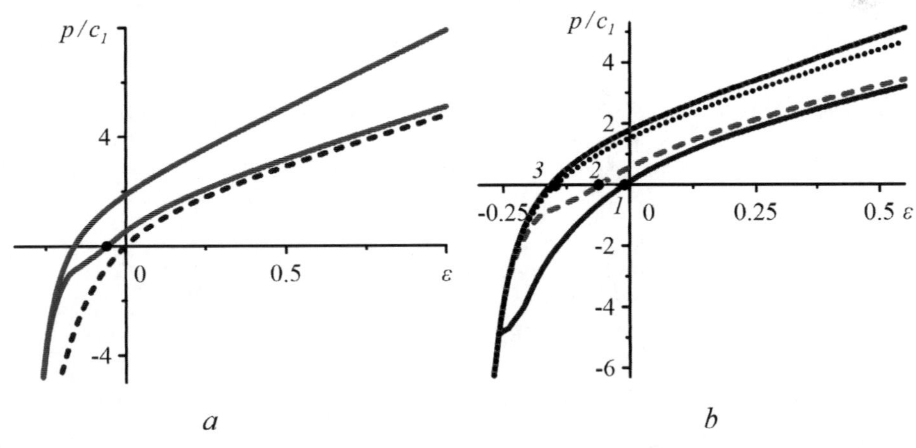

Figure 6. Stress-strain diagrams for the model MRE sample with relative compliance $\beta = 2500$: a) in the absence of magnetic field (dashes) and under the field $h_0 = 0.015$ (solid line); b) in the field $h_0 = 0.02$ (solid line), 0.04 (dashes), 0.06 (dots). The points mark the respective field-induced deformations under zero external stress.

6. Conclusions

The model of a 2P mesoscopic structure element developed in Refs. [7, 8] was specialized to describe the inner deformations which occur in MREs, which are free of external mechanical stress, under the action of an applied magnetic field. Herein, this model is extended for the case where the element is subjected to a uniaxial load. This allows us to analyze the combined magnetomechanical effects in the mesoscopic model system.

In particular, we show that a pure mechanical load is able to produce a strain hysteresis in a 2P element that otherwise dwells in an equilibrium magnetized state. In a qualitative way this proves that clustering of the particles (occurrence of "magnetic staples"), which changes so much the macroscopic properties of the MREs, could be provoked by mechanical loads as well as by magnetization as such.

An attempt to bridge the built up mesoscopic description with the macroscopic evidence is done. In spite of a very simple approach (Voigt hypothesis), the predictions look reasonable, promising and helpful for interpretation of the experimental data.

Acknowledgments

This work was supported by the Basic Research Program of Ural Branch of RAS No.10 (15-10-1-18), RFBR project 16-31-00503 and RFBR-DFG project 16-51-12001 (PAK 907).

References

[1] Zrínyi M, Barsi L and Büki A 1996 *Journal of Chemical Physics* **104** 8750–8756
[2] Nikitin L V, Mironova L S, Stepanov G V and Samus A N 2001 *Polymer Science, series A* **43** 443–450
[3] Ginder J M, Clark S M Schlotter W F and Nichols M E 2002 *International Journal of Modern Physics B* **16** 2412–2418
[4] Stolbov O V, Raikher Yu L and Balasoiu M 2011 *Soft Matter* **7** 8484–8487
[5] Ivaneyko D, Toshchevikov V P, Saphiannikova M and Heinrich G 2012 *Condensed Matter Physics* **15** 411–424
[6] Han Y, Hong W and Faidley L E 2013 *International Journal of Solids and Structures* **50** 2281–2288
[7] Biller A M, Stolbov O V and Raikher Yu L 2014 *Journal of Applied Physics* **116** 114904
[8] Biller A M, Stolbov O V and Raikher Yu L 2015 *Physical Review E* **92** 023202
[9] Bozorth R M 1993 *Ferromagnetism* (Wiley-IEEE Press)
[10] Oswald P 2009 *Rheophysics: The Deformation and Flow of Matter* (Cambridge University Press)
[11] Stepanov G V, Abramchuk S S, Grishin D A, Nikitin L V, Kramarenko E Yu and Khokhlov A R 2007 *Polymer* **48** 488–495
[12] Melenev P V, Rusakov V V and Raikher Yu L 2006 *Journal of Magnetism and Magnetic Materials* **300** e187–e190
[13] Ponte Castañeda P and Galipeau E 2011 *Journal of Mechanics and Physics of Solids* **59** 194–215
[14] Menzel A 2014 *Journal of Chemical Physics* **141** 194907
[15] Romeis D, Toshchevikov V and Saphiannikova M 2016 *Soft Matter* **12** 9364–9376

IOP Conf. Series: Materials Science and Engineering **208** (2017) 012008 doi:10.1088/1757-899X/208/1/012008

Influence of particle size distribution on nanopowder cold compaction processes

G Boltachev, N Volkov, K Lukyashin, V Markov and E Chingina

Institute of Electrophysics, Ural Branch of RAS, 106, Amundsen St., Ekaterinburg, 620016, Russia

E-mail: grey@iep.uran.ru

Abstract. Nanopowder uniform and uniaxial cold compaction processes are simulated by 2D granular dynamics method. The interaction of particles in addition to wide-known contact laws involves the dispersion forces of attraction and possibility of interparticle solid bridges formation, which have a large importance for nanopowders. Different model systems are investigated: monosized systems with particle diameter of 10, 20 and 30 nm; bidisperse systems with different content of small (diameter is 10 nm) and large (30 nm) particles; polydisperse systems corresponding to the log-normal size distribution law with different width. Non-monotone dependence of compact density on powder content is revealed in bidisperse systems. The deviations of compact density in polydisperse systems from the density of corresponding monosized system are found to be minor, less than 1 per cent.

1. Introduction

Processes of cold compaction of nanosized powders are the grave problem for manufacturing the nanostructured ceramics by powder metallurgy methods. Large forces of interparticle adhesion, which result from intermolecular dispersion forces of attraction, and high strength of nanoparticles require to use very high pressures (of gigapascal range) to achieve densities desired at the compaction stage [1-6]. Theoretical search of most effective conditions of compaction processes takes here on special significance and topicality. However it is hampered by absence of a valid phenomenological theory describing the powder body properties. Thereupon the direct simulations of nanopowders by the granular dynamics method, also known as the discrete element method [4, 5, 7-11], have attractive prospects. Different regularities of nanopowder systems, for example, the size effect in nanopowder compaction processes, have been already investigated in the framework of the granular dynamics method [4, 5, 12, 13]. Heretofore objects for numerical experiments are monodisperse systems [8, 12, 13], which consist of particles of the same size, or systems with very artificial particle size distributions, for example, the uniform distribution into the given narrow range of sizes [9]. This situation differs drastically from real experimental conditions [1-3, 6].

It is unlikely that the influence of particle size distribution on the powder compressibility could be detected by full-scale experiments since here there are too many factors, which have an influence at one time. However some data [6] are rather evidence of presence such influence. Attempts of studying this problem in the frames of numerical experiments have been made earlier, for example, in Refs. [7, 11], where powders of micron sizes have been simulated. Interparticle interactions in such powders

IOP Conf. Series: Materials Science and Engineering **208** (2017) 012008 doi:10.1088/1757-899X/208/1/012008

have essential distinctions from the nanoparticle interactions, namely dispersion attraction of particles is negligible, the plastic deformation of particles play an important role, and so on.

2. Objects of study and investigation method

The aim of the present study is to analyze the influence of particle size distribution on the compressibility of nanopowder systems. To achieve the aim the processes of the uniform (two-sided) and the uniaxial (one-sided) compaction have been simulated by granular dynamics method in 2D geometry. The following systems are simulated: equigranular (monodisperse) systems with particle diameter $d =$ 10, 20 and 30 nm; bidisperse systems with different content of small (the diameter is $d_1 = 10$ nm) and large (the diameter is $d_2 = 30$ nm) particles; polydisperse systems, which are described by the lognormal size distribution law with maximum at the diameter $d_{\text{mp}} = 20$ nm and different widths.

The particle size distribution for polydisperse systems is determined in the form of truncated lognormal law

$$f_{\text{n}}(x) = \frac{f(x)}{I_{\text{n}}}, \qquad f(x) = \frac{1}{x}\exp\left[-\frac{(\mu - \ln x)^2}{2\sigma^2}\right], \qquad \mu = \sigma^2 + \ln(d_{\text{mp}}),$$
$$I_{\text{n}} = \int_{d_{\min}}^{d_{\max}} f(x)dx, \qquad d_{\substack{\max \\ \min}} = d_{\text{mp}} \exp\left(\pm\sigma\sqrt{2\ln 6}\right), \tag{1}$$

where I_{n} is the normalization integral, $d_{\text{mp}} = 20$ nm, and the limits $d_{\min}$ and $d_{\max}$ are determined by the condition of the function $f(x)$ reduction by 6 times with respect to the maximum at $x = d_{\text{mp}}$. Beyond the bounds of ($d_{\min}, d_{\max}$) the normalized function $f_{\text{n}}(x)$ is fixed to zero. The systems with different distribution width are studied: $\sigma = 0$ (monodisperse system), 0.15, 0.30 and 0.47. The

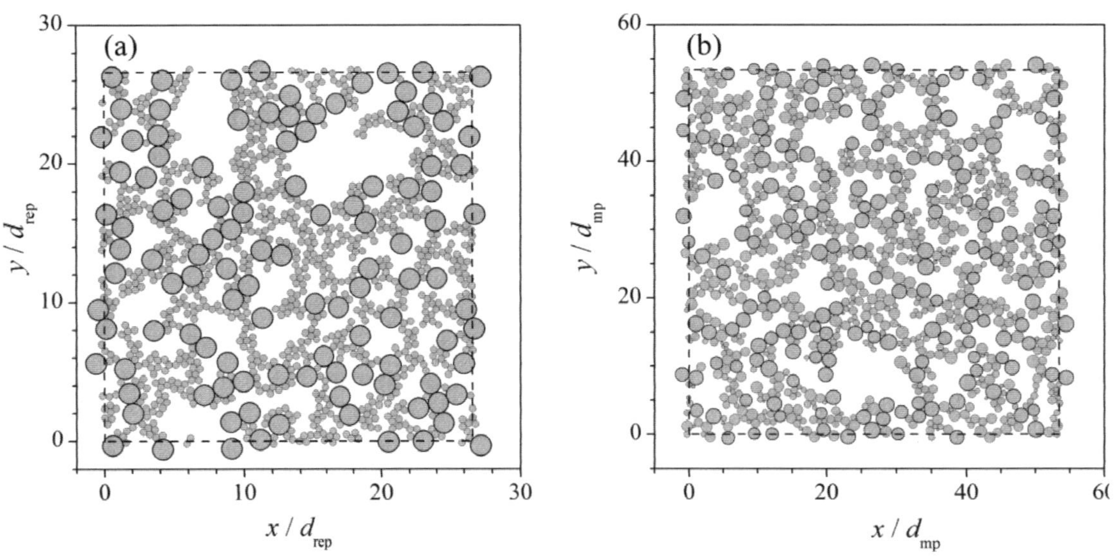

Figure 1. Examples of initial powder structures: (a) a bidisperse system with the relative fraction of small particles $\omega = 0.8$ ($d_{\text{rep}} = 20\,\text{nm}$); (b) a polydisperse system, where the particle size distribution is described by the function of (1) with parameters $\sigma = 0.47$ and $d_{\text{mp}} = 20$ nm. Dashed lines show the model cell boundaries.

maximal value $\sigma = 0.47$ corresponds to the real nanosized powders [5], which are produced by the method of electric explosion of wires [18].

The total number of particles N_p in model cells has been set equal to 1000. At that, as test calculations show, a model cell is a reliable representative volume of the powder body, i.e., the influence of the cell sizes on the model system properties is negligible. The initial density of all model systems has been set equal to $\rho_0 = 0.5$. As the density we have used the relative area of the model cell, which is occupied by the particles, i/.e., $\rho = S_p / S_{cell}$. To generate initial structures the algorithm of Ref. [12] is used, which allows producing isotropic and homogeneous structures in the form of connected 2D-periodical cluster. Figure 1 shows examples of initial structures. It is worth mention that the model polydisperse system with distribution width $\sigma = 0.47$ (see the figure 1-b) looks like real nanopowder photographs obtained by the scanning electron microscopy (see, for example, the figure 1 in Ref. [5]).

The parameters of interparticle interactions correspond to oxide nanosized powders having weak (I type systems) or strong (II type systems) agglomeration tendency [12, 13]. The theoretical model of [12, 13] includes the contact interaction laws, such as the modified Hertz law, which describes the elastic repulsion of particles, the linearized Cattaneo – Mindlin law [14, 15], which describes the "friction" forces at particle shift, the Lurie law [16], which describes the contact elasticity to the particle rolling, as well as modified Hamaker's formula [17], which describes the dispersion attractive forces, and (in the case of II type systems) the possibility of interparticle solid bridges formation [10]. Note that the dispersion attractions and the agglomeration tendency for the nanosized powders, as distinct from powders of micron (or larger) sizes, take on special significance and, in particular, owing to introduction this factors the size effect in compaction processes, well-known from the experimental studies [1-3, 6], is reproduced by the granular dynamics method. All parameters of particle interactions have been described in [12, 13] in detail. An exception is only the utilization in the present 2D simulations, in contrast to 3D calculations of [12, 13], the larger value of the interparticle friction coefficient $\mu = 0.5$.

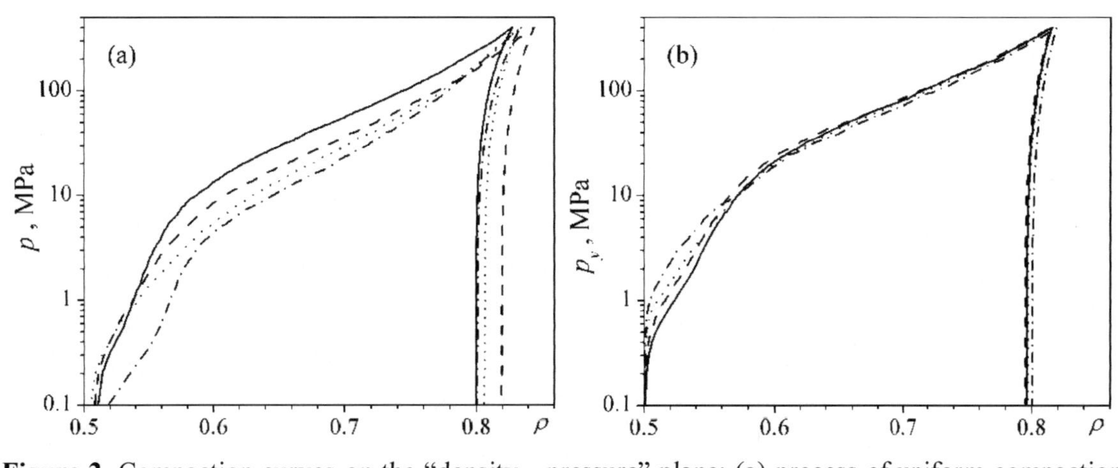

Figure 2. Compaction curves on the "density – pressure" plane: (a) process of uniform compaction of mono- and bidisperse systems of I type with different content of small particles: $\omega = 1.0$ (solid line), 0.9 (dashed line), 0.5 (dotted line), 0.0 (dashed-dotted line); (b) process of uniaxial compaction of the monosystem with the particle diameter $d = 20$ nm (solid line) and polydisperse systems of II type, which are described by the lognormal distribution law (1) with parameters $d_{mp} = 20$ and $\sigma = 0.15$ (dashed line), 0.30 (dotted line), 0.47 (dashed-dotted line).

3. Results and Discussion

Figures 2 – 4 demonstrate the simulation results for bidisperse and polydisperse systems. Figure 2 shows the compaction curves, the curves each constructing on the basis of averaging 20 – 40 independent numerical experiments. The statistical accuracy is shown in figures 3 and 4, where compact densities corresponding to the fixed pressures are presented. As the relative fraction of small particles in the figure 3 we imply the quantity $\omega = N_1 / N_p$. The stress tensor σ_{ij}, averaged over the model cell, is calculated by the known expression [8-10]:

$$\sigma_{ij} = \frac{-1}{V_{\text{cell}}} \sum_{k<l} f_i^{(kl)} r_j^{(kl)}, \tag{2}$$

where the summation is conducted over all pairs (k and l) of interacting particles, $V_{\text{cell}} = S_{\text{cell}} d_{\text{av}}$ and d_{av} is the averaging particle diameter. Note that as the compaction pressure in the figures 2 – 4 we have used the main pressure $p = -\text{Sp}(\sigma_{ij})/2$ for the uniform compaction process and the axial pressure $p_y = -\sigma_{yy}$ (along the loading y-axis) for the uniaxial compaction process.

Figure 2-a shows that for the bidisperse systems at the range of relatively small pressures (1 – 100 MPa) the compact density increases monotonically when passing from a system of small particles (with diameter $d_1 = 10$ nm) to a system of large particles (with diameter $d_2 = 30$ nm). At larger pressures, as the figure 3 shows, dependences $\rho(\omega)$ of bidisperse systems become non-monotone: there is the maximum at the range of ω values about 80 – 90 per cent. In the vicinity of the maximum the density values of bidisperse systems can exceed the densities of corresponding monosized systems by 1-2 per cent. In the systems of II type (see the figure 3-b) this phenomenon is expressed weaker but is on hand too.

The relative simple considerations can explain rather a sharp increase of the compact density when small number of large particles is added to the initial system of small particles, i.e., in the vicinity of $\omega = 1$. Into the area being occupied by the large particles added the density of the initial monodisperse system of small particles ($\rho_1 < 1$) has been increased up to the maximal value equal to 1. At the condition that a large particle added does not effect on the density ρ_1 of surrounding monodisperse

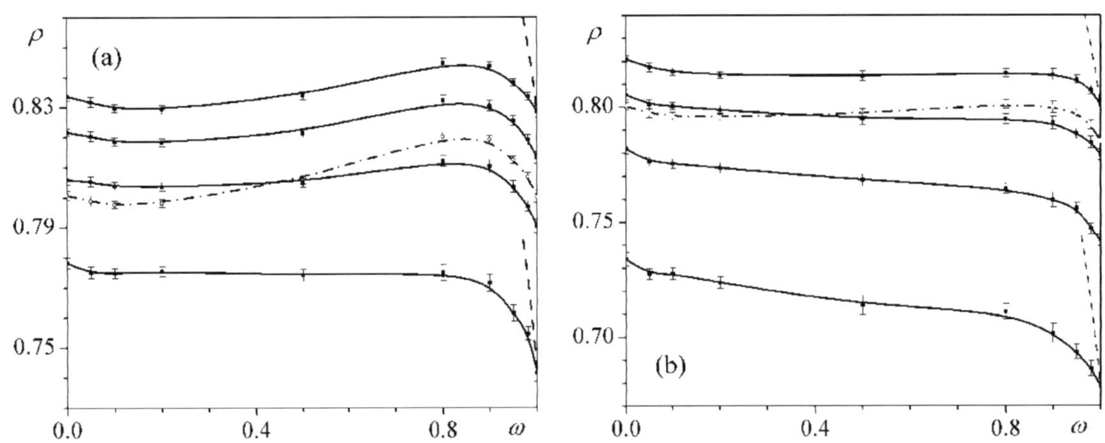

Figure 3. Dependences of the density of bidisperse systems of I type (a) (the uniform compaction) and of II type (b) (the uniaxial compaction) on the relative fraction ω of small particles. Solid lines (from the bottom) correspond to the compaction pressures p = 100, 200, 300 and 400 MPa; dashed-dotted lines show the compact densities after unloading from the maximal pressure (400 MPa); dashed lines are the asymptotes by Eq. (5).

system and also if $d_1 \ll d_2$ and $S_2 \ll S_{\mathrm{cell}}$, where S_2 is the area occupied by the large particles added, it is not difficult to deduce the density change. The density of the initial monodisperse system of small particles and the new density of the bidisperse system are determined by the expressions

$$\rho_1 = \frac{N_{\mathrm{p}}(\pi/4)d_1^2}{S_{\mathrm{cell}}}, \qquad \rho = \frac{N_1(\pi/4)d_1^2 + N_2(\pi/4)d_2^2}{S_{\mathrm{cell}}}, \tag{3}$$

where $N_1 = \omega N$, $N_2 = (1-\omega)N$. The new number of particles $N = N_1 + N_2$, which are situated into the area of the initial model cell, is $N = N_{\mathrm{p}} + N_2 - \Delta N_1$. The number ΔN_1 of "displaced" small particles, which was situated into the area $S_2 = N_2(\pi/4)d_2^2$, is $\Delta N_1 = S_2\rho_1 / \left[(\pi/4)d_1^2\right]$. Combining above listed expression and Eq. (3) we get

$$\Delta\rho = \rho - \rho_1 = \rho_1(1-\rho_1)\frac{d_2^2}{d_1^2}\frac{(1-\omega)}{1+(1-\omega)\left[\rho_1(d_2/d_1)^2 - 1\right]}. \tag{4}$$

In the limit of $\omega \to 1$ this expression gives the asymptotic dependence of the bidisperse system density on the content (ω) in the form of

$$\rho(\omega) = \rho_1 + (1-\omega)\rho_1(1-\rho_1)\frac{d_2^2}{d_1^2}. \tag{5}$$

The asymptotic dependences of (5) are shown by dashed lines in the figure 3. Noticeable deviations of simulation curves $\rho(\omega)$ from the asymptotes (5) in the limit $\omega \to 1$ are connected with weak implementation of the condition $d_1 \ll d_2$ in the bidisperse systems being studied, where we have $d_2/d_1 = 3$.

Figure 4 shows the compaction processes simulation results for polydisperse systems. The uniaxial compaction curves of II type systems, which are demonstrated in the figure 2-b, reveals that visible distinctions of compressibility of the systems simulated are only observed at the range of relatively

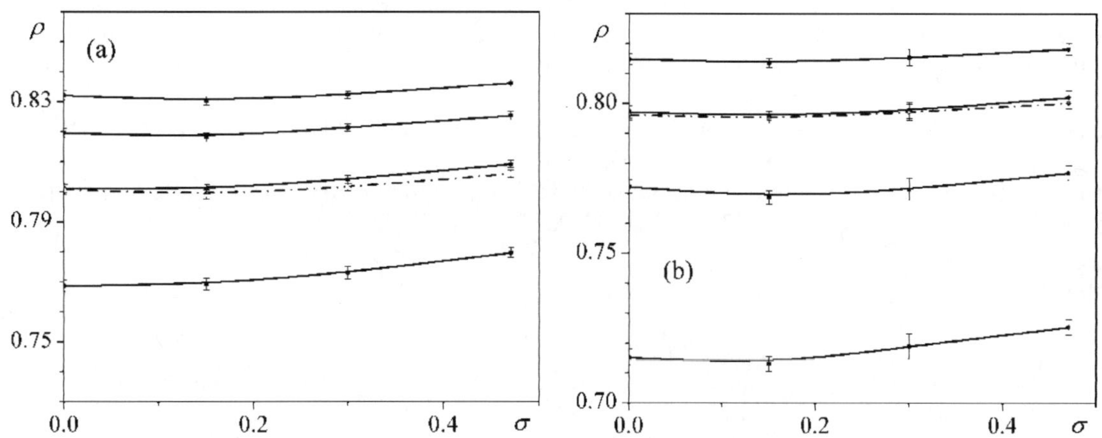

Figure 4. Dependences of the densities of polydisperse systems of I type (a) (the uniform compaction) and of II type (b) (the uniaxial compaction) on the parameter σ, which determines the width of size distribution (1). Solid lines (from the bottom) correspond to the compaction pressures p = 100, 200, 300 and 400 MPa; dashed-dotted lines show the compact densities after unloading from the maximal pressure (400 MPa).

small pressures p_y, namely for pressure values $p_y < 10$ MPa. At larger pressures the density differences become insignificant. Figure 4 demonstrates the density dependences at fixed compaction pressure on the particle size distribution width, which is determined by the parameter σ in Eq. (1). It can be seen that in the range of $\sigma < 0.1$, i.e., in the vicinity of pure monosized system the minor deterioration of compressibility is observed. It is possible that this deterioration is caused by the destruction of quasi-crystallinity regions, which are intrinsic for monodisperse systems [8]. Further increase of parameter σ results in compressibility improvement, i.e., the compact densities begin to go up. However as a whole, as the figure 4 shows, in the range of high compaction pressures ($p_y > 10$ MPa) the transition from a monodisperse system to polydisperse systems have an insignificant influence on the density of the nanosized powder. Density changes do not exceed 1 per cent. This result justifies the application of monodisperse systems in the frameworks of granular dynamics method to describe the experimental properties of real polydisperse nanopowders. It is worth mention here that the rigorous foundation for the last statement requires the fulfillment of numerical experiments in 3D geometry. This remark concerns also the results obtained for bidisperse systems. In particular, the asymptotical dependence (5) in 3D geometry has a form of

$$\rho(\omega) = \rho_1 + (1-\omega)\rho_1(1-\rho_1)\frac{d_2^3}{d_1^3}. \quad (6)$$

It can result in much larger maximums at the dependences $\rho(\omega)$ demonstrated in the figure 3.

4. Conclusion

Compaction processes of mono-, bi- and poly-disperse nanopowder systems are simulated by the granular dynamics method in 2D geometry. The main results of the research are as follows:

1. The influence of the content of bidisperse systems (composed of particles with diameter $d_1 = 10$ nm and $d_2 = 30$ nm) on the powder compressibility has been studied. The non-monotone density dependence on the relative fraction of small particles $\omega = N_1/(N_1 + N_2)$ has been revealed; namely, the dependence $\rho(\omega)$ has a maximum in the vicinity of values $\omega = 0.8 - 0.9$, where the bidisperse system density can exceed by 1-2 per cent the densities of corresponding monodisperse systems.

2. The compressibility of polydisperse systems with lognormal particle size distribution law has been studied. It is revealed that the increase of the distribution width, which is determined by the parameter σ, from zero (a monodisperse system) up to the value $\sigma = 0.47$ that corresponds to real nanopowders, has an insignificant influence on the compact density. This result justifies the application of monodisperse systems in the frameworks of granular dynamics method to describe the experimental properties of real polydisperse nanopowders.

The work has been fulfilled in the frame of state task project No. 0389-2014-0002 and supported by RFBR (project Nos. 16-08-00277-a and 16-33-00826-mol-a).

References

[1] Filonenko V P, Khvostantsev L G, Bagramov R Kh, Trusov L I and Novikov V I 1992 *Powder Metallurgy and Metal Ceramics* **31**(4) 296
[2] Zhao M, Li X, Wang Z, Song L, Xiao L and Xu B 1992 *NanoStructured Materials* **1**(5) 379
[3] Vassen R and Stöver D 1992 *Powder Technology* **72** 223
[4] Boltachev G Sh and Volkov N B 2010 *Technical Physics Letters* **36**(9) 823
[5] Boltachev G Sh, Volkov N B, Kaygorodov A S and Loznukho V P 2011 *Nanotechnologies in Russia* **6**(9-10) 639
[6] Saha B P, Kumar V, Joshi S V, Balakrishnan A and Martin C L 2012 *Powder Technology* **224** 90
[7] Skrinjan O and Larsson P-L 2003 *Proc. of the 2003 European Congress and Exhibition on*

Powder Metallurgy vol 3 (Shrewsbury: European Powder Metallurgy Association) p 335
[8] Agnolin I and Roux J-N 2007 *Phys. Rev.* E **76** 061302
[9] Gilabert F A, Roux J-N and Castellanos A 2008 *Phys. Rev.* E **78** 031305
[10] Balakrishnan A, Pizette P, Martin C L, Joshi S V and Saha B P 2010 *Acta Materialia* **58** 802
[11] Olsson E and Larsson P-L 2011 *Proc. of the EURO PM2011 Congress and Exhibition* vol 3 (Barcelona: European Powder Metallurgy Association) p 265
[12] Boltachev G Sh, Lukyashin K E, Shitov V A and Volkov N B 2013 *Phys. Rev.* E **88** 012209
[13] Boltachev G Sh, Volkov N B, Kochurin E A, Maximenko A L, Shtern M B and Kirkova E G 2015 *Granular Matter* **17** 345
[14] Mindlin R D 1949 *J. Appl. Mech. (Trans ASME)* **16** 259
[15] Jäger J 1995 *Archive of Applied Mechanics* **65** 478
[16] Lurie A I 1964 *Three-Dimensional Problems of the Theory of Elasticity* (NY: Interscience Publishers)
[17] Boltachev G Sh, Volkov N B and Nagayev K A 2011 *J. Colloid Interface Sci.* **355** (2) 417
[18] Kotov Yu A 2003 *Journal of Nanoparticle Research* **5**(5-6) 539

IOP Conf. Series: Materials Science and Engineering **208** (2017) 012009 doi:10.1088/1757-899X/208/1/012009

Aeroelastic stability of cylindrical shells interacting with internal annular fluid flow

S A Bochkarev and S V Lekomtsev

Institute of Continuous Media Mechanics UB RAS, 1, Korolev St., Perm, 614013, Russia

E-mail: bochkarev@icmm.ru

Abstract. This paper is devoted to the analysis of the dynamic behavior of cylindrical shells, containing an internal annular layer of ideal fluid and subject to the external supersonic gas flow. The aerodynamic pressure is calculated based on the quasi-static aerodynamic theory. The behavior of the compressible fluid is described in terms of the perturbation velocity potential. A mathematical formulation of the problem is developed based on the classical theory of shells and virtual displacement principle. A solution of the problem involves computation of complex eigenvalues of the coupled system of equations. The paper presents the results of numerical experiments, which were performed to estimate the influence of the fluid flow velocity on the value of the static pressure in the unperturbed gas flow for shells, interacting with fluid layers of different thicknesses. The numerical simulation shows that a reduction of the fluid layer thickness and increase of the fluid velocity produce a stabilizing effect by virtue of increasing the threshold of aerodynamic stability. However, an essential reduction of the layer thickness can lead, depending on the preset combinations of boundary conditions, to a considerable growth of the stability threshold or to the onset of instability.

1. Introduction

Thin-walled plate and shell structures interacting with fluids, as parts of many industrial applications are used in various technical fields. For a rather long time they have been the focus of numerous theoretical studies [1, 2]. The characteristic feature of these works is that the interaction between the elastic body and liquid or gaseous medium takes place only on the side of one structure surface - outer or inner surface. An exception is coaxial shells, in which the inner shell interacts both with the internal and annular flows. On the one hand, depending on the prescribed boundary conditions and flow velocity a combined action of both flows might be of additive nature, providing a strong destabilizing effect on the examined system and, on the other hand, might enlarge the stability boundaries [3–6]. However, in recent years there have been a number of publications, in which the aeroelastic stability is studied taking into account the influence of a quiescent liquid with a free surface located inside the shell or at its outer surface [7–10]. The panel flutter of isotropic and heated functionally-graded cylindrical shells completely filled with a quiescent or flowing ideal liquid has been analyzed in our recent papers [11, 12]. The objective of this study is to investigate the aeroelastic characteristics of cylindrical shells interacting with the internal annular fluid flow.

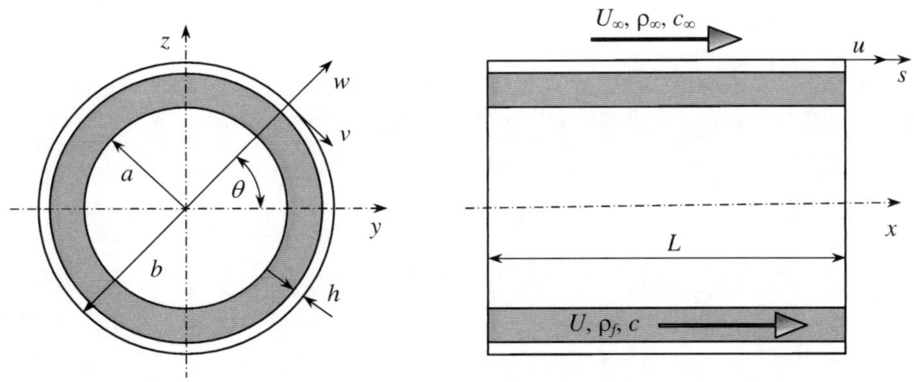

Figure 1. Cylindrical shell, containing an internal annular fluid flow and subject to the external supersonic gas flow.

2. Statement of the problem

Let us consider an elastic cylindrical shell of length L (figure 2), thickness h and radius b. Inside, it interacts with a layer of ideal compressible liquid of radius a, and outside it is subject to a supersonic gas flow with velocity U_∞. Here we will concern ourselves with the influence of the velocity U of the fluid flowing in the annular channel of different heights on the boundary of aeroelastic stability of the shell, at the edges of which we prescribe different combinations of boundary conditions.

The components of the strain vector $\varepsilon = \{\varepsilon_1, \varepsilon_2, \varepsilon_{12}, \kappa_1, \kappa_2, 2\kappa_{12}\}^{\mathrm{T}}$ in the curvilinear coordinate system (s, θ, z) are determined in terms of the classical theory of shells, which is based on the Kirchhoff–Love hypotheses, and are written as [13]

$$
\begin{aligned}
\varepsilon_1 &= \frac{\partial u}{\partial s}, \quad \varepsilon_2 = \frac{1}{b}\left(\frac{\partial v}{\partial \theta} + w\right), \quad \varepsilon_{12} = \frac{1}{b}\frac{\partial u}{\partial \theta} + \frac{\partial v}{\partial s} \\
\kappa_1 &= -\frac{\partial^2 w}{\partial s^2}, \quad \kappa_2 = \frac{1}{b^2}\left(\frac{\partial v}{\partial \theta} - \frac{\partial^2 w}{\partial \theta^2}\right), \quad \kappa_{12} = \frac{1}{b}\left(\frac{\partial v}{\partial s} - \frac{\partial^2 w}{\partial s \partial \theta}\right)
\end{aligned} \tag{1}
$$

Here u, v and w are the meridional, circumferential, and normal components of the displacement vector of the shell.

The physical equations relating the vector of forces and moments $\boldsymbol{T}$ to the vector of strains $\boldsymbol{\varepsilon}$, can be written in the matrix form as

$$
\boldsymbol{T} = \{T_{11}, T_{22}, T_{12}, M_{11}, M_{22}, M_{12}\}^{\mathrm{T}} = \mathbf{D}\boldsymbol{\varepsilon} \tag{2}
$$

The non-zero elements of the stiffness matrix $\mathbf{D}$ for isotropic material are generally determined in terms of the elasticity modulus E, Poisson's ratio ν and shear modulus [13].

A mathematical formulation of the problem is based on the virtual displacements principle, which in the matrix form is written as

$$
\int_{S_s} \delta\boldsymbol{\varepsilon}^{\mathrm{T}}\boldsymbol{T}\mathrm{d}S + \int_{S_s} \delta\boldsymbol{d}^{\mathrm{T}}\rho_0\ddot{\boldsymbol{d}}\mathrm{d}S - \int_{S_s} \delta\boldsymbol{d}^{\mathrm{T}}\boldsymbol{P}\mathrm{d}S = 0 \tag{3}
$$

Here $\boldsymbol{d}$ and $\boldsymbol{P} = \{\ 0 \quad 0 \quad p_a + p_f\ \}$ are the vectors of the generalized displacements and surface loads; p_a and p_f are the aerodynamic and hydrodynamic pressures acting on the external and

IOP Conf. Series: Materials Science and Engineering **208** (2017) 012009 doi:10.1088/1757-899X/208/1/012009

internal surfaces of the shell; $\rho_0 = \int_h \rho_s dz$; ρ_s is the specific density of the material of the shell; S_s is the surface that bounds the body of the shell.

The aerodynamic pressure exerted by the gas flow on the elastic surface is calculated in the framework of quasi-static aerodynamic theory using the approximate formula [14]

$$p_a = -\left(q\frac{\partial w}{\partial s} + q_1\frac{\partial w}{\partial t} - q_2 w \right) \tag{4}$$

where

$$q = \rho_\infty U_\infty^2/\beta = \kappa p_\infty \mathrm{M}_\infty^2/\beta, \quad q_1 = q(\mathrm{M}_\infty^2 - 2)/(U_\infty \beta^2), \quad q_2 = 1/2qR\beta$$

Here, $\mathrm{M}_\infty = U_\infty/c_\infty$ is the Mach number in a gas medium; ρ_∞, p_∞ and c_∞ are the density, static pressure and sound speed in the unperturbed gas flow; q is the modified dynamic pressure; κ is the adiabatic index; $\beta = \left(\mathrm{M}_\infty^2 - 1\right)^{1/2}$.

The hydrodynamic pressure exerted by the liquid, occupying the space V_f, on the wetted surface of the shell $S_\sigma = S_f \cap S_s$ is calculated using the linearized Bernoulli equation [15]

$$p_f = -\rho_f \left(\frac{\partial \phi}{\partial t} + U\frac{\partial \phi}{\partial x} \right) \tag{5}$$

Here ρ_f is the fluid density, S_f is the bounding surface of the fluid volume, ϕ is the perturbation velocity potential, which in the cylindrical coordinates (r, θ, x) is described by the wave equation

$$\nabla^2\phi = \frac{\partial^2 \phi}{\partial r^2} + \frac{1}{r^2}\frac{\partial^2 \phi}{\partial \theta^2} + \frac{\partial^2 \phi}{\partial x^2} + \frac{1}{r}\frac{\partial \phi}{\partial r} = \frac{1}{c^2}\left[\frac{\partial}{\partial t} + U\frac{\partial}{\partial x}\right]^2 \phi \tag{6}$$

where c is the speed of sound in the fluid.

On the shell-fluid interface we impose the impermeability condition

$$\frac{\partial \phi}{\partial n} = \frac{\partial w}{\partial t} + U\frac{\partial w}{\partial s} \tag{7}$$

where n is the normal to the boundary. It is assumed that the internal surface of the annular channel comes into contact with the rigid cylinder, which corresponds to the boundary condition $\partial\phi/\partial n|_{r=a} = 0$.

The perturbation velocity potential at the shell inlet and outlet obeys the following boundary conditions

$$\phi|_{x=0} = 0, \quad \partial\phi/\partial x|_{x=L} = 0 \tag{8}$$

For numerical implementation of the problem, the equation of perturbation velocity potential (6) and boundary conditions (7-8) are transformed using the Bubnov–Galerkin method [16].

3. Numerical implementation

For numerical implementation we used a semi-analytical variant of the finite element method (FEM) [17], in which the solution is represented as a Fourier series with respect to the circumferential coordinate θ

$$(u, w, \phi) = \sum\nolimits_{j=0}^{\infty} (u_j, w_j, \phi_j)\cos j\theta, \quad v = \sum\nolimits_{j=0}^{\infty} v_j \sin j\theta \tag{9}$$

where j is the number of harmonic.

Expressing the unknown variables in (9) in terms of their node values, we obtain the known matrix relations

$$U = \{u, v, w\}^{\mathrm{T}} = \mathbf{N}\boldsymbol{d}^e, \quad \boldsymbol{\phi} = \mathbf{F}\boldsymbol{f}^e, \quad \varepsilon = \mathbf{B}\boldsymbol{d}^e \tag{10}$$

Here $\mathbf{N}$ and $\mathbf{F}$ are the matrices of the shape functions for the shell finite element and the perturbation velocity potential, $\boldsymbol{d}^e$ and $\boldsymbol{f}^e$ are the vectors of the nodal values, $\mathbf{B}$ is the matrix of the relation between the strains and nodal displacements.

Discretization of the shell domain was carried out with the use of high-precision finite element in the form of truncated cone and approximation of the meridional and circumferential components of the displacement vector by a cubic polynomial and the normal component — by the 7th degree polynomial [18]. Discretization of the fluid domain was performed with the use of triangular finite element with linear approximation of the perturbation velocity potential [17].

Using the standard FEM procedures and taking into account equations (2,4,5,9,10), we obtain from the transformed wave equation for the perturbation velocity potential and virtual displacement principle (3) a coupled system of two equations, which

$$\mathbf{M}\{\ \ddot{\boldsymbol{d}}\ \ \ddot{\boldsymbol{\phi}}\ \}^{\mathrm{T}} + \mathbf{C}\{\ \dot{\boldsymbol{d}}\ \ \dot{\boldsymbol{\phi}}\ \}^{\mathrm{T}} + (\mathbf{K} + \mathbf{A})\{\ \boldsymbol{d}\ \ \boldsymbol{\phi}\ \}^{\mathrm{T}} = 0 \tag{11}$$

where

$$\begin{aligned}
\mathbf{M} &= \begin{bmatrix} \mathbf{M}_s & 0 \\ 0 & \mathbf{M}_f \end{bmatrix}, \quad \mathbf{C} = \begin{bmatrix} \mathbf{C}_s & \mathbf{C}_{sf} \\ \mathbf{C}_{fs} & \mathbf{C}_f \end{bmatrix}, \quad \mathbf{K} = \begin{bmatrix} \mathbf{K}_s & 0 \\ 0 & \mathbf{K}_f \end{bmatrix}, \quad \mathbf{A} = \begin{bmatrix} \mathbf{A}_s & \mathbf{A}_{sf} \\ \mathbf{A}_{fs} & \mathbf{A}_f \end{bmatrix} \\
\mathbf{K}_s &= \sum_{m_s} \int_{S_s} \mathbf{B}^{\mathrm{T}}\mathbf{D}\mathbf{B}\mathrm{d}S, \quad \mathbf{M}_s = \sum_{m_s} \int_{S_s} \mathbf{N}^{\mathrm{T}}\rho_0\mathbf{N}\mathrm{d}S, \quad \mathbf{M}_f = \sum_{m_f} \int_{V_f} \frac{1}{c^2}\mathbf{F}^{\mathrm{T}}\mathbf{F}\mathrm{d}V \\
\mathbf{K}_f &= \sum_{m_f} \int_{V_f} \left(\mathbf{F}^{\mathrm{T}}_{,r}\mathbf{F}_{,r} + \frac{1}{r^2}\mathbf{F}^{\mathrm{T}}_{,\theta}\mathbf{F}_{,\theta} + \mathbf{F}^{\mathrm{T}}_{,x}\mathbf{F}_{,x} \right) \mathrm{d}V, \quad \mathbf{C}_{fs} = -\sum_{m_s} \int_{S_\sigma} \mathbf{F}^{\mathrm{T}}\bar{\mathbf{N}}\mathrm{d}S \\
\mathbf{C}_{sf} &= \rho_f \sum_{m_s} \int_{S_\sigma} \bar{\mathbf{N}}^{\mathrm{T}}\mathbf{F}\mathrm{d}S, \quad \mathbf{A}_{sf} = \rho_f \sum_{m_s} \int_{S_\sigma} U\bar{\mathbf{N}}^{\mathrm{T}}\mathbf{F}_{,x}\mathrm{d}S \\
\mathbf{A}_f &= -\sum_{m_f} \int_{V_f} \mathrm{M}^2\mathbf{F}^{\mathrm{T}}_{,x}\mathbf{F}_{,x}\mathrm{d}V, \quad \mathbf{C}_f = \sum_{m_f} \int_{V_f} \frac{2U}{c^2}\mathbf{F}^{\mathrm{T}}_{,x}\mathbf{F}\mathrm{d}V, \quad \mathbf{C}_s = q_1 \sum_{m_s} \int_{S_s} \bar{\mathbf{N}}^{\mathrm{T}}\bar{\mathbf{N}}\mathrm{d}S \\
\mathbf{A}_{fs} &= -\sum_{m_s} \int_{S_\sigma} U\mathbf{F}^{\mathrm{T}}\bar{\mathbf{N}}_{,s}\mathrm{d}S, \quad \mathbf{A}_s = \sum_{m_s} \int_{S_s} \left(q\bar{\mathbf{N}}^{\mathrm{T}}\bar{\mathbf{N}}_{,s} + q_2\bar{\mathbf{N}}^{\mathrm{T}}\bar{\mathbf{N}} \right) \mathrm{d}S
\end{aligned}$$

Here, m_f and m_s is the number of finite elements used to decompose the fluid and shell domains, $\bar{\mathbf{N}}$ is the shape function matrix for a normal component of the displacement vector. Representing the disturbed motion of the shell and the fluid as $(\boldsymbol{d}, \boldsymbol{\phi}) = (\boldsymbol{q}, \boldsymbol{f}) \exp(\mathrm{i}\lambda t)$, we finally arrive at

$$(-\lambda^2\mathbf{M} + \mathrm{i}\lambda\mathbf{C} + \mathbf{K} + \mathbf{A})\{\ \boldsymbol{q}\ \ \boldsymbol{f}\ \}^{\mathrm{T}} = 0 \tag{12}$$

where $\boldsymbol{q}$ and $\boldsymbol{f}$ are some functions of coordinates, $\mathrm{i} = \sqrt{-1}$ and $\lambda = \lambda_1 + \mathrm{i}\lambda_2$ is the characteristic index. The eigenvalue problem was solved using the iterative algorithm, which is based on the Muller method [19].

4. The results of computations

For numerical simulation we chose a cylindrical shell ($h = 3\times10^{-4}$ m) made of aluminum ($E = 70$ GPa, $\nu = 0.3$, $\rho = 2707$ kg/m^3) with the following variants of kinematic boundary conditions prescribed at the ends: simply supported ($v = w = 0$, S), clamped ($u = v = w = \partial w/\partial s = 0$, C) and free-end support (F) conditions. The gas and fluid flows have the following characteristics $\rho_f = 1000$ kg/m^3, $c = 1500$ m/s, $\kappa = 1.4$, $\mathrm{M}_\infty = 3$, $T_\infty = 48.89$°C is the drag temperature in

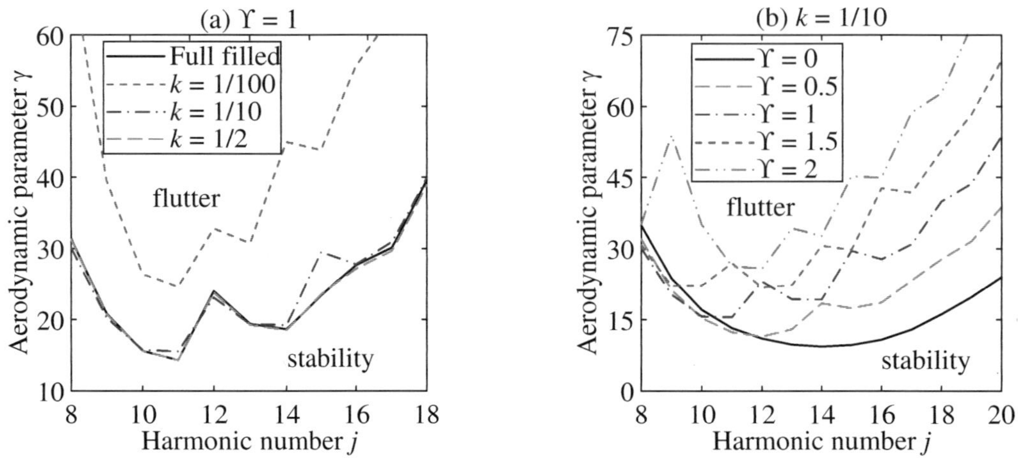

Figure 2. The non-dimensional parameter of the dynamic pressure γ versus the number of harmonics j at different values of the annular channel width k (a) and annular flow velocity Υ (b) for clamped shells.

the unperturbed flow. The numerical experiments have shown that the appropriate accuracy of computation is achieved in the case when the coupled system of equations has about 880 degrees of freedom. The obtained results are interpreted in terms of dimensionless fluid velocity Υ, parameter of dynamic pressure γ and the width of the annular channel k

$$\Upsilon = [U/\xi] \times 10^{-3}, \quad \gamma = [qL^3/D] \times 10^{-4}, \quad k = [b-a]/a$$
$$\xi = \sqrt{E/[\rho(1-\nu^2)]}, \quad D = Eh^3/[12(1-\nu^2)]$$

During the study of a combined action of aerodynamic and hydrodynamic loads, we operate on the assumption that the fluid velocity U is fixed and we search for the static pressure in the unperturbed flow p_∞, at which the system loses its stability.

Figure 4 shows the plots of non-dimensional parameter of the dynamic pressure γ as a function of the harmonic number j, obtained for rigidly clamped shells at different values of flow velocity Υ, and channel width k. For clarity, we also presents in the figure the results of computations made for a similar shell completely filled with the fluid. With increasing fluid velocity ($\Upsilon \neq 0$), the monotonic curve of the non-dimensional aerodynamic parameter changes to the non-monotonic curve. Such behavior is explained by the fact that at different harmonics the flowing fluid may have different effects. In this case, the critical number of the harmonic, i.e., the number of the harmonic with the lowest value of parameter γ, passes into the region of lower values. The influence of the thickness of the annular flow on the size of the instability region for different harmonics is shown in figure 4(a). Here, to represent the non-dimensional size of the fluid layer, we use notation proposed in [1]. From the data displayed in the figures, it follows that the width of the annular channel has only a qualitative impact on the instability boundaries, which is most pronounced in the case of very narrow annular channel.

At a certain height of the annular channel and prescribed boundary conditions the fluid flow exerts a stabilizing effect, which is demonstrated in figure 4(a), depicting the stability boundaries at three values of parameter k. It is evident from the figure, that actually, the critical dynamic pressure in the narrowest channel experiences an exponential growth. This can be attributed to the growth of the hydrodynamic pressure acting on the elastic surface, which strengthens it and prevents thereby the loss of stability.

IOP Conf. Series: Materials Science and Engineering **208** (2017) 012009 doi:10.1088/1757-899X/208/1/012009

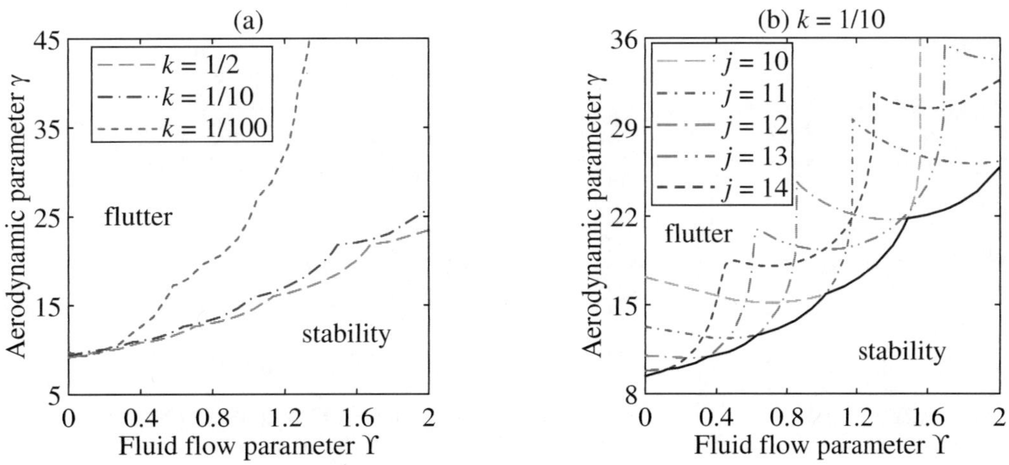

Figure 3. The non-dimensional parameter of the dynamic pressure γ versus the non-dimensional fluid velocity Υ for clamped shells at different values of the annular channel width k (a), and different harmonics j (b).

The non-monotonic behavior of curves in figure 4(a) suggests that with increase of the fluid velocity the critical harmonic number changes. In figure 4(b), this effect is illustrated in greater detail. It is seen that for some harmonics an increase in the velocity of the fluid flow can lead either to an abrupt growth or gradual reduction of the critical parameter γ, violating thereby the smoothness of stability threshold variation.

A combined action of the gas and fluid flows leads to a change in the instability type. In the case of a quiescent fluid the loss of stability occurs in the form of the coupled-mode flutter. Whereas an increase in the velocity of annular flow leads to a change in the type of instability to the single-mode flutter. Note that this form of instability remains invariable at any combinations of boundary conditions.

The dependence of the stability boundary on the thickness of the annular fluid layer is analyzed in detail in figure 4. The data shown in figure 4(a) indicate that a common feature of all harmonics, beginning with a certain value of b/a, is a variation of the parameter γ, which in the case of a thin fluid layer occurs in a jump-wise manner. On the other hand, from the results given in figure 4(b) we can conclude that in the specified parameter range for the examined combination of boundary conditions take place a rise of the stability threshold, which becomes more pronounced with increasingly growing velocity of the fluid.

The analysis of the behavior patterns found for other combinations of boundary conditions has shown that at a certain thickness of the annular fluid layer, the internal fluid flow has a stabilizing effect on the system, which is found to be more pronounced in cantilever shells (figure 4(a)). However, for simply supported shells there is a threshold value of the fluid layer thickness, beyond which even an inessential value of the hydrodynamic pressure parameter leads to the loss of stability (figure 4(b)). It should be noted that according to [11] for thin and long shells completely filled with liquid and subject to similar boundary conditions, a combined action of the aerodynamic and hydrodynamic loads is of the additive nature and leads to an abrupt decrease of the stability threshold.

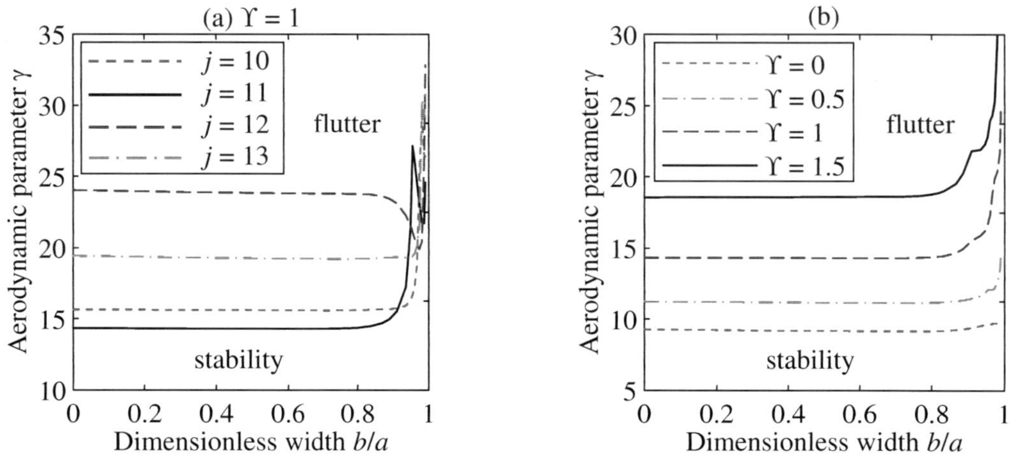

Figure 4. The non-dimensional parameter of the dynamic pressure γ versus the non-dimensional width of the annular channel b/a at different harmonics j (a) and fluid flow velocities Υ (b) for clamped shells.

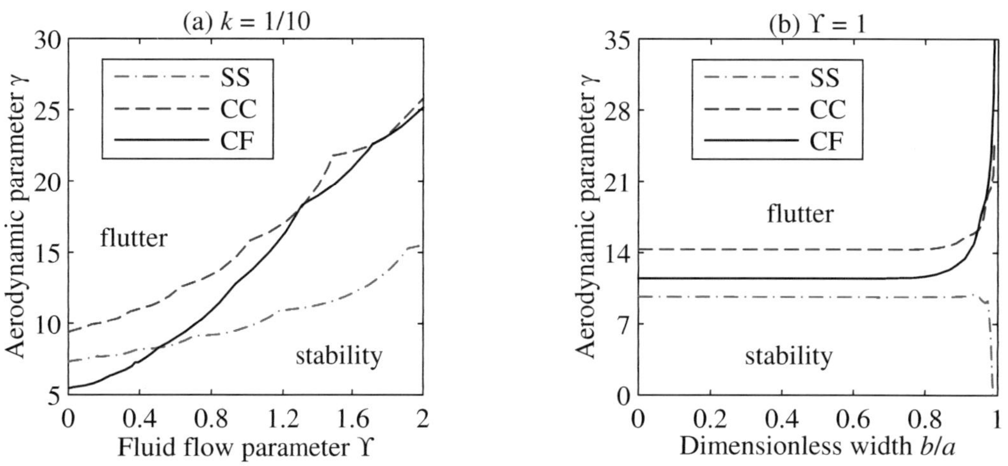

Figure 5. The non-dimensional parameter of the dynamic pressure γ versus the non-dimensional fluid velocity Υ (a) and non-dimensional width of the annular channel b/a (b) for shells under different combinations of boundary conditions.

5. Conclusion

We have considered the results of numerical simulation made to investigate the dynamic behavior of circular cylindrical shells of revolution, containing an annular layer of flowing, ideal, compressible fluid and subject to the external supersonic gas flow. The influence of the velocity of the annular fluid flow on the boundaries of aeroelastic stability has been analyzed for shells under different boundary conditions. It has been found that an increase of the fluid velocity, irrespective of the prescribed combination of the boundary conditions, leads to a change in the type of stability loss — a coupled mode flutter changes to a single-mode flutter. The results of computation allowed us to conclude that there exists a critical thickness of the annular fluid layer, at which the annular fluid flow in shells under any of the examined combinations of the boundary conditions produces a stabilizing effect and increases the aerodynamic stability threshold. For simply supported shells a decrease in the thickness of the annular fluid layer to some critical minimum causes a reverse reaction, namely, a destabilizing effect.

Acknowledgments
The work was supported by the Russian Foundation for Basic Research (grant 15-01-05254) and Ural Branch of RAS program (project 15-10-1-18).

References
[1] Païdoussis M P 2004 *Fluid-Structure Interactions: Slender Structures and Axial Flow* vol 2 (London: Elsevier Academic Press) p 1040
[2] Algazin S D and Kijko I A 2015 *Aeroelastic Vibrations and Stability of Plates and Shells* (Studies in Mathematical Physics vol 25) (Berlin: Walter de Gruyter GmbH) p 231
[3] Païdoussis M P, Chan S P and Misra A K 1984 Dynamics and stability of coaxial cylindrical shells containing flowing fluid *J. Sound Vib.* **97** 201-35
[4] Païdoussis M P, Nguyen V B and Misra A K 1991 A theoretical study of the stability of cantilevered coaxial cylindrical shells conveying fluid *J. Fluids Struct.* **5** 127-64
[5] Bochkarev S A and Matveenko V P 2010 The dynamic behaviour of elastic coaxial cylindrical shells conveying fluid *J. Appl. Math. Mech.* **74** 467-74
[6] Bochkarev S A, Lekomtsev S V and Matveenko V P 2014 Parametric investigation of the stability of coaxial cylindrical shells containing flowing fluid *Eur. J. Mech. A Solids* **47** 174-81
[7] Farhat C, Chiu E K-Y, Amsallem D, Schotté J-S and Ohayon R 2013 Modeling of fuel sloshing and its physical effects on flutter *AIAA J.* **51** 2252-65
[8] Sabri F and Lakis A A 2010 Hybrid finite element method applied to supersonic flutter of an empty or partially liquid-filled truncated conical shell *J. Sound Vib.* **329** 302-16
[9] Sabri F and Lakis A A 2011 Effects of sloshing on flutter prediction of liquid-filled circular cylindrical shell *J. Aircr.* **48** 1829-39
[10] Noorian M, Haddadpour H and Firouz-Abadi R 2015 Investigation of panel flutter under the effect of liquid sloshing *J. Aerosp. Eng.* **28** 04014059
[11] Bochkarev S A and Lekomtsev S V 2015 An aeroelastic stability of the circular cylindrical shells containing a flowing fluid *Vestn. Samar. Gos. Tekhn. Univ. Ser. Fiz.-Mat. Nauki* **19** 750-67 (in Russian)
[12] Bochkarev S A, Lekomtsev S V and Matveenko V P 2017 Aeroelastic stability of heated functionally graded cylindrical shells containing fluid *Mech. Adv. Mater. Struct.* **24** DOI: 10.1080/15376494.2016.1232457
[13] Biderman V L 1977 *Mechanics of Thin-walled Structures* (Moscow: Mashinostroenie) p 488 (in Russian)
[14] Bismarck-Nasr M N 1992 Finite element analysis of aeroelasticity of plates and shells *Appl. Mech. Rev.* **45** 461-82
[15] Vol'mir A S 1979 *Shells in Fluid and Gas Flow. Problems of Hydroelasticity* (Moscow: Nauka) p 320 (in Russian)
[16] Bochkarev S A and Matveyenko V P 2008 Numerical study of the influence of boundary conditions on the dynamic behavior of a cylindrical shell conveying a fluid *Mech. Solids* **43** 477-86
[17] Zienkiewicz O C 1971 *The Finite Element Method in Engineering Science* (London: McGraw-Hill) p 521
[18] Shivakumar K N and Krishna Murty A V 1978 A high precision ring element for vibrations of laminated shells *J. Sound Vib.* **58** 311-8
[19] Matveenko V P, Sevodin M A and Sevodina N V 2014 Applications of Muller's method and the argument principle to eigenvalue problems in solid mechanics *Comput. Continuum Mech.* **7** 331-6 (in Russian)

IOP Conf. Series: Materials Science and Engineering **208** (2017) 012010 doi:10.1088/1757-899X/208/1/012010

Exact solutions for layered large-scale convection induced by tangential stresses specified on the free boundary of a fluid layer

N.V. Burmasheva[1,2,a], **E.Yu. Prosviryakov**[2,3,b]

[1] Ural Federal University, 19 Mira Street, Ekaterinburg, 620002, Russia

[2] Institute of Engineering Science, 34 Komsomolskya Street, Ekaterinburg, 620049, Russia

[3] Kazan National Research Technical University, 10 K.Marx Street, Kazan, 420111, Russia

E-mail: [a]nat_burm@mail.ru, [b]evgen_pros@mail.ru

Abstract. A new exact solution for layered convection of a viscous incompressible fluid is found in this paper. A fluid flow in an infinite layer is considered. Convection in the fluid is induced by tangential stresses specified on the upper non-deformable boundary. Temperature corrections are given on the both boundaries of the fluid layer. The analysis of hydrodynamic fields allows us to state the presence of two stagnant points in the flow of a fluid. It is shown that, in the case of thermocapillary convection in a fluid, only one stagnation point can exist.

1. Introduction

A number of exact solutions describing nonisothermal fluid flows have been obtained to date. A great majority of exact nonisothermal solutions of the Navier-Stokes equations in the Boussinesq approximation are formulated in terms of velocity, pressure, and temperature fields that depend linearly on horizontal coordinate [1, 2]. In refs. [1, 2, 3, 4, 5, 6, 7] one can find the most important papers and reviews that follow fundamental studies authored by Ostroumov [8] and Birikh [9] and deal with physically meaningful exact solutions for advective and convective flows. In this paper, we will consider new exact solutions for layered convection under induced by tangential forces on the upper boundary of an infinite fluid layer.

2. Boundary value problem formulation

We consider convective large-scale motions of a viscous incompressible fluid in an infinite horizontal strip. Mathematical simulation of large-scale motions in a fluid is based on the use of layered flows ($V_z \equiv 0$) [1, 2, 3, 4, 5, 6]. Convective flows of a viscous incompressible fluid are described by the Oberbeck-Boussinesq equation system [7, 10], which consists of the Navier-Stokes equations

$$(\mathbf{V} \cdot \nabla)\mathbf{V} = -\nabla P + \nu \triangle \mathbf{V} + g\beta T \mathbf{k}, \tag{1}$$

the heat equation

$$\mathbf{V} \cdot \nabla T = \chi \triangle T, \tag{2}$$

and the incompressibility equation

$$\nabla \cdot \mathbf{V} = 0. \tag{3}$$

Here $\mathbf{V}(x, y, z) = (V_x, V_y, V_z)$ is a velocity vector; P is the deviation of pressure from hydrostatic, divided by constant average fluid density ρ; T is the deviation from the average temperature; ν and χ are the coefficients of the kinematic viscosity and thermal diffusivity of the fluid, respectively; $\mathbf{k}$ is the unit vector of the z axis directed vertically upwards; ∇ is the Hamiltonian operator; $\triangle$ is the three-dimensional Laplace operator.

Note that, although the terms of the convective derivative from the Navier-Stokes equation and the incompressibility equation are identically equal to zero, they remain present in the heat equation. Therefore, strictly speaking, a nonlinear problem is solved.

To find the hydrodynamic fields, we represent the velocities in the following form [1, 2]:

$$V_x = u(z), \quad V_y = v(z). \tag{4}$$

This representation allows us to find a solution for an overdetermined system. The system (1) - (3) consists of 5 equations with respect to four unknowns (velocity components V_x, V_y and physical fields T and P). The solution (4) identically satisfies the incompressibility equation (3). Thus, the velocity field (4) is solenoidal.

The pressure and the temperature can be represented in special linear forms depending on the horizontal coordinates as

$$T = T_0(z) + T_1(z)x + T_2(z)y, \quad P = P_0(z) + P_1(z)x + P_2(z)y. \tag{5}$$

Then we substitute the selected class of solutions (4) - (5) into the Navier-Stokes equation (1) and the heat equation (2). Equating the coefficients at the identical powers of the horizontal coordinates x and y, we obtain the following system of ordinary differential equations:

$$T_1'' = 0, \qquad T_2'' = 0. \tag{6}$$

$$\frac{\partial P_1}{\partial z} = g\beta T_1, \qquad \frac{\partial P_2}{\partial z} = g\beta T_2 \tag{7}$$

$$\nu u'' = P_1, \qquad \nu v'' = P_2, \tag{8}$$

$$\chi T_0'' = uT_1 + vT_2, \tag{9}$$

$$\frac{\partial P_0}{\partial z} = g\beta T_0. \tag{10}$$

In this paper, we assume that the lower boundary of the fluid layer is absolutely rigid and fixed. The upper boundary is assumed to be free and undeformed. We consider the following boundary conditions: at the lower boundary ($z = 0$) of the fluid layer the adhesion condition is satisfied and the temperature is given by the function

$$T = Ax + By. \tag{11}$$

On the upper boundary ($z = h$), constant pressure S acts and the temperature is given by the function

$$T = \vartheta + Cx + Dy. \tag{12}$$

In addition, stresses are given on the free boundary $z = h$ as

$$\eta \frac{du}{dz} = \xi_1, \quad \eta \frac{dv}{dz} = \xi_2. \tag{13}$$

Thus, we obtain the following system of boundary conditions:

$$u(0) = v(0) = 0,$$

$$
\begin{array}{lll}
T_0(0) = 0, & T_1(0) = A, & T_2(0) = B, \\
T_0(h) = \vartheta, & T_1(h) = C, & T_2(h) = D, \\
P_0(h) = S, & P_1(h) = 0, & P_2(h) = 0, \\
& \eta \dfrac{du}{dz}(h) = \xi_1, & \\
& \eta \dfrac{dv}{dz}(h) = \xi_2. &
\end{array}
\tag{14}
$$

Without loss of generality, we can assume that $S = 0$, thereby counting the reduced pressure from the level specified at the upper boundary.

3. Equation system solution

The case of temperature and stresses simultaneously set on the upper boundary was considered earlier in [11]. Therefore, another particular case of the above-mentioned boundary value problem is considered here, namely, when the temperature perturbation is specified at the lower boundary ($C = D = 0$).

Integrating the system of equations (6) - (10) in view of the boundary conditions, we obtain the exact solution

$$u(z) = \frac{z(24h\xi_1\nu + 6B\beta\eta g h^2 z - 8B\beta\eta g h z^2 + 3B\beta\eta g z^3)}{24\eta h\nu}, \tag{15}$$

$$v(z) = \frac{z(24h\xi_2\nu + 6A\beta\eta g h^2 z - 8A\beta\eta g h z^2 + 3A\beta\eta g z^3)}{24\eta h\nu}, \tag{16}$$

$$
\begin{aligned}
P_0(z) = {} & \frac{168h\nu(60\chi\eta(2hS - \beta g h^2\vartheta + \beta g\vartheta z^2) + B\beta g\xi_2(2h^5 - 5h^3z^2 + 5hz^4 - 2z^5))}{20160\chi\eta h^2\nu} + \\
& + \frac{A\beta g(168h\xi_1\nu(2h^5 - 5h^3z^2 + 5hz^4 - 2z^5))}{20160\chi\eta h^2\nu} + \\
& + \frac{B\beta\eta g(35h^8 - 80h^6z^2 + 168h^3z^5 - 196h^2z^6 + 88hz^7 - 15z^8)}{20160\chi\eta h^2\nu},
\end{aligned}
\tag{17}
$$

$$P_1(z) = -\frac{A\beta g(h - z)^2}{2h}, \tag{18}$$

$$P_2(z) = -\frac{B\beta g(h - z)^2}{2h}, \tag{19}$$

$$
\begin{aligned}
T_0(z) = -z \Bigg(& \frac{210h\nu(-12\chi\eta\vartheta + B\xi_2(h^3 - 2hz^2 + z^3))}{2520\chi\eta h^2\nu} + \\
& + \frac{A(210h\xi_1\nu(h^3 - 2hz^2 + z^3) + B\beta\eta g(20h^6 - 105h^3z^3 + 147h^2z^4 - 77hz^5 + 15z^6))}{2520\chi\eta h^2\nu} \Bigg),
\end{aligned}
\tag{20}
$$

$$T_1(z) = A - \frac{Az}{h}, \tag{21}$$

$$T_2(z) = B - \frac{Bz}{h}. \tag{22}$$

Then we formulate the problem of finding the number of stratification points in the velocity, pressure and temperature fields. Note that the case $A = B = 0$ is not considered here due

IOP Conf. Series: Materials Science and Engineering **208** (2017) 012010 doi:10.1088/1757-899X/208/1/012010

to its triviality. Indeed, if $A = B = 0$, then $P_1 = P_2 = T_1 = T_2 \equiv 0$, $u = \dfrac{z\xi_1}{\eta}$, $v = \dfrac{z\xi_2}{\eta}$, $P_0(z) = \dfrac{\beta g\vartheta\chi\eta(-h^2+z^2)}{2h}$, $T_0(z) = \dfrac{z\vartheta}{h}$, and this means that there are no stratifications. Therefore, we assume hereinafter that $A^2 + B^2 \neq 0$.

4. Investigation of velocity components

The velocity components *uandv* are zero at the origin (the factor z is explicit). Let us now ask the question how many more zeros these functions can have. This question is quite important, since the presence of zeros in these functions indicates the existence of counterflows (stratifications) in the convective flow of a viscous incompressible fluid.

We normalize the expressions in (3) for the components $u(z)$ and $v(z)$ of the velocity vector $\mathbf{V}$ by A reducing them to a dimensionless form. Without loss of generality, we set $A \neq 0$.

We introduce the following dimensionless parameters: $\gamma = B/A$, $\delta = h/l$, where h is the characteristic vertical dimension of the layer and l is the characteristic horizontal dimension of the layer. In addition, the dimensionless coordinate z is determined as $z \to z/h$.

The velocities $u(z)$ and $v(z)$ are divided by $\dfrac{g\beta Al^3}{\nu}$. As a result, we finally arrive at the formula

$$u = \frac{z\delta^3}{24}\left(\gamma(6z - 8z^2 + 3z^3) + 24W_1\right), \tag{23}$$

$$v = \frac{z\delta^3}{24}\left(6z - 8z^2 + 3z^3 + 24W_2\right). \tag{24}$$

Here, $W_i = \dfrac{\nu\xi_i}{g\beta C\eta h^2}$ is the Weber number determined for the value ξ_i.

We study the function $u(z)$. It obviously vanishes at the point $z = 0$.

If $\gamma = 0$, then this zero is unique for $W_1 \neq 0$. If $\gamma^2 + W_1^2 = 0$, then the function $u(z) \equiv 0$. Therefore, we set $\gamma \neq 0$ and represent the function $u(z)$ in the form

$$u(z) = \frac{z\delta^3\gamma}{24} f(z, 24W_1/\gamma), \quad f(z,q) = 6z - 8z^2 + 3z^3 + q. \tag{25}$$

The number of the zeros of the function u_1 on the interval $(0;1)$ determines the number of stratifications of the velocity field $u(z)$.

Figure 1 presents the plot of the function $f(z,0)$. This function reaches its maximum at the point $z_0 = (8-\sqrt{10})/9$ and takes the value $f_{max} = 4(68+5\sqrt{10})/243$ at this point. The function $f(z,0)$ has no zeros in the interval $(0;1)$; however, one can change the value of the parameter q so that the function $f(z,q)$ (as well as the function $u(z)$) will have one or even two zeros on the interval $(0;1)$.

IOP Conf. Series: Materials Science and Engineering **208** (2017) 012010 doi:10.1088/1757-899X/208/1/012010

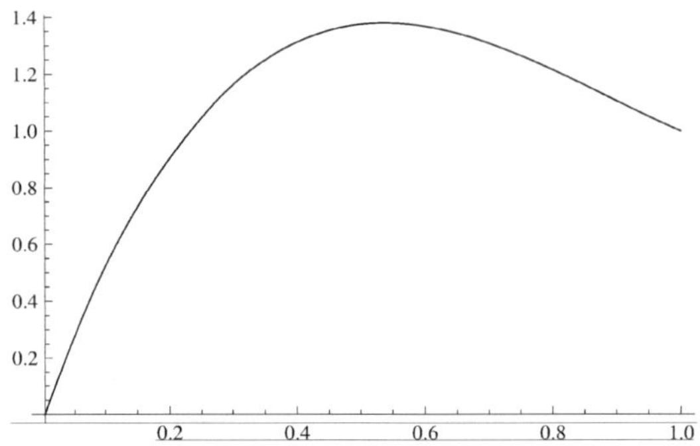

Figure 1. The plot of the function $f(z,0)$.

Thus, returning to the function $u(z)$, we obtain the following estimates:

1) $\frac{24W_1}{\gamma} > f(0,0) = 0$, i. e. the function $u(z)$ does not change its sign;
2) $0 = f(0,0) > \frac{24W_1}{\gamma} \geq -f(1,0) = 1$, i. e. the function $u(z)$ changes its sign once;
3) $1 > \frac{24W_1}{\gamma} > -f(1,0) = -f_{max}$, i. e. the function $u(z)$ changes its sign twice;
4) $-f_{max} = -f(1,0) = \frac{24W_1}{\gamma}$, i. e. the function $u(z)$ changes its sign once;
5) $\frac{24W_1}{\gamma} < -f_{max}$, i. e. the function $u(z)$ does not change its sign.

The behavior of the function $u(z)$ for various values of the ratio W_1/γ is shown in figure 2.

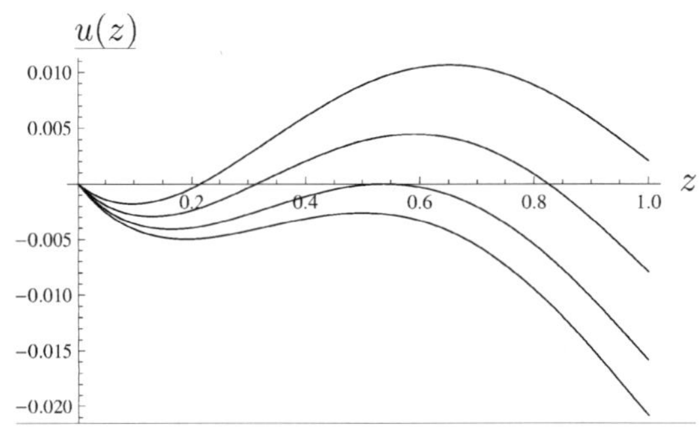

Figure 2. The behavior of the function $u(z)$.

Similar estimates are valid for the function $v(z)$:

1) $24W_2 > f(0,0) = 0$, i. e. the function $v(z)$ does not change the sign;
2) $0 = f(0,0) > 24W_2 \geq -f(1,0) = 1$, i. e. the function $v(z)$ changes its sign once;
3) $1 > 24W_2 > -f(1,0) = -f_{max}$, i. e. the function $v(z)$ changes its sign twice;
4) $-f_{max} = -f(1,0) = 24W_2$, i. e. the function $v(z)$ changes its sign once;
5) $24W_2 < -f_{max}$, i. e. the function $v(z)$ does not change the sign.

Of particular interest in this situation is the study of the velocity vector hodograph. Theoretically, there cannot be more than 25 different forms of the hodograph (5 variants for the $u(z)$ curve and 5 variants for the $v(z)$ curve). However, there are situations that generate one and the same hodograph (for example, for $\gamma = 1$ when $W_1 = W_2 = -1$ or $W_1 = W_2 = 0$, the hodograph curve is the bissectrice of the first quadrant). In addition, we should not forget about the relationship between the parameters of the problem ($W_1\xi_2 = W_2\xi_1$), which leads to additional constraints and, consequently, to a decrease in the number of fundamentally different trajectories of the velocity vector hodograph. As an example, in figures 3 and 4 the velocity vector hodographs are given for two fixed values of the ratio W_1/γ.

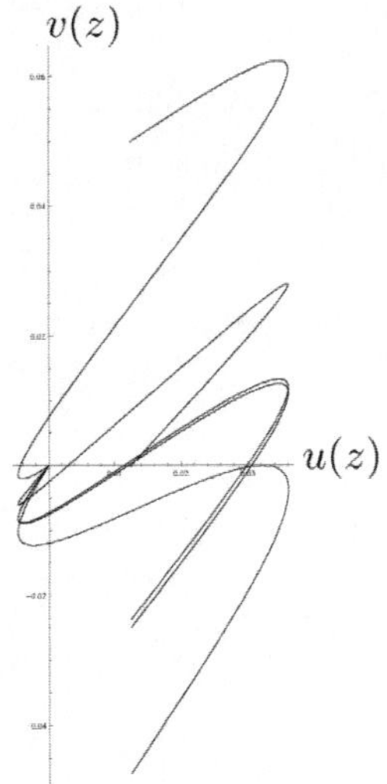

Figure 3. The velocity vector hodograph.

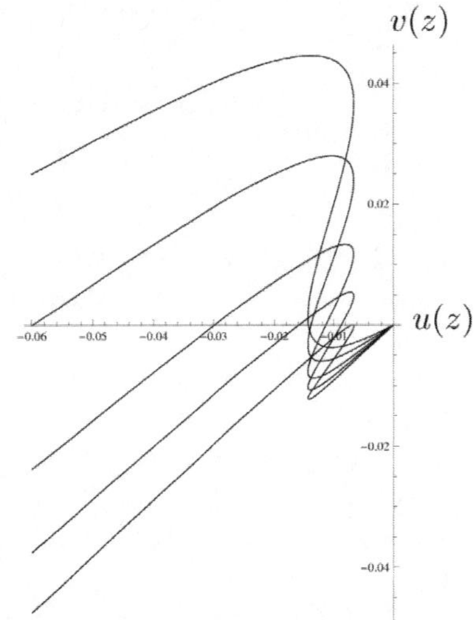

Figure 4. The velocity vector hodograph.

Let us write out the basic quantitative measure of vorticity:

$$\vec{\Omega} = rot\mathbf{V} = \begin{vmatrix} \vec{i} & \vec{j} & \vec{k} \\ \frac{\partial}{\partial x} & \frac{\partial}{\partial y} & \frac{\partial}{\partial z} \\ u & v & 0 \end{vmatrix}, \tag{26}$$

or coordinatewise:

$$\Omega_x = -\frac{\partial v}{\partial z} = -\frac{\delta^3}{2}(2W_2 + z - 2z^2 + z^3), \quad \Omega_y = \frac{\partial u}{\partial z} = \frac{\delta^3}{2}(2W_1 + \triangle(z - 2z^2 + z^3)), \quad \Omega_z = 0. \tag{27}$$

Of particular interest here are combinations of parameters for which the vorticity becomes zero at least at several points (it is not identically equal to zero). It is at these points that the direction of the vortex changes.

It is obvious that, for $\gamma = 0$, there are no such points for the component Ω_y; therefore, we assume that $\gamma \neq 0$. Then Ω_2 can be rewritten as

$$\Omega_y = -\frac{\delta^3}{2}(2W_1 + \triangle(z - 2z^2 + z^3)) = -\frac{\delta^3\gamma}{2}(2W_1/\gamma + \triangle(z - 2z^2 + z^3)). \tag{28}$$

That is, the number of points at which the components $\Omega_x and \Omega_y$ change their direction (one at a time or simultaneously) is governed by the number of zeros of the function$g(z, c) = c + z - 2z^2 + z^3$ (when $c = 2W_1/\triangle\gamma$ for the component Ω_y and when $c = 2W_2$ for Ω_x). With all the further calculations, one can refer to the behavior of the function $g(z, c)$, whose plot for $c = 0$ is presented in figure 5.

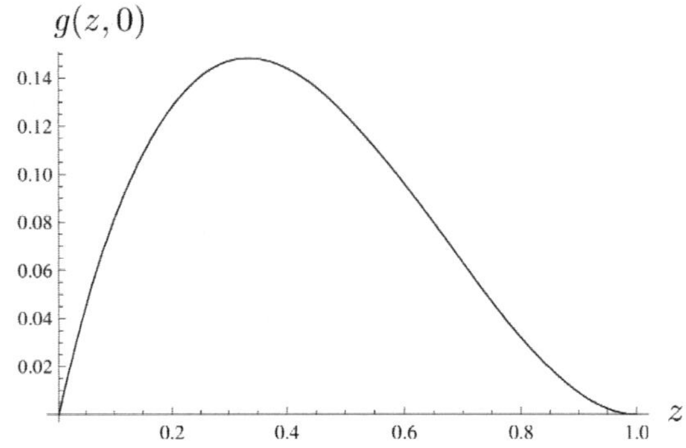

Figure 5. The plot of the function $g(z, 0)$.

5. Conclusion

For the discussed boundary-value problem of simultaneous setting of the boundary temperature and stresses, it has been shown that, in the hydrodynamic fields (in the velocity field in particular), stratifications may arise, depending on certain combinations of the parameters specifying the physical quantities at the boundaries of the region under study. The number of stratification points is always different (0, 1 or 2), and it depends on the values of the constants. The same values also determine the positions of stratification points in the region under investigation. In addition, it has been demonstrated that the motion is vortex (the vorticity is not identically zero); however, under certain combinations of the system parameters, the vortex may change its direction. The conditions for this phenomenon have been discussed.

Acknowledgments

This work was supported by the Foundation for Assistance to Small Innovative Enterprises in Science and Technology (the UMNIK program).

References

[1] Aristov S N and Prosviryakov E Yu 2013 *Rus. J. Nonlin. Dyn.* **9** 651
[2] Aristov S N and Prosviryakov E Yu 2016 *Theor. Found. Chem. Eng.* **50** 286
[3] Sidorov A F 1989 *J. Appl. Mech. Tech. Phy.* **30** 197
[4] Shvarz K G 2014 *Fluid Dynamics* 4 438
[5] Aristov S N, Knyazev D V and Polyanin A D 2009 *Theor. Found. Chem. Eng.* **43** 642
[6] Andreev V K and Bekezhanova V B 2013 *J. Appl. Mech. Tech. Phy.* **54** 171
[7] Gershuni G Z and Zhukhovitskii E M 1976 *Convective stability of incompressible liquid* (Jerusalem: Wiley)
[8] Ostroumov G A 1958 *Free convection under the condition of the internal problem* (NASA TM)
[9] Birich R V 1966 *J. Appl. Mech. Tech. Phy.* **3** 69
[10] Landau L D and Lifshitz E M 1987 *Fluid Mechanics* (Oxford: PergamonPress)
[11] Burmasheva N V and Prosviryakov E Yu 2017 *Vestn. Samar. Gos. Tekh. Univ., Ser. Fiz. Mat. Nauki (J. Samara State Tech. Univ., Ser. Phys. and Math. Sci.)* **21** (*Preprint* doi: 10.14498/vsgtu1527)

IOP Conf. Series: Materials Science and Engineering **208** (2017) 012011 doi:10.1088/1757-899X/208/1/012011

Numerical simulation of helical flow in a cylindrical channel

A Vasiliev, A Sukhanovskii and R Stepanov

Institute of Continuous Media Mechanics UB RAS, 1, Acad. Korolev St., Perm, 614013, Russia

E-mail: vasiliev.a@icmm.ru

Abstract. Numerical simulation of the helical flow in a cylindrical channel with diverter was carried out using open-source software OpenFOAM Extend 4.0. The velocity, vorticity and helicity density distributions were analyzed. It was shown that azimithal contribution of helicity is negative near the wall and positive in the center. In opposite axial helicity contribution is negative in the center and positive near the wall. Analysis of helicity of non-axisymmetric part of the flow showed that it has substantial values near the diverter but than rapidly decreases with y (axial coordinate) and further downstream it can be neglected. Dependencies of integrated values of azimuthal H_ϕ and axial H_y contributions of helicity density on y show a remarkable quantitative similarity. It was found that integral values of H_ϕ and H_y are negative for all y. Magnitudes of H_ϕ and H_y decrease after the diverter up to $y \approx 70$ mm and after that monotonically increase. The flow behind the diverter is characterized by substantial amount of helicity and can be used as a helicity generator.

1. Introduction

The volume integral of hydrodynamic helicity density H is an inviscid invariant of the Euler equation and it is related to the linkage of the vortex lines of the flow [1]. Helicity density H is defined as a scalar product of the local velocity vector $\mathbf{U}$ and the local vorticity vector $\mathbf{W}$

$$H = \mathbf{U} \cdot \mathbf{W}. \tag{1}$$

Helicity density is a pseudoscalar quantity which is a signature of broken mirror symmetry of a flow. It is assumed that helicity can lead to the generation of large-scale magnetic fields (MHD – dynamo [2]) and atmospheric vortices (hydrodynamic alpha effect [3, 4]). The inhomogeneity of turbulent helicity may have an effect on turbulent momentum transport of screw pipe flow [5] and contributes to the sustainment of large-scale vorticity field in a three-dimensional mean flow [6].

The measurement of helicity is extremely difficult as the simultaneous measurements of both the velocity and vorticity vectors are required. Modern optical techniques such as dual or tomographic PIV (Particle Image Velocimetry) make possible reconstruction of helicity density distribution in a plane or even in a volume. Spatial distribution of the helicity density in screw jets was experimentally investigated in [7, 8] where helical flows was generated by screw nozzle. In [7] dual-plane dye laser PIV was used for measurements of all components of vorticity. From the analysis of the results, it was found that regions with high helicity were correlated with regions of high turbulent kinetic energy dissipation.

Tomographic PIV, which is currently under intensive development, makes possible to measure the instantaneous 3D velocity fields in the selected measuring volume of the flow. The helical modes have been studied in low-and high-swirl turbulent jets by using the high-speed tomographic PIV technique [8].

One of the main problems of laboratory studies of helicity is to produce the flow with substantial value of helicity. This problem is far from trivial [9]. One of the promising configuration for generation of such flow is an axial flow in the channel with a diverter [10, 11]. Qualitative observations showed that the flow in a torus with a diverter is very complex and consist of helical motions of different scales. There was no direct measurements of helicity or even estimation of its value. The main goal of the present study is to give a clear answer on the possibility of using such a system as a helicity generator. Special attention is paid to the azimuthal inhomogeneity and anisotropy of flow.

2. Numerical simulation

Numerical simulation of the forced air screw flow in a cylindrical channel was carried out using open-source software OpenFOAM Extend 4.0. For injection of screw flow eight blade diverter was used. Similar diverters were used in studies of screw flows in torus [10, 11]. Inclination angle of the blades was 30°. Figure 1(a) presents the computational domain, where a diverter placed at distance of 20 mm downstream from the inlet boundary. Computational domain is a cylindrical channel of diameter $d = 60\,\mathrm{mm}$ and length $L = 400\,\mathrm{mm}$. The origin of a Cartesian coordinate system (x, y, z) is located at the centre of the diverter. The axis of a cylinder y coincides with the direction of the flow.

The discretization of three-dimensional Navier-Stokes equations for an incompressible flow was carried out by the method of control volumes. The PIMPLE (merged PISO-SIMPLE) algorithm was used for the pressure-velocity coupling. We employed implicit Euler method second order accuracy temporal discretization. For the discretization of the divergence term total variation diminishing (TVD) scheme of second order accuracy was applied. The second order accuracy central difference scheme was applied for discretization of diffusion term. The iterative solvers PCG and PBiCG – the preconditioned conjugate gradient and preconditioned biconjugate gradient methods were used for solving the obtained system of linear algebraic equations. The following boundary conditions are used in calculations: at the inlet, velocity assumed to be uniform $\mathbf{U} = (0, 0.1, 0)$ m/s and normal gradient of pressure is zero; at the outlet, pressure assumed to be uniform $p = 0$ and normal gradient of velocity is zero. For the diverter and cylinder surface no-slip condition was used. Numerical simulation was done using unstructured mesh. Computation mesh had about 4.8 millions elements. All the results have been obtained using initial conditions equal to zero for velocity and pressure in the whole computational domain.

3. Results

Numerical simulation was done for the fixed value of Reynolds number $\mathrm{Re}{=}u_0 d/\nu = 370$, where ν is kinematic viscosity, d is a diameter of cylinder and u_0 is velocity at the inlet. Streamlines shows the three-dimensional structure of the flow in figure 1(a). After the diverter an intensive screw flow is produced.

For analysis of velocity and vorticity fields in a cylindrical channel it is more convenient to use cylindrical (r, ϕ, y) coordinate system instead of Cartesian. Figures 2 and 3 show the spatial distributions of the time-averaged azimuthal and the axial components of the velocity and vorticity in three different cross-sections perpendicular to the y axis. In the present study we did not analysed the radial components of velocity and vorticity, because they (and their gradients) are approximately an order of magnitude smaller than other components. Near the exit from diverter the flow is characterized by eight distinct jets produced by blades of

IOP Conf. Series: Materials Science and Engineering **208** (2017) 012011 doi:10.1088/1757-899X/208/1/012011

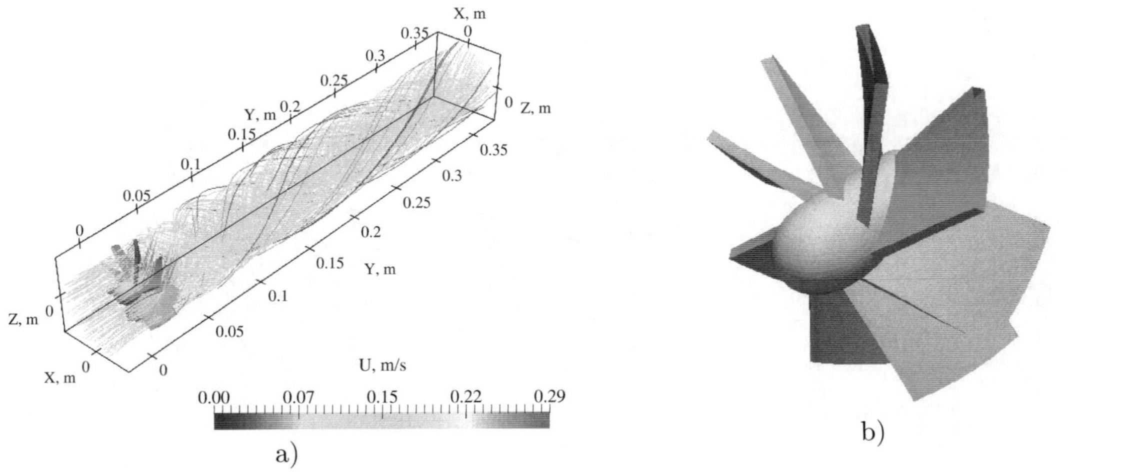

Figure 1. a) – Streamlines of the mean velocity field, b) – the diverter.

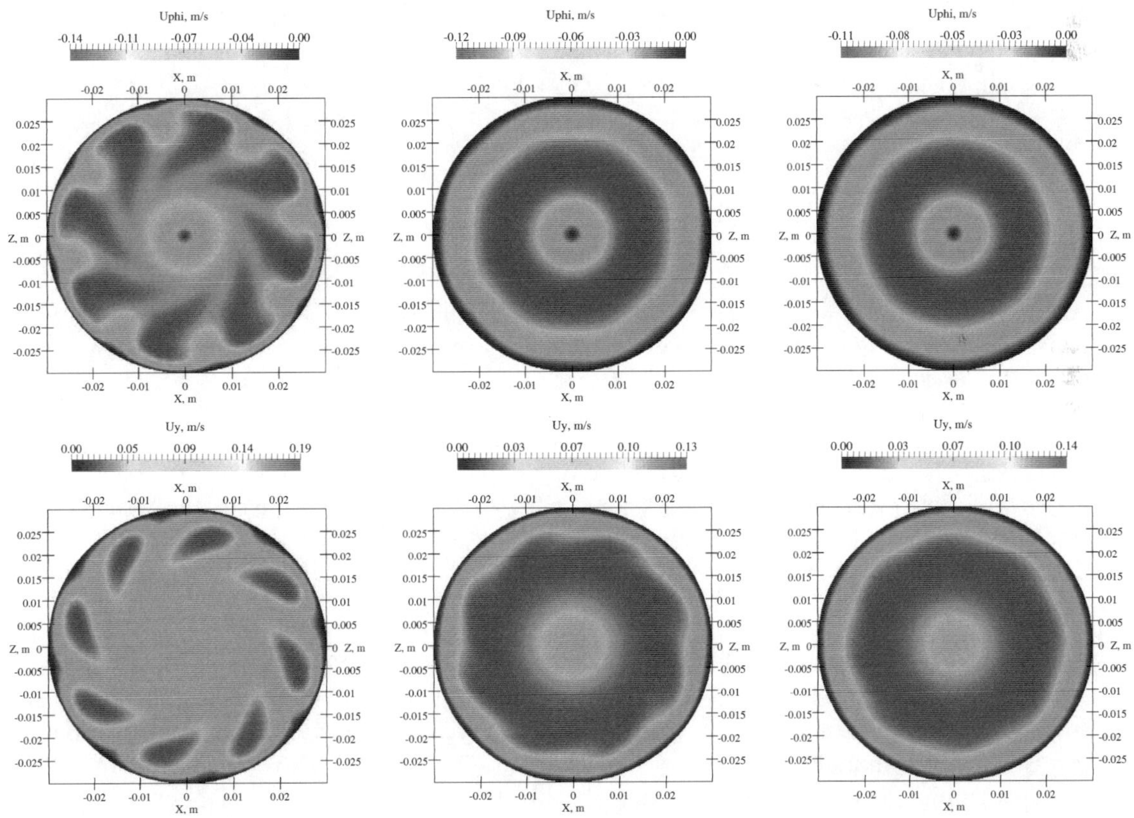

Figure 2. Upper row: time-averaged azimuthal component of the velocity. Lower row – time-averaged axial component of the velocity. Column correspond to cross-sections at $y = 0.05$ m, $y = 0.15$ m and $y = 0.20$ m (left-right).

diverter. Stagnation zone in a central part of the channel is a result of chosen configuration of the diverter (see figure 1(b)). Further from the divertor at $y = 0.15$ m we see that diffusion of

IOP Conf. Series: Materials Science and Engineering **208** (2017) 012011 doi:10.1088/1757-899X/208/1/012011

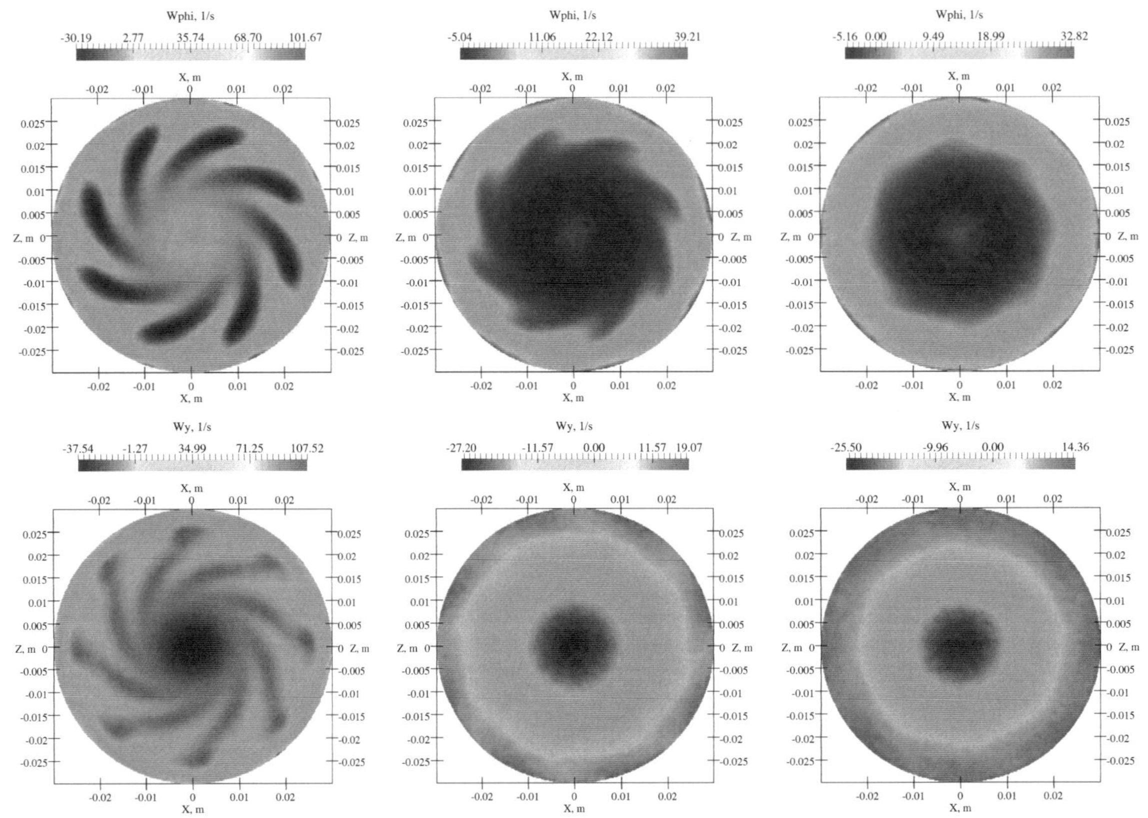

Figure 3. Upper row: time-averaged azimuthal component of the vorticity. Lower row – time-averaged axial component of the vorticity. Column correspond to planes $y = 0.05$ m, $y = 0.15$ m and $y = 0.20$ m (left-right).

momentum by viscosity leads to the rapid axisymmetrization of the flow and instead of eight helical jets the flow took the shape of octagon. Due to non-slip condition on the walls there is a velocity shear near the walls of the channel which has strong influence on the flow evolution. Downstream at $y = 0.2$ m the shape of the flow became almost circular. Helicity describes the correlation between velocity and vorticity vectors so analysis of vorticity distribution is also very important. Blades of the diverter produced mainly two components of vorticity - azimuthal and axial. Similar to the velocity evolution the vorticity distribution became more homogeneous further from the diverter due to diffusion of vorticity. There is evident spatial separation of azimuthal vorticity by sign. Negative azimuthal vorticity is concentrated in the central part of the flow and positive azimuthal vorticity has maximum values near the walls. It is interesting that having similar distributions just after the diverter azimuthal and axial vorticity evolve differently. Negative axial vorticity also concentrated in the center but in a very narrow region in comparison with azimuthal vorticity. Positive axial vorticity is located in a wide band near the walls.

After analysis of velocity and vorticity fields we came to the most interesting and intriguing point - helicity density distribution (see figure 4)

$$\begin{aligned} H &= \mathbf{U}\cdot\mathbf{W} = U_rW_r + U_\phi W_\phi + U_yW_y = H_r + H_\phi + H_y \\ &= U_r(\partial U_y/\partial\phi - r\partial U_\phi/\partial y)/r + U_\phi(\partial U_r/\partial y - \partial U_y/\partial r) + U_y(\partial rU_\phi/\partial r - \partial U_r/\partial\phi)/r. \quad (2) \end{aligned}$$

We consider two main contributions of helicity - azimuthal H_ϕ and axial H_y. Radial contribution H_r has non-zero value near the diverter exit but than rapidly decays. At small values of y distributions of H_ϕ and H_y have eight maximums of helicity density (positive and negative) produced by spiral jets from the diverter blades. Despite similarity of the spatial structure of H_ϕ and H_y distributions of their positive and negative parts are completely different. Azimuthal helicity has positive values in a central part and negative near the walls which is opposite to axial helicity. Helicity density distribution behind the diverter is concentrated in the circular band closer to the walls and characterized by periodic variation of sign. Like evolution of velocity and vorticity helicity density distribution became more axisymmetric but the axisymmetrization of H_y is much faster than of H_ϕ.

At the next step we consider azimuthal modes and its radial profiles. The contributions of different velocity and vorticity components and helicity density is decomposed into the complex Fourier amplitudes:

$$\widehat{f}(r,y) = \frac{1}{2\pi}\int_0^{2\pi} f(r,\phi,y)e^{-im\phi}\,d\phi, \tag{3}$$

where m is the azimuthal wavenumber. Figure 5 shows radial profiles of different velocity components for the axisymmetric mode $m = 0$ at different y. As we mentioned before radial component of velocity is relatively small and can be neglected. Profiles of azimuthal and axial velocity components showed that they evolve along channel in a different way. The viscous stress on the wall supresses azimuthal velocity. The evolution of profile of axial velocity is more complex. Since the total flowrate is fixed the mean (in the transverse cross-section) value of axial velocity is also fixed so the change of radial profile of axial velocity is a result of redistribution without decay. We see that there is a strong change of radial profile of U_y up to $y = 0.15$ m when maximum of U_y is shifted from the wall with increasing of U_y in the central part. After $y = 0.15$ m the profiles became similar with a slight growth in the center.

Radial profiles of azimuthal and axial vorticity components for the axisymmetric mode $m = 0$ at different y are presented in figure 6. Lets consider evolution of the azimuthal vorticity profile for the axisymmetric mode $m = 0$. Behind the diverter ($y = 0.05$ m) near the wall there almost linear dependence of positive values of W_ϕ than closer to the center the sign of W_ϕ is reversed and we see the band of weakly changing negative values of W_ϕ. At $y = 0.15$ m there is remarkable change of the azimuthal vorticity profile. The slope of linear part near the wall strongly decreases and there is almost uniform distribution of negative values of W_ϕ in the central part. The axial vorticity profile has two quasilinear parts with different slopes behind the diverter, negative in the center and positive near the wall. Downstream this difference in slopes is vanished and axial vorticity profile became almost linear at all radii.

Radial profiles of helicity for mode $m = 0$ at different y is a result of both velocity and vorticity evolution. Downstream the profiles of azimuthal and axial contributions of helicity density for mode $m = 0$ became more smooth and stable (see figure 7). Azimuthal contribution of helicity is negative near the wall and positive in the center. In opposite axial helicity contribution is negative in the center and positive near the wall. Total helicity has almost linear profile for $r < 0.2$ m and vanishes at boundary. The screw flow in a described system can be treated as a sum of axisymmetric mode $m = 0$ and non-axisymmetric mode $m = 8$. Analysis of helicity of non-axisymmetric mode of the flow showed that it rapidly decreases with y and further downstream it can be neglected (see figure 8). It means that the main role in generation of helicity in the considered system plays axisymmetric part of the flow.

Finally we present dependences of integrated over ϕ and r values of H_ϕ and H_y on the coordinate y (see figure 8, right). Please note that at $y > 300$ mm the boundary condition at the exit influences on the flow structure and we exclude this region from our analysis. Radial contribution decays very fast because the walls of the channel effectively suppress all

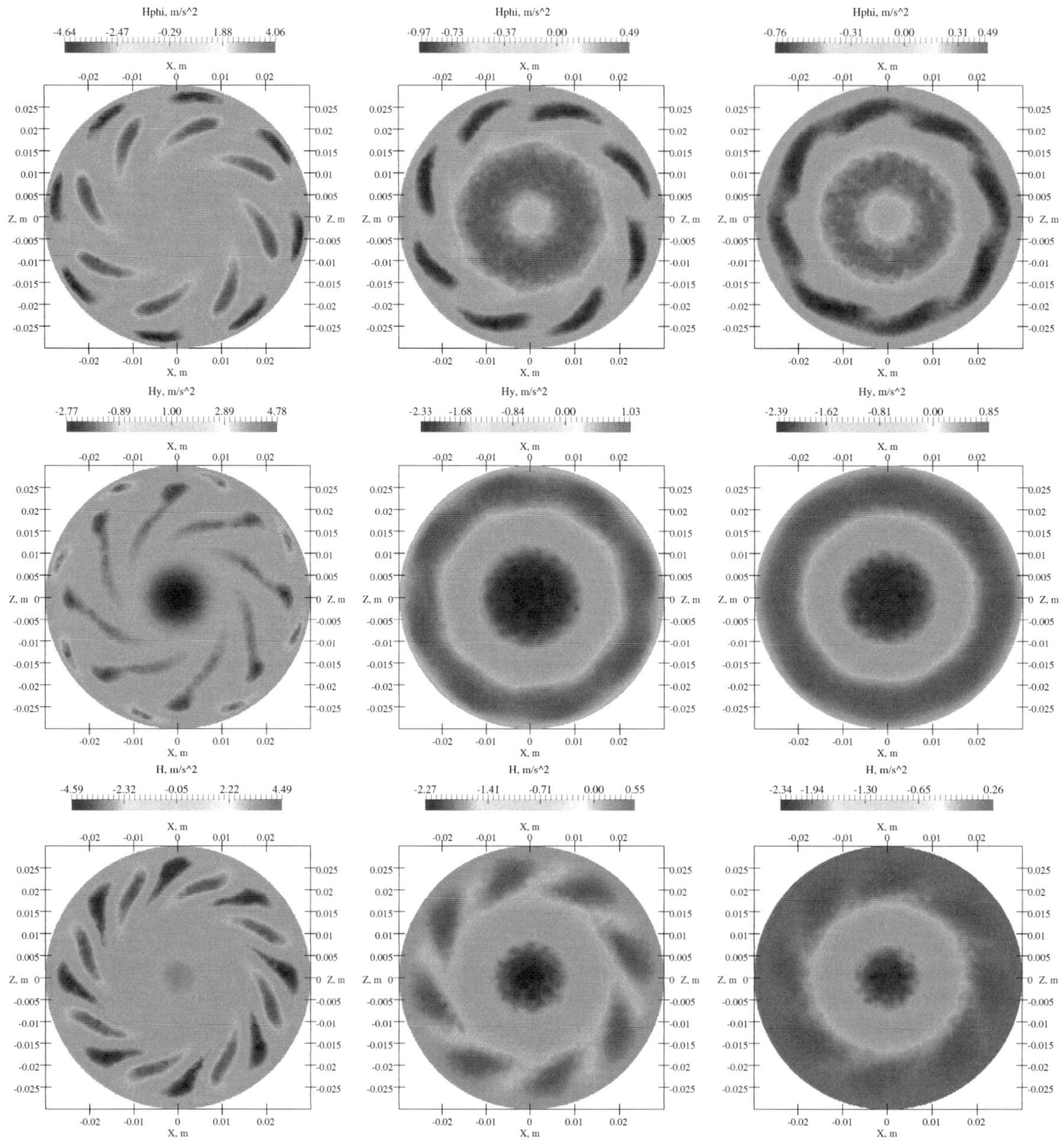

Figure 4. Upper row: time-averaged azimuthal helicity density H_ϕ. Middle row – time-averaged axial helicity density H_y. Lower row – time-averaged helicity density H. Column correspond to planes $y = 0.05$ m, $y = 0.15$ m and $y = 0.20$ m (left-right).

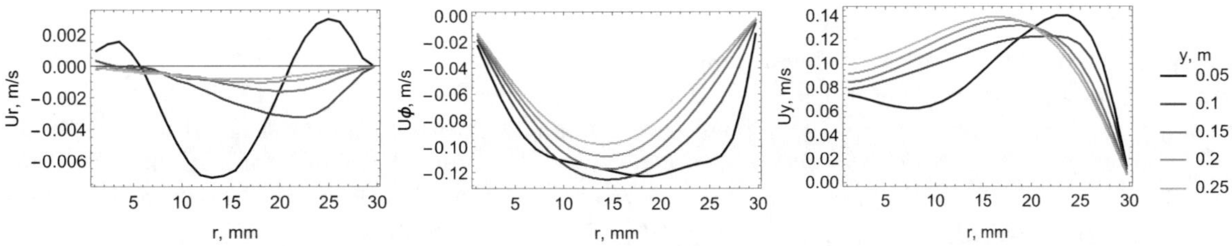

Figure 5. Radial velocity profiles of axisymmetric mode ($m = 0$) at different distance from diverter (y).

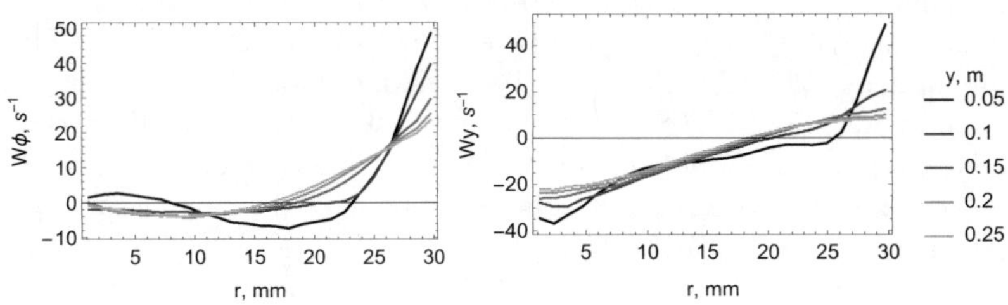

Figure 6. Radial vorticity profiles of axisymmetric mode ($m = 0$) at different distance from diverter (y).

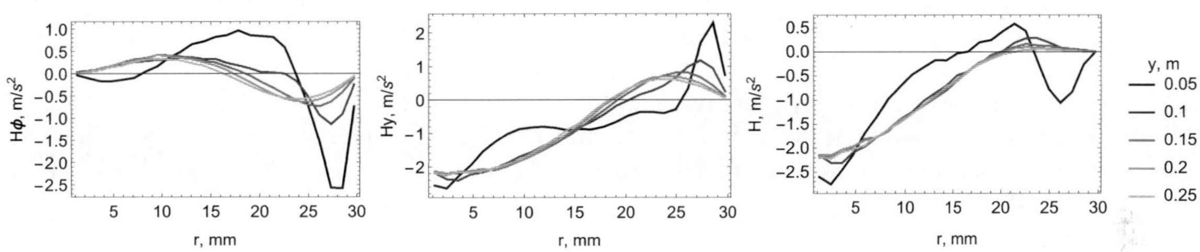

Figure 7. Radial helicity density profiles of axisymmetric mode ($m = 0$) at different distance from diverter (y).

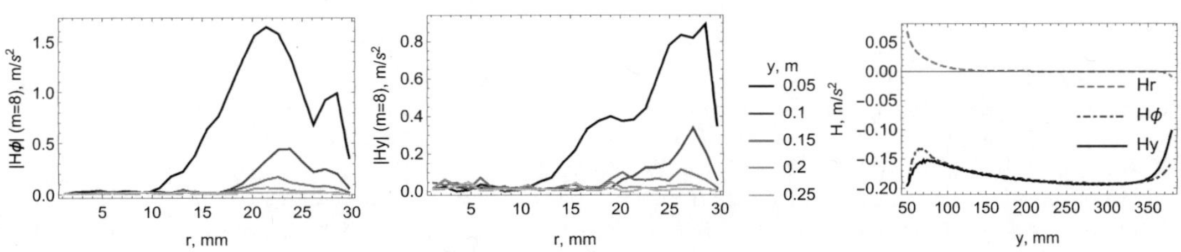

Figure 8. Radial helicity density profiles of non axisymmetric mode ($m = 8$) at different distance from diverter (y) (left and center). Mean density of helicity in cross-section at different distance from diverter (y) (right).

radial motions. Considering azimuthal and axial contribution there is a remarkable quantitative similarity. Despite completely different spatial distribution their integral values belong to the one curve. The integral values of H_ϕ and H_y are negative for all y. Magnitudes of H_ϕ and H_y decrease after the diverter up to $y \approx 70$ mm and after monotonically increase. Evidently this result requires more attention and further study.

4. Conclusions

Numerical simulation of the helical flow in a cylindrical channel with diverter was carried out using open-source software OpenFOAM Extend 4.0. There are several key points about helicity generation in the pipe flow past through fixed diverter: amount of total helicity in outlet flow, scale distribution, axial inhomogeneity and anisotropy. Therefore 3D distributions of velocity, vorticity and helicity density were analyzed in details. Helicity density distribution is characterized by different sign in the central and near wall regions. Azimuthal contribution of helicity is negative near the wall and positive in the center. In opposite axial helicity contribution is negative in the center and positive near the wall. The spiral flow in a described system can be treated as a sum of axisymmetric mode $m = 0$ and non-axisymmetric mode $m = 8$. Analysis of helicity of non-axisymmetric mode of the flow showed that it has substantial values near the diverter but rapidly decreases with y and further downstream it can be neglected. It means that the main role in generation of helicity in the considered system plays axisymmetric part of the flow. Radial contribution of helicity decays very fast downstream because the walls of the channel effectively suppress all radial motions. Despite different spatial distribution variation of mean values of H_ϕ and H_y along the channel has remarkable quantitative similarity. It was shown that integral values of H_ϕ and H_y are negative for all y. Magnitudes of H_ϕ and H_y decrease after the diverter up to $y \approx 70$ mm and after that monotonically increase. Evidently this result requires more attention and further study.

We can conclude that for the first time we demonstrated that the flow behind the diverter is characterized by substantial amount of helicity concentrated in largest scale and can be used as a helical flow injector. In particular the structure of flow develops fast to axisymmetric distribution with well balanced a azimuthal H_ϕ and axial H_y contributions. Axisymmetrization of the flow and rapid decay of radial velocity makes possible to use only radial dependence of azimuthal and axial velocity components to estimate helicity as $H \approx -U_\phi(\partial U_y/\partial r) + U_y(\partial r U_\phi/\partial r)/r$. This result has important outcome for laboratory studies because in case of forced screw flows in channels StereoPIV (measurements three components of velocity in a plane) can be used for direct measurements of the helicity.

Acknowledgments

This work was supported by the Russian Science Foundation (grant RSF-16-41-02012).

References

[1] Moffatt H K and Tsinober A 1992 *Annu. Rev. Fluid. Mech.* **24** 281–312
[2] Steenbeck M, Krause F and Rädler K H 1966 *Z. Naturforsch., Part A* **21** 369
[3] Moiseev S S, Sagdeev R Z, Tur A V, Khomenko G A and Shukurov A M 1983 *Sov. Phys. Dokl.* **28** 926
[4] Levina G V, Moiseev S S and Rutkevich P B 2000 *Adv. Fluid Mech.* **25** 111–162
[5] Yokoi N and Yoshizawa A 1993 *Phys. Fluids* **5** 464–477
[6] Yokoi N and Brandenburg A 2016 *Phys. Rev. E* **93** 033125 (*Preprint* 1511.08983)
[7] Regunath G S, Zimmerman W B, Tesař V and Hewakandamby B N 2008 *Exp. Fluids* **45** 973
[8] Markovich D M, Dulin V M, Abdurakipov S S, Kozinkin L A, Tokarev M P and Hanjali K 2016 *J. Turbulence* **17** 678–698
[9] Eidelman A, Elperin T, Gluzman I and Golbraikh E 2014 *Phys. Fluids* **26** 065103 (*Preprint* 1312.1236)
[10] Denisov S A, Noskov V I, Sukhanovskii A N and Frik P G 2001 *Fluid Dyn.* **36** 734–742 ISSN 1573-8507
[11] Frick P, Noskov V, Denisov S, Khripchenko S, Sokoloff D, Stepanov R and Sukhanovsky A 2002 *Magnetohydrodyn.* **38** 143–162

IOP Conf. Series: Materials Science and Engineering **208** (2017) 012012 doi:10.1088/1757-899X/208/1/012012

The study of the dissipation heat flow and the acoustic emission during the fatigue crack propagation in the metal

A N Vshivkov[1], A Yu Iziumova[1], I A Panteleev[1], A E Prokhorov[1], A V Ilinykh[2], V E Wildemann[2] and O A Plekhov[1,3]

[1] Institute of Continuous Media Mechanics UB RAS, 1, Korolev St., Perm, 614013, Russia
[2] Center of Experimental Mechanics, Perm National Research Polytechnic University, 29, Komsomolskiy Ave., Perm, 614990, Russia
[3] Chair of dynamics and strength of machines, Perm National Research Polytechnic University, 29, Komsomolskiy Ave., Perm, 614990, Russia

E-mail: vshivkov.a@icmm.ru

Abstract. This work is aimed at developing a thermodynamic approach to describing the propagation of fatigue cracks in metals. An attempt is made to explain the change in the character of heat dissipation at different stages of crack propagation: the nucleation, the Paris regime, the critical growth. The studies were conducted on two metal alloy: 304 AISE stainless steel and titanium alloy VT1-0. The investigation of the fatigue crack propagation was carried out on flat samples with stress concentrators. The stress concentrator was the triangular side notch. To monitor the dissipated thermal energy it was used method of infrared thermography and the contact heat flux sensor based on the Seebeck effect. Also the registration system of the acoustic emission was used for more exactly description of the fatigue crack propagation. Analysis of acoustic emission data on the basis of cluster analysis made it possible to classify various mechanisms of the damage process. A correlation was found between the integral dissipated thermal energy and the total energy of acoustic emission during the propagation of a fatigue crack. The joint application of these techniques has made it possible to reveal the moment's activation of failure mechanisms and their relationship to the dissipated heat flux.

1. Introduction

Theoretical and experimental study of the processes accompanying the evolution of the structure of the material during its deformation and fracture is an actual task of modern experimental mechanics. Its solution allows for a deeper understanding of the nature of the destruction processes and the development of new effective methods for assessing the operational resource of construction materials. One of the actual problems in this area is the problem of predicting the rate of propagation of fatigue cracks in metals. Many authors proposed dependencies linking the rate of crack growth and such quantities as the J-integral, the work of plastic deformation, the size of the zone of plastic deformation, the amount of dissipated energy and other [1-4].

It is now well known that the change in the shape of metallic samples during plastic deformation is accompanied by the formation and movement of defects in the structure of the material that cause the energy conversion processes. The classical assumption of an almost complete dissipation of the energy of deformation into heat [5] turns out to be correct only in a limited number of cases. Today, there are techniques that allow real-time monitoring on the temperature change of the sample during the deformation process (the method of infrared thermography) and the number and nature of the defects formed (acoustic emission method). These data allow us to estimate the number of emerging defects and the rate of energy dissipation in the material during the process of damage.

The power of the heat flux can be calculated based on the analysis of evolution of the temperature field (for instance, measured by infrared thermography) on the sample surface, at least for flat samples [6-8]. This calculation is usually associated with the need to differentiate strongly oscillating signals and to determine the parameters responsible for the interaction of the sample with environment.

One of the options for improving the reliability and accuracy of the results is the development of an independent method for measuring the power of heat sources. Such an idea was originally used for studying the energy dissipation under liquid flow [9] as well as the failure of metals [10]. This problem was solved by using a contact heat flux sensor, which was developed based on the Seebeck effect.

This work is devoted to the study of the features of energy dissipation and acoustic emission in stainless steel AISE 304 and titanium alloy OT4-0 during the fatigue crack propagation. The main task of this study is to establish the relationship between the processes of heat dissipation and the evolution of the structure of the material directly during the loading process. At carrying out of experimental investigation simultaneously three methods of nondestructive control were used and their results were compared. In the course of the experiment, the processes accompanying the evolution of the structure of the material were controlled by the infrared scanning, the measurement of the heat flux by the contact sensor and by the acoustic emission method.

2. Experimental setup

A series of samples made from stainless steel AISE 304 and titanium alloy VT1-0 were tested in the servo-hydraulic testing machine Instron 8802. The tests for samples with stress concentrators were carried out in the Center of Experimental Mechanics. The geometry of the samples is shown in Figure 1. During tests the samples were subjected to cyclic loading of 20 Hz with constant stress amplitude and ratio R = 0. The crack length in the course of the experiment was measured by the potential drop method [11, 12]. The electrical potential drop method is accepted as being capable of monitoring the fatigue crack and propagation in steel structures. The size of a crack in a steel member is predicted by applying a constant d.c. (direct current) or a.c. (alternating current) to the member and by measuring an increase in electrical resistance due to the crack. In this case, the potential method is capable of a sensitivity as fine as 0.02 mm for a d.c. 5 A.

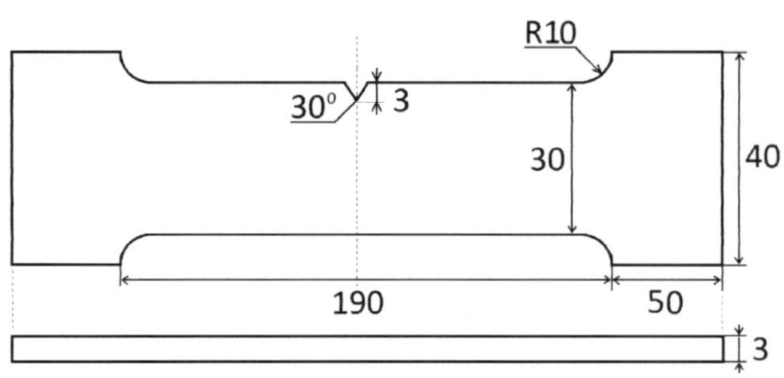

Figure 1. Geometry of samples.

To analyze the dissipated energy at the crack tip a contact heat flux sensor was designed and constructed. The proposed sensor is based on the Seebeck effect, which is the reverse of the Peltier effect.

The Peltier effect is a thermoelectric phenomenon, in which the passage of electric current through conducting medium leads to the generation or absorption of heat at the point of contact (junction) of two dissimilar conductors. The quantity of heat and its sign depend on the type of materials in contact, the direction and the strength of the electric current.

The quantity of heat absorbed or dissipated by the element is directly proportional to the current intensity and the time of its passage.

$$P = \Pi_{AB} I \tag{1}$$

P – the power of heat flux; I – the direct current; Π_{AB} – Peltier coefficient.

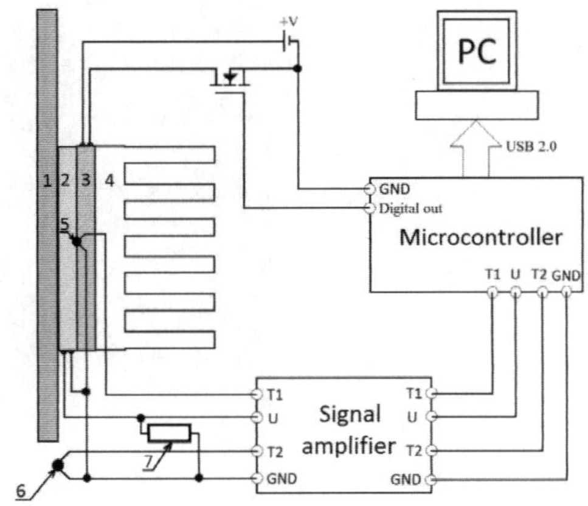

Figure 2. Schematic of the device.1 – testing sample; 2 – "measuring" Peltier element; 3 – "cooling" Peltier element; 4 – radiator; 5, 6 – thermocouple; 7 – resistor.

Figure 1 presents a schematic diagram of the heat flux sensor. The following notation is used in figure 1: sample (1), the heat flux sensor (2). A thermal contact between the sample and the sensor is provided due to the introduction of the thermal paste. Structurally, the sensor comprises two Peltier elements ("measuring" (2) and "cooling" (3)), thermocouples (5), (6) and the radiator (4). The measuring Peltier element is connected to a low-resistance resistor of 1.2 Om (7). To measure the heat flow through the "measuring" Peltier element during the experiment the temperature on its free surface should be a constant. The cooling peltier element caulked with a radiator was connected with the "measuring" Peltier element. This cooling system has feedback and is controlled based on two temperature sensors located between "measuring" and cooling Peltier elements and far from the studied sample in the zone with constant temperature.

The signal from the sensor (voltage at the resistor (7)) is measured by the amplifier and registered in the ADC of the microcontroller. The data are transmitted from the microcontroller to the personal computer for further processing. The "cooling" Peltier element is controlled via pulse width modulation.

These sensors were calibrated using a device with a controlled heat flux. A wire resistor with the known resistance is glued on a plastic plate with a size equal to that of test samples. The heat isolating system provides the heat flux from the resistance to the sensor only. The heat flow was calculated using the values of the resistor voltage and the electric current across the resistor.

The evolution of the temperature field was recorded by infrared camera FLIR SC 5000. The spectral range of the camera is 3-5 μm. The maximum frame size is 320×256 pixels; the spatial resolution is 10-4 meters. The temperature sensitivity is 25 mK at 300 K. Calibration of the camera was made based on the standard calibration table. It was used FLIR SC5000 MW G1 F/3.0 close-up lens (distortion is less than 0.5%) to investigate the plastic zone in details.

The system Amsy-5 Vallen was used for registration of the acoustic emission signal. This system had two broadband M31 sensors with a working range of 300-800 kHz. For amplification, AEP4 preamps (gain 34 dB) were used. The acoustic emission transducers were attached to the surface of the sample with a cyanocrylate adhesive. Continuous acoustic emission was recorded with frequency of

2 MHz. Parameters of acoustic emission (maximum amplitude, energy) were stored every 40 milliseconds (a waveform record also was made after 40 milliseconds with a duration of 4 ms). The location of acoustic emission sensors is shown in figure 3.

The principal scheme of experiments is presented in figure 4, 5.

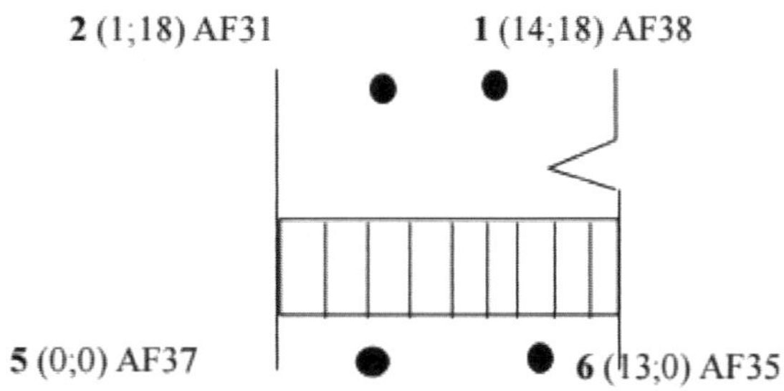

a)

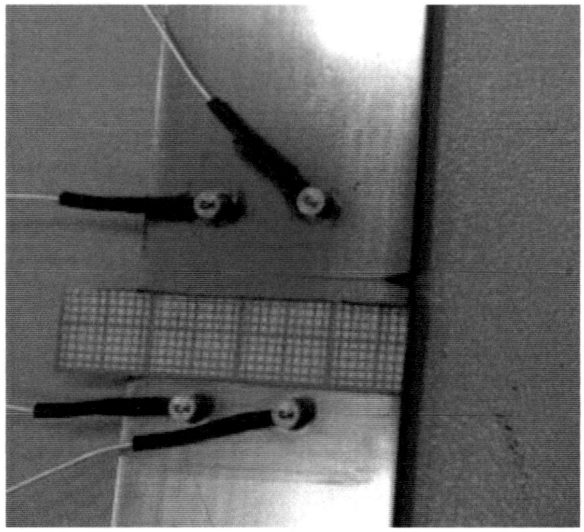

b)

Figure 3. The location of acoustic emission sensors.

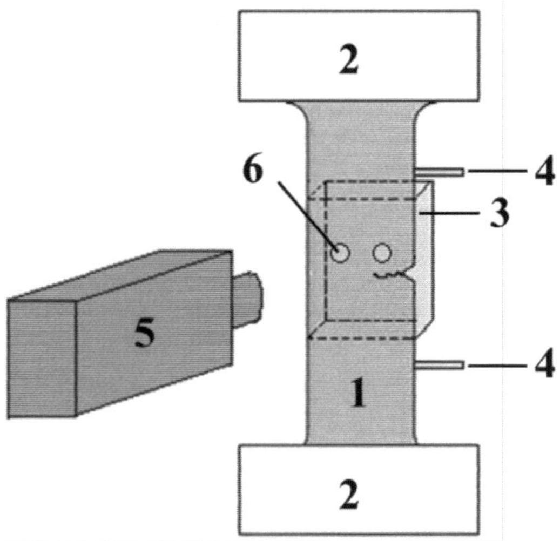

Figure 4. Schematic of the measured equipment.

Figure 5. The sample with the measured equipment.

The following notation are accepted in figure 3: 1 – specimen under testing, 2 – grips of testing machine, 3 – the contsct heat flax sensor, 4 – the potential drop measuring setup to monitor the crack length, 5 – the infrared camera, 6 – sensors of acoustic emission.

3. Results

A series of samples from the titanium alloy VT1-0 and stainless steel AISE 304 was tested as outlined above. The propagation of the fatigue crack corresponded to the Paris law (figures 6).

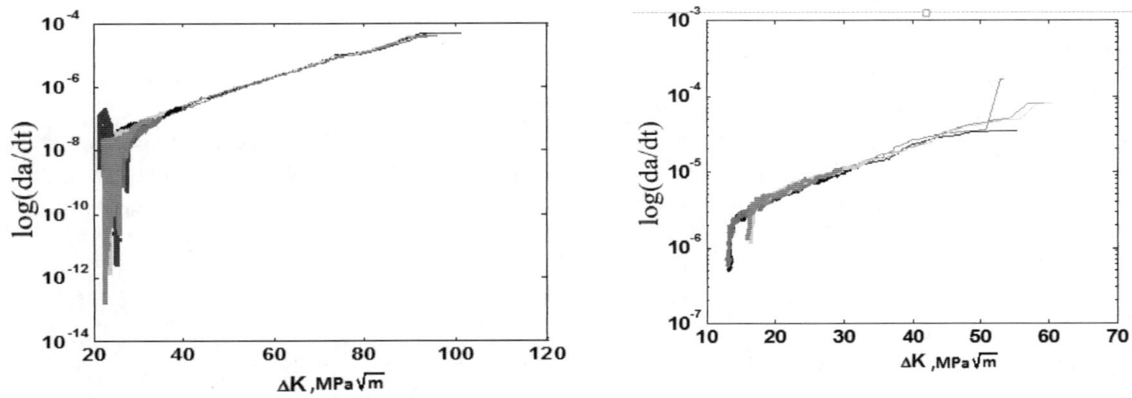

a) b)

Figure 6. The Paris curve for the stainless steel AISE 304 (a) and the titanium alloy VT1-0 (b).

3.1. The dissipated thermal energy

The developed sensor was used to study the heat dissipation caused by growth of fatigue cracks. Typical result of measurements of the heat flux during the experiment is presented in figure 7.

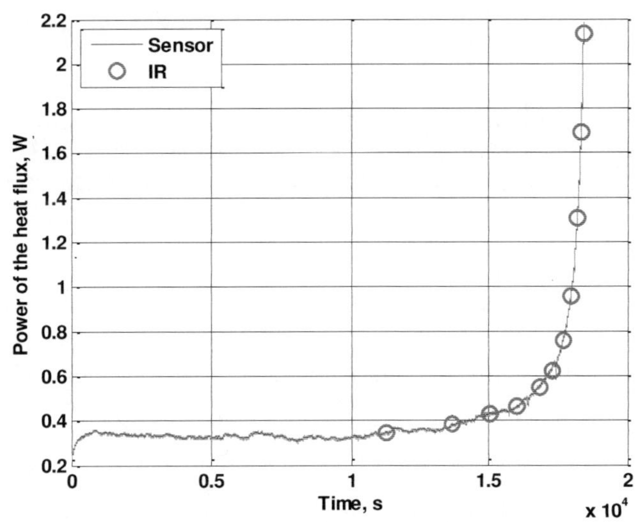

Figure 7. The characteristic dependence of the power of heat flux from the crack tip.

A comparison between contact sensor of the heat flux and the infrared thermography data is shown in figure 7. Infrared camera allows one to register the temperature field on the specimen surface. For calculate the heat flux the heat conductivity equation was used. It has to be averaged on z-coordinate (thickness), and it will have following view:

$$Q_{\text{int}}(x,y,t) = \rho c\left(\frac{\partial\theta(x,y,t)}{\partial t} + \frac{\theta(x,y,t)-T_0}{\tau}\right) - k\left(\frac{\partial^2\theta(x,y,t)}{\partial x^2} + \frac{\partial^2\theta(x,y,t)}{\partial y^2}\right) \quad (2)$$

where τ is time constant which is related to the heat losses[13,14]. The coefficient τ was measured before each test.

The processing of infrared data was made by algorithm of movie compensation and filtering. These algorithms were described in details in [15].

3.2. The acoustic emission

Figure 8 shows the dependence of the energy of the acoustic emission for two channels and the crack length on time. There is a significant and non-monotonic change in the energy of the acoustic emission during the cyclic deformation of the sample (more than 2-time). The sharp growth of the crack at the end of the test is accompanied by a sharp increase in the energy of the acoustic emission.

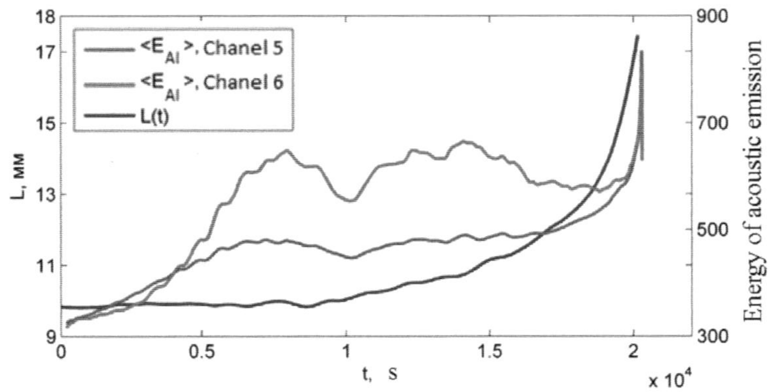

Figure 8. The signal of the acoustic emission and the crack length during the experiment.

For a more detailed analysis of the stage of the acoustic emission variation, the dependences of the first two statistical moments for the accumulated energy of the acoustic emission on time were constructed (figure 9, 10). Analysis of the obtained results showed that the moment of occurrence of the crack t_{2-2}, fixed by the potential drop method, corresponds to a break on the dependences of the first two statistical moments recorded by the more remote sensor (channel 6). At the same time, the same break is observed on the acoustic emission signal from channel 5, but for an earlier time t_{2-1}. The observed difference can be caused by closer positioning of the AE sensor to the crack tip (channel 5), which provides its greater sensitivity in comparison with channel 6. It should also be noted that for data from a closer-lying acoustic sensor, there is a difference in estimates of the moment of appearance of a crack t_{2-1} and t'_{2-1} for first and second statistical moments. This means that the more "sensitive parameter" is the second statistical moment, which begins to change in the first place.

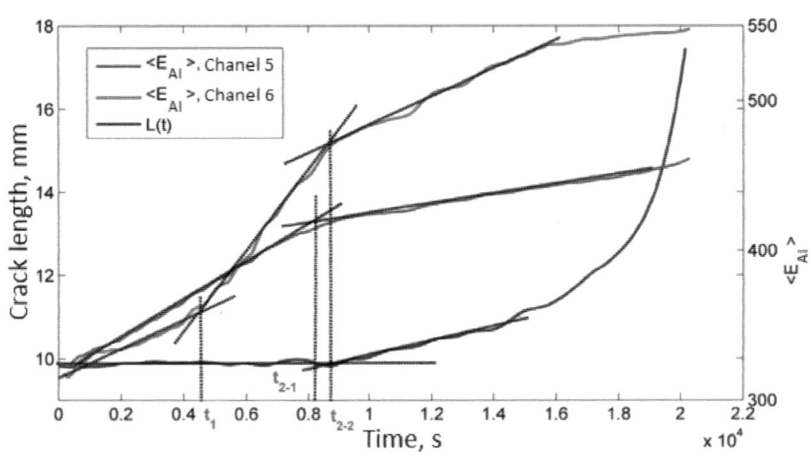

Figure 9. The dependences between the mean energy of the acoustic emission on the time, combined with the time dependence of the crack length.

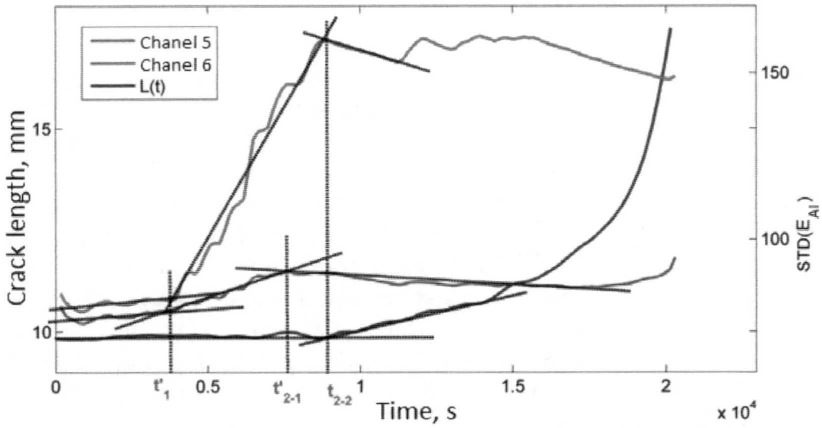

Figure 10. The dependences between the standard deviation of the accumulated energy on the time, combined with the time dependence of the crack length.

On the obtained dependences one can also single out another characteristic time t_1 (t'_1), which, apparently, characterizes the moment of appearance of the local zone of plastic deformation, in which the fatigue crack will appear in the future. This time moment is also differently identified on the time dependences of the statistical moments. Here it is important to note two features:

1. if the time instant t1 is identified on the dependence of the average accumulated energy only for the more remote acoustic sensor, then the dependence of the standard deviation is for both, and for both is the same;
2. The mean level of accumulated energy of the acoustic emission registrated on the closer acoustic sensor only at the initial stage of deformation exceeds the energy level of the sensor from the more remote sensor. This is due to the fact that after the starting the crack propagation, a closer sensor is placing behind the crack tip.

A comparison of the acoustic emission and the heat flux during the propagation of the fatigue crack is also performed. For comparison of the methods, the integral energy is normalized to the maximum value. From the results presented in Figure 11, it can assume a qualitative agreement of the integral thermal energy and cumulative energy of the acoustic emission.

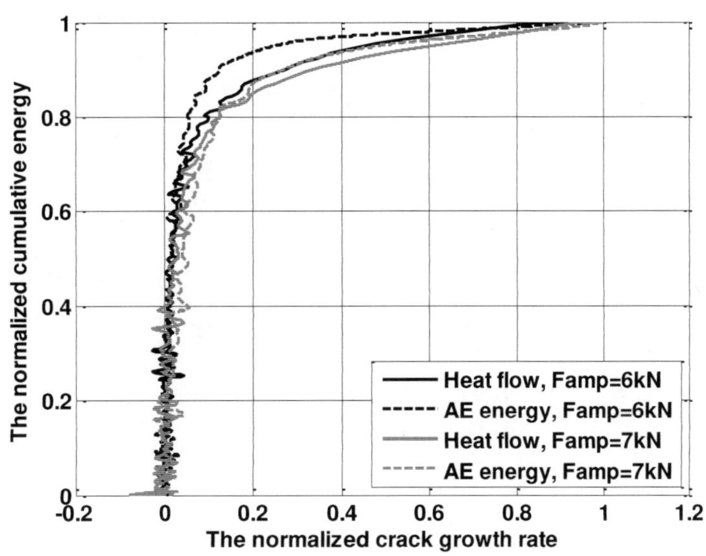

Figure 11. The comparison of the integral dissipated thermal energy and the cumulative energy of the acoustic emission.

4. Conclusion

The paper describes the structure and performance of the heat flux sensor which was designed based on the Seebeck effect. This sensor can be used to analyze the dissipated energy of the material during the deformation process either in combination with the method of infrared thermography (for its verification), or independently. With this technique a comprehensive investigation into the processes of energy dissipation during the propagation of fatigue cracks in metal alloys was done.

As a result of the analysis of the acoustic emission signal, it is shown that the use of the method in a cyclic testing of the metal makes it possible to identify the moment of appearance of the crack. The method has a greater sensitivity compared to the method of the potential drop method or the method based on the power of the dissipated energy. Also, the method of acoustic emission makes it possible to identify the moment of the beginning of preparation of the material (the appearance of local zones of plastic deformation) to the appearance of a fatigue crack. When comparing methods (acoustic emission and energy approach), correlations between integral dissipated thermal energy and the cumulative energy of acoustic emission are revealed.

This work was supported by the grant of the President of Russian Federation for support of young Russian scientists and leading scientific schools [MK-1236.2017.1] and grants of the Russian Foundation for Basic research [grants number 16-31-00130 and 16-31-155].

References

[1] Matvienko Yu G and Morozov E M 2004 *International Journal of Fracture* **125** 249–261
[2] Rosakis P, Rosakis A J, Ravichandran and G Hodowany J 2000 *J. Mech. Phys. Solids.* **48** 581–607
[3] Oliferuk W, Maj M and Raniecki B 2004 *Materials Science and Engineering A* **374** 77–81
[4] Izyumova A and Plekhov O 2014 *FFEMS* **37** 1330–1337
[5] Farren W S and Taylor G I 1925 *Proc. Royal. Soc. of London. Ser. A* **107** 422–451
[6] Iino Y 1979 *Engineering fracture mechanics* **12** 279–299
[7] Chow C L and Lu T J 1991 *Engineering Fracture Mechanics* **39** 1–20
[8] Dowling N E and Begley J A 1976 *ASTM STP 590* 83–104
[9] Pradere C, Joanicot M, Batsale J-C, Toutain J and Gourdon C 2006 *QIRT Journal* **3** 117–135
[10] Izyumova A Yu, Plekhov O A, Vshivkov A N, Prokhorov A A and Uvarov S V 2014 *Technical Physics Letters* **40** 810–812
[11] Nayeb-Hashemi H, Swet D and Vaziri A 2004 *Measurement* **36** 121–129
[12] Hartman G A and Johnson D A 1987 *Experimental Mechanics* **27** 1 106–112
[13] Boulanger T, Chrysochoos A, Mabru C and Galtier A 2004 *Int. J. Fatigue* **26** 221–229
[14] Chrysochoos A and Louche H 2000 *Int. J. Eng. Sci* **38** 1759–1788
[15] Fedorova A, Bannikov M, Terekhina A and Plekhov O 2014 *Quant. InfraRed Thermogr. J.* **11** 2–9

IOP Conf. Series: Materials Science and Engineering **208** (2017) 012013 doi:10.1088/1757-899X/208/1/012013

Experimental study of evaluation of mechanical parameters of heterogeneous porous structure

O Gerasimov, E Koroleva and O Sachenkov[a]

Kazan Federal University, 18, Kremlyovskaya St., Kazan, 420008, Russia

E-mail: [a] 4works@bk.ru

Abstract. The paper deals with the problem of determining the mechanical macroparameters of the porous material in case of knowing the information about it's structure. Fabric tensor and porosity was used to describe structure of the material. Experimental study presented. In research two-component liquid polyurethane plastics of cold curing Lasilcast (Lc-12) was used. Then samples was scanned on computer tomography. Resulting data was analyzed. Regular subvolume was cut out after analyses. Then mechanical tests was performed. As a result we get information about fabric tensor, porosity, Young's modulus and Poisson ratio of the sample. In the abstract presented results for some samples. Taking into account the law of porosity variation, we considered the problem of evaluating the mechanical macro parameters depending on the nature of the porous structure. To evaluate the macroparameters, we built the dependence of the Young's modules and Poisson ratio of the material on the rotation angle α and the pore ellipticity parameter λ. The sensitivity of the deformations to the elastic constants was also estimated.

1. Introduction

When solving the problems of determining the stress-strain state (SSS) of porous structures, it is necessary to take into account structural peculiarities of the material [1-5]. Important problem is evaluation of mechanical parameters using structural data. Nowadays, in biomechanics is popular to use fabric tensor [6-9]. It is assumed that the fabric tensor is a quadratic form that describes the shape of a pore. Some physical relations which connect the stress-strain tensor through the elastic constants, the fabric tensor, and the solid volume fraction in the material is known [10]. In this paper experimental study is presented. The goal of investigation is to show dependence between structure and integral mechanical parameters (Young's modulus and Poisson's ratio).

2. Materials and Methods

In abstract we consider some experiments to show relations between structure and mechanical parameters. It was shown by mathematical models that this relation exist [11-13]. Parametric model which connects mechanical parameters and structural parameters in terms of fabric tensor was built (we use two parameters: λ – ellipticity and α – direction). In presented paper we performed experimental study. We produced parallelepiped model (length was greater than over two dimensions) using casting of plastics. Using the same plastic we produced porous models (using active chemical reaction). After that we scan every sample on computer tomography (CT) and examine received data (see fig. 1). To find regular volume, the received data was meshed to smaller cubes and in every cube porosity and fabric tensor was calculated. Then we analysed distribution of structural parameters and cut out regular sub volume.

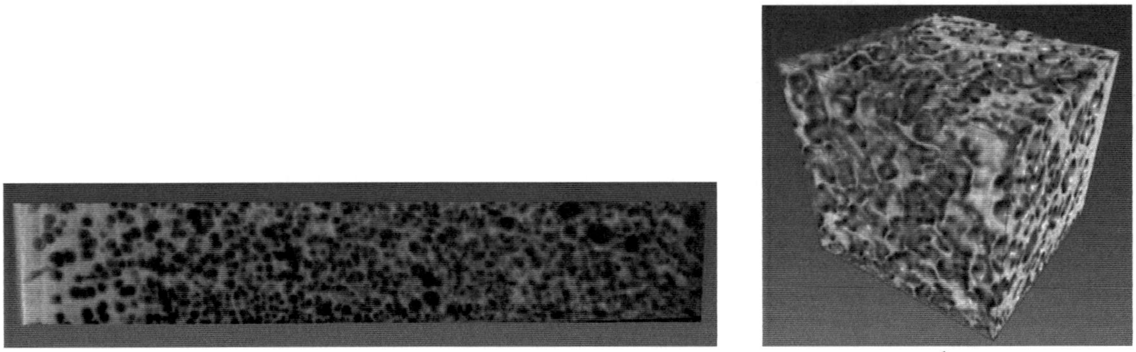

a b

Figure 1. CT scan of sample (a) and sub volume of sample (b).

Extracted sub volume after trimming was tested on test set (see fig. 2). Sample (1 on fig. 2) was setup in support (4 on fig. 2) on cross-arm (5 and 6 on fig. 2). We fixed strain gauge on sample (2 on fig. 2) in transverse direction. To measure longitudinal strain we used test's sensor (3 on fig. 2). During testing motile cross-arm (5 on fig. 2) moved up. During compression of sample we measured longitudinal and transverse strain and applied force.

3. Results and Discussion

In research two-component liquid polyurethane plastics of cold curing Lasilcast (Lc-12) was used. Declared properties: mixture viscosity 105 mPa, Shore hardness – 75, density 1.05 gm/sm^3, Young's modules 500 MPa (for compression). We produced solid and porous samples. For CT scan we used Vatech PaX-I 3D. Raw CT data was analyzed to find regular volumes. In this paper we presents results for sub volumes with porosity 45% and 50%. For the samples eigenvalues of fabric tensor was calculated either. The main of eigenvectors was directed in longitudinal direction. Quotient of two last eigenvalues was 0.54 and 0.62 respectively. Then the samples was extracted and tested (by the method described above). After treatment the data diagrams of changing Young's (E) modules and Poisson's ratio (ν) during the deforming was built (see fig. 3). So, for solid sample Young's modules was 518 MPa, Poisson's ration – 0.24. For sample with porosity 45% and ellipticity 0.54 was received: Young's modules – 267 MPa, Poisson's ration – 0.2. For sample with porosity 50% and ellipticity 0.62 was received: Young's modules – 163 MPa, Poisson's ration – 0.14.

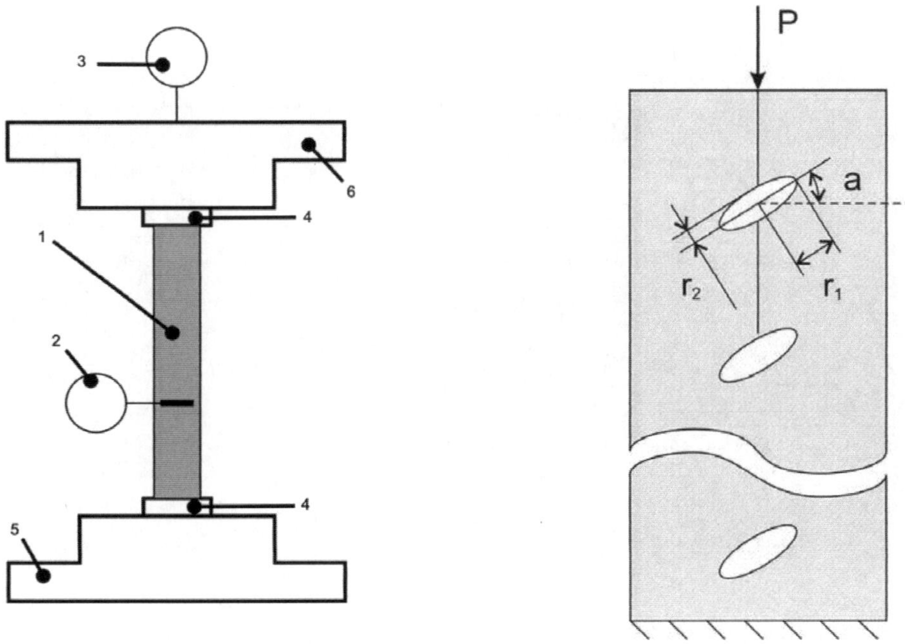

a b

Figure 2. Scheme of test set (a) and computational scheme (b).

In previous paper [10] a block diagram loaded by the compressive force P was considered. It was assumed that the sample uniformly filled with pores having the porosity ν_0. The pores have an elliptical shape with the fixed semi-radii and form an angle $\pi/2$-α with the line of action of the applied force.

For description of the porous structure, we will use the formalism associated with the fabric tensor. Then we will supplement the system by introducing physical relations in the form [10, 14-16]:

$$\tilde{\sigma} = (g_1 + g_2 e) tr\tilde{\varepsilon} \times E + (g_3 + g_4 e)\tilde{\varepsilon} + g_5(\tilde{\varepsilon}\tilde{K} + \tilde{K}\tilde{\varepsilon}) + \\ + g_6\left(tr(\tilde{K}\tilde{\varepsilon}) \times E + tr\tilde{\varepsilon} \times \tilde{K}\right), \bar{x}\in\bar{S}, t \geq 0, \tag{1}$$

here e is the variation of the solid bone volume fraction, $\tilde{K}$ is the fabric tensor deviator, g_i are the elastic constants.

The law of porosity variation was used in the following form:

$$\nu = \frac{\nu_0 + \sigma_i\left(\frac{1}{3K_{sp}} - \frac{1}{3K_s}\right)}{1 + \frac{\sigma_i}{3K_{sp}}}, \tag{2}$$

where K_{sp} is the bulk modulus of elasticity of the skeleton with pores, K_s is the bulk modulus of elasticity of the skeleton, ν and ν_0 are the actual and initial porosities respectively, and σ_i is the compressive stress.

To describe structure was uses two parameters: λ – ellipticity and α – direction. In this case fabric tensor can be presented in the following form:

$$\tilde{H} = Q(\alpha) \cdot \begin{pmatrix} \lambda \cdot h & 0 \\ 0 & h \end{pmatrix} \cdot Q^T(\alpha), \tag{3}$$

where $Q(\alpha)$ – rotational matrix.

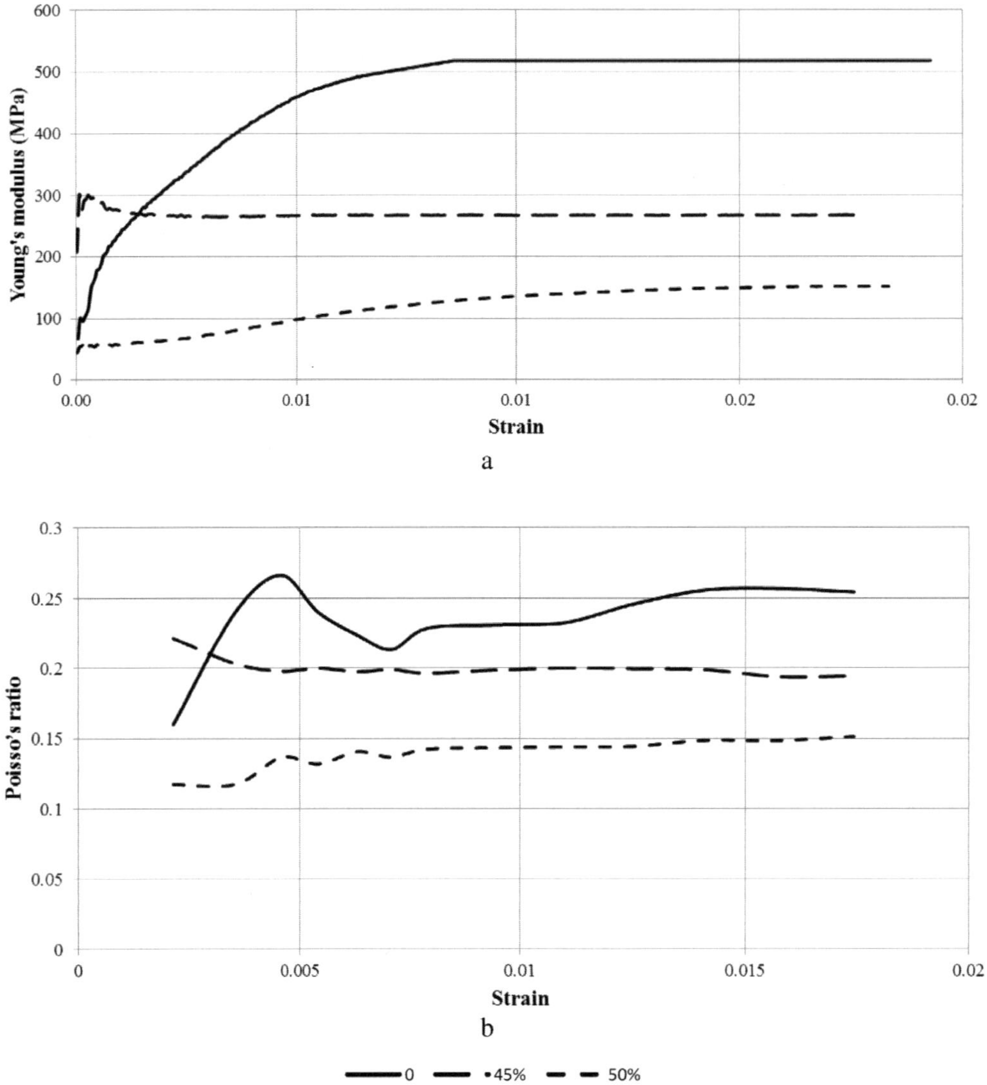

Figure 3. Results of experiments: the Young's modules (a) and Poisson's ratio(b) for different samples.

For evaluating the macroparameters, we built a dependence of the Young's modules and Poisson's ratio of the material on the λ-α axes (see Fig. 4). It was noted that in the case of circular pores, the obtained dependence does not depend on the parameters n, v_0, K_s (but obviously depends on the elastic constants g_i). We used data of solid sample to calculate constants g_i. Using them diagrams for the Young's modules and Poisson's ratio in the λ-α axes was built (see fig. 4). Previously it was shown that sensitivity to the elastic constants showed their underestimation to the value of up to 20% results in the deformation increase by up to 30%. With an increase of the elastic constants by 20%, the deformations decrease by 20%.

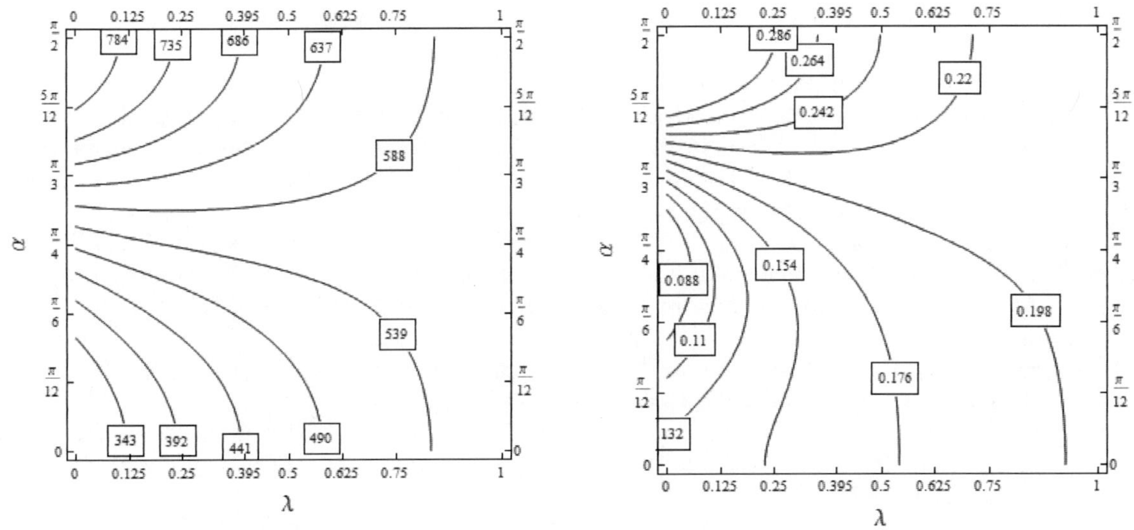

a b

Figure 4. Diagrams of computed Young's modules (a) and Poisson's ratio (b).

4. Conclusion

This study examines the problem of finding dependence between structure and mechanical parameters. Experimental study was presented. Fabric tensor and porosity was used to describe structure of the material. We consider samples which was made of plastic. After producing of samples we scan them on CT and then examine the structure. The first step was to find in the sample regular subvolume and then cut it out. On the second step we produced mechanical text of the sample. As a result we get information about fabric tensor, porosity, Young's modulus and Poisson ratio of the sample. In the abstract presented results for some samples. For the stated law of porosity variation, we considered the problem of evaluating the stress-strain state and get diagrams of the Young's modules and Poisson's. To evaluate the macroparameters, we built the dependence of the Poisson ratio of the material on the rotation angle α and the pore ellipticity parameter λ. The sensitivity of the deformations to the elastic constants was found to be 20–30%

Acknowledgements

The work was partly supported by the Russian Foundation for Basic Research within the scientific project No. 16-04-00772 and by the subsidy allocated to Kazan Federal University for the state assignment in the sphere of scientific activities.

Special thanks to Independent X-ray diagnostic centers Picasso.

References

[1] Cowin S C and Benalla M 2011 Graphical illustrations for the nur-byerlee-carroll proof of the formula for the biot effective stress coefficient in poroelasticity *Journal of Elasticity* **104** 133–141 DOI 10.1007/s10659-011-9324-7

[2] Thompson M and Willis J R 1991 A reformation of the equations of anisotropic poroelasticity *Journal of Applied Mechanics* **58** 612–616 doi:10.1115/1.2897239

[3] Berezhnoi D V, Sachenkov A A and Sagdatullin M K 2014 Geometrically nonlinear deformation elastoplastic soil *Applied Mathematical Sciences* **8** 6341–6348 http://dx.doi.org/10.12988/ams.2014.48672

[4] Berezhnoi D V, Sachenkov A A and Sagdatullin M K 2014 Research of interaction of the deformable designs located in the soil *Applied Mathematical Sciences* **8** 7107–7115 http://dx.doi.org/10.12988/ams.2014.49706
[5] Berezhnoi D V and Sagdatullin M K 2015 Calculation of interaction of deformable designs taking into account friction in the contact zone by finite element method *Contemporary Engineering Sciences* **8** 1091–1098 http://dx.doi.org/10.12988/ces.2015.58237
[6] Sachenkov O, Kharislamova L, Shamsutdinova N, Kirillova E and Konoplev Yu 2015 Evaluation of the bone tissue mechanical parameters after induced alimentary Cu-deficiency followed by supplementary injection of Cu nanoparticles in rats *IOP Conference Series: Materials Science and Engineering* **98** 012015 doi:10.1088/1757-899X/98/1/012015
[7] Shigapova F A, Mustakimova M F, Saleeva G T and Sachenkov O A 2015 Definition of the intense and deformable jaw state under the masseters hyper tone *International Journal of Applied Engineering Research* **10** 44711–44714
[8] Galiullin R R, Sachenkov O A, Khasanov R F and Andreev P S 2015 Evaluation of external fixation device stiffness for rotary osteotomy *International Journal of Applied Engineering Research* **10** 44855–4860
[9] Akulich A Yu, Akulich Yu V, Denisov A S, Ivanov D V and Shulyatev A F 2016 Analysis of main reasons of nonunions and methods of osteosynthesis of the femoral neckfractures *Russian J. of Biomechanics* **20** 145–151
[10] Sultanov L U 2015 Numerical Modelling of Deformation of Soil Foundations *Procedia Earth and Planetary Science* **15** 119–124 doi:10.1016/j.proeps.2015.08.029
[11] Gerasimov O, Shigapova F, Konoplev Yu and Sachenkov O 2016 Evaluation of the stress-strain state of a one-dimensional heterogeneous porous structure *IOP Conference Series: Materials Science and Engineering* **158** 012036
[12] Duarte R J, Marques A S, Ramos A, Nyashin Y I and Mesnard M 2016 Numerical evaluation of the mechanical behavior of the ulna in dogs when using an intramedullary pin *Russian J. of Biomechanics* **20** 152–60
[13] Zhuravkov M A and Romanova N S 2016 Determination of physical and mechanical properties of biomaterials on base of the nanoindentation technologies and fractional order models *Russian J. of Biomechanics* **20** 5–22
[14] Davydov R L, Sultanov L U and Kharzhavina V S 2015 Elastoplastic Model of Deformation of Three- Dimensional Bodies in Terms of Large Strains *Global Journal of Pure and Applied Mathematics* **11** 5099–5108
[15] Rho J Y, Roy M E, Tsui T Y and Pharr G M 1999 Elastic properties of microstructural components of human bone tissue as measured by indentation *J. Biomed. Mater. Res.* **45** 45–48
[16] Kichenko A A, Tverier V M, Nyashin Y I, Osipenko M A and Lokhov V A 2012 On application of the theory of trabecular bone tissue remodeling *Russian J. of Biomechanics* **16** 46–64

IOP Conf. Series: Materials Science and Engineering **208** (2017) 012014 doi:10.1088/1757-899X/208/1/012014

Investigation of epoxy matrix viscosity in the initial stage of its formation

V G Gilev[1], A V Kondyurin[1, 2] and S V Rusakov[1]

[1] Perm State National Research University, 15, Bukirev St., Perm, 614990, Russia
[2] School of Physics, University of Sydney, Sydney, New South Wales 2006, Australia

E-mail: gvg@psu.ru, rusakov@psu.ru

Abstract. The present study includes experimental investigations of rheological properties of the epoxy composite matrix based on epoxy resin L and hardener EPH 161. It is shown that the temperature dependences of the oligomer viscosity and the hardener can be satisfactorily described by the Arrhenius equation. Dependence of epoxy resin viscosity on temperature and volume concentration of the hardener at the initial stage of the polymerization process is investigated. It is shown that the experimental results can be generalized with one universal dependence of concentration and temperature.

1. Introduction

Due to their unique properties, epoxy-based polymer materials are widely used in industrial processes and in creating various polymer constructions and structures. They have high dielectric and strength properties, radiation resistance, so these materials can be used both on Earth and in space [1, 2]. A number of studies suggest that epoxy compositions are to be used for bulky space structures, which are supposed to get cured directly in space in high vacuum, cosmic radiation and extreme temperatures [3-5]. The curing process, as well as end-use properties of the curable composites, largely depend on the evaporation process of epoxy active components. Experiments in space flights are complicated and expensive. Therefore, the terrestrial experiments in vacuum and plasma, as well as theoretical modeling of the processes of epoxy compositions are more realistic.

In the literature there are many studies devoted to the curing processes of epoxy resins (see for example [6-8]). Physical and chemical properties of structural materials are also quite well investigated.

At the same time, the studies of the processes occurring in the initial stage of polymerization (i.e. immediately after the components are mixed) are not numerous. Among the studies covering these problems is [9]. By the methods of NMR and dielectrometry it investigates the kinetics of macromolecule formation and their mobility in the mixture of epoxy resin and polyethylene polyamine immediately after mixing at temperatures of 20 °C and 100 °C, and makes an assumption that the final properties of the product are formed at the initial stage of formation of the macromolecules occurring in the reaction of the oligomer and the hardener. Calorimetric, chemical and spectroscopic methods are widely used as experimental methods for studying epoxy oligomer structuring processes, as they allow us to obtain analytical expressions describing the kinetics of hardening of epoxy resins. Strictly speaking, the immediate registration of varying concentrations of reactive groups is possible only with IR spectroscopy and NMR. However, in some cases the use of these methods leads to ambiguity in the interpretation of the obtained results due to the complexity of spectra of real multicomponent compositions. At the same time, for observation of macroscopic properties of curable compositions we can apply rheological investigations which allow us not only to determine the location of the gel point at which the evaporation of the binder active components decreases, but also to connect macrokinetic description of the reaction with physical and mechanical properties of the reacting system such as binder viscosity and prepreg elasticity modulus directly affecting the ability of the uncured composite to unfold from a shipping container in the space orbit. Of course, viscosimetric flows may affect the process of composite polymerization, but any technological process at its initial stage is connected to shear deformations.

In the present study we measured the effective viscosity of solutions of the oligomer and its hardener depending on the temperature and hardener concentration at the initial stage of the polymerization process.

2. The object of study and preparation of samples

In the experiments we used the epoxy composition - epoxy resin L and EPH 161 hardener certified for use in constructional composites applied in aircraft industry. The mixture was prepared in the required proportions in a weight ratio of epoxy resin and curing agent (hardener). After preparations, the mixture was thoroughly mixed for 1-2 minutes with the electromechanical mixer and further for 1–2 minutes in the ultrasonic bath Digital Ultrasonic Cleaner CD 4820 (at 40 kHz). The last operation degases the mixture. It took 3–4 minutes to prepare the samples – quite a short period – as the rate of polymerization processes is rather high under normal conditions, and the composition properties change rapidly after its components get mixed.

Measurements of the mass of solutions and their components necessary to calculate the concentration of the mixture were performed using the analytical balance LV-210A (accuracy class 2, absolute measurement error ± 0.4 mg).

To measure the density of samples necessary for calculation of the volume concentration of solutions we used the digital vibration densimeter DMA 5000 M with an absolute measurement error not more than 0.000005 g/cm^3. The densimeter has a built-in Pelletier thermostat to study temperature dependences of the density.

Rheological measurements were performed using a rotary rheometer Physica MCR 501. This device gives rheological data in a wide range of fixed and dynamic tests in a controlled shear stress (CS) and controlled shear rate (CR) modes from measurements of flow curves to dynamic analysis of complex fluids.

The "cone-plate" system was used in all the measurements (the cone diameter d = 25 mm and the angle α = 1°). The used geometry provides shear rate gradient uniformity in the measurement gap. To verify the data, several measurements were performed using a cone with d = 50 mm (cone angle 1°). The temperature of the sample was controlled by a Peltier temperature controller with accuracy less then 0.1 °C. The header and measuring system were kept inside a protective cover for uniformity of the sample temperature. During the experiment, about 0.07 ml of the solution was located on the rheometer work surface. The thickness of the solution layer on its outer radius was 0.047 mm, which ensured the sample to achieve the working temperature quickly even in conditions of heat release due to the polymerization reaction.

Preliminary experiments showed that at temperatures below 15 °C the flow curves of the oligomer and hardener are similar to pseudoplastic materials: the viscosity decreases slightly as the shear rate increases, and then becomes constant – its value becomes correspondent to the second Newtonian viscosity. Non-Newtonian behavior is observed for the resin and hardener solutions.

When measuring the viscosity of Newtonian liquids the CR and CS modes have equal opportunities and give identical results. This statement is no longer valid for rheological tests of non-Newtonian fluids. Given that the CS-rheometers provide higher sensitivity at low shear rates, the controlled shear stress mode was selected as the main research method, as it is more consistent with the actual application conditions.

Comparison at constant shear stress has another advantage: although different classes of fluids may have different viscosity curves, the analysis of curves obtained at constant shear stress, enables us to compare them with each other, and the character of their flow is similar to the flow in capillaries. In all experiments, we used the value of shear stresses of 2 Pa, which provides an egress of the flow curves to the second Newtonian viscosity throughout the range of all the investigated temperatures and solution concentrations. The realizable shear rate range was from 0.015 to 150 s^{-1}, depending on the solution concentration and medium temperature.

3. The measurement results and their analysis

Temperature dependences of resin density and hardener in the temperature range of 10 to 80 ^{0}C is very well approximated by expressions of the form: $\rho_0 = -0.0007135\,T + 0.9783$ and $\rho_c = -0.0007777\,T + 1.1656$ for the hardener and resin respectively. If the temperature is presented in °C, then the density dimension is shown in g/cm^3. The approximation error does not exceed 3%.

The results of measurements of the temperature dependences between the resin effective viscosity and the hardener are shown in Fig. 1. The inset shows large-scaled sections of these dependences (ordinate direction).

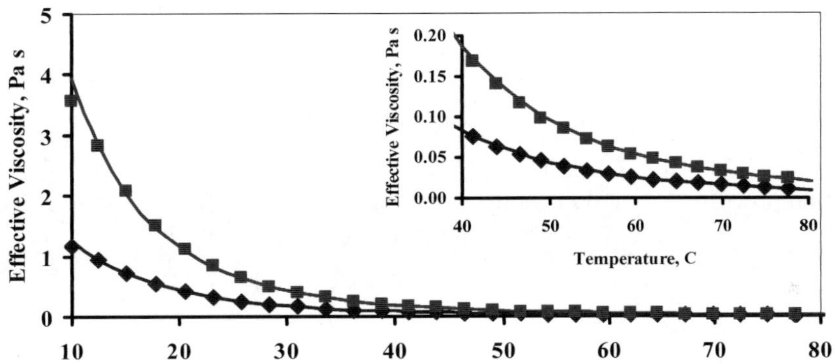

Figure 1. Dependence of effective viscosity on temperature: ■ – epoxy resin; ◆ – hardener; Solid lines – approximations of dependencies

Solid lines in Fig. 1 present experimental data approximations by the Arrhenius equation: $\eta_O = 1.277*10^{-5}\,Exp\left(\dfrac{1120.3}{87.42+T}\right)$ for the hardener; $\eta_C = 4.512*10^{-5}\,Exp\left(\dfrac{937.6}{72.42+T}\right)$ for the resin. If the temperature T is presented in °C in these formula, the viscosity η is shown in Pa•s. The approximation error does not exceed 3%.

The measured values of the epoxide compound viscosity (hardener concentration is 20%) are greater than the corresponding values of resin viscosity. This means that the polymerization starts in the process of mixing.

It is well known that viscosity variations occurring in the course of epoxy hardening have quite a complicated character, but at the initial stages of the reaction the linear growth of macromolecules prevails. In this case, the viscosity of the reaction system with time t usually conforms to the power law $\eta = k\,t^a$, where a takes values ranging from 1 to 3.5 [10].

Actually, the measurements showed that the dependences of the effective viscosity of resin and hardener solutions on time at the initial stage of polymerization are linear (a = 1) with the coefficient k which depends on the concentration of the hardener (Fig. 2). This allows us to extrapolate them to the value $t \to 0$ in order to bring all the viscosity values measured at different concentrations to the same time, since in experiment it is not possible to get the same time in sample preparation. These values of viscosity are used in all experimental results hereinafter.

In Fig. 3 shows the results of measurements of the solution viscosity depending on the volume concentration of the hardener C for 17 °C.

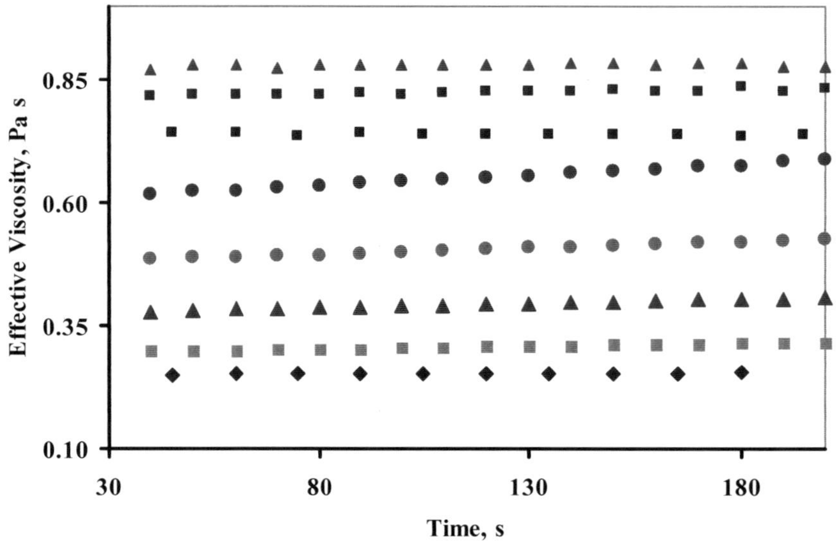

Figure 2. Dependences of solution effective viscosity on time at the initial stage of the polymerization process. From top to bottom hardener solutions C: 0.025; 0.02; 0 (resin); 0.3; 0.4; 0.6; 0.8; 1 (hardener)

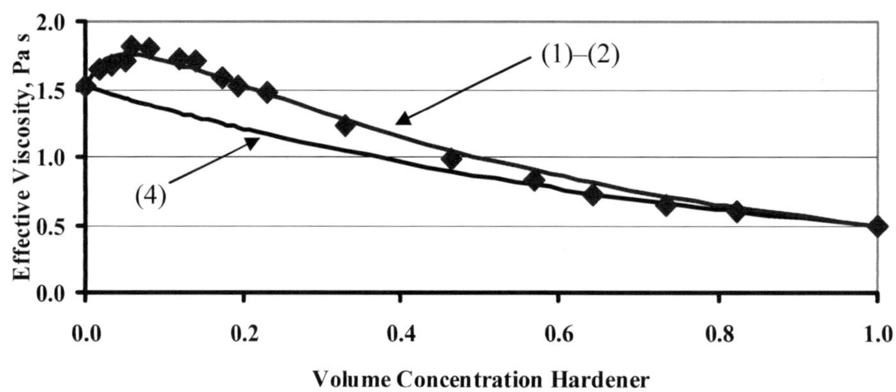

Figure 3. Dependence of epoxy composition effective viscosity on hardener concentration. Solid lines - calculation by formulas (1) - (2) and (4). Details are given in the text

The viscosity curve has a characteristic maximum which value is bigger than the oligomer viscosity, but after reaching its maximum the solution viscosity decreases so that within the limit $C \to 0$ it corresponds to the resin viscosity. It should be noted that the measurement results are well reproduced both at different time points, and at other – bigger value of tangential stresses.

To assess the effect of the polymerization process at the initial stage of epoxide compound formation we use a simplified version of the model of kinetics of the process proposed in [11]. Curing of the epoxy composition occurs inhomogeneously so at any time there are reacted macromolecules with chemical bonds and also there are initial macromolecules, which did not yet react. Assume that at the initial stage of the process all molecules are initial. In this case, the Cauchy model of the corresponding dynamic system takes the form:

$$\begin{cases} \dfrac{dn_0^{ep}(t)}{dt} = -\left(K_{0,0}n_0^{am}(t) + K_{1,0}^{am}n_1^{am}(t)\right)n_0^{ep}(t), \\ \dfrac{dn_1^{ep}(t)}{dt} = \left(K_{0,0}n_0^{am}(t) + K_{1,0}^{am}n_1^{am}(t)\right)n_1^{ep}(t), \\ \dfrac{dn_0^{am}(t)}{dt} = -\left(K_{0,0}n_0^{ep}(t) + K_{0,1}n_1^{ep}(t)\right)n_0^{am}(t), \\ \dfrac{dn_1^{am}(t)}{dt} = \left(K_{0,0}n_0^{ep}(t) + K_{0,1}n_1^{ep}(t)\right)n_1^{am}(t). \end{cases} \quad (1)$$

$$n_0^{ep}(0) = n_0^{ep,0}, \quad n_1^{ep}(0) = 0, \quad n_0^{am}(0) = n_0^{am,0}, \quad n_1^{am}(0) = 0. \quad (2)$$

Here the index "*ep*" denotes the resin concentration, and "*am*" – the hardener concentration. The assumption made above leads to the following balance equation:

$$n_0^{ep}(t) + n_1^{ep}(t) = n_0^{ep.,0}, \quad n_0^{am}(t) + n_1^{am}(t) = n_0^{am,0}, \quad n_0^{ep,0} + n_0^{am,0} = 1. \quad (3)$$

To account the viscosity variations the Chong formula was used in the form [11];

$$\eta_{solution}(t) = \eta_{solution}(0)\left(1 + \frac{1.25\varphi(t)}{1 - \varphi(t)/\varphi_{max}}\right)^2,$$

where $\varphi(t) = \dfrac{V_{ep}n_1^{ep}(t) + V_{am}n_1^{am}(t)}{V_{ep}n_0^{ep,0} + V_{am}n_0^{am,0}}$.

The parameters of equations (2) were $K_{0,0} = 4 \cdot k_a$, $K_{0,1} = K_{1,0} = 2 \cdot k_a$, and the constants k_a and $\dfrac{V_{ep}}{V_{am}}$ were determined by the least squares method with the use of experimental data (see Fig. 3.). The result of numerical solution of the problem at t = 3 min is also presented in Fig. 3. Here we also can see the curve calculated by the known relation for determining the viscosity of the mixture:

$$\eta_S = \eta_0^{n_0^{am,0}} \cdot \eta_c^{n_0^{ep,0}} \quad (4)$$

Thus, we can conclude that the observed effect (see. Fig. 3) is actually the result of the polymerization initial stage.

On the other hand, dimensions of macromolecule aggregates formed at the initial stage of polymerization depend on the concentration of components in the solutions. To compute the intrinsic viscosity of such an aggregate we can use the hypothesis that the ratio of characteristic viscosities is proportional to the ratio of aggregated molecules volume to the total volume of individual unreacted molecules. In other words, from the macroscopic point of view the solutions can be regarded as heterogeneous systems with two different medium fractions. This approach is widely used to analyze the viscosity of disperse systems taking into account the interaction of particles [12] and is successfully applied in [11] to approximate the dependence of the epoxy composition viscosity on the polymerization time. This gives grounds for using the obtained experimental data (Fig. 3) to determine the empirical viscosity dependence on the hardener concentration. At that, one of the main requirements is the need for the selection of a simple expression, enabling an adequate description of the experimental data and the ability to generalize the results to other parameters of the medium.

The regression analysis yields

$$\eta_S = \frac{1.329}{C}\eta_O \, Exp\left(-\frac{0.179}{0.0398+C}\right), \tag{5}$$

where η_S and η_O are the viscosities of the solution and hardener, respectively; C is the hardener volume concentration. Substituting the dependence of hardener viscosity on temperature to this expression, we obtain

$$\eta_S = \frac{1.697\cdot 10^{-5}}{C} Exp\left(-\frac{0.179}{0.0398+C}+\frac{1120.3}{87.42+T}\right) \tag{6}$$

which is the empirical relationship allowing estimation of the solution viscosity in relation to the medium temperature and hardener concentration at $C > 0.015$, as at lower concentrations approximation dependence has a special feature: at $C < 0.015$ the estimated value of the viscosity increases and tends to infinity. In order to verify the obtained expression, effective viscosity measurements were carried out at different temperatures. The results of the comparative analysis of the experimental and calculated data are shown in Fig. 4.

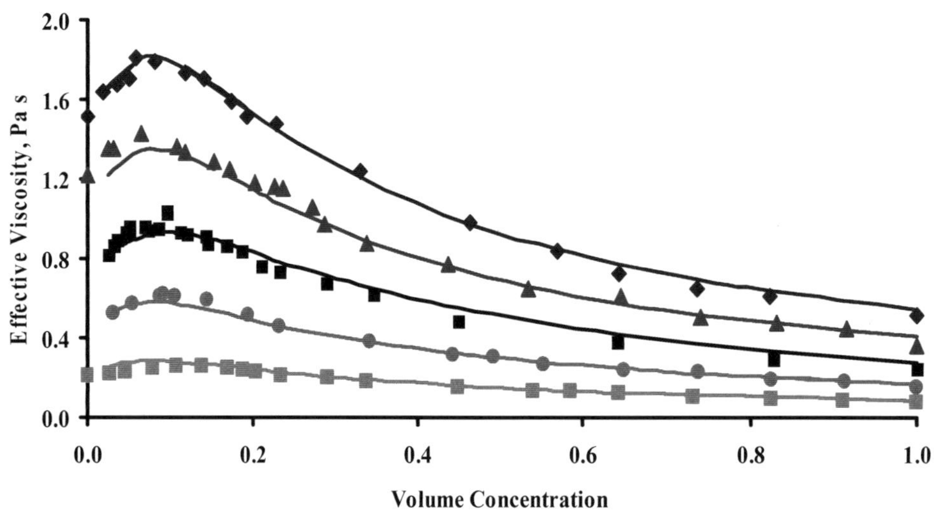

Figure 4. Dependence of effective viscosity on hardener concentration and medium temperature. *From top to bottom*: T = 17; 20; 25, 30; 40 ^{0}C. Markers show the experimental results, solid lines present the formula (6) evaluation

The relative error of approximation in the range $C > 0.015$ is 5% on the average. As can be seen from the figure, the empirical relationship adequately describes the experimental data in relation to the volume concentration of solutions and medium temperature.

4. Conclusion

The physical properties of epoxide compound (epoxy resin L, hardener EPH 161) are investigated. Empirical dependences of resin density and viscosity and hardener on temperature within the range of 10 to 80 °C are obtained. Dependences of composite solution viscosity in relation to hardener volume concentration and medium temperature are measured. Finally, the investigation presents the empirical relationship adequately describing the experimental results, enabling estimation of the corresponding variation in a wide temperature range according to measurements of the viscosity concentration dependence measured at a certain temperature.

Acknowledgments

This study was supported by RFBR Grant №14-08-96011, №15-01-07946, №16-48-590844.

References

[1] Gunyaev G M, Dushin M I, Ivonin Y N, Kvacheva L A, Mikhailov V V *et al* 1983 *Mech. of composite materials* **2** 211
[2] Startzev O V, Rudnev V P, Ivonin Y N, Nikishin E F, Barbashev E A et al 1987 *High-Molecular Compounds* **29 (12)** 2577
[3] Kondyurin A V 1997 Plastic mixtures 8 25
[4] Briskman V A, Yudina T M, Kostarev K G, Kondyurin A V, Leontyev V B et al 2001 *Acta Astronautica* **48 (2-3)** 169
[5] Kondyurin A V, Nechitailo G S 2009 *Space and Rocket Science* **3 (56)** 182
[6] Irzhak V I Rozenberg B A and Enikolopyan N S 1979 *Cross-linked polymer*s (Moscow: Nauka) p 250
[7] Chernin I Z, Smehov F M, Zherdev Y 1982 *Epoxide polymers and compounds* (Moscow: Chemistry) p 232
[8] Mitzel E, Koenig J 1986 *Epoxy resins and composites* (Berlin) **2** 73
[9] Izmestjev I V, Solovyev M N 2010 *Bulletin of Perm State University. Physics* **1 (36)** 54
[10] Malkin A Ya, Kulichikhin S G 1985 *Rheology in the processes of formation and transformation of polymers* (Moscow: Chemistry) p 240
[11] Kondyurin A V, Komarov, L A, Svistkov A L 2010 *Mechanics of Composite Materials and Structures* **16 (4)** 597
[12] Happel J, Brenner H 1976 *Hydrodynamics at low Reynolds numbers* (Moscow: Mir) p 630

IOP Conf. Series: Materials Science and Engineering **208** (2017) 012015 doi:10.1088/1757-899X/208/1/012015

Conjecture on reflectionlessness of blood-vascular system as a wave-conducting medium

D S Goldobin[1,2] **and I A Mizeva**[1]

[1]Institute of Continuous Media Mechanics UB RAS, 1 Akademika Koroleva street, 614013 Perm, Russia

[2]Department of Theoretical Physics, Perm State University, 15 Bukireva street, 614990 Perm, Russia

E-mail: Denis.Goldobin@gmail.com, mizeva@icmm.ru

Abstract. Our research is related to the employment of photoplethysmography (PPG) and laser Doppler flowmetry (LDF) techniques (measuring the blood volume and flux, respectively) for the peripheral vascular system. We derive the governing equations of the wave dynamics for the case of extremely inhomogeneous parameters. We argue for the conjecture that the blood-vascular system as a wave-conducting medium should be nearly reflection-free. With the reflectionlessness condition, one can find the general solution to the governing equation and, on the basis of this solution, analyse the relationships between PPG- and LDF-signals.

1. Introduction

Currently, the filtration of blood through microvessels and tissues of the human body is a subject of growing interest of researchers. Two of the most popular noninvasive tools for experimental studies on the subject are photoplethysmography (PPG) and the laser Doppler flowmetry (LDF) [1]. The PPG signal is proportional to the volume of blood in microvessels, while the LDF signal is proportional to the flow velocity multiplied by the number of erythrocytes in the scanned tissue volume. In figure 1, one can see sample signals measured on human hand fingers. In particular, for a deep inspiratory gasp manoeuvre [2, 3], experimental observations reveal that, under certain conditions maintained during functional tests, a phase shift between the 1 Hz-components of PPG- and LDF-signals appears [4], which illustrates how combined implementation of these two tools may be seminal for diagnostics of the state of the blood-vascular system. It is therefore imperative to improve our understanding of the relationships between PPG- and LDF-signals and of their dependance on the state of the vessels.

In this paper we focus on the consideration of the propagation of the waves of flux/pressure through the blood-vascular system as a wave-conducting medium with highly inhomogeneous properties. Specifically, the net cross-section area of vessels varies from 4.5 to 200 cm^2 along the system, which is as much as 2 orders of magnitude. With such a strong inhomogeneity of parameters, one can, as an approximation, neglect minor effects such as viscoelasticity of the vessel walls [5, 6, 7] (the viscoelasticity effect is treated as "minor" here only against the background of a very strong effect of inhomogeneity) and consider the walls to be merely elastic.

Further, for the time scale of a heart pulse $T = 0.2\,\mathrm{s}$ and the characteristic value of blood kinematic viscosity $\nu = 4\cdot 10^{-6}\,\mathrm{m}^2/\mathrm{s}$ (as the blood is a non-Newtonian liquid, one can speak of a

IOP Conf. Series: Materials Science and Engineering **208** (2017) 012015 doi:10.1088/1757-899X/208/1/012015

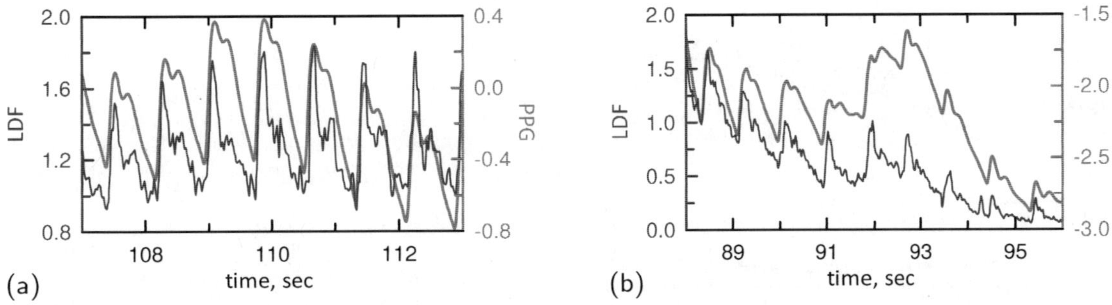

Figure 1. Sample readings from the plethysmograph measuring the blood volume in vessels (PPG, red thick line) and from the Doppler measuring the product of the flow velocity and the blood volume (LDF, blue thin line) are presented for the normal conditions (a) and for the functional testing conditions (b)

reference value of the viscosity coefficient for estimates but not of a rigorously defined coefficient), one can evaluate the thickness of the viscous boundary layer: $\delta_{\text{vis}} = \sqrt{\nu T/\pi} \approx 0.5\,\text{mm}$. Hence, for the vessels of inner diameter $d_{\text{in}} > 1\,\text{mm}$, one can assume an inviscid blood flow. In this paper we will restrict our consideration to the propagation of waves through the part of the blood-vascular system, where the vessel inner diameter $d_{\text{in}} > 1\,\text{mm}$. For the thinner vessels the system is significantly more stiff (which is known from experiments and will be also apparent with the results of our derivation at the end of the paper) and the wave propagation through this remaining part is very fast. With this fast propagation of waves and the conditions of the mass conservation and the flux conservation, one can approximately assume the blood flux and volume to be nearly immediately transferred from the edge of the part of the blood-vascular system where $d_{\text{in}} > 1\,\text{mm}$ to the microvascular system. To summarise, we will consider an inviscid blood flow and associate the blood flux and vessel volume oscillations at the distant-from-heart end with measured LDF- and PPG-signals.

Generally, in a wave-conducting medium, the inhomogeneity of parameters produces reflection of waves. The absence of reflection is possible but requires mutually consistent variation of inhomogenous parameters or their specific dependence on coordinates [8, 9, 10, 11]. Moreover, a waveguide with as strong inhomogeneity of parameters as in the blood-vascular system would be generally nearly impenetrable for waves, reflecting the major part of the energy of incident waves. Meanwhile, in practice, one observes reflection of pulses from the aorta bifurcation point and similar reflections, but they are not as strong as could generally be. This suggests a plausible conjecture that in a healthy state the system is optimised in a way to minimise the reflections. The latter is actually wide-spread for natural systems with different sorts of adaptation dynamics [8, 9]. The approximation of the reflectionlessness of the system will provide us with the opportunity to derive the general solution for the equations of wavy dynamics we will derive in this paper.

The paper is organised as follows. In Sec. 2, we consider the possibility of a phase shift between PPG- and LDF-signals for the stationary-profile waves without any restrictions on mechanical properties of the vessel walls. In Sec. 3.1, we derive governing equations for the wave dynamics in the case of strong inhomogeneity of the system parameters. In Sec. 3.2, the general solution to these equations is obtained for the reflection-free case. In Secs. 3.4 and 3.5, on the basis of this solution, the relationships between the waves of the blood volume (PPG) and the flux (LDF) are established. Finally, in Sec. 4, we summarise our findings.

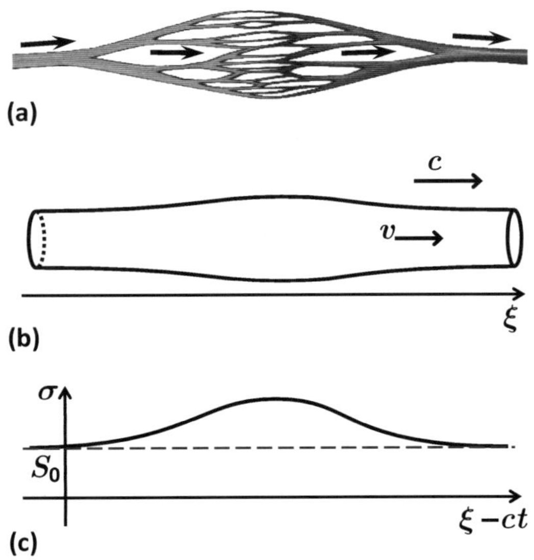

Figure 2. (a) Sketch of the branching vascular network, (b) wave of deformation of an elastic tube containing liquid, (c) schematic picture of a volume wave propagating without dispersion

2. Absence of the phase shift for stationary-profile waves

Prior to considering the relationships between the waves of volume and flux for the general case of inhomogeneous parameters, it is suitable to answer a preliminary question: Why the phase shift is NOT to be considered as typical?

One can show that in an elastic tube with laminar flow of inviscid liquid the linear (small-amplitude) waves of pressure *or* widening of the tube propagate without dispersion and are governed by an equation of the type of the string vibration equation [12].

Let us consider the relationship between the wave of volume of an elastic tube and the wave of flux in it, without imposing any additional assumptions except for the stationarity of wave profile. We make a natural choice for the spatial coordinate; we use the coordinate ξ which measures the length along the unstretched vessel. This coordinate is tied to the vessel wall, *i.e.* it becomes stretched in space when the vessel is stretched. To characterise both widening and stretching of a vessel we introduce the following parameters: S which is the cross-section area of the vessel and

$$\sigma = \frac{\delta V}{\delta \xi} \tag{1}$$

which measures the volume per elementary length $\delta\xi$ of the vessel. Notice, $\sigma = S$ in the case of widening of a vessel without longitudinal stretching—for a stretched state σ and S are different. The average over cross-section velocity of the blood flow in the vessel will be denoted by v, and the wave propagation speed will be c (see figure 2b). Accordingly, the liquid flux

$$q = v\,S\,. \tag{2}$$

Note, LDF measures the flow velocity multiplied by the number of erythrocytes, which is proportional to

$$q_{\mathrm{LDF}} = v\,\sigma \tag{3}$$

and generally differs from the flux $q = v\,S$. However, specifically for the site of measurements in [4], one can neglect the elongation of capillaries and associate LDF-readings with flux q. In this paper we associate LDF for the microcirculatory system with the flux.

As the liquid is incompressible, one can write the condition of the volume conservation:

$$\frac{\partial}{\partial t}\sigma(\xi,t) = -\frac{\partial}{\partial \xi}\Big(v(\xi,t)\,S(\xi,t)\Big)\,. \tag{4}$$

We assume the wave propagation speed c to be constant along the vessel and consider the volume wave of stationary profile (figure 2c)

$$\sigma(\xi, t) = \sigma(\xi - ct)\,. \tag{5}$$

For such a wave $(\partial/\partial t) = -c(\partial/\partial\xi)$. Then equation (4) can be integrated with respect to ξ;

$$c\,(\sigma - \sigma_0) = q - q_0\,, \tag{6}$$

where σ_0 and q_0 correspond to the unperturbed state of the vessel and the flux in it. As the vessel is not stretched in the unperturbed state, $\sigma_0 = S_0$ and equation (6) can be recast as

$$q = c\sigma - cS_0 + q_0\,. \tag{7}$$

According to the latter equation, the flux and the vessel volume vary in-phase.

3. Wave propagation through an inhomogeneous vascular network

As shown in the previous section, in a vessel with homogeneous properties, the phase shift is impossible. Nonetheless, the question whether the phase shift can appear due to variation of parameters (elasticity and inner diameter) along the vessel or not remains unanswered. The construction of mathematical description of the system with inhomogeneous parameters is our task in the current section.

3.1. Vessel with inhomogeneous parameters

Equation (4) describes the liquid incompressibility and holds valid for the case where the diameter and the elasticity of the vessel vary along its length. The time-derivative of this equation yields

$$\frac{\partial^2}{\partial t^2}\sigma(\xi, t) = -\frac{\partial}{\partial\xi}\left[\frac{\partial}{\partial t}\Big(v(\xi, t)\,S(\xi, t)\Big)\right]. \tag{8}$$

We consider perturbations against the background of a static state of a vessel with a time-independent flow; $\sigma(\xi, t) = \sigma_0(\xi) + \sigma_1(\xi, t)$, $S(\xi, t) = S_0(\xi) + S_1(\xi, t)$, $v(\xi, t) = v_0(\xi) + v_1(\xi, t)$ and pressure $p(\xi, t) = p_0(\xi) + p_1(\xi, t)$, which can be used for calculation of the time-derivative in the r.h.s. of equation (8). The Bernoulli's equation for unsteady potential flow

$$\frac{\partial v}{\partial t} = -\nabla\left(\frac{1}{\rho}p + \frac{v^2}{2}\right) \tag{9}$$

yields for perturbations

$$\frac{\partial v_1}{\partial t} = -\nabla\left(\frac{1}{\rho}p_1 + v_0 v_1\right). \tag{10}$$

Then the time-derivative

$$\frac{\partial}{\partial t}(vS) = v_0(\xi)\frac{\partial}{\partial t}S_1(\xi, t) + S_0(\xi)\frac{\partial}{\partial t}v_1(\xi, t) = v_0\frac{\partial}{\partial t}S_1 - S_0\nabla\left(\frac{1}{\rho}p_1 + v_0 v_1\right)$$

$$= v_0\left(\frac{\delta S}{\delta\sigma}\right)\frac{\partial}{\partial t}\sigma_1 - S_0\nabla\left(\frac{1}{\rho}\left(\frac{\delta\sigma}{\delta p}\right)^{-1}\sigma_1 + v_0 v_1\right). \tag{11}$$

The ratio of variations $(\delta S/\delta\sigma)$ under varying pressure in the vessel is a dimensionless quantity of the order of the magnitude of 1 and depends on the ratio of the elasticity moduli and their possible anisotropy.

IOP Conf. Series: Materials Science and Engineering **208** (2017) 012015 doi:10.1088/1757-899X/208/1/012015

Let us estimate the orders of the magnitude of three contributions in $(\partial/\partial t)(vS)$ in equation (11). Fields σ_k, S_k, v_k, p_k (where $k = 0, 1$) are of the same order of magnitude as in the case of a homogeneous vessel. From equation (6), one finds $c\sigma_1 \sim v_0 S_1 + S_0 v_1$. As $S_1 \sim \sigma_1$ and $v_0 \ll c$, the velocity perturbation $v_1 \approx c\sigma_1/S_0$. For wave fields, $(\partial/\partial t) \sim -c(\partial/\partial\xi)$. Hence, three contributions in $(\partial/\partial t)(vS)$ are of the following orders of magnitude:

$$v_0 \frac{\partial\sigma_1}{\partial t}\,, \qquad \frac{S_0}{\rho c}\left(\frac{\delta\sigma}{\delta p}\right)^{-1}\frac{\partial\sigma_1}{\partial t}\,, \qquad v_0 \frac{\partial\sigma_1}{\partial t}\,, \tag{12}$$

in the order of their appearance. For a *vessel with homogeneous parameters* the relation $(\partial/\partial t) = -c(\partial/\partial\xi)$ is accurate, and from equations (8) and (11) for $v_0 \ll c$ one can obtain

$$c^2 \approx \frac{S_0}{\rho}\left(\frac{\delta\sigma}{\delta p}\right)^{-1}. \tag{13}$$

Thus, the second term in sum (11) is of the order of magnitude of $c(\partial\sigma_1/\partial t)$ and large compared to the first and third terms, which $\sim v_0(\partial\sigma_1/\partial t)$.

Taking only the leading term of (11) into account, one can recast equation (8) as

$$\frac{\partial^2}{\partial t^2}\sigma_1 = \frac{\partial}{\partial\xi}\left[S_0(\xi)\frac{\partial}{\partial\xi}\left(\beta(\xi)\,\sigma_1\right)\right], \tag{14}$$

where $\beta := [\rho(\delta\sigma/\delta p)]^{-1}$ characterises the elasticity of the vessel walls.

Noteworthy, with proper functions $S_0(\xi)$ and $\beta(\xi)$, equation (14) is valid for an approximate description of propagation of waves through a bunch of parallel vessels with hierarchy of branchings.

3.2. Solutions of equation (14) in the reflection-free case

Equation (14) admits reflectionless wave propagation, if it can be recast in the form

$$\frac{\partial}{\partial\zeta_1}\frac{\partial}{\partial\zeta_2}F = 0\,. \tag{15}$$

Introducing operator

$$\hat{Q}F := \frac{S_0(\xi)}{\alpha(\xi)}\frac{\partial}{\partial\xi}\left(\alpha(\xi)\,F\right), \tag{16}$$

one can rewrite equation (14) in the form

$$\left(\frac{S_0(\xi)}{\beta(\xi)}\frac{\partial^2}{\partial t^2} - \hat{Q}^2\right)\frac{\beta(\xi)\sigma_1}{\alpha(\xi)} = 0\,. \tag{17}$$

The latter equation can be factorised,

$$\left(\sqrt{\frac{S_0(\xi)}{\beta(\xi)}}\frac{\partial}{\partial t} - \hat{Q}\right)\left(\sqrt{\frac{S_0(\xi)}{\beta(\xi)}}\frac{\partial}{\partial t} + \hat{Q}\right)F = 0\,, \tag{18}$$

if

$$\frac{\partial}{\partial\xi}\sqrt{S_0(\xi)/\beta(\xi)} = 0\,. \tag{19}$$

Equation (19) is the condition of the reflectionlessness of the system. We keep the square root in this equation on purpose, since in reality the ratio (S_0/β) is not perfectly constant along

IOP Conf. Series: Materials Science and Engineering **208** (2017) 012015 doi:10.1088/1757-899X/208/1/012015

the vessel network and with equation (19) one can have a quantitative estimate how accurate the assumption of the reflectionlessness is.

For the waves of our interest, equation (18) yields

$$\left(\frac{\partial}{\partial t} + \sqrt{\beta(\xi)\, S_0(\xi)}\frac{\partial}{\partial \xi}\right) \beta(\xi)\, \sigma_1(\xi, t) = 0\,, \tag{20}$$

and thus

$$\sigma_1(\xi, t) = \frac{1}{\beta(\xi)} f\left(\int^{\xi} \frac{d\xi_1}{\sqrt{\beta(\xi_1)\, S_0(\xi_1)}} - t\right). \tag{21}$$

The wave travel time for ξ_0 to ξ is

$$\tau(\xi, \xi_0) = \int_{\xi_0}^{\xi} \frac{d\xi_1}{\sqrt{\beta(\xi_1)\, S_0(\xi_1)}}\,. \tag{22}$$

Expression (21) is a solution to equation (14) for arbitrary continuous function f. From equation (21) for homogeneous elasticity β and net cross-section area S_0 of vessels, one would find the wave propagation speed $c = \sqrt{\beta S_0}$. Substituting solution (21) into equation (4), one can obtain the flux

$$q(\xi, t) = q(\xi_0, t) + \int_{\xi_0}^{\xi} \frac{f' d\xi_1}{\beta(\xi_1)}\,, \tag{23}$$

where the prime denotes the derivative of a function with respect to its argument.

As the first step of analysis of solutions (21) and (23), we qualitatively consider the issue of the phase shift between the waves of volume and flux.

3.3. Qualitative analysis of the phase shift: wave extrema

Let us consider propagation of a solitary wave through the blood-vascular system, choosing the reference point ξ_0 behind the wave so, that $q(\xi_0, t) = q_0(\xi_0)$. Physiologically, the location of point ξ_0 can be chosen immediately after the heart. At certain point ξ, where

$$\left.\frac{\partial \sigma_1}{\partial t}\right|_{\xi} = 0\,, \tag{24}$$

differentiating equation (23) with respect to time, one finds

$$\left.\frac{\partial q}{\partial t}\right|_{\xi} = -\int_{\xi_0}^{\xi} \frac{f'' d\xi_1}{\beta(\xi_1)} = f'(\xi_1)\sqrt{S_0(\xi_1)/\beta(\xi_1)}\Big|_{\xi}^{\xi_0} + \int_{\xi_0}^{\xi} f' \frac{d}{d\xi_1}\left(\sqrt{\frac{S_0(\xi_1)}{\beta(\xi_1)}}\right) d\xi_1 = 0\,, \tag{25}$$

because where $\partial\sigma_1/\partial t = 0$, f' also becomes zero (which can be explicitly seen from the time-derivative of expression (23)), and the reflectionlessness condition (19). For the case of a train of waves, the situation is the same.

Thus, for inhomogeneous parameters of the system, the extrema of flux and volume are attained simultaneously (figure 1). Moreover, one can notice that this property is accurately preserved even for a transient processes, when system parameters are changing with time (in figure 1b after approximately 90 sec. a deep inspiratory gasp manoeuvre begins).

3.4. The relation between measurable signals

Let us now derive the relation between measurable signals of the blood volume and flux. From equation (21), one can find

$$\beta(\xi_1)\,\sigma_1(\xi_1,t)=\beta(\xi)\,\sigma_1(\xi,t+\tau(\xi,\xi_1)) \tag{26}$$

and evaluate the integrand of equation (23) from a measurable signal $\sigma_1(\xi_{\rm m},t)$ (where $\xi_{\rm m}$ is fixed to the observation site, and $\sigma_1(\xi_{\rm m},t)$ is measured as a function of time);

$$\frac{f'\Big(\int^{\xi_1}[\beta(\xi_2)\,S_0(\xi_2)]^{-1/2}d\xi_2-t\Big)}{\beta(\xi_1)}=-\dot\sigma_1(\xi_1,t)=-\frac{\beta(\xi_{\rm m})}{\beta(\xi_1)}\dot\sigma_1(\xi_{\rm m},t+\tau(\xi_{\rm m},\xi_1))\,, \tag{27}$$

where the dot stands for the time-derivative. Hence, equation (23) takes the form

$$q(\xi_{\rm m},t)=q(\xi_0,t)-\beta(\xi_{\rm m})\int\limits_0^{\tau(\xi_{\rm m},\xi_0)}\dot\sigma_1(\xi_{\rm m},t+\tau)\sqrt{\frac{S_0[\xi_1(\tau)]}{\beta[\xi_1(\tau)]}}\,d\tau\,, \tag{28}$$

where $\xi_1(\tau)$ is implicitly determined by equation (22) or, which is the same but can be more convenient technically in some situations, by the problem

$$\xi_1(\tau=0)=\xi_{\rm m}\,,\qquad d\xi_1=-\sqrt{\beta(\xi_1)\,S_0(\xi_1)}\,d\tau\,. \tag{29}$$

For the time instances, *when the blood flow and the vessel walls near the heart are at the ground state (i.e., between heart pulses)*, one can set $q(\xi_0,t)=q_0(\xi_0)$ in equation (28) and find the relation involving only measurable signals $q(\xi_{\rm m},t)$ and $\sigma_1(\xi_{\rm m},t)$;

$$q(\xi_{\rm m},t)=q_0(\xi_0)-\beta(\xi_{\rm m})\int\limits_0^{\tau(\xi_{\rm m},\xi_0)}\dot\sigma_1(\xi_{\rm m},t+\tau)\sqrt{\frac{S_0[\xi_1(\tau)]}{\beta[\xi_1(\tau)]}}\,d\tau\,. \tag{30}$$

4. Conclusion

We have analytically demonstrated, that elastic waves of volume and flux in the vessels of the blood-vascular system attain extrema simultaneously even in the case where the inhomogeneity of parameters is significant. The experimental observations also suggests nearly no shift of principal extrema of PPG- and LDF-signals for unsteady conditions (see figure 1b). The observed phase shift between the extracted 1 Hz-components of PPG- and LDF-signals [4], which is not detected for steady conditions and appears for unsteady ones, is presumably associated with an integral change in the pulse forms.

The relation between waveforms of volume and flux has been derived (equations (21) and (23) *or* equations (28) and (30), where the latter equation is for specific time instances). These relations provides opportunity for extraction of diverse information on the variation of the system parameters (elasticity and inner diameter of vessels along the network) in response to different external conditions and functional tests from the readings of PPG and LDF.

For the system under consideration the spacial inhomogeneity of parameters is essential as they change by two orders of magnitude along the system and the spatial extent of the wave pulses is commensurable to the characteristic scale of the parameter inhomogeneity. A waveguide with such a strong inhomogeneity of parameters would be generally nearly impenetrable for waves, reflecting the major part of the energy of incident waves. In practice, the system seems to be optimised from the view point of reflectionlessness; the reflection in a healthy state is possibly minimal. The approximation of the reflectionlessness of the system has allowed us to derive the reported relations between waveforms of blood volume and flux.

As an important implication of the reflectionlessness conjecture, one can expect that an increasing pathological variation of the vessel inner diameters and elasticity properties should result in the growing violation of the reflectionlessness conditions and thus also in an increasing stress impact on both the heart and large vessels by reflected waves.

Acknowledgments

The work has been financially supported by RFBR (grant no. 17-41-590560 r_a).

References

[1] Allen J, Frame J R, Murray A 2002 *Physiol. Meas.* **23** 365–73
[2] Laude D, Goldman M, Escourrou P and Elghozi J L 1993 *Clin. Exp. Pharmacol. Physiol.* **20** 619–26
[3] Di Marco L Y, Zheng D and Murray A 2012 *Comput. Cardiol.* **39** 745–8
[4] Podtaev S, Mizeva I and Alan J 2012 *J. Vasc. Res.* **48** 130
[5] Formaggia L, Lamponi D and Quarteroni A 2003 *J. Eng. Math.* **47** 251–76
[6] Valdez-Jasso D, Haider M A, Banks H T, Santana D B, German Y Z, Armentano R L and Olufsen M S 2009 *IEEE Trans. Biomed. Eng.* **56** 210–9
[7] Valdez-Jasso D, Bia D, Zocalo Y, Armentano R L, Haider M A and Olufsen M S 2011 *Ann. Biomed. Eng.* **39** 1438–56
[8] Didenkulova I and Pelinovsky E 2009 *Phys. Lett.* A **373** 3883–7
[9] Didenkulova I and Pelinovsky E 2012 *Nonlin. Processes Geophys.* **19** 1–8
[10] Petrukhin N S, Pelinovsky E N and Batsyna E K 2011 *JETP Lett.* **93** 564–7
[11] Petrukhin N S, Pelinovsky E N and Batsyna E K 2012 *Astron. Lett.* **38** 388–93
[12] Pedley T J 1980 *The Fluid Mechanics of Large Blood Vessels* (Cambridge University Press)

IOP Conf. Series: Materials Science and Engineering **208** (2017) 012016 doi:10.1088/1757-899X/208/1/012016

Influence of geometrical parameters on convective flows in non-uniformly heated cylindrical fluid layers

A. Evgrafova[1,2], A. Sukhanovskii[1], M. Kuchinskii[2], E. Popova[1]
[1]Institute of Continuous Media Mechanics, Academ. Korolyov, 1, Perm, 614013, Russia
[2]Perm State University, Bukireva St. 15, Perm, 614990, Russia
E-mail: eav@icmm.ru

Abstract. Convection in a fluid layer with a free surface in a cylindrical container non-homogeneously heated from below was studied experimentally. The heater has a circular form and was placed in the center of the vessel. Such a system in the presents of rotation is very promising for studying the nature of tropical cyclones. In current paper we considered the influence of different geometric parameters on the dynamics of convection flows. For it two experimental setups of different sizes were used. Measurements were done for fluids with different values of Prandtl number and heating powers. It was shown that the structure and intensity of mean flow producing by a horizontal temperature gradient are defined by Rayleigh number Ra. The basic flow leads to unstable temperature stratification over the heating area and the formation of a system of secondary flows. The formation of secondary flows depends on characteristics of thermal boundary layer and is described by local Rayleigh number Ra_δ.

1. Introduction

Convective flows produced by the non-uniformly heating exist in a large variety of systems of different nature and scale. In particular the very interesting problem is natural convection produced by a localized heat source. In technological processes such flows are commonly observed in heat exchangers, cooling systems of modern electronic equipment and nuclear reactors [1,2]. Concerning atmospheric boundary layer some studies were motivated by processes of impurity propagation from a strong thermal source (for example oil ignition)[3]. Development of urban climatology leads to the problem of convection from so-called urban heat islands and studies of large-scale convective circulation from extended local heat sources with ambient temperature stratification [4,5,6].

The specific of the convection driven by local heat source is producing both vertical and horizontal temperature gradients. As a result of the applied horizontal temperature gradient, a large-scale meridional circulation appears. In the presence of rotation meridional circulation transports angular momentum from the periphery to the center and produces cyclonic vortex. Such system was proposed as a laboratory model of tropical cyclone [7]. PIV measurements showed that the general structure of mean radial and azimuthal flows in the proposed model is similar to the typical structure of a hurricane [8]. In the boundary layer of cyclonic vortex secondary convective structures in the shape of horizontal rolls were observed [8]. There is remarkable similarity of these secondary flows and rainbands in real tropical cyclones [9,10]. Secondary flows may significantly influence on heat and mass transfer in the boundary layer [11,12]. The nature and characteristics of secondary flows over heating area in a non-rotating cylindrical layer were studied in [13]. Various types of visualization revealed co-existence of radial and transverse convective rolls. Radial rolls are very robust and similar to the longitudinal rolls studied in rectangular tank [14,15]. Transverse rolls in a cylindrical layer are formed periodically and move to the center along with a basic flow. The characteristic frequency of transverse rolls depends on Rayleigh number for a wide range of governing parameters.

It is not clear how strong is the influence of geometrical parameters on the structure and characteristics of basic and secondary flows in cylindrical fluid layer with localized heat source. As a first step to solve this problem we consider two experimental setups. The paper is organized as follows. In section 2 we describe the experimental setup and measurement technique. Experimental results are presented in section 3 and conclusions are given in section 4.

2. Experimental setup

Two cylindrical vessels were used. The first vessel (I) has diameter $D = 300$ mm (figure 1 a). The heater is a brass cylindrical plate mounted flush with the bottom. The diameter of the plate $2R$ is 104 mm, and its thickness is 10 mm. The second vessel (II) has diameter $D = 690$ mm (figure 1 b). The bottom of the larger vessel is the textolite plate with thickness of 20 mm. In the center of plate the brass heater of diameter $2R = 195$ mm was placed flush with the bottom. The temperature of the heaters was controlled by computer. Silicon oils with different values of kinematic viscosity, PMS-20, PMS-10 and PMS-5 (20, 10 and 5 cSt at T = 25 ^{0}C) were used as working fluids. The surface of the fluid was always open. The depth of the layer for the vessel I was fixed $h = 30$ mm. For the vessel II two series of experiments were carried out with $h = 30$ mm and $h = 60$ mm. The room temperature was kept constant by air-conditioning system, and cooling of the fluid was provided mainly by the heat exchange with surrounding air on the free surface and some heat losses through sidewalls. Temperature inside the fluid layer was measured by copper-constantan thermocouple. The velocity field measurements were made with a 2D particle image velocimetry (PIV) system Polis. The measurement, and the processing of the results were performed using the software package Actual Flow.

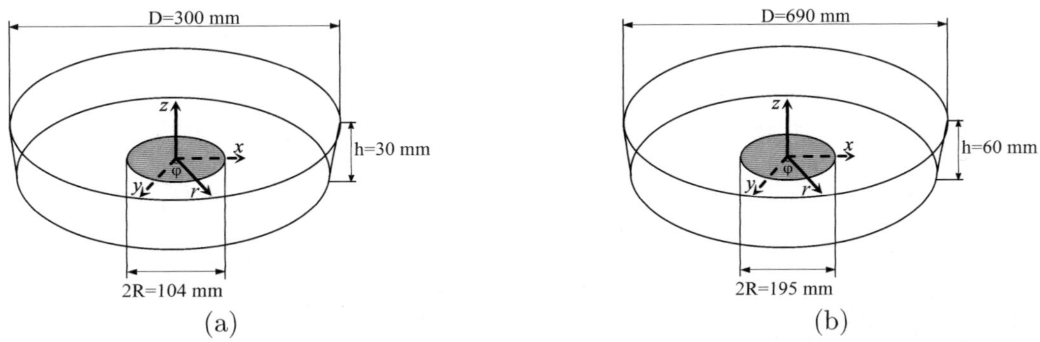

Figure 1. Experimental models, dimensions and location of the coordinate system.

We use the set of the non-dimensional parameters which are commonly used for similar problems. These are Prandtl number Pr and Rayleigh number Ra:

$$Pr = \frac{\nu}{\kappa} \tag{1}$$

$$Ra = \frac{g\beta h^3 \Delta T}{\nu^2} \tag{2}$$

where g is the gravitational acceleration, h is the layer depth, β is the coefficient of thermal expansion, ΔT - temperature difference between temperatures of the heater and the room, ν is the coefficient of kinematic viscosity.

3. Results

Detailed description of the basic flow structure can be found in [13] but brief description of the general structure of the large-scale flow is necessary for a better understanding of the results. The heat flux in the central part of the bottom is a source of the intensive upward motion above the heater. Warm fluid cools at the free surface and moves toward the periphery where the cooled fluid moves downward along the side wall. Large-scale advective flow occupies the whole vessel (figure 2). Experimental measurements of velocity fields in a vertical cross-section over the heating area showed that instantaneous fields are irregular and asymmetric. Along with the main updraft in the center there are less intensive but pronounced upgoing convective flows close to the periphery of the heater.

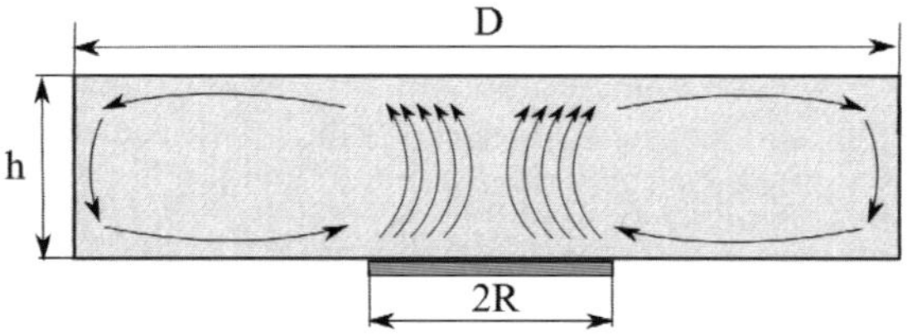

Figure 2. Large-scale circulation.

At first we wanted to compare the large-scale flow structure for vessels I and II. If the structure of the flow does not depend on the size of the vessel (for the same ratio of the layer depth to the heater diameter and value of Prandtl number) than non-dimensional velocity field should be the same for the same values of Rayleigh number. Non-dimensional radial velocity fields are shown in figure 3. Velocities are averaged in the azimuthal direction and in time. Positive values describe divergent flow, and negative values convergent flow. As a characteristic velocity we chose ratio of kinematic viscosity and characteristic size (depth of the layer) $v* = \nu/h$. One can see that structures of mean flow have a good qualitative agreement. For the quantitative comparison we present mean radial velocity profiles (figure 4). Profiles in both case are plotted for r/R=0.5. The agreement is satisfactory, the discrepancy (about 20%) can be explained by different ratio of the heater size and vessel diameter which also important geometric parameter. We can conclude that for the described parameters the variation of the geometric sizes of the vessel for fixed Rayleigh number does not substantially change the structure of the main large-scale flow.

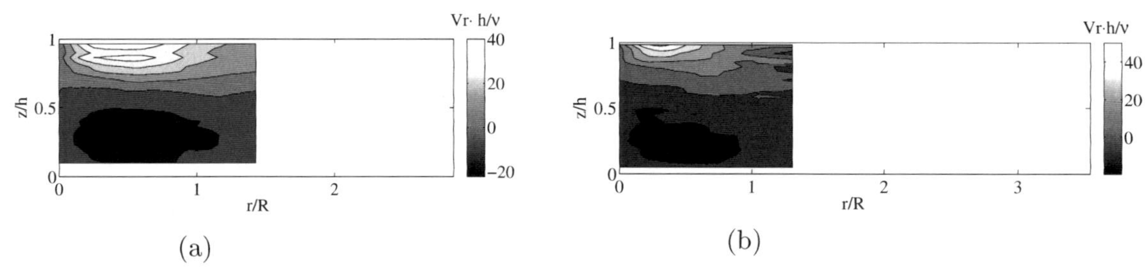

Figure 3. Mean radial velocity fields for Ra = $1.1 \cdot 10^7$, Pr=66 : a - D = 300 mm, h = 30 mm; b - D = 690 mm, h = 60 mm.

The large-scale advective flow in the lower part of the layer leads to the formation of boundary layer with potentially unstable temperature stratification above the heater and makes possible the generation of the secondary convective flows. The structure and specifics of secondary

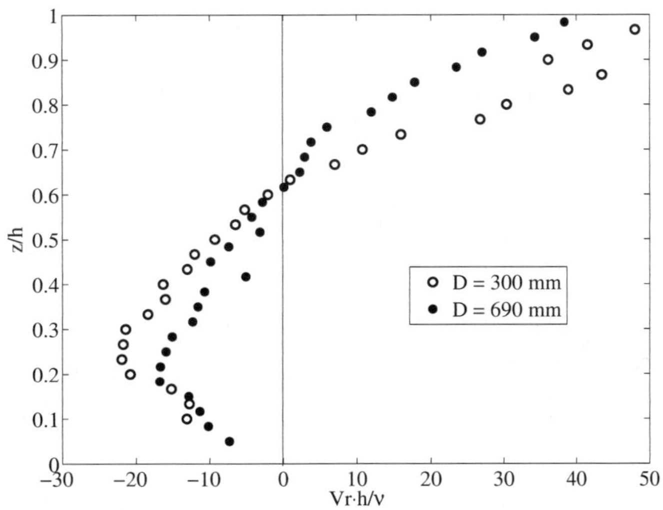

Figure 4. Mean radial velocity profiles at r/R = 0.5. $Ra = 11{\cdot}10^6$. $Pr = 66$.

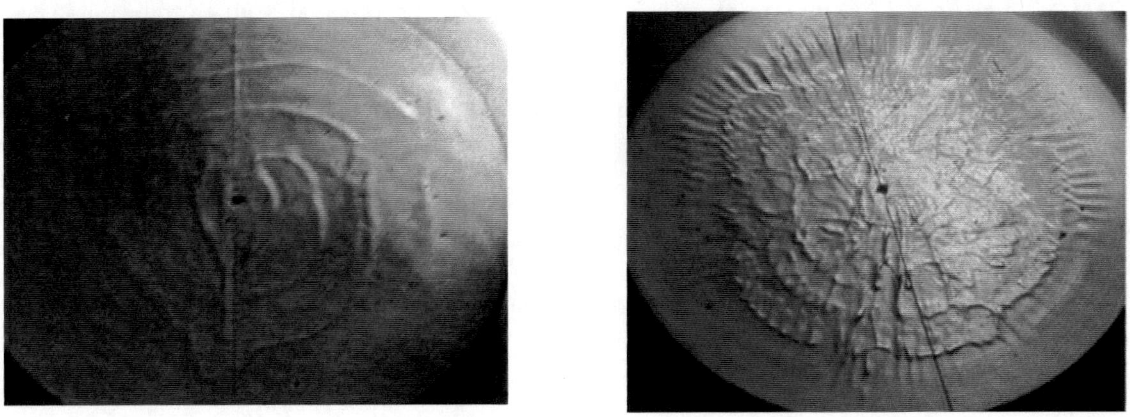

(a) (b)

Figure 5. Visualization of secondary flows: a - $Ra = 0.2 \cdot 10^5$, b - $Ra = 1.2 \cdot 10^6$.

flows over the heater in the case of non-rotating cylindrical layer are described in detail in [13]. Secondary flows visualized by shadowgraph technique are presented in figure 5. Regimes for the weak heating are characterized by appearance of ring-like rolls. An increase of the heat flux leads to more complex convective patterns - the superposition of the spiral and a system of radial rolls on the periphery. The authors in [13] showed that radial rolls are robust and stationary. Ring-like rolls and spirals are orientated in transverse direction to the main flow and appear periodically on the heater periphery.

Dependence of transverse roll formation frequency f for on Rayleigh number for different fluids and aspect ratios is presented in figure 6a. One can see that frequency f grows with an increase of the Rayleigh number for both vessels but there is significant quantitative discrepancy between measurement for different values of the layer depth. For a correct description of

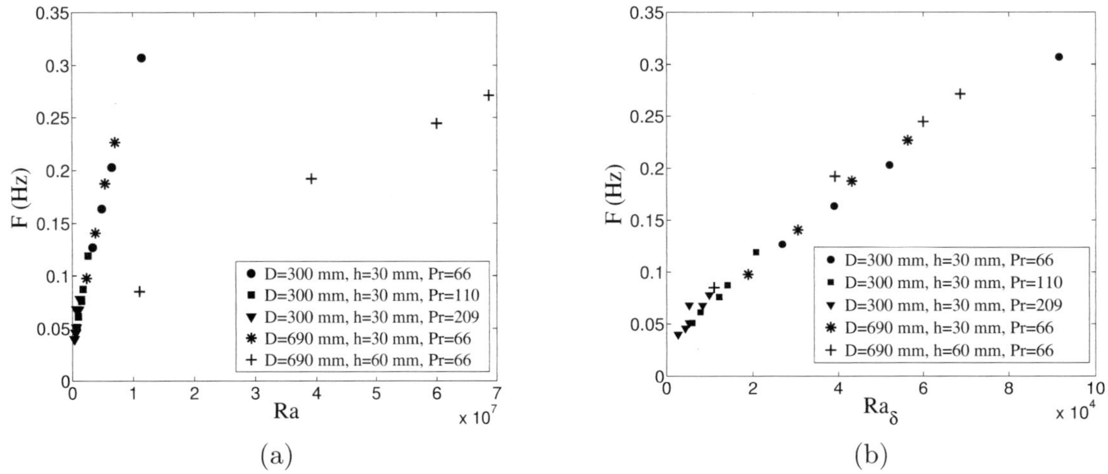

Figure 6. Characteristic frequence f versus Rayleigh number (a) and local Rayleigh number (b).

secondary flow regimes, the local governing parameter should be used, for example the local Rayleigh number, defined through the characteristics of the thermal boundary layer, because processes in a boundary layer with unstable temperature stratification are connected but do not directly depend on global parameters. From the temperature measurements for $Ra = 1.1 \cdot 10^7$ (vessel II) we estimated that height of thermally unstable layer has value $\delta \approx 6$ mm. Replacing h by fixed value of δ (6 mm) for calculation of local Rayleigh number Ra_δ we see that all experimental points lie on the same curve. This result needs to be discussed. There are two possible explanations. The first one is that characteristic frequency f does not depend on δ and is defined only by physical properties of the fluid and applied temperature difference. The second one required that δ is constant or changes weakly for all experimental realisations presented in figure 6b. In order to check that we plan to study the properties of thermal boundary layer for a wide range of experimental parameters.

It is also found that variation in fluid viscosity for fixed model does not change the structure and dynamics of secondary flows. In figure 7 instantaneous radial velocity fields over the heater at the height $z = 3mm$ are presented for different Prandtl numbers but constant D, h and Ra_δ. The values of velocity and the shape of fields are very similar.

4. Conclusions

The influence of geometrical parameters on the structure and characteristics of basic and secondary flows from a localized heat source in a cylindrical layer was studied experimentally. Two experimental setups of different sizes were used. Measurements were done for fluids with different values of Prandtl number and heating powers. It was shown that the structure and intensity of mean flow producing by a horizontal temperature gradient are defined by Rayleigh number Ra. The basic flow leads to unstable temperature stratification over the heating area and the formation of a system of secondary flows. The formation of secondary flows depends on characteristics of thermal boundary layer and are described by local Rayleigh number Ra_δ. For calculation of Ra_δ we used the assumption that for the described experiments δ is constant or has only slight deviation from the constant value. In order to check that we plan to study the properties of thermal boundary layer for a wide range of experimental parameters.

The financial support of the grant RFBR 16-31-00150 is gratefully acknowledged.

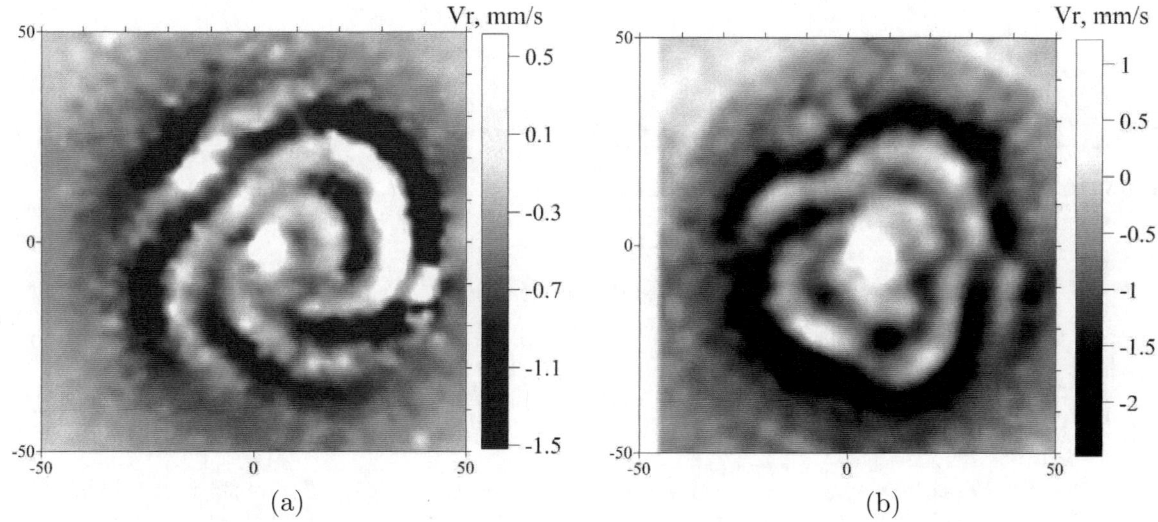

Figure 7. Instantaneous radial velocity fields at $z = 3$ mm ($Ra_\delta = 3.1{\cdot}10^4$, D = 300 mm, h=30 mm): a - $Pr = 110$, b - $Pr = 209$.

References

[1] Chiu WKS et al. 2000 *Phys. Fluids* **12** 2128
[2] Aydin O and Yang WJ 2000 *Int. J. Numerical Methods for Heat and Fluid Flow* **10 (5)** 518
[3] Shvarts FG and Shklyaev VA 2009 *Computational continuum mechanics* **1** 96
[4] Lu J et al 1997 *J. Ap. Met.* **36** 1377
[5] Kurbatskii AF 2001 *J. App. Met.* **40** 1748
[6] Arnfield AJ 2003 *Int. J. Climatol.* **23** 1-26
[7] Bogatyrev GP 1990 *Letters of JETP* **51** 557-559
[8] Sukhanovskii et al. 2016 *Quart J.R.Met.Soc.* **316** 23-33
[9] Moon Y, Nolan DS. 2009 *J. Atmos. Sci.* **72** 164-190
[10] Moon Y, Nolan DS. 2009 *J. Atmos. Sci.* **72** 191-215
[11] Morrison I et al. 2005 *J. Atmos. Sci.* **62** 2662
[12] Zhang JA et al. 2008 *Boundary-Layer Meteor.* **128** 173
[13] Sukhanovskii et al. 2016 *Physica D* **316** 23-33
[14] Batalov et al. 2007 *Fluid Dyn.* **42** 540-549
[15] Sukhanovsky et al. 2012 *Eur. Phys. J. B* **85** 12

Spatial localization of nonlinear waves spreading in materials in the presence of dislocations and point defects

V I Erofeev, A V Leontieva and A O Malkhanov

Mechanical Engineering Research Institute of RAS, 85, Belinskogo St., Nizhny Novgorod, 603024, Russia

E-mail: erof.vi@yandex.ru

Abstract. Within the framework of self consistent dynamic problems, the impact of dislocations and point defects on the spatial localization of nonlinear acoustic waves propagating in materials has been studied.

1. Introduction

The experimental data accumulated to this date allows us to state that dislocations and point defects have a significant influence on the propagation of acoustic waves in a solid. The role of dislocations is particularly important in the propagation of waves in a deformed or cyclically loaded body [1 - 3]. Theoretical description of the propagation of an acoustic wave in a solid with a changing density of dislocations would allow us to approach the problem of estimating the real state of a material and predicting its residual resource.

When the laser radiation or particle flux (e.g. in ion implantation) are exposed to the material it leads to creation of point defects (vacancies, interstices) [4]. The transmitting of an intense longitudinal acoustic waves facilitates the change in areas of tension and compression, the activation energy of formation of point defects, leading to their spatial redistribution. Defects migrating in material, recombine at various centers. The role of such centers can play a dislocation, impurity introduction, etc.

The aim of the paper is to study nonlinear self-consistent problems about the propagation of acoustic waves, taking into account their interaction with dislocations and point defects which are existed in the material.

2. Dislocation displacement solitons.

To take into account the interaction of dislocations with the lattice of the crystal, as well as the interaction of dislocations with one another, a mathematical model was proposed in [5, 6]:

$$\frac{\partial^2}{\partial t^2}U - c^2\frac{\partial^2}{\partial x^2}U = \frac{\beta}{\rho}\frac{\partial}{\partial x}\xi, \tag{1}$$

$$A\frac{\partial^2}{\partial t^2}\xi + B\frac{\partial}{\partial t}\xi = -\beta\frac{\partial}{\partial x}U. \tag{2}$$

Here U – acoustic displacement, ξ – dislocation displacement, A – mass of dislocations, B – frictional force per unit length of dislocation, ρ – density of a material, $\beta = \beta_{jkl} \cdot b_j$ – coefficient of

IOP Conf. Series: Materials Science and Engineering **208** (2017) 012017 doi:10.1088/1757-899X/208/1/012017

acoustic dislocation interaction (where β_{ijlk} – tensor of acoustic dislocation interaction, b_j – Burgers' vector). In this case, the pulsation component will be assumed to be proportional to the square of the dislocation displacement ξ [5]: $A = A_0\left(1 + A_1\xi^2\right)$, $B = B_0\left(1 + B_1\xi^2\right)$, then the equation (2) will take the following form:

$$\begin{gathered} \frac{\partial^2}{\partial t^2}U - c^2\frac{\partial^2}{\partial x^2}U = \frac{\beta}{\rho}\frac{\partial}{\partial x}\xi, \\ A_0(1 + A_1\xi^2)\frac{\partial^2}{\partial t^2}\xi + B_0\left(1 + B_1\xi^2\right)\frac{\partial}{\partial t}\xi = -\beta\frac{\partial}{\partial x}U \end{gathered} \tag{3}$$

Assuming, at the first stage of the investigation, the dislocation subsystem to be conservative $\left(B_0 = 0\right)$, let us rewrite (3) in the form of the one equation:

$$\begin{gathered} \frac{A_0}{\beta}\frac{\partial^4}{\partial t^4}\xi - \frac{c^2 A_0}{\beta}\frac{\partial^4}{\partial x^2 \partial t^4}\xi - \frac{\beta}{\rho}\frac{\partial^2}{\partial x^2}\xi + \\ + \frac{A_0 A_1}{\beta}\left(\frac{\partial^2}{\partial t^2} - c^2\frac{\partial^2}{\partial x^2}\right)\left(\xi^2\frac{\partial^2}{\partial t^2}\xi\right) = 0 \end{gathered} \tag{4}$$

which describes the propagation of the dislocation displacement wave.

Equation (4) has solutions in the form of traveling nonlinear stationary waves propagating with a constant velocity $(V = const)$.

If the material possesses positive nonlinearity $(A_1 > 0)$, then dislocation displacements solitons might form in it.

In the subsonic range of the velocity $(V < c)$ the form of the soliton is described with hyperbolic secant:

$$\xi = A_c \sec h\left(\frac{x - Vt}{\Delta}\right) \tag{5}$$

and represents localized bell-shaped perturbation having a positive polarity. The width of the soliton is determined by the effective mass per unit dislocation length $\left(A_0\right)$ and by the coefficient of acoustic dislocation interaction (β), i.e.

$$\Delta \sim \sqrt{A_0}, \Delta \sim 1/\beta, \tag{6}$$

while the amplitude is defined by the nonlinearity parameter $A_c = \sqrt{2/A_1}$.

In the supersonic range of velocity $\left(V < c\right)$ the form of the soliton is described with hyperbolic tangent:

$$\xi = A_c th\left(\frac{x - Vt}{\Delta}\right) \tag{7}$$

and represents the step between two constant values $\xi_{(+)}$ and $\xi_{(-)}$. Such a soliton is called "kink", it is analogous to a shock wave whose front width is determined by the relations (5), and the amplitude (the magnitude of the drop) is equal to $A_c = \sqrt{1/A_1}$.

In materials which possess a negative dislocation nonlinearity $(A_1 < 0)$ solitons cannot be formed, only periodic nonlinear stationary waves can propagate in such materials.

Let us estimate the influence of the nonconservativity of the dislocation subsystem on the wave processes.

If the dislocations are assumed to be free-inertial $(A_0 = 0)$, then the system (3) can be reduced to the equation:

$$\frac{d\xi}{d\eta} = \frac{\beta^2 \xi}{\rho^2 V(V^2 - c^2) B_0 (1 + B_1 \xi^2)} \tag{8}$$

where $\eta = x - Vt$, which take the following form when $(B_1 \xi^2 << 1)$:

$$\frac{d\xi}{d\eta} + \alpha\xi - \alpha B_1 \xi^3 = 0, \tag{9}$$

where $\alpha = \dfrac{\beta^2}{\rho^2 B_0 V (c^2 - V^2)}$.

The solution to the equation (9) we will look in the form $\xi = \xi_0 e^{ik\eta} + \xi_0^* e^{-ik\eta}$, where $k = k' + ik''$ – complex wavenumber, ξ_0 – complex amplitude, ξ_0^* – complex conjugate amplitude.

Neglecting the higher harmonics and taking into account the contribution of the nonlinearity only to the fundamental harmonic, we obtain:

$$k' = 0,$$
$$k'' = \alpha(1 - 3B_1 |\xi_0|^2). \tag{10}$$

During the propagation of a supersonic wave *(c<V)* exponential damping occurs, besides a "soft" nonlinearity, i.e. when $(B_1 < 0)$ the nonlinear wave decays faster than the linear wave. With a "hard" nonlinearity, i.e. when $(B_1 > 0)$ the linear wave decays faster than the nonlinear one.

During the propagation of a subsonic wave *(c>V)* the dislocation viscosity leads to an increase in the amplitude of the wave, i.e. the system behaves as an active one. For a "soft" nonlinearity, a nonlinear wave grows faster than a linear wave, for a "hard" nonlinearity, a linear wave grows faster than a linear wave.

3. Point defects

In [7, 8] it is shown that the problem of propagation of acoustic wave in the material with defects, should be considered as self-consistent, comprising along with a dynamic equation of elasticity theory, the kinetic equation for the density of defects:

$$\frac{\partial^2 u}{\partial t^2} - c_l^2 \frac{\partial^2 u}{\partial x^2} - \frac{\beta_N}{\rho} \frac{\partial u}{\partial x} \frac{\partial^2 u}{\partial x^2} = -\frac{\left(\lambda + \frac{2}{3}\mu\right)\Omega_j}{\rho} \frac{\partial n_j}{\partial x} \tag{11}$$

$$\frac{\partial n_j}{\partial t} = q_0 + q_\varepsilon \frac{\partial u}{\partial x} + D_j \frac{\partial^2 n}{\partial x^2} - \beta_j n_j \tag{12}$$

Here $u(x,t)$ – longitudinal dislocation material particles (the wave is considered flat); $n_j(x,t)$ – volume concentration of point defects; ($j = v$ – for vacancies, $j = i$ – for interstices); $c_l = \sqrt{(\lambda + 2\mu)/\rho}$ – the velocity of the longitudinal wave propagation in the material, in the case of absence of defects; λ, μ – Lame modulus; ρ – density of the material; β_N – nonlinearity coefficient: $\beta_N = 3\lambda + 6\mu + 2A + 6B + 2C$; A, B, C – Landau modulus of the third order; Ω_j – dilatational parameter which characterizes the volume change when a single point defect appears (for vacancies $\Omega_j < 0$, for interstices $\Omega_j > 0$).

Equation (12) is written under the assumption that the main processes determining the behavior of the defects are the processes of generation, recombination and diffusion. Through q_0 it is denoted the rate of generation of point defects in the absence of deformation; the second term in the right part of (12) is a deformation correction to the generation of defects; D_j – diffusion coefficient of defect of type j; β_j– recombination velocity on the effluent. Volume of mutual recombination of unlike defects is not taken into account.

The system (11), (12) can be reduced to a single nonlineary equation with respect to displacement:

$$\frac{\partial^2 u}{\partial t^2} - \left(c_\ell^2 - \frac{q\left(\lambda + \frac{2}{3}\mu\right)\Omega_j}{\rho \cdot \beta_j} \right) \frac{\partial^2 u}{\partial x^2} - \frac{D_j}{\beta_j}\left(\frac{\partial^4 u}{\partial x^2 \partial t^2} - c_\ell^2 \frac{\partial^4 u}{\partial x^4} \right) + \frac{1}{\beta_j}\left(\frac{\partial^3 u}{\partial t^3} - c_\ell^2 \frac{\partial^3 u}{\partial x^2 \partial t} \right) =$$
$$= \frac{\beta_N}{2\rho}\left[\frac{\partial}{\partial x}\left(\frac{\partial u}{\partial x}\right)^2 - \frac{D_j}{\beta_j}\frac{\partial^3}{\partial x^3}\left(\frac{\partial u}{\partial x}\right)^2 + \frac{1}{\beta_j}\frac{\partial^2}{\partial x \partial t}\left(\frac{\partial u}{\partial x}\right)^2 \right] \tag{13}$$

Let us introduce dimensionless displacement, coordinate and time:

$$\tilde{U} = \frac{U}{U_0}, \tilde{x} = x\sqrt{\frac{\beta_j}{D_j}}, \tilde{t} = t\sqrt{\frac{\beta_j\left[c_l^2 - q_\varepsilon(\lambda + \frac{2}{3}\mu)\Omega_j \right]}{D_j}} \tag{14}$$

Then equation (13) can be transformed to the following form:

$$\frac{\partial^2 \tilde{U}}{\partial \tilde{t}^2} - \frac{\partial^2 \tilde{U}}{\partial \tilde{x}^2} - \frac{\partial^4 \tilde{U}}{\partial \tilde{x}^2 \partial \tilde{t}^2} + \frac{1}{a}\frac{\partial^4 \tilde{U}}{\partial \tilde{x}^4} + b\sqrt{a}\left(\frac{\partial^3 \tilde{U}}{\partial \tilde{t}^3} - \frac{1}{a}\frac{\partial^3 \tilde{U}}{\partial \tilde{x}^2 \partial \tilde{t}} \right) =$$
$$= \frac{\beta_N}{2\rho}\left\{ \frac{\sqrt{\beta_j} U_0}{\sqrt{D_j} c_l^2 a}\left[\frac{\partial}{\partial \tilde{x}}\left(\frac{\partial \tilde{U}}{\partial \tilde{x}}\right)^2 - \frac{\partial^3}{\partial \tilde{x}^3}\left(\frac{\partial \tilde{U}}{\partial \tilde{x}}\right)^3 \right] + \frac{U_0}{D_j c_l \sqrt{a}}\frac{\partial^2}{\partial \tilde{x} \partial \tilde{t}}\left(\frac{\partial \tilde{U}}{\partial \tilde{x}}\right)^2 \right\} \tag{15}$$

Here we introduced the following designations $a = 1 - \frac{q_\varepsilon\left(\lambda + \frac{2}{3}\mu\right)\Omega_j}{\beta_j \rho c_l^2}$, $b = \frac{c_l}{\sqrt{\beta_j D_j}}$.

In the linear approximation, $\beta_N = 0$ the equation (15) has the following form:

$$\frac{\partial^2 \tilde{U}}{\partial \tilde{t}^2} - \frac{\partial^2 \tilde{U}}{\partial \tilde{x}^2} - \frac{\partial^4 \tilde{U}}{\partial \tilde{x}^2 \partial \tilde{t}^2} + \frac{1}{a}\frac{\partial^4 \tilde{U}}{\partial \tilde{x}^4} + b\sqrt{a}\left(\frac{\partial^3 \tilde{U}}{\partial \tilde{t}^3} - \frac{1}{a}\frac{\partial^3 \tilde{U}}{\partial \tilde{x}^2 \partial \tilde{t}}\right) = 0 \quad (16)$$

Looking for the solution of equation (16) in the form of traveling harmonic wave:

$$U = A_0 e^{i(\omega t - kx)} + A_0^* e^{-i(\omega t - kx)}, \quad (17)$$

where ω - the frequency; $k = 2\pi/\Lambda$ - the wave number (Λ - the wavelength); A_0 - complex amplitude (A_0^* - its complex-conjugate value), we get the following dispersion equation:

$$k^4 - k^2(a - a\omega^2 + ib\sqrt{a}\omega) + a(\omega^2 + ib\sqrt{a}\omega^3) = 0. \quad (18)$$

Note that in equation (18), spatial and time scales of the longitudinal wave, contains complex coefficients, which implies that the wave will not only spread in the medium, but also to fade along with propagation.

Firstly, consider the special case, when $b = 0$, this is possible if the diffusion coefficient to tend to infinity $(D_j \to \infty)$. Then the coefficients of the dispersion equation (18) will be real and it will look like:

$$k^4 - k^2(a - a\omega^2) + a\omega^2 = 0. \quad (19)$$

The frequency and wave number are related by:

$$\omega = \pm k\sqrt{\frac{1 - \frac{1}{a}k^2}{1 + k^2}}, \quad (20)$$

i.e., the presence of point defects leads to the dispersion of longitudinal elastic waves. The graph of (20) determines the dispersion curve in the plane (ω, k). The curve at small wave numbers has the asymptote $\omega = k$, and for large – asymptotically approaches a straight line $\omega = k/\sqrt{a}$.

Phase velocity of the wave is determined by the formula:

$$V_{ph} = \frac{\omega}{k} = \pm\sqrt{\frac{1 - \frac{1}{a}k^2}{1 + k^2}}, \quad (21)$$

and the group velocity is associated with it by the ratio (Rayleigh):

$$V_{gr} = V_{ph} + k\frac{dV_{ph}}{dk}. \tag{22}$$

When $k \to 0$ phase velocity $V_{ph} \to 1$, when $k \to \infty$ $V_{ph} \to 1/\sqrt{a}$. If point defects are vacancies $(j = v)$, then $\Omega_j < 0$, consequently $1/\sqrt{a} < 1$ and dispersion in this case is normal $(V_{ph} > V_{gr})$. If point defects are interstices $(j = i)$, then $\Omega_j > 0$, consequently $1/\sqrt{a} > 1$ and dispersion is abnormal $(V_{ph} < V_{gr})$.

When $b \neq 0$, i.e. when the values of the diffusion coefficients (D_j) are finite, the wave number in the study of the dispersion equation (18) can be written in the form $k = k' + ik''$, where k'- characterizes the propagation constant $\left(V_{ph} = \frac{\omega}{k'}\right)$- phase velocity of the wave, but $k'' = \alpha(\omega)$ characterizes the wave attenuation.

Longitudinal wave in a medium with point defects is dispersive and damped along its propagation. Attenuation has an extreme character.

Returning to the study of nonlinear equation (15), let us assume that the dispersion and attenuation of waves due to the presence of point defects, and the nonlinearity of the material is the value of the same order of smallness. The solution of equation (15) is looking in the form of the asymptotic expansion of displacement in the small parameter:

$$\tilde{U} = U_0 + \varepsilon U_1 + \ldots \tag{23}$$

And introduce new variables:

$$\xi = \tilde{x} - c\tilde{t}; \eta = \varepsilon \tilde{x} \tag{24}$$

This choice of variables is explained by the fact that the perturbation propagating with the velocity along the axis $\tilde{x}$, slowly evaluates in space due to the nonlinearity, dispersion and dissipation. After substituting (23) and (24) in (15) in the zero approximation of ε it is obtained the expression for the velocity $c = 1$ that matches the velocity c_l in dimensional variables.

The first approximation on ε gives the evolutionary equation with respect to axial deformation $W = \partial U_0 / \partial \xi$:

$$\frac{\partial W}{\partial \eta} + \frac{(a-1)}{2a}\frac{\partial^3 W}{\partial \xi^3} + \frac{b(a-1)}{2\sqrt{a}}\frac{\partial^2 W}{\partial \xi^2} + qW\frac{\partial W}{\partial \xi} = 0, \tag{25}$$

where $q = \beta_N \sqrt{\beta_j} U_0 / 2\rho\sqrt{D_j} c_l^2 a$. Equation (25) is called the Korteweg-de Vries-burgers. It has a solution in the form of a localized wave (a kink):

$$W = A_0 \exp\left(\tilde{\xi}\right) \sec h^2\left(\frac{\tilde{\xi}}{2}\right), \tilde{\xi} = \frac{2}{\Delta}(\xi - V\eta) \tag{26}$$

Here A_0 - the amplitude of the wave, V - its velocity, Δ - the characteristic width. Their values are determined by the expressions:

$$A_0 = \frac{3c_l^2\left[-q_\varepsilon\left(\lambda + \frac{2}{3}\mu\right)\Omega_j\right]}{25\beta_j^2\sqrt{\beta_j}(-\beta_N)U_0\sqrt{D_j}} \cdot \left[1 - \frac{q_\varepsilon\left(\lambda + \frac{2}{3}\mu\right)\Omega_j}{\beta_j\rho c_l^2}\right], \tag{27}$$

$$V = \left|\frac{3q_\varepsilon\left(\lambda + \frac{2}{3}\mu\right)\Omega_j}{25\beta_j^2 D_j\rho}\right|, \Delta = \frac{10\sqrt{\beta_j D_j}}{c_l\sqrt{1 - \frac{q_\varepsilon\left(\lambda + \frac{2}{3}\mu\right)\Omega_j}{\beta_j\rho c_l^2}}}. \tag{28}$$

In the relations (27) – (28) $\frac{q_\varepsilon\left(\lambda + \frac{2}{3}\mu\right)\Omega_j}{\beta_j\rho c_l^2} < 1$, consequently for estimations the following formulas are valid:

$$A_0 \approx \frac{-3c_l^2 q_\varepsilon\left(\lambda + \frac{2}{3}\mu\right)\Omega_j}{25\beta_j^2\sqrt{\beta_j D_j}(-\beta_N)U_0}, \quad \Delta \approx \frac{10\sqrt{\beta_j D_j}}{c_l}. \tag{29}$$

Equation (26) describes the stationary wave profile. Such waves are formed by a joint of a balance between nonlinearity, dispersion and dissipation. Consequently, the presence of point defects in the material affects the process of localizing the propagation of strain waves. If point defects are vacancies $(j = v)$, then the wave profile represents the difference ("step") from highest to lowest. If point defects are the interstices $(j = i)$, then the wave profile is the difference from lowest to highest. For both types of defects the amplitude of the wave decreases with increasing recombination rate at drains as $A_0 \sim 1/\beta_j^{\frac{5}{2}}$, its velocity decreases as $V \sim 1/\beta_j^2$, but width increases as $\Delta \sim \sqrt{\beta_j}$. This behavior is the same wave parameters and with the growth of the diffusion coefficient: the amplitude and speed decrease $A \sim 1/\sqrt{D_j}$, $V \sim 1/D_j$, but width increases $\Delta \sim \sqrt{D_j}$.

4. Conclusions

Expressions are obtained that describe the dependence of the dislocation displacement soliton width on the effective mass per unit length of the dislocation, and the coefficient of acoustic dislocation interaction.

It is shown that the interaction of a nonlinear deformation wave with the field of concentration of point defects (vacancies, interstices) leads both to wave scattering and to a change in the activation energy of defect formation and their spatial redistribution.

The research was carried out under the financial support of the Russian Foundation for Basic Research (projects NN 15-08-01836-a, 16-08-00971-a, and 16-08-00776-a).

References

[1] Lebedev A B 1996 *Philos.Mag.* B**74** 545
[2] Eiras J A 2000 *J. Alloy and Compounds* **310** 68
[3] Hirao M *et al* 2000 *Acta Mater.* **48** 517
[4] Mirzoev F Kh *et al* 1996 *Physics-Uspekhi* **39** 1
[5] Burlak G N and Ostrovskii I V 1997 *Technical Physics Letters* **23** 725
[6] Erofeev V I and Romashov V P 2002 *Technical Physics Letters* **28** 218
[7] Mirzoev F and Shelepin L A 2001 Technical *Physics. The Russian J. Appl. Phys.* **46** 952
[8] Belubekyan M V *et al* 2015 *Mater. Phys. and Mech.* **23** 20

IOP Conf. Series: Materials Science and Engineering **208** (2017) 012018 doi:10.1088/1757-899X/208/1/012018

Numerical studies of the polymer melt flow in the extruder screw channel and the forming tool

S V Ershov[a] and N M Trufanova

Perm National Research Polytechnic University, 29, Komsomolsky ave., Perm, 614990, Russia

E-mail: [a] Ershov_sv@bk.ru

Abstract. To date, polymer compositions based on polyethylene or PVC is widely used as insulating materials. These materials processing conjugate with a number of problems during selection of the rational extrusion regimes. To minimize the time and cost when determining the technological regime uses mathematical modeling techniques. The paper discusses heat and mass transfer processes in the extruder screw channel, output adapter and the cable head. During the study were determined coefficients for three rheological models based on obtained viscosity vs. shear rate experimental data. Also a comparative analysis of this viscosimetric laws application possibility for studying polymer melt flow during its processing on the extrusion equipment was held. As a result of numerical study the temperature, viscosity and shear rate fields in the extruder screw channel and forming tool were obtained.

1. Introduction

Investigation of the polymer flow in the extruder screw channel is essential, during the insulating polymer processing, because the limit of melt maximum temperature is one of the main requirements. In the plasticizing extruder metering zone, where the material is molten, take place pressure buildup, intense mixing, and homogenization of the melt, which can lead to the local material overheating. However, problems arise not only in that part of the extrusion equipment. When exiting the extruder screw channel, material enters the forming device, which also has a significant influence on the polymer processing. A sharp change in channel geometry makes a significant contribution to the character of the melt flow, causing local overheating. Therefore, determination the ways of polymer materials dissipation energy and the melt overheating volume and level reduction is one of the main problems pursued by the researchers. The aim of this work was to study flow and heat transfer processes in the extruder screw channel and forming device during production of plastic insulation and comparison of different rheological laws and geometric models.

Currently, there are a number of mathematical models describing the flow and heat transfer in extruders screw channels [1-5]. All known approaches, based on the laws of conservation, use the principle of inverse movement and turns the screw channel on a plane, that is an assumption.

A schematic illustration of the extruder is represented in figure 1. It consists of the screw, body, heating elements and a forming device. The extruder is divided into four zones: the feeding zone, a delay melting zone, a melting and metering zone. In this paper, a study was conducted with the last 5 turns of the extruder, output adapter and forming tool, because this is where the greatest overheating takes place.

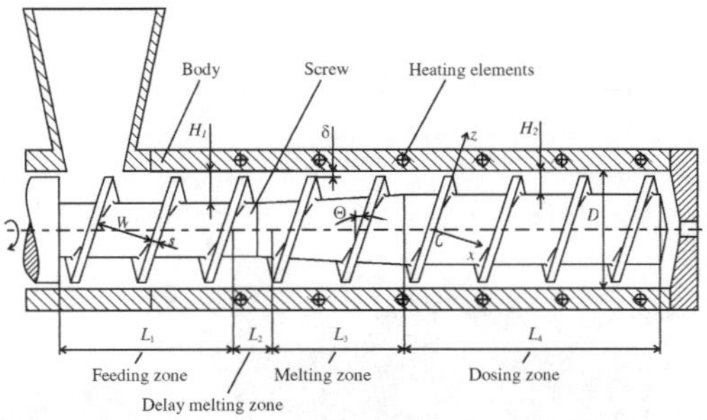

Figure 1. A schematic illustration of the extruder.

2. Mathematical statement and initial data

In the construction of spatial mathematical models of flow in the extruder metering zone following assumptions were made: the process is stationary and steady at constant mass flow, the polymer melt is considered to be a purely viscous, incompressible liquid, mass forces are equal to zero.[6] In view of these assumptions, the system of differential equations has the form:

$$\frac{\partial V_i}{\partial x_i} = 0; \tag{1}$$

$$\rho_m V_i \frac{\partial V_j}{\partial x_i} = -\frac{\partial P}{\partial x_i} + \frac{\partial \tau_{ij}}{\partial x_j}; \tag{2}$$

$$\rho_m C_m V_i \frac{\partial T}{\partial x_i} = \frac{\partial}{\partial x_i}\left(\lambda_m \frac{\partial T}{\partial x_i}\right) + \tau_{ij}\frac{\partial v_i}{\partial x_j}, \tag{3}$$

where ρ_m, C_m, λ_m – respectively, density, heat capacity and thermal conductivity of the polymer melt; $V_i = v_x, v_y, v_z$— velocity component of the polymer melt movement; $x_i = x, y, z$ – rectangular coordinates; τ_{ij} – stress tensor deviator components; P – pressure; T – temperature.[6].

The connection between the deviator stress tensor and the strain rate tensor is given by:

$$\tau_{ij} = \mu_э \left(\frac{\partial v_i}{\partial x_j} + \frac{\partial v_j}{\partial x_i}\right), \tag{4}$$

where $\mu_э$ – effective viscosity of the polymer melt[1].

Viscosity dependence on temperature was determined by the Reynolds equation:

$$\mu_0 = \mu_{0|_{T_0}} \exp\left(-\beta\left(T - T_0\right)\right) \tag{5}$$

The paper considered with the different rheological laws describing the dependence of the effective viscosity versus shear rate:

power law [7]

$$\mu_э = \mu_0 \dot{\gamma}^{\frac{n-1}{2}} \tag{6}$$

Carreau model [8]

$$\mu_э = \left[1 + \lambda_k^2 \dot{\gamma}\right]^{\frac{n-1}{2}} \left(\mu_0 - \mu_\infty\right) + \mu_\infty \tag{7}$$

Cross model [7]

$$\mu_э = \mu_\infty + \frac{\mu_0 - \mu_\infty}{1 + \lambda \dot{\gamma}^m} \tag{8}$$

where $\dot{\gamma}$ – shear rate; μ_0 – viscosity at zero shear rate; μ_∞ – viscosity at infinite shear rate; n –

viscosity anomaly index; λ_k, λ, m – constants of the models,

To study the processes in the extruder channel, models presented in Figure 2 were built.

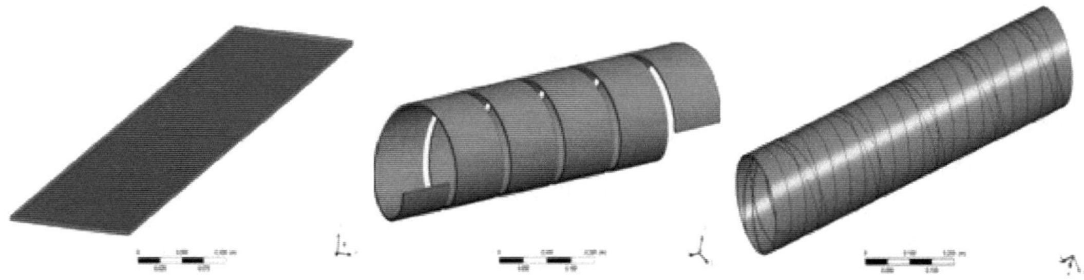

a b c

Figure 2. Geometric models of the extruder channel: a – Flat model; b – Cylindrical model without a gap; c – Cylindrical model with a gap.

The geometry of the channel is presented in Table 1. The body temperature was set at 200 ˚C, the temperature of the screw is 210˚C, the temperature of the input material is 225˚C . Calculations were carried out for the screw rotation speed of 60 revolutions / min.

Table 1. The geometry of the channel

Parameter	Value
Inner diameter of a barrel, mm	160.0
Outer diameter of a screw, mm	159.4
Twist, mm	160.0
Channel width, mm	137.3
Flight Wight, mm	15.3
Length of the dosing geometric zone, turn	5
Channel depth in the metering zone, mm	4
Helix angle, ˚	17°39′
Radial clearance between the flight and the barrel, mm	0.3

Study of the polymer melt rheological characteristics was carried out using a rotary rheometer Discovery HR-2 «TA Instruments». During the experiment, the analyzed polymer was subjected oscillating deformations in the «plate – plate» system. Thermophysical characteristics of the material and rheological models coefficients are shown in Table 2.

Table 2. Polymer properties

Material	Density, ρ (kg/m^3)	Specific heat capacity, C $(J/kg{\cdot}K)$	Thermal conductivity, λ $(W/m{\cdot}K)$
PE 153-02K	779	2500	0.182
The coefficients, used in the rheological laws μ_0 (Pa·s)			
Power law: μ_0=23336.9, β=0.0124, n=0.34, T_0=433			
Cross model: μ_0=188912, β=0.0131, n=0.64, λ=25.8548, μ_∞=51.1269, T_0=433			
Carreau model: μ_0=88064.2, β=0.0124, n=0.40424, λ_K=11.5052, μ_∞=124.83, T_0=433			

Figure 3 shows the curves of viscosity change versus shear rate, built according to the studied laws.

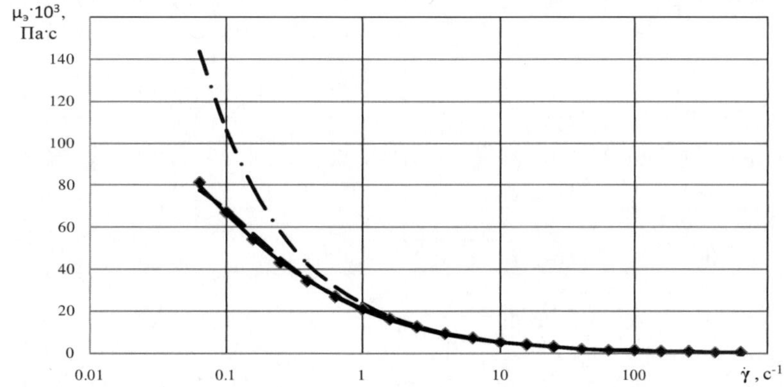

Figure 3. Dependence of the polymer melt effective viscosity by shear rate
◆ – the experimental values; ▬ ▬ – Carreau model; ▬▬ – Cross model; ▬ · ▬ — power low

Figure 3 shows, that all shear flow curves are quite similar, with the exception of low shear rate values, where power law is significantly different. The average values deviation obtained by a power law and experimental data was 14.6%, whereas the deviation for Kerry and Cross models was 2.9 and 0.9 (%), respectively.

The equations solution (1-8) was carried out in engineering calculations package ANSYS CFX by finite element method. Construction of finite element mesh was made in ICEM CFD system.

To determine the optimal number of elements, grids with different number of elements for different channel geometries were constructed and then divergence of the maximum temperature at the channel exit was evaluated. For the deployed channel and cylindrical model without gap, a grid with a number of elements equal to 334125 was chosen. Further increase in the number of elements leads to a longer counting time, while the discrepancy becomes less than 0.2%. To solve the problem of polymer melt flow in a helical channel with a gap, a grid with a number of elements equal to 1 757 000 was required. This is due to the small size of the gap between the crest of the screw and the inner surface of the cylinder. To establish the solutions 30,000 iterations was required.

3. Results and Discussion

As a result of a numerical study of the polymer melt flow processes in the extruder dosing zone, the shear rate, temperature and viscosity fields along the channel length were received.

At the first stage of numerical studies, the fields of temperature, velocity, viscosity, and shear rate were obtained for a flat and screw channel model under conditions of reversed motion and with a screw rotation. Also a comparison of three mathematical models for different channel geometries was held. The investigations were carried out for the power law of viscosity.

The maximum and average values of temperature, viscosity and shear rates across the channel volume are shown in Table 3.

Table 3. Average values of temperature, shear rates and viscosity for power law, for various geometric models.

Variant of calculation №	Geometric model	Movement type	Temperature, (°C)		Shear rate, $[s^{-1}]$	Viscosity, [Pa·s]
			Max.	Avg.	Avg.	Avg.
1	Flat channel	Reversed	254.7	233.8	134.5	578.0
2		Unconverted	302.8	263.8	257.8	757.1
3	Helical without a gap	Reversed	250.1	233.5	134.3	531.3
4		Unconverted	289.3	262.3	247.7	780.5
5	Helical with a gap	Reversed	247.7	231.6	166.2	482.5
6		Unconverted	254.1	235.0	246.5	1162.0

In the case of reversed motion (variants 1,3,5), the temperature values are close, the difference in the maximum temperature does not exceed 3%, and in the average temperature is 1%, in spite of the fact that for variant 5 there are significant differences in the maximum values of the shear rate and viscosity. This is due to the presence of a gap between the crest of the screw and cylindrical body, where the highest shear rates are arise. But in view of the small channel region above the crest of the screw, this is significantly less pronounced on the temperature fields.

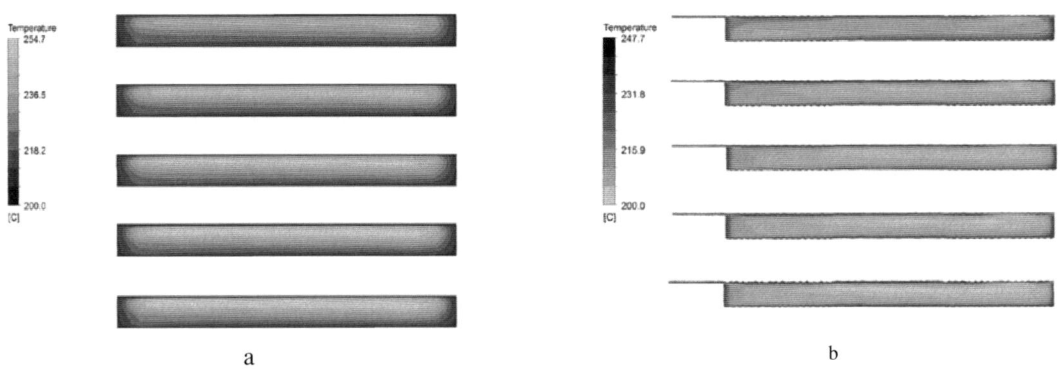

Figure 4. Temperature fields using the principle of reversed motion: a – Flat channel; b – Helical channel with gap.

The difference in the temperature values obtained for the flat channel model using the inverse motion principle and the helical channel model with a gap in the case of a screw movement is less than 1%, in spite of the fact that the nature of the polymer melt flow in these channels is different. In addition, the channel model with a gap in the case of a moving screw is the closest to the real conditions of polymer flow in the extruder dosing zone and allows more complete consideration of heat and mass transfer processes during polymer materials processing.

The important factor in the study of polymer flow processes is the choice of rheological laws that describe the change in viscosity from shear rate and temperature. Since the viscometric curves constructed according to the described above models (Fig. 3) have a significant discrepancy in the region of low shear rates, it was necessary to analyze the effect of the selected rheological law on the polymer flow in the extruder screw channel.

Shear rates values realized in the channel for all three rheological laws under equal conditions and assumptions are differ by not more than 3% (see Table 4). Using the power law in describing the polymer flow in a channel is possibil, because there are no shear rates in the dosing zone at which this model does not correctly describe the viscosity change. When considering models with regard to the rotation of the screw, shear rates of less than 1 s^{-1} are arise. This, certainly, goes beyond the range of

IOP Conf. Series: Materials Science and Engineering **208** (2017) 012018 doi:10.1088/1757-899X/208/1/012018

adequate description of the viscosity change, however, regions with such values of shear rates are small and do not significantly influence on the investigations results. The difference between the maximum and average temperatures at the exit from the extruder channel does not exceed 2°C. Thus, the use of a power law to describe the polymer flow in an extruder channel is possible.

Table 4. Average values of temperature, shear rate and viscosity in terms of volume for the flat channel.

Rheological law	Movement type	Shear rate, s^{-1}			Viscosity, Pa·s		
		Avg.	Max.	Min.	Avg.	Max.	Min.
Power law	Reversed	134.5	317.7	35.7	578.0	1172.0	165.5
	Unconverted	257.8	1030.0	1.1	757.1	6219.0	40.8
Carreau model	Reversed	134.4	324.1	36.1	606.3	1229.0	165.1
	Unconverted	244.1	1170.0	0.9	810.8	6170.0	34.0
Cross model	Reversed	134.5	320.2	35.8	577.9	1178.0	157.7
	Unconverted	253.0	1078.0	1.0	745.6	5331.0	36.1

In order to more fully study the processes which occur during the polymer compositions processing on extrusion equipment, a full model was constructed (see Figure 3).

Figure 5. A geometric model of the investigated extruder channel and forming tool.

Since the helical channel model with a gap in the case of screw rotation yields the most adequate results, it was chosen to study the polymer melt flow processes in a full model which consists of dosing zones, output adapter and cable heads.

Heat and mass transfer processes in the output adapter are the most interesting in studying the polymer flow within the framework of the described above model, because in this area the polymer melt flow is reconstructed. This occurs because channel geometry is changing and material is heating due to the energy dissipation. In the course of numerical investigation, the temperature field in the adapter was obtained (see Figure 5). It should be noted that in the middle of this area the maximum temperature value increases by 10°C. Depending on the geometric features of the output adapter, in particular the angle of the conical part, overheating can reach 30°C, which can be critical for a number of polymer materials. Such increase in the temperature of the melt is associated with the appearance of a circulation movement at the end of the screw. Figure 6 shows the flow functions that allow you to see the path of the material movement and identify the areas of the vortex flow in the output adapter.

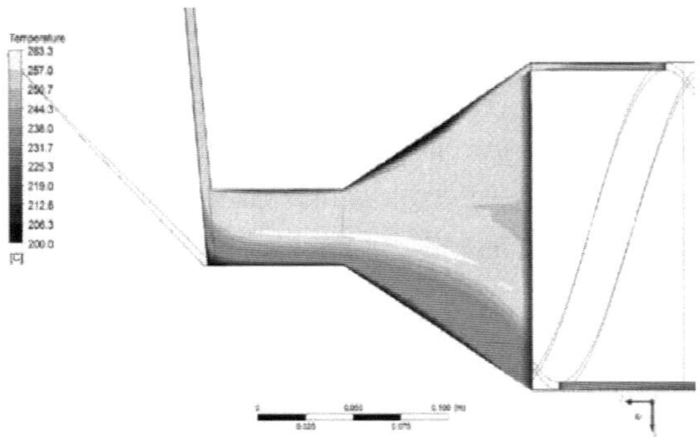

Figure 6. Temperature field in the output adapter.

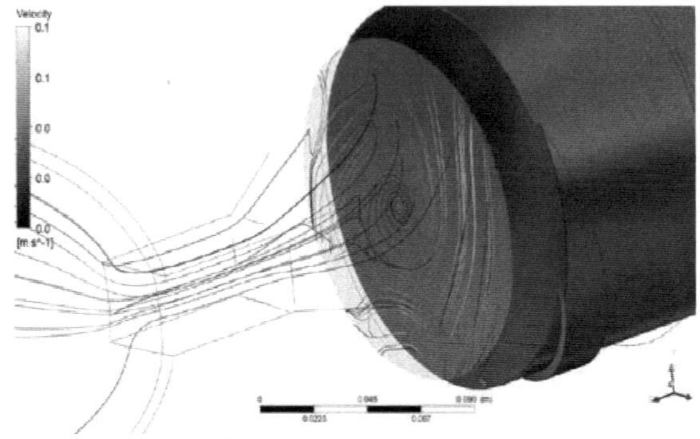

Figure 7. Material flow functions.

Pressure is the most important parameter in choosing the rational operation mode of the extrusion equipment. Figure 7 shows the pressure and flow rate characteristics of the extruder and forming tool. The pressure values obtained by calculating the model, taking into account the output adapter and the forming tool (marked with dots), quite well coincide with the operating points, which are determined on the basis of the flow-rate characteristics separately for the extruder and cable head. The difference does not exceed 4%. The complete model allows to obtain pressure values directly at the manometer installation site in a real extruder.

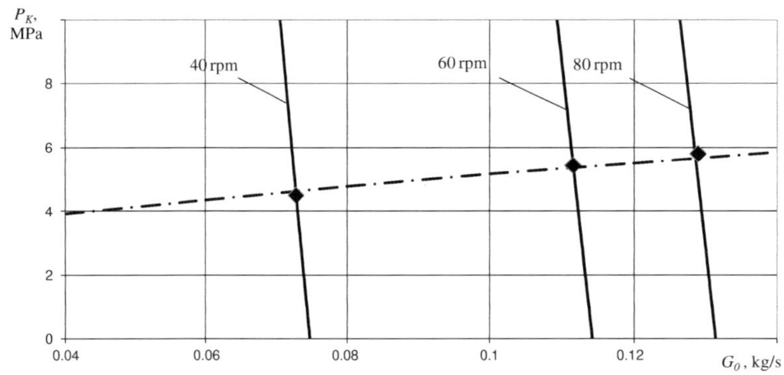

Figure 8. Pressure and flow rate characteristic of an extruder and forming tool: ♦ - data obtained by calculating the full model; ▬▬ - pressure and flow rate characteristic of the extruder at different speeds of rotation of the screw; ▬ ▪ ▬ - pressure and flow rate characteristic of the forming tool.

4. Conclusion

As a result of the study, a number of conclusions can be done. Since for a number of processed polymers the value of the maximum melt temperature is essential, it is important to adequately describe the polymers viscosity dependence on the shear rate and temperature during calculations. Despite the fact that the viscometric flow curves built from the Kerry and Kross models differ quite strongly from the power law, the temperature, viscosity and shear rate fields obtained in the study are quite close. In addition, the power law has a significant advantage, which lies in the simplicity of its use and the small number of empirical coefficients.

An analysis of the mathematical model with different channel geometries of the dosing zone allows us to conclude that they are valid for the case of reversed motion. Thus, the shape of the channel (flat or screw) has little effect on the flow process, while the choice of the movable wall (rotation of the body or screw) has a significant effect on the calculation results.

Using the principle of reversed motion leads to a changes in the polymer flow nature, which accordingly affects the temperature, viscosity and shear rates in the channel of the extruder. The possibility of using the principle of inverted motion depends on the chosen mathematical model (channel geometry).

The calculation of the full model makes it possible to consider the heat and mass transfer processes, which are close to reality, both in the main channel of the extruder and in the forming tool. In addition, it becomes possible to determine the rational mode of extrusion equipment operation by means of one model calculation.

References

[1] Trufanova N M, Shcherbinin A G and Yankov V I 2009 *Melting polymers in extruders*.(Moscow–Izhevsk: SIC "Regular and Chaotic Dynamics", Institute of Computer Science) p 336
[2] Rauwendaal C 2008 *Extrusion of polymers*.(St. Petersburg: Profession) pp 295–353
[3] Rauwendaal C 2004 Finite element studies of flow and temperature evolution in single screw extruders *Plastics, Rubbers and Composites* **33 (9/10)** 390–396
[4] Subbotin E V, Shherbinin A G and Trufanova N M 2015 Numerical analysis of polymers heat and mass transfer processes in extruder screw channel, taking into account the thermal conductivity of the screw *Computational Continuum Mechanics* **8 (3)** 329–339
[5] Syrjälä S 1999 On the analysis of fluid flow and heat transfer in the melt conveying section of a single-screw extruder *Numerical Heat Transfer* Part **A 35** 25–47
[6] Tadmor Z, Gogos C 1984 *Theoretical Foundations of polymer processing*. (Moscow: Chemistry) pp 96–159
[7] Malkin A Y and Isaev A I 2007 *Rheology: concepts, methods, and applications*. (St. Petersburg: Profession) p 184
[8] Carreau P J 1972 Rheological Equations from molecular network Theories *Transactions of the society of Rheology* **16** 99–127

IOP Conf. Series: Materials Science and Engineering **208** (2017) 012019 doi:10.1088/1757-899X/208/1/012019

The numerical study of the coextrusion process of polymer melts in the cable head

M V Kozitsyna and N M Trufanova

Perm National Research Polytechnic University, 29, Komsomolsky ave., Perm, 614990, Russia

E-mail: Bachurinamaria@gmail.com

Abstract. The process of coextrusion consists in a simultaneous creation of all necessary insulating layers of different polymers in the channel of a special forming tool. The main focus of this study is the analysis of technological, geometrical and rheological characteristics on the values of the layer's thickness. In this paper are considered three geometries of cable head on the three–dimensional and two–dimensional representation. The mathematical models of separate and joint flow of polymer melts have been implemented by the finite element method in Ansys software package. The velocity fields, temperature, pressure in the cross–sections of the channel and by the length have been obtained. The influence of some thickness characteristics of insulation layers has been identified.

1. Introduction

Today the coextrusion is the most effective method for the manufacture of cable products composed of two or more layers of polymer insulation, the result of which is the high–quality multilayer coating. In this case, the melts of polymers came from different extruders to the cable head, thereby reduced the time of the process, the costs of production, at the same time the quality of the products is increased. In the research papers devoted to coextrusion, the main attention is the influence of technological and rheological parameters on the character of the joint flow of polymer melts [1–4]. However the influence of geometrical parameters isn't investigated enough, because a conical–cylindrical channel of the cable head was considered without adapter [5–6]. The main aim of this paper is to study the stratified flow of two anomalously viscous fluids in the channel of the extrusion head, which has various spatial geometries, and it is presented in Figure. 1. At the same time the paper focuses on the evaluation of the effect of rheological characteristics on fields of the temperature, pressure and velocity materials and technological characteristics on thicknesses of the layers.

2. Mathematical statement and initial data

In this study, we consider three geometries of a cable head in three–dimensional and two–dimensional representations with channels of different length. Figure 1 shows the configuration of channels of the forming tool, (a) – is a complete three–dimensional model of the cable head with an adapter, the total length is 590 mm; (b) – is the full model, in the flat view, without the adapter, the length of the model is 308 mm; (c), (d) – are truncated models in a three–dimensional and flat view, the lengths of the model are 190 mm; (e), (f) – are truncated models in a three–dimensional and flat view, the lengths of the model are 104 mm. In the following analysis the short notation for models (a–e) are used. The streams of polymers moving along the channels 1–3 create multilayer covering in the form of 1–screen; 2–insulation; 3–screen. The obtained cable product is shown in Figure 2.

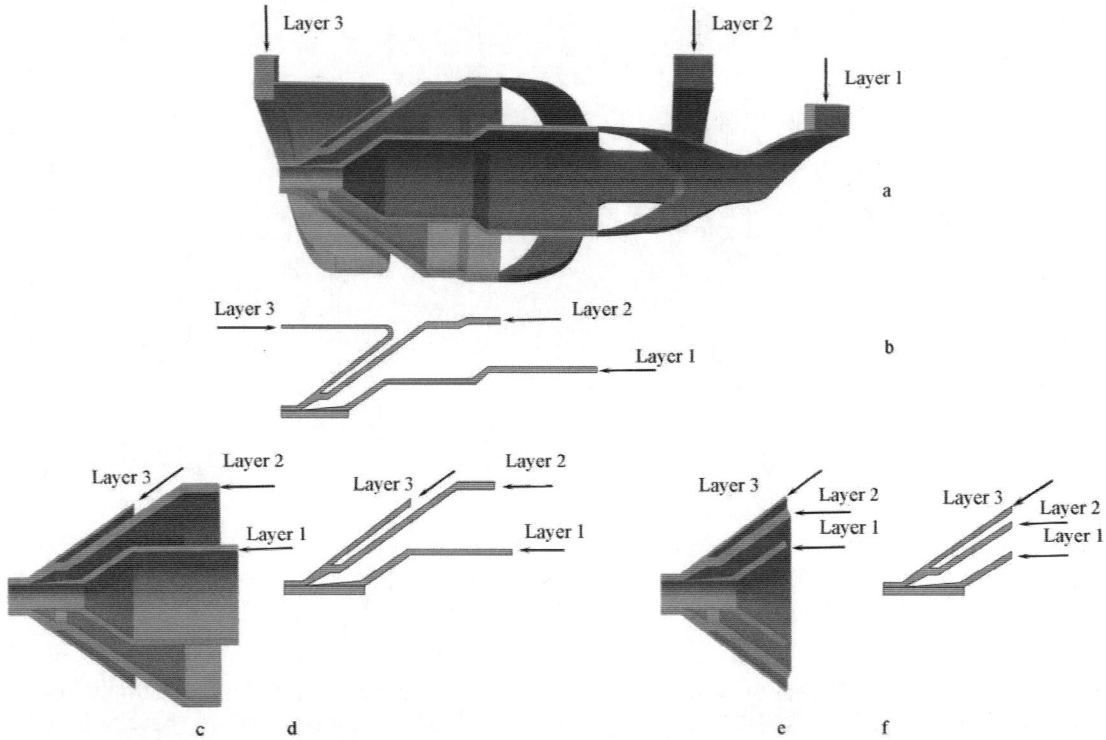

Figure 1. The configurations of cable head of the models (a) (e).

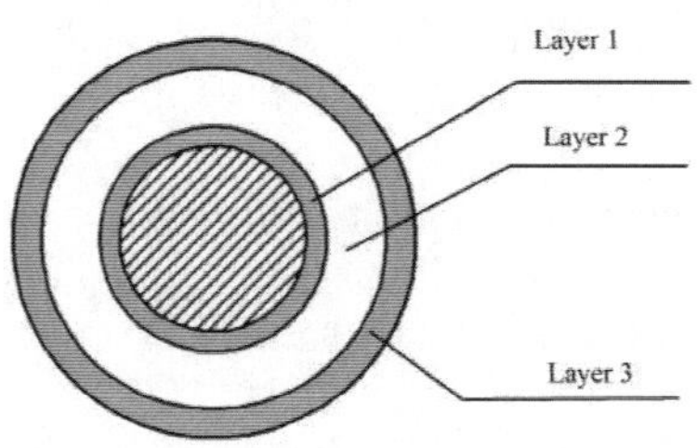

Figure 2. The configuration of cable.

The mathematical assumptions of heat and mass transfer processes in stratified flow the polymer melts are based on the conservation equations. To make the analysis feasible, the process has to be simplified, so thus the following assumptions are made:
– the process is stationary;
– environment is incompressible and there are no elastic properties [7];
– surface forces exceed the mass forces;
– are axisymmetric properties of the flow;
– slip and impermeability conditions are defined at the channel boundaries;
– thermophysical characteristics are constant;

As a result, the mathematical model is a following set of differential equations [8], where each one relates to every layer in the flow:

$$\frac{\partial(\rho_m V_x)}{\partial x}+\frac{\partial(\rho_m V_y)}{\partial y}+\frac{\partial(\rho_m V_z)}{\partial z}=0 \qquad (1)$$

$$\rho_m\left(V_x\frac{\partial V_x}{\partial x}+V_y\frac{\partial V_x}{\partial y}+V_z\frac{\partial V_x}{\partial z}\right)=-\frac{\partial P}{\partial x}+\frac{\partial \tau_{xx}}{\partial x}+\frac{\partial \tau_{xy}}{\partial y}+\frac{\partial \tau_{xz}}{\partial z} \quad (2)$$

$$\rho_m\left(V_x\frac{\partial V_y}{\partial x}+V_y\frac{\partial V_y}{\partial y}+V_z\frac{\partial V_y}{\partial z}\right)=-\frac{\partial P}{\partial y}+\frac{\partial \tau_{yx}}{\partial x}+\frac{\partial \tau_{yy}}{\partial y}+\frac{\partial \tau_{yz}}{\partial z} \quad (3)$$

$$\rho_m\left(V_x\frac{\partial V_z}{\partial x}+V_y\frac{\partial V_z}{\partial y}+V_z\frac{\partial V_z}{\partial z}\right)=-\frac{\partial P}{\partial z}+\frac{\partial \tau_{zx}}{\partial x}+\frac{\partial \tau_{zy}}{\partial y}+\frac{\partial \tau_{zz}}{\partial z} \quad (4)$$

$$\rho_m C_m\left(V_x\frac{\partial T}{\partial x}+V_y\frac{\partial T}{\partial y}+V_z\frac{\partial T}{\partial z}\right)=\frac{\partial}{\partial x}\left(\lambda_m\frac{\partial T}{\partial x}\right)+\frac{\partial}{\partial y}\left(\lambda_m\frac{\partial T}{\partial x}\right)+\tau_{xx}\frac{\partial V_x}{\partial x}+$$
$$+\tau_{yy}\frac{\partial V_y}{\partial y}+\tau_{zz}\frac{\partial V_z}{\partial z}+\tau_{xy}\left(\frac{\partial V_x}{\partial y}+\frac{\partial V_y}{\partial x}\right)+\tau_{yz}\left(\frac{\partial V_y}{\partial z}+\frac{\partial V_z}{\partial y}\right)+\tau_{zx}\left(\frac{\partial V_z}{\partial x}+\frac{\partial V_x}{\partial z}\right) \quad (5)$$

$$\tau_{xx}=2\mu_э\frac{\partial V_x}{\partial x},\tau_{yy}=2\mu_э\frac{\partial V_y}{\partial y},\tau_{zz}=2\mu_э\frac{\partial V_z}{\partial z},$$
$$\tau_{yz}=\tau_{zy}=\mu_э\left(\frac{\partial V_y}{\partial z}+\frac{\partial V_z}{\partial y}\right),\ \tau_{xy}=\tau_{yx}=\mu_э\left(\frac{\partial V_x}{\partial y}+\frac{\partial V_y}{\partial x}\right),\ \tau_{zx}=\tau_{xz}=\mu_э\left(\frac{\partial V_z}{\partial x}+\frac{\partial V_x}{\partial z}\right) \quad (6)$$

$$\mu_э=\mu_0\exp(-\beta(T-T_0))\left(\frac{I_2}{2}\right)^{\frac{(n-1)}{2}} \quad (7)$$

Where P – is a pressure; T– is a temperature; ρ – is a density; C – is a heat capacity; λ –is a thermal conductivity; $\mu_э$ – is an effective viscosity, μ_0– is the consistency coefficient at temperature T_0; V_x, V_y, V_z– the components of the velocity vector; I_2 – quadratic invariant of the tensor of the deformation rates; β – the temperature coefficient of viscosity; n– an anomaly of viscosity; τ_{xx}, τ_{xy}, τ_{xz}, τ_{yx}, τ_{yy}, τ_{yz}, τ_{zx}, τ_{zy}, τ_{zz} – are the normal and tangential stresses [2].

The following boundary conditions are applied on the set of differential equations from (1) to (7):
– Velocity components on fixed walls are equal to zero;
– At the contact boundary of moving conductor, the longitudinal component of velocity is equal to velocity of conductor (0,4m/s);
– The velocity profile is defined (calculated from flow rate) and set at the inlet of the channels;
– The boundary condition of second type of speed and temperature is defined at the output;
– The stationary walls temperature of the channel is 170 °C;
– The temperature of polymer melts at the extruder exit is 150 °C;
– The temperature of the moving conductor is 110 °C;
–At the interfaces of flows polymers is equality of velocities and surface forces;
The properties of polymer liquids are presented in Table 1.

Table 1. The rheological and thermophysical characteristics of the materials

Material	consistency factor μ_0 at T=443 K, Pa	Temperature coefficient of viscosity в, 1/K	Anomaly index, n	Density c, kg/m3	Coefficient of thermal conductivity λ, W / (m K)	Specific heat capacity C, J / (kg K)
Channels material–1,3 Borealis LE0540	38523	0.0027	0.251	1080	0.182	2500
Channels material–2 Borealis LE4421M	14946	0.0168	0.542	779	0.182	2500

The system of differential equations (1) – (7), supplemented boundary conditions was solved numerically, by the finite element method in the ANSYS software package. The convergence of the solution of the problem with respect to the number of countable iterations and the number of elements in the model has been made.

Table 2. The nominal thickness.

Number of channels	1	2	3
Nominal thickness, mmc	0.6	3.6	0.7

3. Results and Discussion

The first part of the study focuses on determination of the cable head adapter effect on the heat and mass transfer processes in the forming tool and on the thickness of the resulting multilayer coating. In all the variants of the calculation in the channels the flow rates were set like that the thicknesses of the coatings were created equal to the nominal values. They value are given in Table 3. According to the results presented in Table 4, the values of the thickness layers in models are different : in models (a) the value of thickness completely are consistent nominal significances, by the way in the case by models (b) and (d) the maximal difference is equal 14%. At the same time, for the models (b), (d), (e), this difference is 33%. Thus, the results obtained for the plane models aren't adequate, enough. This makes it difficult to use them in modeling the real processes.

Table 3. The value of flow rates in channels

Number of channels	1	2	3
Flow rate (Q), kg/c	0.021	0.09	0.017

Table 4. The results of a numerical study for models (a–f)

	Model (a)			Model (b)		
Number of channel	1	2	3	1	2	3
Layer thickness, mm	0.6	3.6	0.7	0.7	3.3	0.9
Maximum temperature, °C		206			188	
Average temperature in the channels, °C		168			157	
Maximum speed, m / s		2.2			1.3	
Pressure drop, Pa	67	53	47	35	12	30
	Model (c)			Model (d)		
Number of channel	1	2	3	1	2	3
Layer thickness, mm	0.7	3.5	0.7	0.7	3.3	0.9
Maximum temperature, °C		196			180	
Average temperature in the channels, °C		155			154	
Maximum speed, m / s		1.5			1.27	
Pressure drop, Pa	35	28	22	30	10	13
	Model (e)			Model (f)		
Number of channel	1	2	3	1	2	3
Layer thickness, mm	0.6	3.5	0.8	0.7	3.4	0.8
Maximum temperature, °C		191			174	
Average temperature in the channels, °C		154			152	
Maximum speed, m / s		1.3			1.2	
Pressure drop, Pa	30	11	18	26	7	14

Figure 3 shows the velocity fields of materials for models (a–e). The maximum speed is developed in the narrowest part of channel 1 in the model (a). Its value is 5 times higher, than the speed of the conductor's movement. The maximum difference for axisymmetric and flat models is two times to model (a) and (e); in 1.6 times by model (a) and (b), it's related with into account the effect braking polymer near of the walls. Comparison of the velocities for axisymmetric models (a) (c) and (e) was showed the maximum difference in the maximum speed to 1.6 times for models (a) and (e), which is determined by the different length of the channels. The average speed for axisymmetric models (a, c, e) is approximately 0.75 m/s, for flat models this parameter is 0.62 m/s, the difference is 16%, consequently.

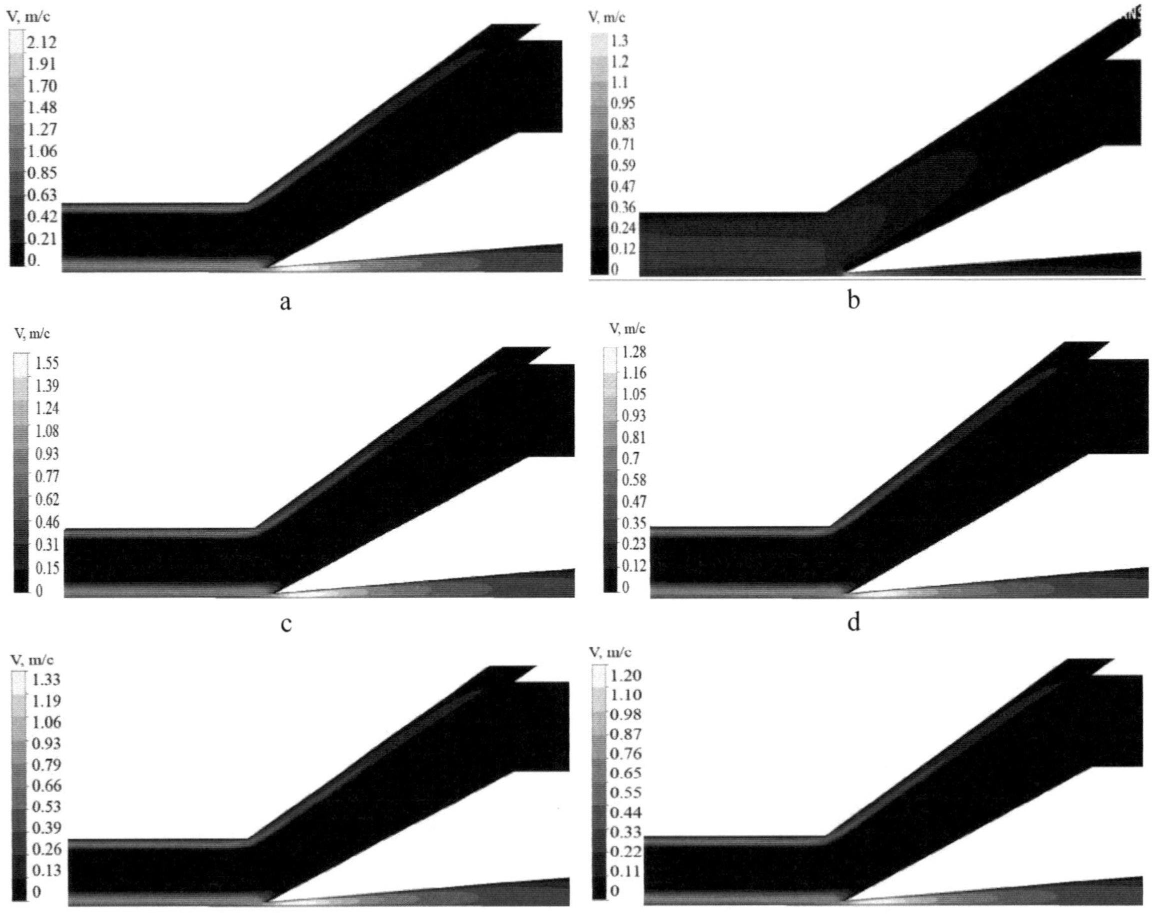

Figure 3. The fields of speeds material for model (a–f).

The images presented in Figure 4 show the steady heating of the material. Nevertheless, it is important to note that the contribution of the forming tool adapter is well illustrated in the analysis of maximum temperatures. The highest temperature is achieved in model (a) and it is equal to 206 °C, while at the same time for model (e), this value was 191 °C. As a result, an increase in the temperature of the materials by 7% is due to contribution of the cable head adapter. Under the similar comparison for models (b) and (f), the difference was 7%. At the same time, while comparing the maximum temperatures in the models (a) – (b), (c) – (d), (e) – (f), the average difference was 8%. It should be noted that for the model (f) the maximum temperature is close to the temperature of the supplied melt, while for model (a) it is 18% higher. This difference comes from the contribution of viscous dissipation, which is well observed for model (a), where the length is almost 6 times higher than in model (f).

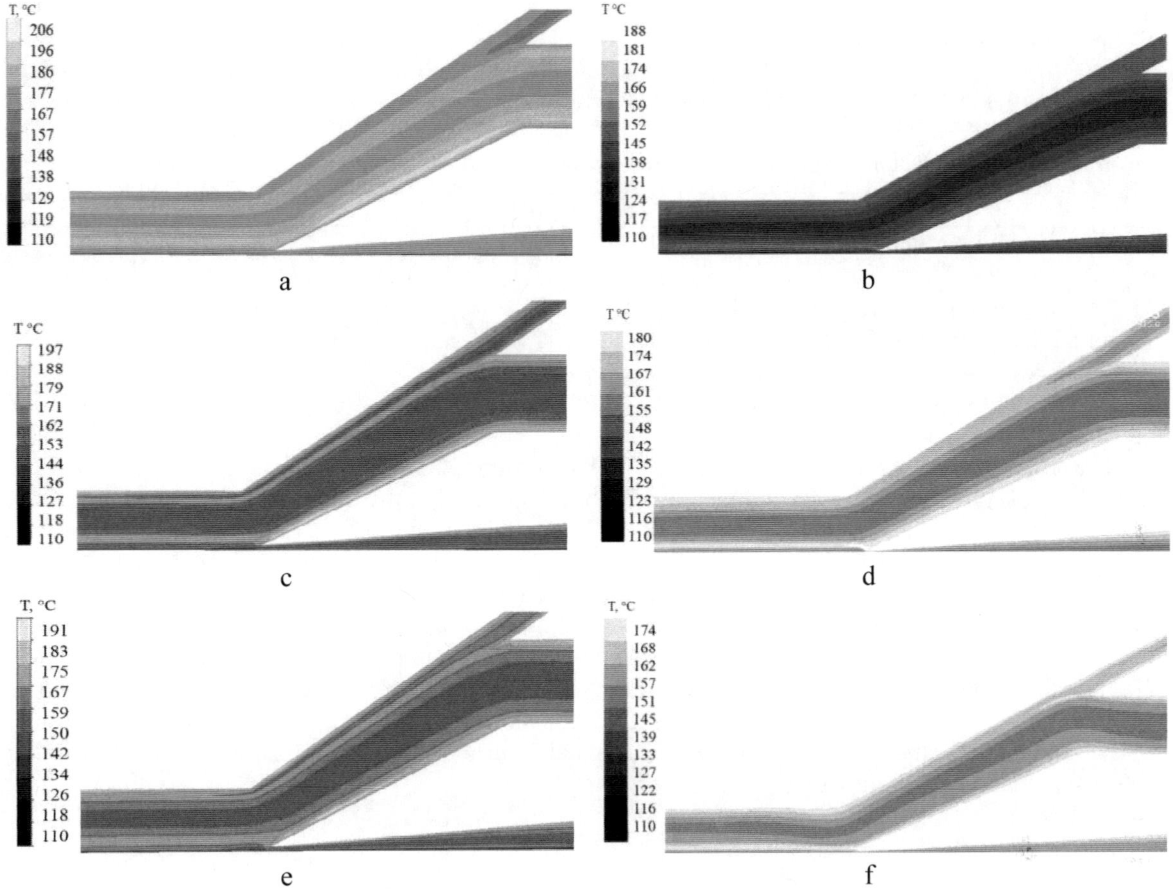

Figure 4. The temperature fields in the channels of the forming tool for models (a) – (e).

Since in the polymers processing the rheological properties of the same material can be different significantly, from batch to batch, that the second part of this study focuses on the evaluation of the effect of rheological properties, viscosity in this case, on the thickness of insulation layers. The case was considered when the material of layer 2, was replaced by a material with a viscosity less than twice, at constant flow rates and linear velocity of the conductor. The results for model (a) – (b) are presented in Table 5. In these parts was observed a slight change in the temperature and velocity of the material. However, a significant change was observed for the values of the pressure drop in layer 2. For the model (a), the value decreased by 47% and 41% for model (b), due to the decrease in the viscosity of the material of the 2 layer, the contribution of viscous dissipation decreased, respectively, the polymer material moves along the channel with less force.

Table 5. The results of simulation for layer 2 with viscosity μ_0= 6642 Pa

	Model (a)			Model (b)		
Number of channel	1	2	3	1	2	3
Layer thickness, mm	0.6	3.4	0,9	0.8	3.2	0.9
Maximum temperature, °C		202			185	
Average temperature in the channels, °C		162			156	
Maximum speed, m / s		1.98			1.3	
Pressure drop, Pa	65	28	36	35	7	28

In the second case, when the viscosity of layers 1 and 3 increased 2 times, the results are presented in Table 6. The maximum temperature is increased in model (a) by 13%, and by 4% for model (b), as well as an increase of value pressure drop in layer 1 by 48% and in layer 3 by 30% for model (a), in layer 1 by 50% and in layer 3 by 15% for the model (b).

Table 6. The results of simulation for layer 2 with viscosity1,3 μ_0=74730 Pa

	Model (a)			Model (b)		
Number of channel	1	2	3	1	2	3
Layer thickness, mm	0.6	3.4	0.9	0.6	3.5	0.8
Maximum temperature, °C		236			195	
Average temperature in the channels, °C		154.9			152	
Maximum speed, m / s		2			1.2	
Pressure drop, Pa	128	53	68	71	45	35

The final part of this work focuses on the determination of the influence of technological parameters, such as the flow rate and the velocity of the material to the thickness of the applied layers.

Data dependence the thickness of the layers and mass flow are presented in Figure 5. After the flow rate in channel 1 is increased at 5 times (Figure 5.a), the thickness of layer 1 is increased at 3 times, and the thickness of layer 2 and 3 is decreased accordingly by 30% and 14%. Then the flow rate in channel 2 is increased at 2.5 times, the layer 1 was thinner by 33%, layer 2 was thickened 1.1 times, layer 3 was thinned 57% (Figure 5.b). It the last case the flow rate in channel 3 is increased at 5 times, (Figure 5.c), the thickness of layer 1 isn't changed, layer 2 is decreased by 27%, and layer 3 is increased by 2.4 times. The results are presented for model (a) and (b). The analysis of the results allows to conclude that the effect of flow in channel 3 on the thickness of the layers formed in the other two channels is minimized.

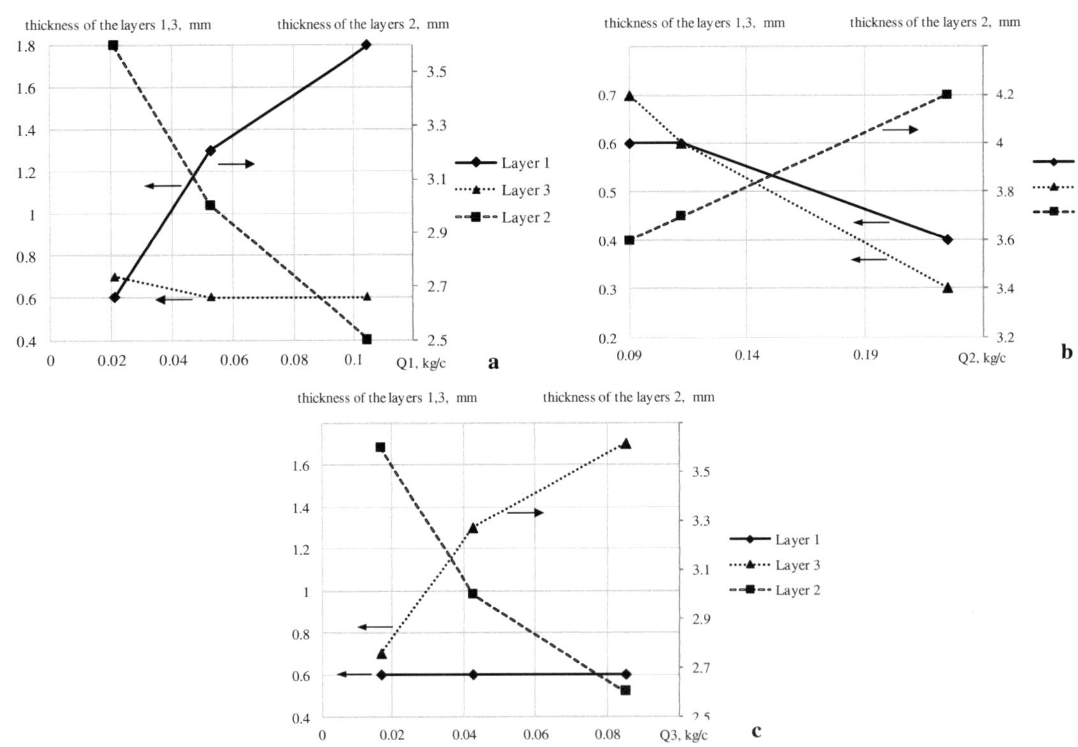

Figure 5. The influence the flow rates on thicknesses of layer.

In industry, is the question of the possibility increasing the production capacity, therefore, the cases of increasing the moving velocity were considered, and it was necessary to achieve constant values of the nominal thicknesses (Table 2). Figure 6 shows the influence of the linear velocity on the thickness of the layers. We see that the required result is possible at the velocity of the moving conductor is increasing to 0.6 m/s. If the velocity of moving conductor increases by 2.25 times, the thickness of layer 1 is decreased by 1.2 times, the thickness of layer 2 is increased by 2%, and layer 3 is not changed.

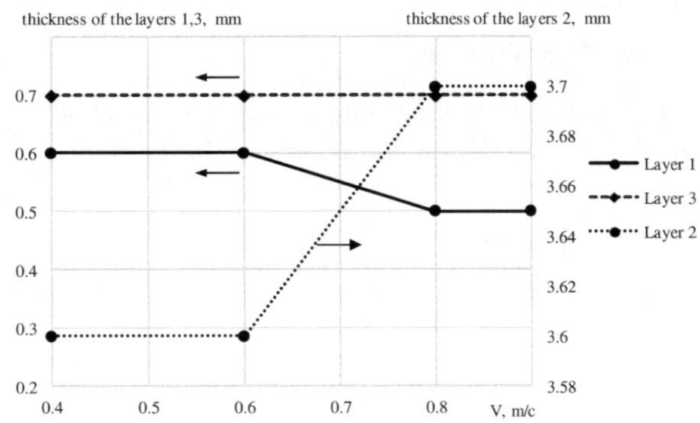

Figure 6. The influence of the linear velocity on the thickness of the layers.

4. Conclusion

The results of this study can be applied in the process of analyzing multilayers flows of polymer materials in a forming tool [6]. The main attention should be directed to the contribution of the cable head adapter, the analysis of the results allows to consider the significant contribution of the latter to the processes of heat and mass transfer of anomalous liquids. The obtained data indicates insufficient adequacy of the results of flat moles and, as a result it limits the application. The difference on field's temperatures and velocities has not exceeded 8%. However, it is necessary to take into account, that the change in the value of the pressure drop in the channel of the forming tool can reach 50%, when changing from one material to another, and all other conditions being equal.

References

[1] Han C D and Chin H B 1979 *Polymer engineering and science* **19 (16)** 1156
[2] Khan A A and Han C D 1977 *Trans. Soc. Rheol.* **21** 101
[3] Martyn M T, Spares R, Coates P D and Zatloukal M 2004 *Plastics, Rubbers and Composites* **33 (1)** 27
[4] Martyn M T, Spares R, Coates P D and Zatloukal M 2009 *J. Non–Newtonian Fluid Mech.* **156** 150
[5] Bachurina M V, Kazakov A V and Trufanova N M 2015 *Computational continuum mechanics* **8 (3)** 299
[6] Kazakov A V and Trufanova M V 2012 *Electrical Engineering* **83 (11)** 641
[7] Tadmor Z and Gogos C 1984 *Theoretical Foundations of polymer processing* (Moscow: Chemistry) pp 96–178
[8] Michaeli W 2007 *Extrusion dies for plastics and rubber* (St Petersburg: Profession) pp 91–151

IOP Conf. Series: Materials Science and Engineering **208** (2017) 012020 doi:10.1088/1757-899X/208/1/012020

Numerical modeling of mechanical behavior of multilayered composite plates with defects under static loading

V V Korepanov[a], **G S Serovaev**

Institute of Continuous Media Mechanics UB RAS, 1, Acad. Korolev St., 614013, Perm, Russia

[a]E-mail: kvv@icmm.ru

Abstract. Evaluation of the mechanical state of a structure or its components in the process of operation based on detection of internal damages (damage detection) becomes especially important in such rapidly developing spheres of production as machine building, aerospace industry, etc. One of the most important features of these industries is the application of new types of materials among which polymer based composite materials occupy a significant position. Hence, they must have sufficient operational rigidity and strength. However, defects of various kinds may arise during the manufacture. Delamination is the most common defect in structures made from composite materials and represents a phenomenon that involves the complex fracture of layers and interlayer compounds. Among the reasons of delamination occurrence are: disposition of anti-adhesive lubricants, films; insufficient content of binder, high content of volatile elements; violation of the molding regime; poor quality of anti-adhesive coating on the surface of the tooling. One of the effective methods for analyzing the influence of defects is numerical simulation. With the help of numerical methods, it is possible to track the evolution of various parameters when the defect size and quantity change. In the paper, a multilayered plate of an equally resistant carbon fiber reinforced plastic was considered, with a thickness of each layer equal to 0.2 mm. Various static loading cases are studied: uniaxial tension, three and four-point bending. For each type of loading, a numerical calculation of the stress-strain state was performed for healthy and delaminated plates, with different number and size of the defects. Contact interaction between adjacent surfaces in the zone of delamination was taken into account.

1. Introduction

Evaluation of the mechanical state of a structure or its components in the process of operation based on detection of internal damages (damage detection) becomes especially important in such rapidly developing spheres of production as machine building, aerospace industry, etc. One of the most important features of these industries is the application of new types of materials among which polymer based composite materials occupy a significant position. Hence, they must have sufficient operational rigidity and strength. However, defects of various kinds may arise during the manufacture. To solve such problems, complex theoretical and experimental studies aimed at finding the reasons for their occurrence and investigating the nature of mechanical processes, both in structural elements and in material under possible operational loads are needed.

A layered composite material with different options of microstructure that takes into account the presence of a certain type of defects is considered in this paper. Delamination is the most common among them and represents a phenomenon that involves the complex fracture of layers and interlayer

IOP Conf. Series: Materials Science and Engineering **208** (2017) 012020 doi:10.1088/1757-899X/208/1/012020

compounds. The appearance of a delamination leads to the fact that in the region of separation the object is divided into several parts, which begin to absorb the loads independently of each other. In this case, the stiffness of each of the parts is much lower than the stiffness of the object working as a single unit [1]. Among the reasons of delamination occurrence are: disposition of anti-adhesive lubricants, films; insufficient content of binder, high content of volatile elements; violation of the molding regime (high temperature, high cooling rate, unregulated thermal or mechanical impact); poor quality of anti-adhesive coating on the surface of the tooling.

The appearance of delamination can lead to a loss of the bearing capacity of the structure, its breakdown and decommissioning, which makes the task of detecting this defect at an early stage of development highly relevant.

In order to simulate delamination processes, layered material in [2] is modeled as a laminate stacking sequence of homogeneous layers and interlayer interfaces. This simulation is used for predicting the start of delamination and its growth in the case of static loads without buckling. Development of experimental methods for determining the sustainability to delamination and crack resistance of reinforcing fibers and polymer matrix of composites is an active area of research. [3] provides an overview on the standardization of test methods. The latest researches leading to new standardized test methods, complementing and updating earlier reviews are presented. The causes of delamination and its effect on the structural characteristics, as well as, analytical and experimental methods for delamination detection are discussed in [4]. Delamination in layered composite plates under compression loads is analyzed in [5]. The method includes an analytical approach based on fracture mechanics and numerical analysis, assuming that the adhesive layers may be modeled using elastic springs with a finite tensile strength. Analytical and numerical results are compared and discussed. In [6], uniaxial loading tests of laminates with large number of layers which are prone to delamination were performed. The onset of delamination is based on acoustic emission.

Since the use of visual inspection methods is difficult due to the fact that delamination is an internal defect, diagnostic methods aimed at obtaining information about the appearance of delamination type defect and its propagation are of great importance [7-10]. A number of issues need to be answered before the organization of the monitoring of the mechanical behavior of the structure. And first of all what are the mechanical characteristics of the controlled object that can be measured and are most influenced by the appearance of a defect. An effective method for solving this problem is mathematical modeling. The analysis of stress-strain state change of the simulated object under static loading and the appearance and development of delamination type defect in composite materials is studied in this paper.

2. Modeling of defects in composite multilayered plates

Generally, a numerical study of the response of static parameters to the presence of a defect of different sizes can be considered for three models of delamination in a multilayered composite structure.

Free mode model in which the adjacent volumes, simulating the delamination, are not connected (and thus allowing their interpenetration). The free model allows interpenetration of the surfaces, therefore surface S_1 and S_2 are free from strains.

Constrained model in which the matching nodes in the delamination zone are bound by the same component of displacement. For the constrained model the matching nodes, belonging to surfaces S_1 and S_2, have the same displacement in the z direction, but in the other directions the displacements are independent.

A model, taking into account contact interaction in delamination zone.

The following states of contact are considered in simulation:

1) the state of open contact (Figure 1) when the surfaces are separated;

2) the state of closed contact (Figure 2). In this case, the surfaces contact with each other and their relative behavior is determined by the contact stiffness, which is equal to the elastic modulus of the material multiplied by the size of the contact element adjacent to the surface.

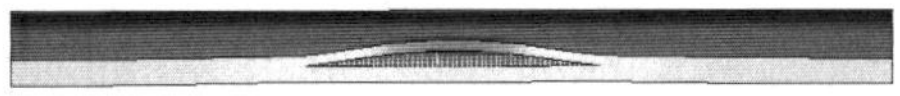

Figure 1. Contact state: open contact.

Figure 2. Contact state: closed contact.

3. Numerical calculation of the composite multilayered plate with defects under static loading

As the object of the study a 15-layer plate of equal strength CFRP UWC-39, with the thickness $h1c$=0.2 mm of each layer was considered. Each layer of the plate was simulated within a three-dimensional model of elasticity theory with orthotropic effective characteristics for each layer. The effective characteristics for each layer were made equal and are shown in Table 1.

The following variants of the arrangement of defects at the boundary of the layers are considered: one central defect (Figure 3, Figure 5); two symmetrically (with respect to the Y axis) located defects (Figure 4, Figure 6). The number of delaminations between the layers along the thickness of the plate was set from one to six, as schematically shown in Figure 5 and Figure 6, the linear size of the defects Ld varied from 10 to 40 mm.

Table 1. Orthotropic effective characteristics of composite

Material	E_{xx}, GPa	E_{yy}, GPa	E_{zz}, GPa	G_{xy}, GPa	G_{xz}, GPa	G_{yz}, GPa	ν_{xy}	ν_{xz}	ν_{yz}
UWC-39	58.0	58.0	10.0	4.5	3.0	3.0	0.05	0.3	0.18

Elements with quadratic displacement approximation were used to create a finite element model of the plate. The construction of the finite element model was performed in ANSYS.

The virtual work principle is used for mathematical formulation of the problem

$$\int_V \sigma_{ij}\delta\varepsilon_{ij}dV = \int_{S_\sigma} F_i\delta u_i dS \qquad (1)$$

u_i - components of the displacement vector, $\varepsilon_{ij} = \frac{1}{2}(\frac{\partial u_i}{\partial x_j} + \frac{\partial u_j}{\partial x_i})$ - components of strain tensor, $\sigma_{ij} = C_{ijkl}\varepsilon_{kl}$ - components of stress tensor, C_{ijkl} - stiffness tensor, F_i - components of the body forces vector.

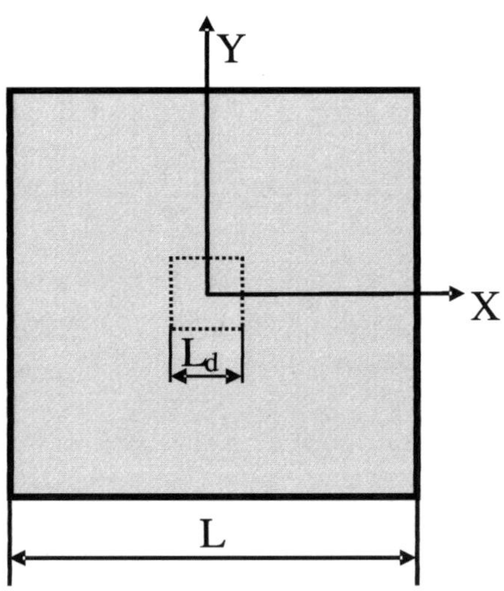

Figure 3. Plate with one central defect.

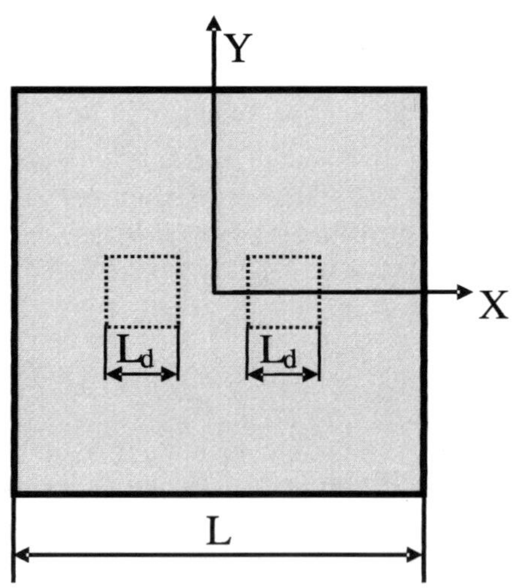

Figure 4. Plate with two defects.

IOP Conf. Series: Materials Science and Engineering **208** (2017) 012020 doi:10.1088/1757-899X/208/1/012020

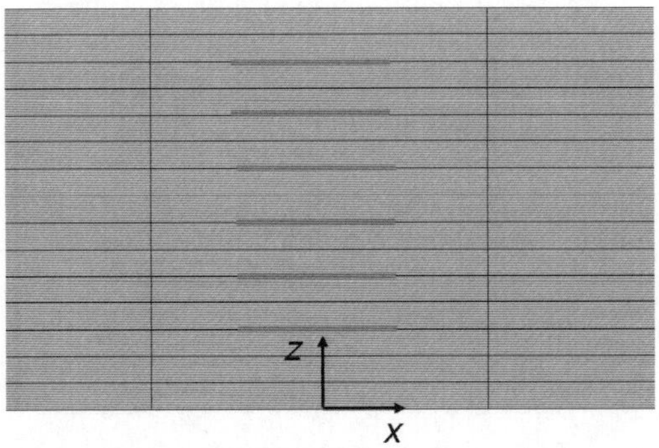

Figure 5. The scheme of the arrangement of the delaminations along the thickness of the plate, corresponding to Figure 3.

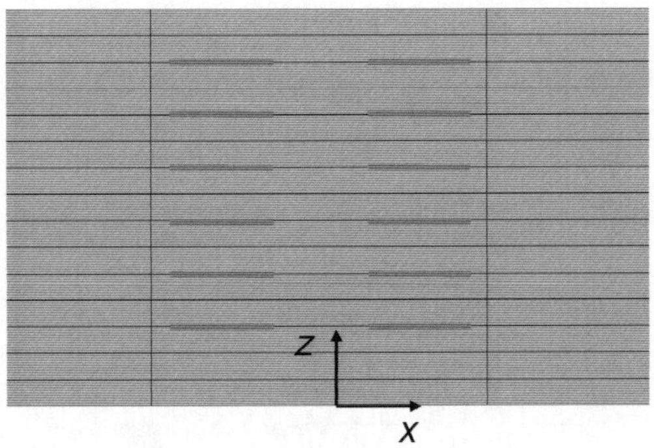

Figure 6. The scheme of the arrangement of the delaminations along the thickness of the plate, corresponding to Figure 4.

The following load cases are considered:
- uniaxial tension;
- 3 point bending (Figure 7);
- 4 point bending (Figure 8).

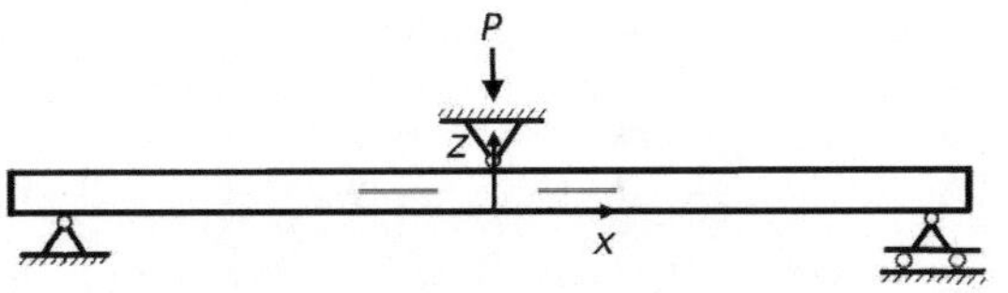

Figure 7. 3 point bending scheme.

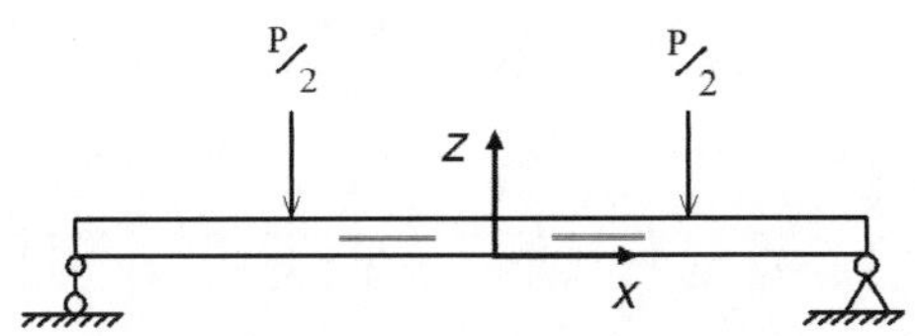

Figure 8. 4 point bending scheme.

For each of the considered loading schemes, the change in the parameters of the stress-strain state as a function of the number of defects and their dimensions in a multilayer plate was analyzed. The corresponding components of stress and strain tensors can be chosen as such parameters. For the problem of uniaxial tension of a plate, these parameters are longitudinal strain and stress, for 3 and 4 point bending problems, tensile strain and stress are examined. The sensitivity of the selected parameters will be estimated by comparing the results on a homogeneous plate and a plate with defects.

The results of the calculations showed that the stress-strain state remains practically unchanged for uniaxial tension and four-point bending (the difference in the respective strain and stress components in the defect zones is no more than 2%). The three-point bending is of the greatest interest from the strain response point of view (Figure 7).

The dependences of the tensile strain ε_{xx} on the surface along the Y=0 line for a homogeneous plate and a plate with defects for one and 6 delaminations along the thickness of the plate with one and two defects in the layer (Figure 3 and Figure 4) and for different defect sizes are shown in Figures 9-11. Six delaminations along the thickness of the plate were set as shown in Figure 5 and Figure 6, and one delamination over the plate thickness was set in the section $Z = h/2$, where h is the thickness of the plate.

Dependences of the tensile strain ε_{xx} on the surface of the plate at the point (*X*=0, *Y*=0), depending on the size of the delamination *Ld* for the case of one and six delaminations with one and two defects in the layer are represented in Figure 12.

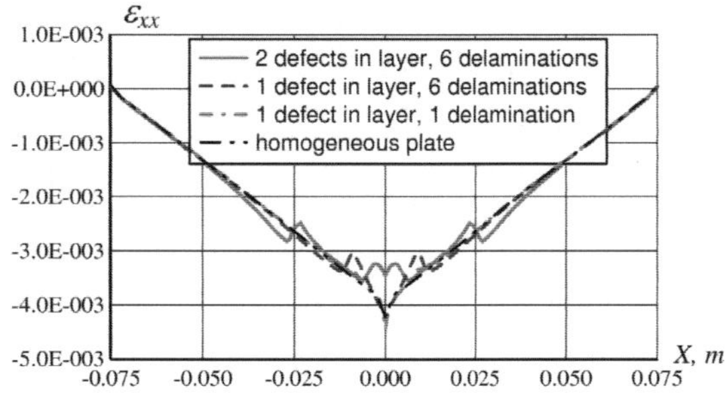

Figure 9. Tensile strain distribution along the *Y*=0 line, *Ld*=20 mm.

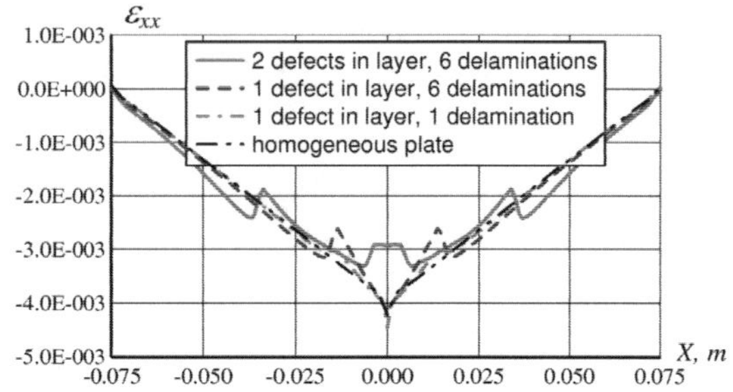

Figure 10. Tensile strain distribution along the *Y*=0 line, *Ld*=30 mm.

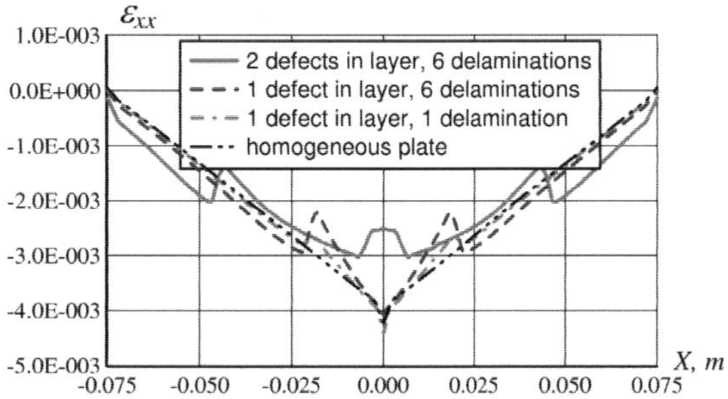

Figure 11. Distribution of tensile strain along the *Y*=0 line, *Ld*=40 mm.

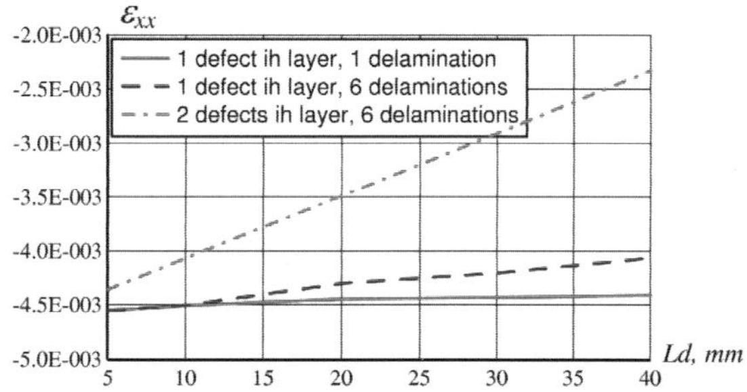

Figure 12. Dependences of the tensile strain on the size of the delamination *Ld*.

The analysis of the obtained numerical results showed that when measuring the strain fields with the purpose of recording the development of delaminations in composite materials, it is necessary to locate sensors in the zone of 3 or 4 sizes of the expected defect. Therefore, in this case, two options are possible:

- In the first variant it is necessary to provide a dense sensor arrangement in the volume of the monitored defect;
- In the second variant, the sensors are located in the zones of stress concentration, where defects are expected to appear.

Zones of stress concentration can be determined for a specific construction using methods of mathematical modeling.

Acknowledgments
This research was done at the expense of the Basic Research Complex program of Ural Branch of RAS No 10 (grant № 15-10-1-18).

4. References

[1] Zhang Z., Shankar K., Ray T., Morozov E.V., Tahtali M. 2013 *Composite Structures* **102**
[2] Allix O., Ladeveze P. 1992 *Composite Structures* **22**
[3] A.J. Brunner, B.R.K. Blackman, P. Davies 2008 *Engineering Fracture Mechanics* **75**
[4] Amar C. Garg 1988 *Engineering Frachue Mechanics* **29** 5
[5] D. Bruno 1988 *Theoretical and Applied Fracture Mechanics* **9**
[6] R.Y. Kim, S.R. Soni. 1984 *Journal of Composite Materials* **18**

[7] Stepinski T., Uhl T., Staszewski W. 2013 *Advanced Structural Damage Detection: From Theory to Engineering Applications* (John Wiley & Sons)
[8] Adams D.E. 2007 *Health Monitoring of Structural Materials and Components* (John Wiley & Sons)
[9] Morassi A., Vestroni F. 2008 *Dynamic methods for damage detection in structures* (NewYork: Springer)
[10] Zou Y., Tong L., Steven G.P. 2000 *Journal of Sound and Vibration* **230** 2

IOP Conf. Series: Materials Science and Engineering **208** (2017) 012021 doi:10.1088/1757-899X/208/1/012021

Some models of viscous gas dynamics

I I Lipatov

Central Aerohydrodynamic Institute,1, Zhukovsky St., Zhukovsky, Moscow reg., 140180, Russia

E-mail: igor_lipatov@mail.ru

Abstract. In this paper, the problems containing two or more small parameters are investigated. In the analysis of such problems, a diagrammatic method is used to visualize the structure of perturbed flow regions, and to estimate the similarity parameters and mathematical models depending on the limiting transitions. The interaction between the laminar flows and transonic flows is considered in the context of the problem containing a small parameter inverse to the Reynolds number, the Mach number other than 1 and the pressure perturbation amplitude of the initiating viscous-inviscid interaction processes. The results of the flow analysis in a laminar boundary layer under the assumption of discontinuity of boundary conditions, for example, discontinuity of the surface velocity, are presented. For such problems a diagram of possible embedded perturbed flow regions is plotted. In addition, the results of studying the nonstationary processes of interaction between the flow in the boundary layer and the external supersonic flow are discussed.

1. Introduction

It is assumed that at a distance l from the leading edge of a plane surface there is a region of local interactions initiated by a certain perturbation, for example, an obstacle on the surface or a shock wave. It is assumed that the origin of the Cartesian coordinate system coincides with the leading edge of the plate subject to a transonic gas flow. The notation adopted for time, Cartesian coordinates, components of the velocity vector, pressure, density, and dynamic viscosity coefficient, is as follows

$$tl/u_\infty, xl, yl, uu_\infty, vu_\infty, p\rho_\infty u_\infty^2, \rho\rho_\infty, \mu\mu_\infty$$

Despite the fact that the Reynolds number $\mathrm{Re} = \rho_\infty u_\infty l / \mu_\infty$ is supposed to be large, a laminar flow is considered. This suggestion is justified for the analysis of unsteady problems, including linear stability problems.

2. Methodology

In addition to a small parameter, characterizing the thickness of the boundary layer $\varepsilon = \mathrm{Re}^{-1/2}$, the problem contains another two small parameters $\beta = \sqrt{(M^2 - 1)}$ and the amplitude of the pressure perturbation Δp. In principle, the perturbation can also be characterized by other parameters, such as the characteristic spatial and temporal scales, because, as for a turbulence of the external flow, the scale and spectral composition of the perturbations are of considerable significance. In what follows,

IOP Conf. Series: Materials Science and Engineering **208** (2017) 012021 doi:10.1088/1757-899X/208/1/012021

the discussion will be limited to the analysis of regimes, for which the viscous-inviscid interaction processes are important.

3. Results for transonic flows analysis

For further analysis, it is essential to distinguish a three-layer structure in the laminar boundary layer [1-3]. The asymptotic analysis shows that the following relation is satisfied

$$\Delta p \sim \varepsilon^{1/2}\beta^{-1/2} \tag{1}$$

In the near-wall region, the influence of the forces of viscosity, inertia, and induced pressure has the same order, provided that in the external flow the boundary layer induces linear perturbations.
For the subsequent analysis, this relationship is convenient to represent in the following form:

$$\ln(\Delta p)/\ln\varepsilon = f(\ln\beta/\ln\varepsilon).$$

Then, relation (1) is represented by the line AM in figure 1.

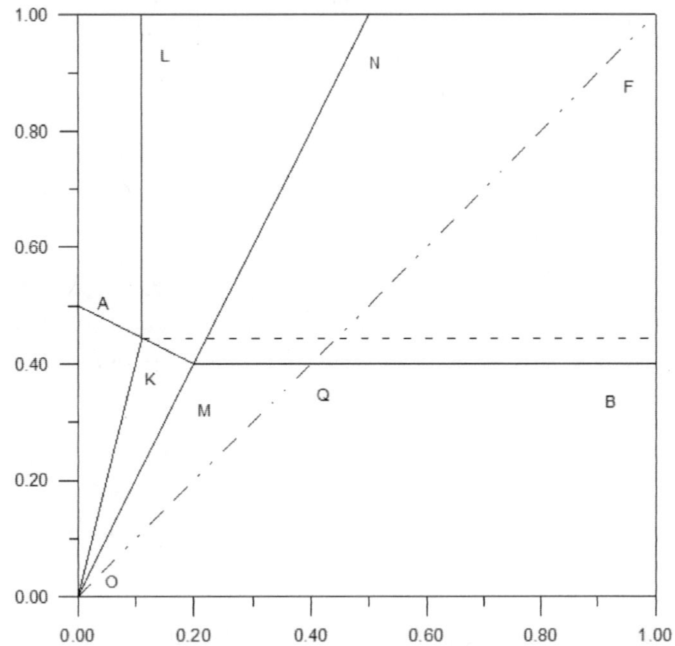

Figure 1. Transonic flow regimes diagram

From a comparison between the characteristic times in the outer region and in the near-wall region, it follows that their ratio is

$$\frac{T}{t} \sim \Delta p^{1/2}\beta^{-2} \tag{2}$$

Then, the condition of equality of the orders of characteristic times can be represented by the line OK in figure 1. For linear flow regimes in the near-wall region, this estimate is given as

$$\frac{T}{t} \sim \varepsilon^{1/4}\beta^{-9/4} \tag{3}$$

The times for this case are described by the line KL in figure 1 [4].

IOP Conf. Series: Materials Science and Engineering **208** (2017) 012021 doi:10.1088/1757-899X/208/1/012021

It is also important to evaluate the effect of the transonic parameter $K = \beta^2 / \delta^{2/3}$. For linear and nonlinear flow regimes in the near-wall region, the following relationship leads to a significant effect of the transonic parameter

$$\Delta p \sim \beta^2 \tag{4}$$

The last relationship is represented by the OMN line in figure 1.

Thus, using the results obtained above and presented in figure 1, for each limiting procedure, corresponding to the lines in the diagram (or the relation between the small parameters of the problem), there is a corresponding boundary-value problem (or model) and the similarity parameters. It should be noted that the most common models correspond to the intersection points of the lines in the diagram [5].

4. Results for problems with discontinuous boundary conditions

As an example, we consider a flow near the velocity discontinuity region on a plate having a moving section at a distance l from the leading edge, whose velocity is equal to u_w (figure 2).

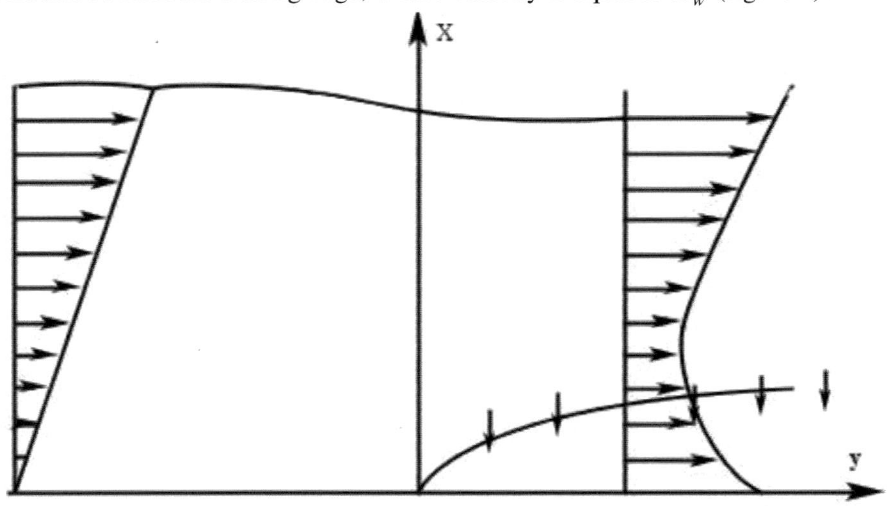

Figure 2. Flow structure near the point of velocity discontinuity

The notation used for the Cartesian coordinates measured along the surface and normal to it, time, velocity vector components, density, pressure, viscosity coefficient, and total enthalpy is similar to the notation introduced in the previous Section.

We consider the structure of a perturbed stationary flow, for which

$$u(x < 1{,}0) = 0, \quad u(x > 1{,}0) = u_w > 0. \tag{5}$$

A change in the velocity of the streamlines located near the surface moving with a velocity u_w at $x > 1$, and streamlines with almost zero velocities for $x < 1$, can lead to the formation of a new boundary layer downstream from the point of discontinuity of the boundary condition.

Using the momentum variation equations, one can find the distance, at which the friction in the resulting boundary layer becomes compatible with the friction in the main boundary layer

$$u_w / \varepsilon \sim 1/\varepsilon \ , \quad x_1 \sim u_w^{\,3} \tag{6}$$

IOP Conf. Series: Materials Science and Engineering **208** (2017) 012021 doi:10.1088/1757-899X/208/1/012021

The relations obtained can be conveniently depicted graphically in the form of the relationship $\ln x_i / \ln \varepsilon = f(\ln u_w / \ln \varepsilon)$, and then relation (6) is represented by OB line (figure 3).

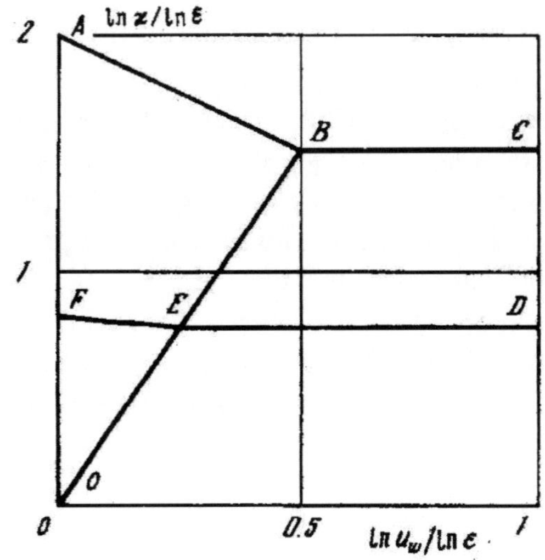

Figure 3. Disturbed flows scales as a function of the wall velocity

One can also find the distance x_2 from the discontinuity point of the boundary condition, at which the thickness and length of the region of nonlinear perturbations have the same order of magnitude and where, in essence, the assumptions of the boundary layer theory are violated

$$x_2 \sim \varepsilon^2 / u_w \tag{7}$$

The relation (7) is represented by the line AB in figure 3.
An estimate of the pressure perturbation allows us to find the distance x_3, at which the induced pressure gradient exerts a nonlinear effect on the near-wall region of the main boundary layer. Here, it should be noted that a change in the thickness of this region determines the total change in the displacement of the boundary layer

$$\Delta p \sim u_3^2 , \quad x_3 \sim \varepsilon^{4/5} u_w^{-1/5} \tag{8}$$

The last of the relations (8) is represented by the line EF in figure 3. In the case of linear perturbation regime, its realization can be associated with greater relative influence of the viscosity forces. The equality of the orders of magnitude of the terms, describing the influence of the forces of viscosity and inertia, leads to an estimate

$$x_4 \sim \varepsilon^{3/2} \tag{9}$$

The relation (9) is described by the line BC in figure 3. For linear interaction modes the following estimate holds true:

$$x_5 \sim \varepsilon^{3/4} \tag{10}$$

In Figure 3, it corresponds to ED line.

The diagram of disturbed flow regions depicted in figure 3 can be used to determine the dimensions of these regions and the nature of the flow, occurring in these regions at given amplitude of the parameter u_w. Thus, the perturbations with the amplitude $O(\varepsilon^{1/4}) \le u_w \le O(1)$ lead to the appearance of the region near the discontinuity with the dimensions determined by the line AB, in which the arising flow is described by the system of the Navier-Stokes equations for an incompressible fluid. The next in length is the region whose longitudinal dimension is determined by the line EF, where the flow is described in the first approximation by the Burgers equation. At the intermediate distances, when the parameters are located in the region between the lines AB and EF, the influence of viscosity on the flow in the region of nonlinear perturbations is insignificant and a compensation interaction mode is realized [7]. The absence of viscous terms in the equations, describing the perturbed flow, necessitates the introduction of a sub- region, in which the influence of viscous forces is of the same order as the influence of inertia forces. At the same time, there is a region with a length defined by the line, in which the effect of viscosity is significant and the surface friction is of the same order of magnitude as the friction at the initial boundary layer. The point E, as noted above, corresponds to the general case, when the nonlinear processes of interaction with the external flow occur in the "free interaction" region [1-3].

When the parameter u_w changes in the range $O(\varepsilon^{1/4}) \le u_w \le O(1)$ other than the above mentioned range, in which the flow is described by the Navier-Stokes equations, another region appears. Its length is determined by the line BE. In this case, the initiated flow is described by a system of boundary layer equations with a compensation interaction condition. In this region, the skin friction on the surface is comparable with the skin friction in the undisturbed flow. Finally, for $u_w \sim \varepsilon^{1/2}$, the friction tends to undisturbed value directly in the region where the flow is described by the Navier-Stokes system of equations.

5. Results for unsteady viscous-inviscid interaction

The flow of supersonic viscous gas is considered. Assume that at the distance l from the leading edge of the plate there is a flap or base region. The flap angle changes with time or the base pressure changes. The analysis is performed for such flows regimes, in which the amplitude of the flap deflection or the amplitude of pulsations of the bottom pressure are small.

Using the estimates that follow from the equations of motion, we can show that if the relation

$$\Delta p \tau \sim \varepsilon \tag{11}$$

holds true, the regime of nonlinear nonstationary viscous-inviscid interaction is realized. In the case of realization of the linear regime, the following relations are satisfied

$$\Delta p \sim \tau \sim \varepsilon^{1/2} \tag{12}$$

For clarity, it is useful to represent the asymptotic relations between the obtained parameters in the form of a diagram, as in the previous sections.

Then relation (11) is represented by the line AB, and relation (12) by the line AD. Then the BAD line separates the regions of variation of the parameters, corresponding to the steady-state and nonsteady flow regimes. The EAB line separates the regions of parameter variation corresponding to the linear and nonlinear flow regimes, and the condition determining the boundary between the linear and nonlinear regimes in the case of steady flows, (the EA line was obtained in [7-8]). The region of

IOP Conf. Series: Materials Science and Engineering **208** (2017) 012021 doi:10.1088/1757-899X/208/1/012021

variation of the parameters locates to the left of the line AD and its continuation corresponds to stationary modes of viscous-inviscid interaction. In turn, the parameters to the right of this line correspond to the nonstationary flow interaction regimes in the laminar boundary layer with an external supersonic flow. In this case, the AB line separates the regions of parameters, for which the influence of viscosity (regions of nonlinear and linear regimes) is significant or insignificant.

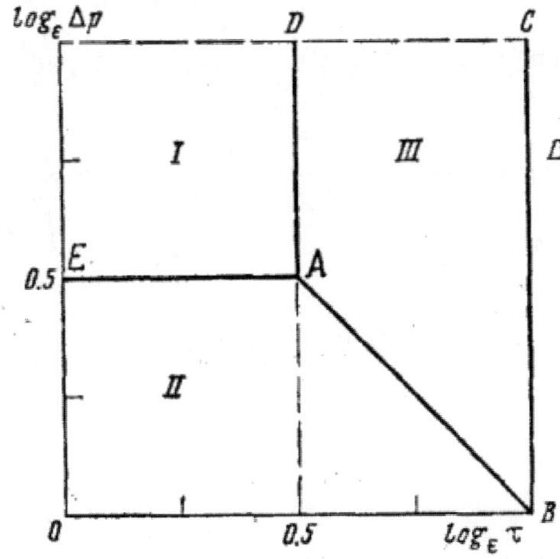

Figure 4. Disturbed flow regimes diagram

6. Conclusions

Thus, the above diagram and also the diagrams obtained in the previous sections have proved to be useful for visualization of possible flow regimes and estimation of the corresponding mathematical models along with similarity parameters.

Acknowledgments
Part of this study was financially supported by the Grant of the Ministry of Education and Science of the Russian Federation (Contract No. 14.G39.31.0001 of 13.02.2017) and by the Grant of the Russian Foundation for Basic Research (No. 17-01-00129).

References

[1] Neyland V 1969 *Fluid Dynamics* **4** 53
[2] Stewartson K and Williams P 1969 *Proc. Roy. Soc., A* **312** (1509) 181
[3] Messiter A, Feo A and Melnik R 1971 *AIAA J.* **9**(6) 1197
[4] Ryzhov O 1977 *Dokl. AN SSSR* **236** 1091
[5] Diesperov V and Lipatov I I 2003 *Fluid Dynamics* **5** 60
[6] Lipatov I 1999 *Appl. Math. Mech.* **2** 32
[7] Neyland V 1974 *Tr TsAGI* **1529**
[8] Ryzhov O 1980 *Appl. Math. Mech.* **44** 739

IOP Conf. Series: Materials Science and Engineering **208** (2017) 012022 doi:10.1088/1757-899X/208/1/012022

UDV study of a liquid metal vortex flow

G Losev[1], **R Khalilov**[1], **I Kolesnichenko**[1,2]

[1] Institute of Continuous Media Mechanics UB RAS, 1, Acad. Korolev St., Perm, 614013, Russia

[2] Perm National Research Polytechnic University, 29, Komsomolsky ave., Perm, 614990, Russia

E-mail: Losev_g@mail.ru

Abstract. The characteristics of an oscillating vortex flow of liquid metal were studied experimentally using an ultrasound Doppler velocimeter (UDV). The flow was generated by a local alternating magnetic field induced in a rectangular thin cell filled with gallium eutectic. The influence of medium temperature change and stirring on the UDV measurements was considered. The best set of parameters providing the reliability of long-term measurements were determined. The non-monotonic behavior of dependence of the local kinetic energy on the external alternating magnetic field intensity was found.

1. Introduction

The studied flow is initiated by an external alternating magnetic field generating a vortex electric current in the metal. An electromagnetic force is cosequently generated by the interaction of the current with the magnetic field. This force gives rise to apperance vortex flows. In most cases the area of magnetic field generation is localized, which is associated with restriction to inductor size. Generation of a vortex flow with a few largescale eddies was described in [1]. Such a flow is able to keep its pattern only at small force values. When the intensity of force action increases and exceeds some critical value, the flow becomes unstable [2]. In this case the eddies are moving in the layer plane. In our previous work [2], we used the potential probe (PP) for velocity measurements. This type of the probe can determine only the local characteristics of the flow [3, 4]. It is also interesting to measure the spatial characteristics of the flow and the flow patterns. However, the existing optical methods (e.g. PIV) cannot be used for this purpose because of the optical opacity of liquid metals. For this reason, application of an ultrasound doppler velocimeter (UDV) is justified. This device has been widely used in similar resaerch studies (e.g. [5, 6, 7]).

The study focuses on the study of a liquid metal flow in a plane rectangular layer (figure 1). The layer thickness is much less than its planar sizes. The region under the influence of an alternating magnetic field is located in the center of the plane. The value of the vector component transveral to the layer magnetic induction exceeds the value of the plane component. The influence of an alternating magnetic field generates the vortex electric current. The electromagnetic force is directed from the periphery of the region to its center. At the initial time, four-eddies flow of liquid metal are initiated.

The main purpose of this work is to investigate the velocity of the examined liquid metal (gallium eutectic). To this end, it is necessary to determine the dependence of the mean kinetic

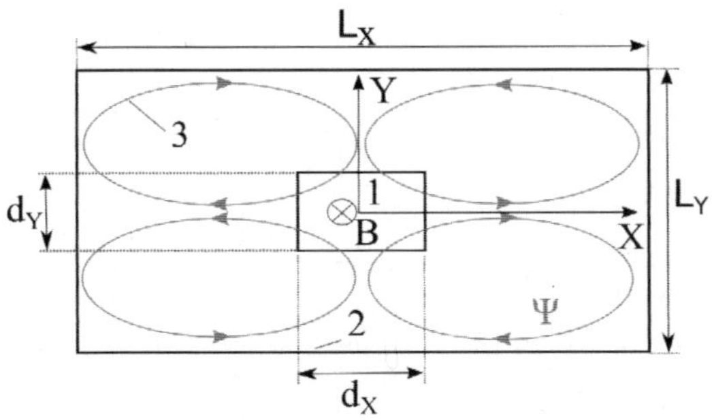

Figure 1. Sketch of layer and flow initial pattern: 1 – area of magnetic field localization, 2 – flow of liquid metal, 3 – streamlines of initial flow. Dimensional parameters $L_X = 200$ mm, $L_Y = 100$ mm, $d_X = 67$ mm, $d_Y = 56$ mm.

energy and oscillation frequency on the strength and frequency of an vortex current in the inductor coil. To describe the flow properties, four parameters were used. The first is the mean local kinetic energy of the flow W_y defined as the time average value of the velocity squared. The subscript y shows the projection of the velocity component onto the coordinate axis. The second is the dominant frequency of velocity profile oscillations f_d defined as the frequency which corresponds to the Fourier transform energy maximum of the local velocity. The third is the spectrum filling coefficient k_S defined as the frequency averaged over the spectral energy density normalized to unity. The maximum value of 1024 matches the so called white noise, and the minimum value of 1 matches harmonic oscillations. The latter is the relative number of lost velocity profile points N_l. The lost points are defined as the points whose value does not exceed 5% of the mean velocity value.

Ultrasound doppler velocimetry includes emitting ultrasound wave packages by an ultrasound doppler sensor, their consequent adoption, and calculation of the Doppler frequency shift between the emitted and received wave packages. The ultrasonic silicon gel has to be used to provide sound contact between the probe and the wall. However, the presence of sound-reflecting particles moved by the flow in the medium is necessary. The gallium oxide particles reflect ultrasound waves in gallium eutectic. These particles are formed by oxidation of gallium in the atmosphere [8]. This means that the UDV measures not exactly the velocity of the medium, but the velocity of solid particles. Thus, the possible lack of repeatability of energy dependencies can be explained not only by the random behavior of eddies or the changes in the properties of media subjected to heating caused by Joule losses, but also by the changes in the number of sound-reflecting particles. The stirring of dispersion medium by the intensive flow leads in this case to sedimentation of sound-reflecting particles or, on the contrary, to washing the particles away from the walls of the cell. This affects the qualitative and quantitative characteristics of a reflecting ultrasound signal (e.g. [9]). Furthermore, the variation of medium temperature as well the drying of an ultrasound gel can influence the results of velocity measurements. Therefore, one more purpose of this work is to develop a measurement method that would allow getting high-accuracy and repeatable results.

The existing method of electromagnetic force generation has found the use in industrial applications. Travelling and rotating magnetic fields are generally used for contactless transport of the molten metal through technological channels (e.g. [10]), as well as for mixing flows generated in enclosed volumes (e.g. [11]). Contactless generation of the molten metal flow is useful for continuous casting. Besides, the application of stirrers for manifacturing cylindrical ingots in aluminium semicontinuous casting machines is being extensively investigated (e.g. [12]). The supporting rollers of a steel continuous casting machine can, after small update, generate an alternating magnetic field. This makes it possisible to generate a stirring flow (e.g. [13]). There is no traveling magnetic field, this machine design assures proper mixing of a hardened ingot core.

A liquid metal flow passing through the channel can also be obtained in the absence of a traveling magnetic field method. Bearing this in mind, a channnel of specific configuration was developed [14]. An electromagnet in this channel creates a local electromagnetic force and generates thus a large-scale vortex flow. The resulting centrifugal force causes the pressure drop in the pump channel. The use of one or several inductors allows one to create a stirriing flow in an enclosed rectangular channel filled with molten metal [2]. In the production of solid ingots the stirrring has a strong impact on the crystallization process [15]. The simple device described in this paper may find application in different reserach laboratories to produce metals and alloys in comparatively small quantities. Stirring provides homogenisation of inclusions and temperature distribution, grain refining and smoothing of the crystallization front (e.g. [16, 17, 18]).

As can be seen from above, the liquid metal flow generated in plane layers by an external alternating magnetic field is important problem. The instability of eddies is desirable for stirring the liquid metal [2] and undesirable for pumping the flow through the channel [14]. This provokes the interest in a detailed study of such flows.

2. Methods

The experimental setup (fig. 2) consists of the plexiglass cell 1 filled with liquid gallium eutectic (density 6256 kg/m^3, kinematic viscosity $3.1 \cdot 10^{-7}$ m^2/s, conductivity $3.56 \cdot 10^6$ Sm, sound speed 2828 m/s). Dimensions of the cell are $200 \times 100 \times 15$ mm^3. The cell is placed in the gap between the poles of the C-core inductor 2. The AC power source is used as the power supply 3 (Pacific Smart Source). The non-contact ampere-meter FLUKE i2000 Flex AC Current probe controls amperage with accuracy to second decimal places of ampere. The cover of the cell is equipped with two PP 4 with diameters of 10 and 18 mm and the chromel-aluminum thermocouple 5. The data acquisition card 6 Nation Instruments c-DAQ receives the signals from PP and a thermocouple. The UDV probe 7 allows measuring velocity component in any cross section of cell.

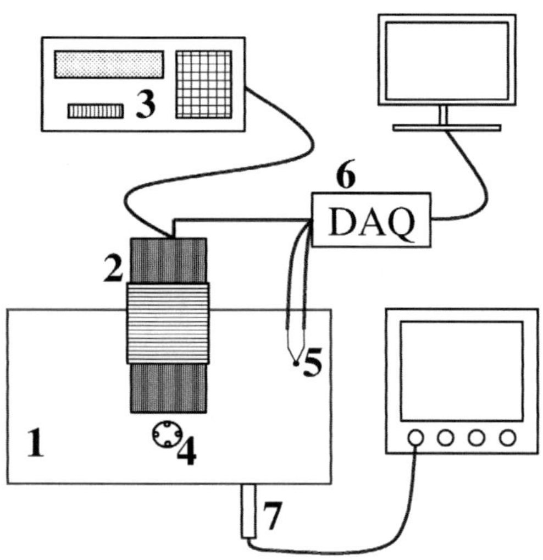

Figure 2. Experimental setup: 1 – plexiglass cell filled with gallium eutectic, 2 – C-core inductor, 3 – AC power supply, 4 – potential probes, 5 – chromel-aluminum thermocouple, 6 – data acquisition card, 7 – UDV probe.

The amperage measurement is much easier than the magnetic induction measurement. Therefore the dependence of induction in the gap on the current in coils can be found. Such measurements were made by Lake Shore 421 Gaussmeter. This dependence governed by the linear law allows one to present results in terms of current, but not magnetic field induction.

The velocity profiles were measured for different values of the external magnetic field by UDV in the cell section chosen for the study. To check the quality of the obtained data, it is more convenient to make a Fourier analysis of velocity in a single point of the profile instead

IOP Conf. Series: Materials Science and Engineering **208** (2017) 012022 doi:10.1088/1757-899X/208/1/012022

of the full-profile analysis . Moreover, comparison of energy parameters in a single point of the velocity profile with the available data is more adequate than the spatial average of the velocity profile. Since the velocity profile oscillations are quite similar to harmonic oscillations, the single dominant frequency f_d exists in the velocity spectrum. This frequency influences the eddy shape and oscillating movement. The time interval between velocity profile records was 124.9 ms. This imposed limitations of 4 Hz on the maximum resolvable frequency of a spectrum. The specific filter was used to determine f_d. If k_S had a value more than 240, the velocity signal was considered as chaotic. Hence it can be concluded that the dominant frequency does not exist, and f_d can be defined as zero.

3. Results

The first step toward eliminating factors that affect negatively velocity measurements included study of W_y varying with time under constant ambient conditions. The velocity profile evolution is recorded for 66 minutes with UDV (about 2^{15} profiles). In addition, the velocity profile and the PP and thermocouple readings were recorded simultaneously. This provided the opportunity to compare the nature of energy dependencies obtained by different methods. When the recording was completed, every record was divided into 30 separate parts. The mean values were found by time averaging of these parts.

The analysis of UDV measurements indicates that W_y declines significantly after approximately 30 minutes of the experiment (figure 3). To investigate this feature, the dependence of N_l on time has been found (figure 4). It is evident that the valid signal of the UDV velocity is being absorbed by the background noise. Thereby we have arrived at the conclusion that the long-duration velocity measurements carried out with UDV (more than 2000 s) do not provide high-quality results.

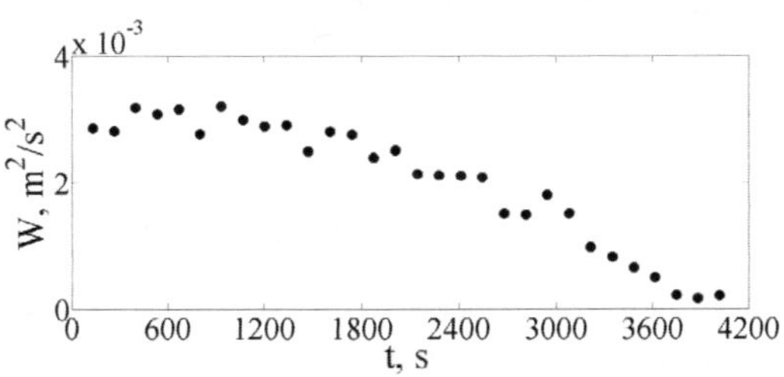

Figure 3. Evolution of mean local kinetic energy.

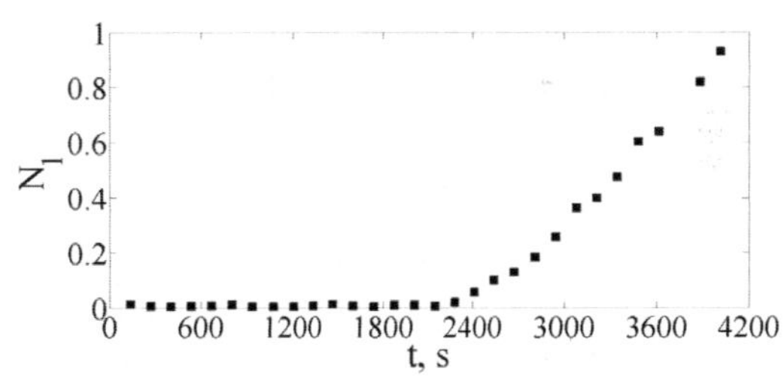

Figure 4. Evolution of the number of lost point on the velocity profile.

Mixing and heating of the liquid metal, as well as the drying of ultrasound gel, can influence velocity measurements. During the next step, an attempt was made to eliminate the influence of these factors. The velocity measurements of low-intensity flow (coil current was 4.0 A, current frequency was 50.0 Hz) were made using the UDV. Each experiment consisted of 12 records of velocity profiles. The time interval of a single record was 256 s (2048 velocity profiles), whereas the stirring time in the presence of a strong magnetic field (coil current was 8.0 A, current frequency was 50.0 Hz) was 180 s. In the first experiment, measurements were made without renovation of the drying ultrasound gel within the time intervals between velocity profile recordings (exp. 1). The second experiment included renovation of the gel after every record (exp. 2). In the third experiment (exp. 3), the liquid metal in the cell was preheated to 50 degrees using a water temperature regulator before measurements started. Figure 5 demonstrates how W_y varies with time. It is obvious that in all the experiments the process of stirring improves the stability of mean energy. At the same time, the time dependence of f_d demonstrates similarity and stability. However, the f_d tends to decrease slightly in experiments 1 and 2. The pulsations of local kinetic energy increase. The PP indicates the constancy of kinetic energy. This fact suggests that the change in the gallium eutectic temperature has a negative influence on the UDV velocity measurement accuracy, and the stirring affects positively.

The time dependence of N_l is even a more important characteristic than W_y. Figure 6 illustrates the improvement in stability measurements for metal preheating. The N_l do not exceed 7% of the mean velocity. When the liquid metal was stirred at varying temperature, N_l was about 60%. Therefore, the preheating of liquid metal and the constancy of the medium temperature during measurements improve the stability of velocity and energy characteristics of the flow measured by UDV.

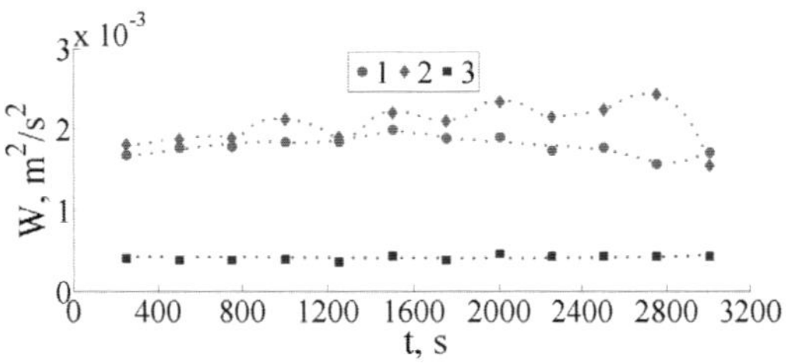

Figure 5. Evolution of mean kinetic energy during three experiments: 1 – stirring after each series of velocity recordings, 2 – renovation of the ultrasound gel and stirring after each recording series, 3 – preheating of liquid metal before the experiment and stirring after each recording series.

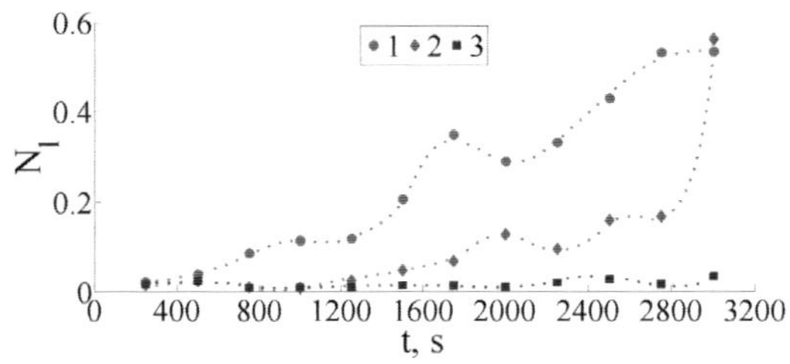

Figure 6. The number of lost points of the velocity profile for three experiments: 1 – stirring after each series of velocity recording, 2 – renovation of the ultrasound gel and stirring after each recording series, 3 – preheating of liquid metal before the experiment and stirring after each recording series.

The most appropriate system relaxation time was found at the following stage. It is the shortest time at which the system can recover. To determine it, we performed a number of additional experiments. The indicated values of PP and thermocouple allow one to control the state of the system and UDV measurements as well. The current in the inductor coils had a value of 8.0 A. The high flow intensity provided good stirring during the experiment. The measurement time was 1956 s (16384 velocity profiles). Each of the signals was divided into 20 parts. The data obtained was averaged and provided the dependencies similar to described above. After each measurement the power supply was shut down and the system was returned to relaxation condition during the set time of 15, 30, 45 and 60 minutes between each of the following two measurements.

Using the described method the time dependencies of W_y and N_l were established for five measurements (figure 7). It is obvious that the initial condition and 60 minutes after the relaxation period are most similar. At the same time it is seen that the dependencies of 30 and 45 minutes relaxation periods also show almost similar results. It is necessary to note that the PP data suggest a 7 fold increase in kinetic energy. Furthermore, the time dependence of N_l increases slightly (less than 5% in all measurements). The best result was obtained for the 15 minute relaxation time. In this case, the growth of N_l is less than half percent.

The results allowed us to develop the measuring procedure for getting high-quality and repeatable velocity and energy dependencies using UDV. Figure 9 demonstrates the velocity evolution in two ways. The first implies application of the proposed method, and the second includes direct UDV measurements without any specific procedures.

The determined difference between velocity evolutions makes it possible to distinguish between the evolutions of kinetic energies. In the case of direct velocity profile measurements,

Figure 7. Evolution of mean kinetic energy for different relaxation standstill periods: 1 – initial condition, 2 – 15 minute relaxation, 3 – 30 minute relaxation, 4 – 45 minute relaxation, 5 – 60 minute relaxation.

Figure 8. Changes in the number of lost velocity profile points for different relaxation standstill periods: 1 – initial condition, 2 – 15 minute relaxation, 3 – 30 minute relaxation, 4 – 45 minute relaxation, 5 – 60 minute relaxation.

it was impossible to obtain repeatable results for the mean energy W_y. On the other hand, our measurement method gives the repeatable dependencies of W_y on the current intensity and frequency of the coil. In addition, the time dependencies of mean energy calculated using UDV and PP are close (figure 10).

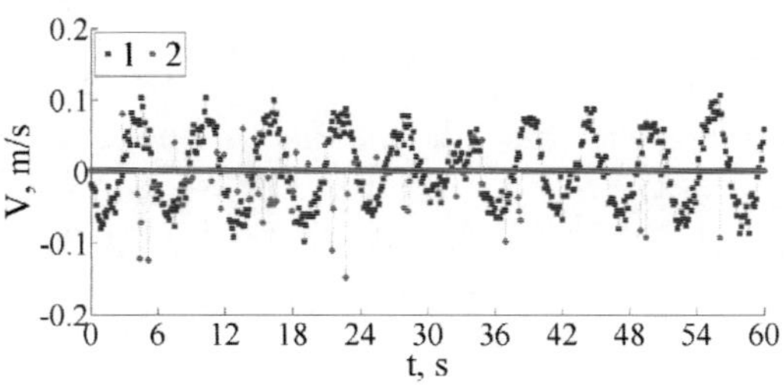

Figure 9. Evolution of the velocity at a fixed point: 1 – measured using the developed method, 2 – directly measured.

Figure 10. Mean kinetic energy evolution: 1 – calculated from UDV measurements, 2 – calculated from PP measurement.

The dependencies of W_y (figure 11) and f_d (figure 12) on the current frequency of the coil were determined. The mean kinetic energy demonstrates a non-monotonic dependence on the alternating magnetic field frequency.

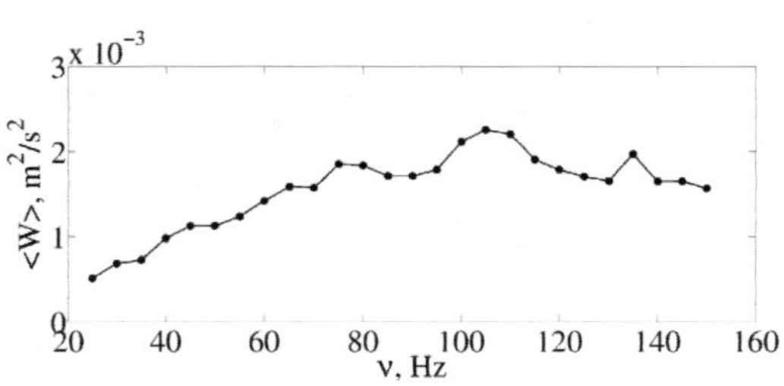

Figure 11. Mean local kinetic energy vs. magnetic field frequency.

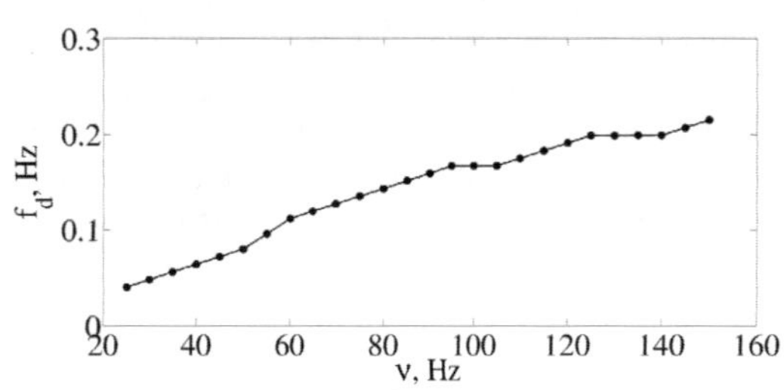

Figure 12. Vortexes oscillation frequency vs. magnetic field frequency.

Figure 11 shows a well-marked oscillation together with a background increase of W_y in the frequency range 25 to 105 Hz. Thereafter the energy gradually decreases when the magnetic field frequency exceeds 110 Hz. This provides insights into qualitative changes in flow patterns. The frequency of eddy oscillations increases linearly with the frequency of the magnetic field. Therefore the intensity of the stirring can be improved with growing coil current frequency under the same amperage.

4. Conclusions

It was found that application of UDV for long-duration measurements essentially differs from short-duration measurements. The preliminary recording of the velocity of gallium eutectic showed the lack of repeatability of mean kinetic energy dependencies on the external force intensity. It was found that the UDV recordings are becoming unreliable during long-time profile recordings. The investigation of parameters which influence liquid metal flow velocity measurements (stability of medium temperature, sound-reflecting particles of gallium oxide distribution, acoustic properties of ultrasound gel) showed that temperature variations have a strong negative effect on the velocity signal quality. In some cases this can leads to a complete loss of the profile. The stability of energy characteristics of the flow measured by UDV under temperature regulation was found. The high intensity stirring of eutectic also affects positively the quality of recordings. It causes a decrease in the number of profile lost points. In addition the relaxation standstill between the experiments improves the data quality. The data recorded with relaxation time periods more than 15 minutes are in good agreement with the results of several tests. As can be seen from above, we have found the experimental parameters best suited for getting accurate data.

The use of the described method in the flow parameters investigation allows us to determine the non-monotonic dependence of local kinetic energy on the external alternating magnetic field value and frequency. This proves the qualitative changes of the flow structure. The frequency of the velocity profile oscillation increases almost linearly with increasing magnetic field frequency. Therefore the vortex flow in a rectangular cell generated by the local alternating magnetic field is unsteady, and its behavior significantly depends in the magnetic field frequency.

References

[1] Kirko I, Almukhametov V and Khripchenko S 1988 *Dokl. Acad. Nauk USSR (in Russian)* **302** 845–847
[2] Kolesnichenko I, Khalilov R and Khripchenko S 2007 *Magnetohydrodynamics* **43** 45–52
[3] Ricou R and Vives C 1982 *Int. J. of Heat Mass Transfer* **25** 1579–1588
[4] von Weissenfluh T 1985 *Int. J. of Heat Mass Transfer* **28** 1563–1574
[5] Andreev O, Kolesnikov Y and Thess A 2009 *Exp. Fluids* **46** 77–83
[6] Ben-David O, Levy A, Mikhailovich B and Azulay A 2013 *Int. J. of Heat Mass Transfer* **67** 260–271
[7] Franke S, Räbiger D, Galindo V, Zhang Y and Eckert S 2016 *Flow Meas. Instrum.* **48** 64–73
[8] Nowak M 2002 *Exp. in fluids* **31** 249–255
[9] Brito D, Nataf H C, Cardin P, Auber J and Masson J P 2001 *Exp. fluids* **31** 653–663
[10] Lielpeter Y 1969 *Liquid metal induction MHD machines* (Zinatne, Riga (in Russian))
[11] Moffatt H K 1991 *Phys. Fluids A* **3** 1336–1343
[12] Mapelli C, Gruttadauria A and Peroni M 2010 *J. Mater. Process. Technol.* **210** 306–314
[13] Zelenetskij A, Khripchenko S and Tsaplin A 1992 *Magnitnaya Gidrodinamika (in Russian)* **1** 96–100
[14] Khripchenko S, Khalilov R, Kolesnichenko I, Denisov S, Galindo V and Gerbeth G 2010 *Magnetohydrodynamics* **46** 85–97
[15] Oborin P and Kolesnichenko I 2013 *Magnetohydrodynamics* **49** 231–236
[16] Willers B, Eckert S, Nikrityuk P A, Räbiger D, Dong J, Eckert K and Gerbeth G 2008 *Metall. Mater. Trans. B* **39** 304–316
[17] Ščepanskis M, Jakovičs A and Nacke B 2010 *Magnetohydrodynamics* **46** 413–423
[18] Denisov S, Dolgikh V, Khripchenko S, Kolesnichenko I and Nikulin L 2014 *Magnetohydrodynamics* **4** 249–265

IOP Conf. Series: Materials Science and Engineering **208** (2017) 012023 doi:10.1088/1757-899X/208/1/012023

Vibration effect on the Soret-induced convection of ternary mixture in a rectangular cavity heated from below

T P Lyubimova[1,2] and N A Zubova[1]
[1] Institute of Continuous Media Mechanics UB RAS, 1, Acad. Korolev St., Perm, 614013, Russia
[2] Perm State National Research University, 15, Bukirev St., Perm, 614990, Russia

E-mail: lubimova@psu.ru

Abstract. This paper presents the results of numerical simulation of the Soret-induced convection of ternary mixture in the rectangular cavity elongated in horizontal direction in gravity field. The cavity has rigid impermeable boundaries. It is heated from the bellow and undergoes translational linearly polarized vibrations of finite amplitude and frequency in the horizontal direction. The problem is solved by finite difference method in the framework of full unsteady non-linear approach. The procedure of diagonalization of the molecular diffusion coefficient matrix is applied, allowing to eliminate cross-diffusion components in the equations and to reduce the number of the governing parameters. The calculations are performed for model ternary mixture with positive separation ratios of the components. The data on the vibration effect on temporal evolution of instantaneous and average fields and integral characteristics of the flow and heat and mass transfer at different levels of gravity are obtained.

1. Introduction

The Soret effect can make a significant effect on the convection in mixtures caused by the density inhomogeneity in the gravity field. The thermodiffusion properties of a mixture consisting of n components are described by the vector of $n-1$ the elements called the separation ratios vector:

$$\boldsymbol{\psi} = -\beta_T^{-1}\mathbf{B}\mathbf{D}^{-1}\boldsymbol{D}_T, \tag{1}$$

where $\mathbf{B}$ is the diagonal matrix of the coefficients of the concentration dependence on the density, β_T is the thermal expansion coefficient, $\mathbf{D}$ is the matrix of the molecular diffusion coefficients, $\boldsymbol{D}_T$ is the vector of thermal diffusion coefficients. When subjected to the temperature gradient, the molecules of the components with negative separation ratio move to more heated domain of the cavity and the molecules of the components with a positive separation ratio move to the cold domain.

Linear stability analysis carried out for convection of ternary mixtures with fixed positive separation ratio of one component in horizontal layer and square cavity in static gravity field has shown the existence of monotonic instability at the positive separation ratio of other component and the oscillatory instability at the negative separation ratio in the case of heating from the bellow. The monotonic instability occurs at the negative separation ratio of the other component in the case of heating from above [1, 2].

In the presence of vibrations, convection may exist even in zero gravity conditions. In gravity field vibrational and gravitational mechanisms of convection interact, intensifying or reducing each other.

IOP Conf. Series: Materials Science and Engineering **208** (2017) 012023 doi:10.1088/1757-899X/208/1/012023

For example, the longitudinal high-frequency vibrations make destabilizing effect on the horizontal layer of binary mixtures with positive separation ratios and stabilizing effect on binary mixtures with negative separation ratios [3]. Vibrations applied in the direction perpendicular to the layer play stabilizing role independently of the separation ratio sign [4]. In square cavity at side heating in conditions of low gravity the residual flow can be damped by applying an appropriate vibration intensity in the appropriate direction [5]. Direct numerical simulation of convection in the water - isopropanol mixtures in cubic cavity subjected to the gravity and vibrations perpendicular to the temperature gradient is carried out in [6].

Linear stability analysis [7] for horizontal layer of ternary liquid mixtures with the negative separation ratios of the first component has shown that longitudinal low-amplitude high-frequency vibrations reduce the area of the convectionless state stability in the case of heating from below and to extend stability area in the case of heating from above. Numerical simulations [8] have demonstrated that for the ternary mixture in the rectangular cavities in zero gravity conditions at the increasing the vibration intensity there is a transformation of the average flow structure associated with the instability of quasi-equilibrium state in the central part of the cavity.

The present paper presents the results of the direct numerical simulation of the Soret-induced convection of ternary mixture in the rectangular cavity subjected to the heating from below, horizontal vibrations and gravity.

2. Formulation of the problem

Let us consider Soret-induced convection of ternary mixture in the horizontally elongated rectangular cavity with the rigid impermeable boundaries. The cavity performs the longitudinal harmonic vibration with amplitude a and angular frequency ω. The vertical boundariess of the cavitye are adiabatic and on the horizontal boundaries the constant different temperatures corresponding to the heating from below are maintained.

Unsteady non-linear equations of the Soret-induced convection of ternary in the Boussinesq approximation in the reference frame of oscillating cavity are [9]:

$$\frac{\partial \vec{u}}{\partial t}+\vec{u}\nabla\vec{u}=-\nabla p+\nabla^2\vec{u}+\frac{\mathrm{Ra}}{\mathrm{Pr}}\left(T+\boldsymbol{I}\cdot\boldsymbol{C}\right)\vec{\gamma}+\frac{\mathrm{Ra_v}}{\mathrm{Pr}}cos\left(\Omega t\right)\left(T+\boldsymbol{I}\cdot\boldsymbol{C}\right)\vec{n}\,, \tag{2}$$

$$\frac{\partial \boldsymbol{C}}{\partial t}+\vec{u}\nabla\boldsymbol{C}=\frac{1}{\mathbf{SC}}\left(\nabla^2\boldsymbol{C}-\boldsymbol{\psi}\nabla^2 T\right), \tag{3}$$

$$\frac{\partial T}{\partial t}+\vec{u}\nabla T=\frac{1}{\mathrm{Pr}}\nabla^2 T\,, \tag{4}$$

$$\nabla\cdot\vec{u}=0\,. \tag{5}$$

Here $\vec{u}$ is the velocity vector, p is the pressure, $\boldsymbol{C}=\left(C_1,\ \ldots,\ C_{n-1}\right)^T$ is the transpose vector of concentrations, $\boldsymbol{I}=\left(1,\ \ldots,\ 1\right)$ is the unit vector, $\vec{\gamma}$ is the unit vertical vector and $\vec{n}$ is the unit vector in the direction of vibrations.

The equations are written in the dimensionless form using the following quantities as the scales: the height of the cavity H for the length, ν/H for the velocity, H^2/ν for the time, $\rho_0\nu/H$ for the pressure, temperature difference between the lower and upper boundaries ΔT for the temperature, and $\beta_T\Delta T\mathbf{B}^{-1}$ for the vector of concentrations. Here ν is the viscosity of the mixture and ρ_0 is the initial mixture density. Equations (2)-(5) contain the following dimensionless parameters: $\mathrm{Pr}=\nu/\chi$ is the Prandtl number (χ is the thermal diffusivity), $\boldsymbol{\psi}$ is the separation ratios vector, $\mathbf{SC}=\nu^{-1}\mathbf{BDB}^{-1}$ is the matrix of Schmidt numbers, $\mathrm{Ra}=g\beta_T\Delta TH^3/\left(\nu\chi\right)$ is the Rayleigh number, $\Omega=\omega H^2/\nu$ is the dimensionless vibration frequency, $\mathrm{Ra_v}=a\omega^2\beta_T\Delta TH^3/\left(\nu\chi\right)$ is the vibrational analogue of the

Rayleigh number (equals to the usual Rayleigh number multiplied by the ratio of vibrational acceleration to gravity acceleration and related to the vibrational Rayleigh number used in the average approach as $\mathrm{Ra_v} = \Omega\left(2\,\mathrm{Pr}\,\mathrm{Ra_{va}}\right)^{1/2}$).

The boundary conditions are:

$$x = 0, L: \vec{u} = 0, \frac{\partial T}{\partial x} = \frac{\partial \boldsymbol{C}}{\partial x} = 0, \tag{6}$$

$$y = 0, H: \vec{u} = 0, T = \pm 1/2, \frac{\partial}{\partial y}(C - \psi T) = 0. \tag{7}$$

3. Numerical method

The problem (2)-(7) was rewritten in terms of stream function and vorticity and solved numerically by finite difference method. The spatial derivatives were approximated by central differences. Unsteady equations were solved using explicit finite difference scheme with a constant time step $h^2/8$, where h is the constant spatial step (the grid 200×40 was used in main calculations). The Poisson equation for the stream function was solved by successive over-relaxation method. The vorticity values at the boundaries of the cavity were calculated using the Thom formula [10].

For ternary mixtures the matrix of molecular diffusion coefficients **D** consists of four elements two nondiagonal elements of which are responsible for the cross-diffusion effect. This leads to the appearance of additional terms in the equation and complicates the calculations. The diagonalization procedure of matrix **D** allows to eliminate the cross diffusion terms in the equations and to decrease the number of governing parameters, in this case the equations structure does not change [11]. Such transformation applied to the problem (2)-(7) in dimensionless form can be written as:

$$\boldsymbol{C} = \mathbf{BV}(\mathbf{BQ})^{-1}\widehat{\boldsymbol{C}}, \ \boldsymbol{\psi} = \mathbf{BV}(\mathbf{BQ})^{-1}\widehat{\boldsymbol{\psi}},$$

where **V** is the matrix whose columns are the eigenvectors $v_i = \left(v_{i1}, \ldots, v_{in-1}\right)^T$ of the matrix **D**, $\mathbf{Q} = diag\left\{q_1, \ldots, q_{n-1}\right\}$, $q_i = \beta_i^{-1}\sum_{j=1}^{n-1}\beta_j v_{ij}$. After the transformation for dimensionless problem the matrix of the Schmidt numbers becomes diagonal.

The calculations were carried out for rectangular cavities with aspect ratio 5:1. The initial conditions corresponded to the uniform vertical gradients of temperature and solutes concentrations:

$$T = 1/2 - y \text{ and } \boldsymbol{C} = \boldsymbol{\psi}(1/2 - y).$$

The calculations were carried out for ternary mixtures with the following dimensionless parameters: the Prandtl number $\mathrm{Pr} = 10$, the Schmidt numbers $\mathrm{Sc_1} = 100$ and $\mathrm{Sc_2} = 1000$, separation ratios $\psi_1 = 0.3$ and $\psi_2 = 0.1$. The chosen parameters are typical for the liquid mixtures. The dimensionless frequency of vibrations was fixed $\Omega = 445$. For the cavity of 1 cm height filled by ternary liquid mixture with the viscosity $\nu \sim 10^{-6}$ m^2/s this value corresponds to the dimensional vibration frequency approximately equal to 1 Hz. The Rayleigh number was varied in the range $\mathrm{Ra} = 10^2 - 10^4$ and the vibrational Rayleigh number in the range $\mathrm{Ra_v} = 10^2 - 2\cdot 10^5$.

4. Results

4.1. Soret-induced convection in gravity field in the absence of vibrations

As known [12], in the case of single-component fluid in the horizontal layer heated from below with rigid perfectly conductive boundaries the crisis of conductive state occurs at the Rayleigh number equal to $\mathrm{Ra} \approx 1.7 \cdot 10^3$, the wave number of the most dangerous perturbations is approximately equal to 3.1. The calculations carried out in the present work for ternary mixture with the parameters specified above, in section 3, in the absence of vibrations, have shown that in this case the onset of convection takes place at $\mathrm{Ra} \approx 2.5 \cdot 10^2$ and the spatial period of the emerging convective structures is $\lambda \approx 2.5$ (the corresponding wave number $k \approx 2.5$), (see fig. 1a, $\mathrm{Ra} \approx 3 \cdot 10^2$).

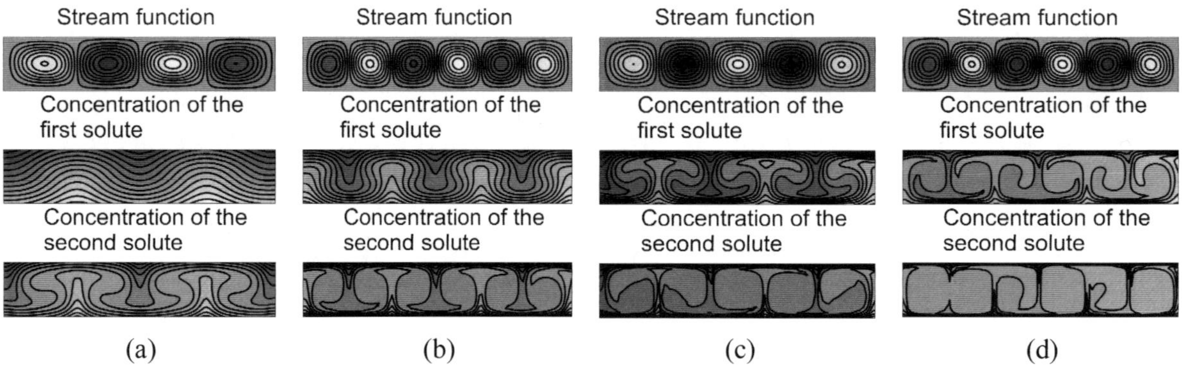

Figure 1. Streamlines of flow and solute concentration fields in the absence of vibration ($\mathrm{Ra}_v = 0$): (a) – $\mathrm{Ra} = 3 \cdot 10^2$, (b) – $\mathrm{Ra} = 1 \cdot 10^3$, (c) – $\mathrm{Ra} = 2 \cdot 10^3$, (d) – $\mathrm{Ra} = 5 \cdot 10^3$.

With the increase of the Rayleigh number the spatial period of the convective structures decreases (see Fig. 1b, for $\mathrm{Ra} = 1 \cdot 10^3$, where $\lambda \approx 1.7$ ($k \approx 3.8$)). When the Rayleigh number reaches the value $\mathrm{Ra} \approx 1.7 \cdot 10^3$, the thermogravitational mechanism of instability takes the floor and the spatial period of the convective structures increases to $\lambda \approx 2.0$ ($k \approx 3.1$) (Fig. 1b), which is accompanied by the sharp increase in the intensity of the flow and heat and mass transfer and the change of the inclination of the curve Nu/L(Ra) (Fig. 2). With further increase of the Rayleigh number the spatial period again decreases (Fig. 1d).

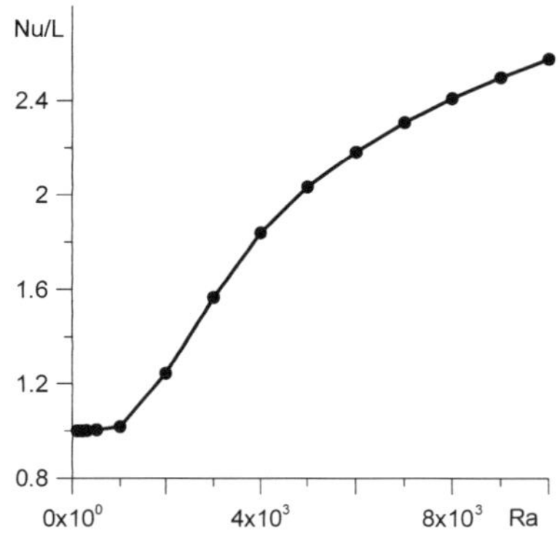

Figure 2. Dimensionless heat flux per unit length versus Rayleigh number in the absence of vibrations ($\mathrm{Ra}_v = 0$).

4.2. Vibrational convection of the mixture in the absence of gravity

Figure 3 shows the structure of the average flow and solute concentration fields under vibrations in the absence of gravity. The average values were obtained by averaging the instantaneous values over the vibration period. For small values of the vibrational Rayleigh number (Fig. 3a, $\mathrm{Ra}_v = 1 \cdot 10^3$) the average flow in the form of the four vortexes located in pairs near the sidewalls of the cavity is generated and in the central part of the cavity the flow is nearly absent, the flow intensity is rather weak and the deformation of the solute concentration fields by the convective flow is not visible. When the vibrational Rayleigh number approaches $5\cdot10^4$, the qualitative transformation of the average flow structure occurs due to the crisis of the quasi-equilibrium state in the central part of the cavity. The new regime corresponds to the multi-vortex steady average flow (Fig. 3b, $\mathrm{Ra_v}=5\cdot10^4$), the spatial period of the vibrational convective structures is approximately equal to 2.0 ($k\approx3.1$). This flow transformation is accompanied by a sharp increase in the intensity of the heat transfer (Fig. 4).

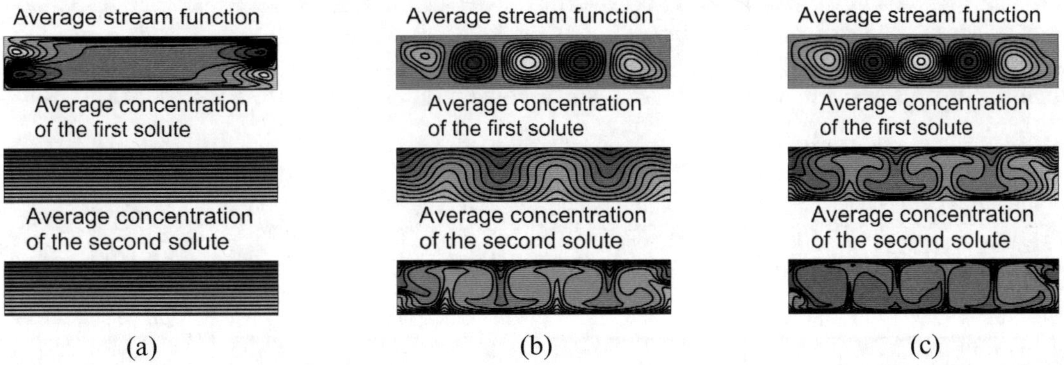

Figure 3. Streamlines of average flow and solute concentration fields in zero gravity conditions ($\mathrm{Ra}=0$): (a) – $\mathrm{Ra_v}=1\cdot10^3$, (b) – $\mathrm{Ra_v}=5\cdot10^4$, (c) – $\mathrm{Ra_v}=1\cdot10^5$.

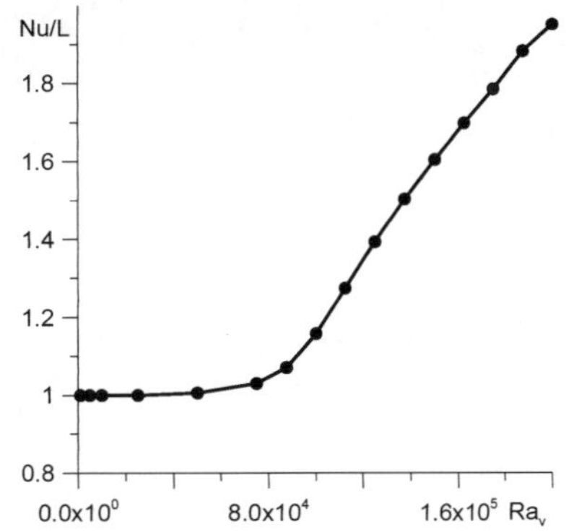

Figure 4. Dimensionless heat flux per unit length versus vibrational Rayleigh number in zero gravity conditions ($\mathrm{Ra}=0$).

4.3. The effect of vibration on Soret-induced convection in gravity field

Figures 5, 6 present the average fields of stream function and solute concentrations for steady convective regimes in the presence of gravity and vibrations. As one can see, in the range of vibrational Rayleigh numbers $3\cdot10^2-1\cdot10^5$ the spatial period of steady average flow is constant and equal to approximately 2.5 ($k\approx2.5$) (fig. 5a-c, fig. 6a-c). Further growth of the vibration intensity leads to decrease of the spatial period (fig. 5d, fig. 6d).

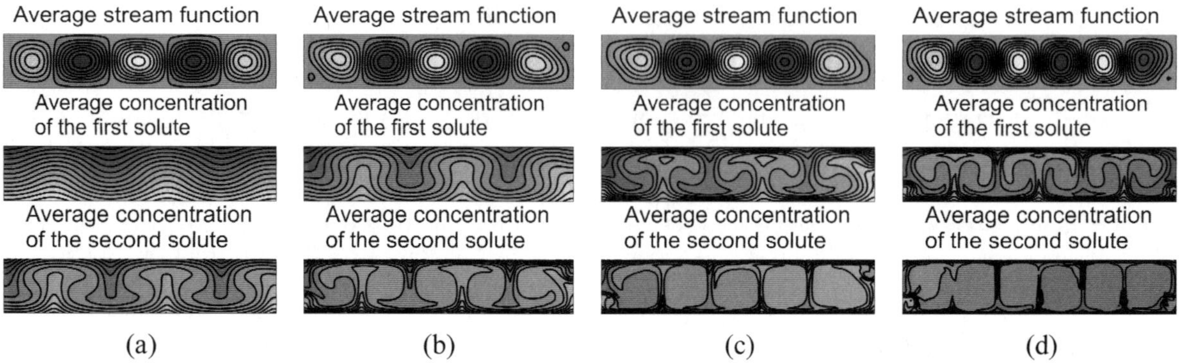

Figure 5. Streamlines of average flow and solute concentration fields at $\mathrm{Ra} = 3{\cdot}10^2$: (a) – $\mathrm{Ra_v} = 1{\cdot}10^3$, (b) – $\mathrm{Ra_v} = 5{\cdot}10^4$, (c) – $\mathrm{Ra_v} = 1{\cdot}10^5$, (d) – $\mathrm{Ra_v} = 2{\cdot}10^5$.

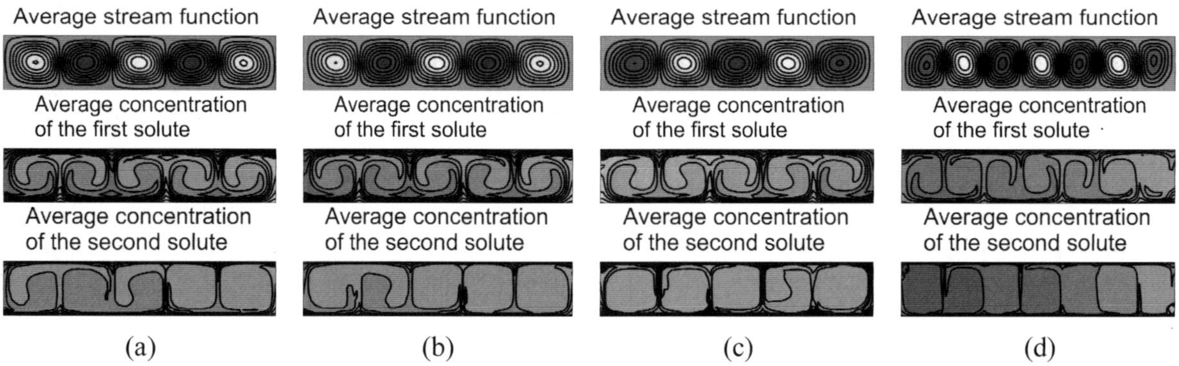

Figure 6. Streamlines of average flow and solute concentration fields at $\mathrm{Ra} = 5{\cdot}10^3$: (a) – $\mathrm{Ra_v} = 1{\cdot}10^3$, (b) – $\mathrm{Ra_v} = 5{\cdot}10^4$, (c) – $\mathrm{Ra_v} = 1{\cdot}10^5$, (d) – $\mathrm{Ra_v} = 2{\cdot}10^5$.

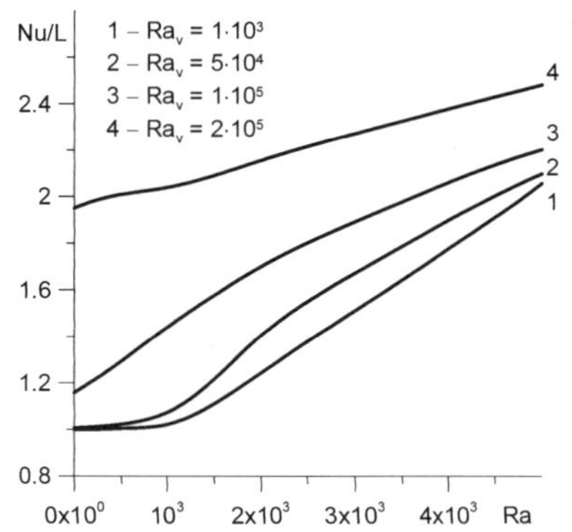

Figure 7. Dimensionless heat flux per unit length versus Rayleigh number at different vibrational Rayleigh numbers.

The dimensionless heat flux per unit length through the lower boundary of cavity monotonously increases with the increase of vibration intensity (fig. 7). At low vibration intensity (fig. 7, curve 1), as in absence of vibrations (see section 4.1), at $\mathrm{Ra{=}Ra^*} \approx 1\cdot 10^3$ the sharp increase of the flow intensity and heat and mass transfer related to the activation of thermogravitational mechanism of instability is observed; it is accompanied by the sharp change of inclination of the curve $\mathrm{Nu/L(Ra)}$. In this case the destabilizing effect of vibrations leads to the fact that this change occurs at lower Rayleigh number than in the absence of vibrations (at $Ra_v = 10^3$ the critical Rayleigh number for the onset of thermogravitational convection in horizontal layer of single-component fluid is approximately 907). The increase of Ra_v leads to the further decrease Ra* (Fig. 7, curve 2). Further growth of the vibration intensity leads to a smoothing of curve Nu/L(Ra) (the change of the curve Nu(Ra) inclination associated with the activation of thermogravitational mechanism becomes less pronounced (fig. 7, curves 3, 4)).

5. Conclusions

The horizontal vibration effect on the Soret-induced convection of ternary mixture with positive separation ratios of both solutes in horizontally elongated rectangular cavity with rigid boundaries heated from below is studied numerically. It is found that in the absence of vibrations in the cavity the multi-vortexes flow with a spatial period decreasing with increasing gravity (Rayleigh number) occurs in a threshold manner. When the Rayleigh number reaches the value which corresponds to the onset of thermogravitational convection, the flow transformation occurs which is accompanied by the sharp increase in intensity of flow and change the inclination of the Nusselt number dependence on the Rayleigh number curve. Vibrations make destabilizing effect, as a result, the Rayleigh number, where the sharp increase in the intensity of the convective flow of ternary mixture is observed, decreases.

Acknowledgments

The study was supported from the grant of President of Russian Federation for the support of Leading Scientific Schools of Russian Federation (grant NSh-9176.2016.1).

References

[1] Ryzhkov I I 2013 *Fluid Dynamics* **48(4)** 477–490
[2] Lyubimova T P and Zubova N A 2015 *The European Physical Journal E* **38** 19
[3] Gershuni G Z, Kolesnikov A K, Legros J C and Myznikova B I 1997 *Journal of Fluid Mechanics* **330** 251–269
[4] Gershuni G Z, Kolesnikov A K, Legros J C and Myznikova B I 1999 *International Journal of Heat and Mass Transfer* **42** 547–553
[5] Lyubimova T, Shklyaeva E, Legros J C, Shevtsova V and Roux B 2005 *Advances in Space Research* **36** 70-74
[6] Shevtsova V, Melnikov D, Legros J C, Yan Y, Saghir Z, Lyubimova T, Sedelnikov G and Roux B 2007 *Physics of Fluids* **19** 01711
[7] Lyubimova T P 2015 *The European Physical Journal E* **38** 43
[8] Lyubimova T P, Zubova N A and Shevtsova V M 2016 *Journal of Physics: Conference Series* **681** 012041
[9] Gershuni G Z and Lyubimov D V 1998 *Thermal Vibrational Convection* (New York: Wiley)
[10] Thom A and Apelt C J 1961 *Field computations in engineering and physics* (London: Van Nostrand)
[11] Ryzhkov I I and Shevtsova V M 2009 *Physics of Fluids* **21** 014102
[12] Silveston P L 1958 *Forsch Ing Wes* **24** 29-32 59-69

IOP Conf. Series: Materials Science and Engineering **208** (2017) 012024 doi:10.1088/1757-899X/208/1/012024

Numerical modelling of meniscus deformations and flows in a liquid bridge subjected to axial vibrations

T P Lyubimova[1,2], Ya N Parshakova[1], R V Skuridin[1]
[1] Institute of continuous media mechanics UB RAS, 1, Acad. Korolev St., Perm, 614013, Russia
[2]Perm State University, 614990, Bukireva st, 15, Perm, Russia

E-mail: lubimova@psu.ru

Abstract. Flows and deformations of free surface of isothermal liquid bridge under the influence of axial vibrations of finite amplitude and frequency in microgravity conditions are studied numerically by Volume of Fluid method. Investigation is carried out in the framework of full non-average approach. Numerical data on instantaneous and average velocity fields and instantaneous and average shapes of free surface of fluid at different vibration frequencies and amplitudes are obtained.

Vibrations are simple and low-energy means of control of inhomogeneous fluids behavior. One of important manifestation of vibration influence on fluids is generation of average flows. Investigation of generation of average flows was started by the work [1] of Rayleigh, where generation of average flow in the Stokes boundary layer near rigid surface during excitation of standing acoustic wave in the channel was studied. Schlichting obtained ([2]) effective boundary conditions for the tangential component of average velocity on rigid boundary based on the analysis of viscous boundary layer near oscillating rigid surface. In work [3] by Longuet-Higgins the generation of average flows near free surface due to the one-dimensional waves propagating on the free surface was studied. Generalization of the results of Longuet-Higgins for the case of two-dimensional wave fields was carried out in [4]. Theoretical, numerical and experimental study of flows in isothermal fluid near vibrating surface is carried out in [5], where method of averaging of vibrational flow during theoretical analyze is suggested, averaged solution of momentum equations is obtained, comparison with experimental results is performed. Zharikov et al. [6] performed an experimental study of the liquid motion near vibrating bodies imitating different techniques of growth from the liquid phase; in particular, the growth of $NaNO_3$ crystal subject to axial vibrations, in Cz-configuration, with small amplitudes (from 0.005 to 0.5 mm) and frequencies up to 100 Hz. This experimental study showed that vibration-driven convection has an important effect on the shape of the liquid-solid interface, and demonstrated that the vibrational flow would be able to counteract the buoyancy-driven flow. Works [7-10] are devoted to the investigation of the vibration effect on flows and heat mass transfer during crystal growth out of melt by floating zone method. In these papers pulsational and average flows, induced by small-amplitude high-frequency axial vibrations of one or both rods and the interaction of this flows with thermocapillary flow induced by the dependence of surface tension on temperature were studied on the basis of average approach. In the present work the flows and deformations of free surface of isothermal liquid bridge under the influence of axial vibrations of finite amplitude and frequency in microgravity conditions are investigated numerically by Volume of Fluid method.

1. Problem statement

Let us consider the interface deformations and flows in a cylindrical liquid bridge, surrounded by a cylindrical layer of air, situated between two horizontal rigid plates (Fig. 1). The system is subjected to periodic forcing with finite amplitude a and finite circular frequency ω along the bridge axis (periodically changing gravity force or synchronous vibrations of both rigid plates). Gravity force is absent. Isothermal problem is under consideration.

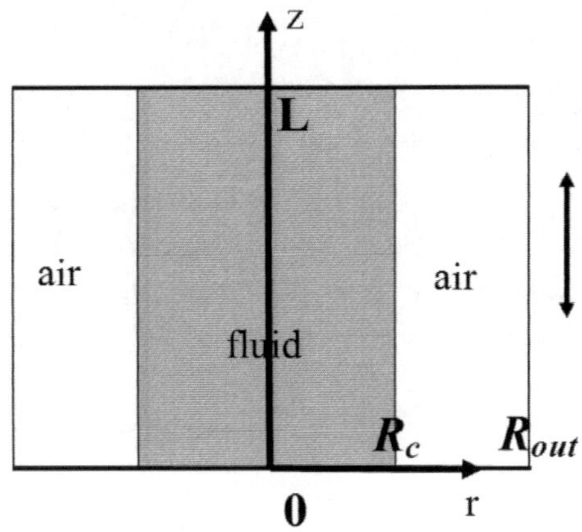

Figure 1 Problem configuration

2. Governing equations

Equations, describing the motion of isothermal viscous uncompressible fluid under the influence of axial vibrations in cylindrical coordinate system have the following form:
for radial component of velocity u:

$$\frac{\partial}{\partial t}(\rho u)+\frac{1}{r}\frac{\partial}{\partial z}(r\rho \mathrm{v}u)+\frac{1}{r}\frac{\partial}{\partial r}\left(r\rho u^2\right)-\rho\frac{w^2}{r}=-\frac{\partial p}{\partial r}+\frac{1}{r}\frac{\partial}{\partial z}\left[r\mu\left(\frac{\partial u}{\partial z}+\frac{\partial \mathrm{v}}{\partial r}\right)\right]+\frac{2}{r}\frac{\partial}{\partial r}\left(r\mu\frac{\partial u}{\partial r}\right)-2\mu\frac{u}{r^2} \quad (1)$$

for azimuthal velocity w:

$$\frac{\partial}{\partial t}(\rho w)+\frac{1}{r}\frac{\partial}{\partial z}(r\rho u w)+\frac{1}{r}\frac{\partial}{\partial r}(r\rho \mathrm{v}w)=\frac{1}{r}\frac{\partial}{\partial z}\left[r\mu\frac{\partial w}{\partial z}\right]+\frac{1}{r^2}\frac{\partial}{\partial r}\left[r^3\mu\frac{\partial}{\partial r}\left(\frac{w}{r}\right)\right]-\rho\frac{\mathrm{v}w}{r} \quad (2)$$

for axial component of velocity v:

$$\frac{\partial}{\partial t}(\rho \mathrm{v})+\frac{1}{r}\frac{\partial}{\partial z}\left(r\rho \mathrm{v}^2\right)+\frac{1}{r}\frac{\partial}{\partial r}(r\rho u\mathrm{v})=-\frac{\partial p}{\partial z}+\frac{2}{r}\frac{\partial}{\partial z}\left(r\mu\frac{\partial \mathrm{v}}{\partial z}\right)+\frac{1}{r}\frac{\partial}{\partial r}\left[r\mu\left(\frac{\partial \mathrm{v}}{\partial r}+\frac{\partial u}{\partial z}\right)\right]-a\omega^2\rho\sin\omega t \quad (3)$$

Here ρ is the density, μ is the dynamic viscosity of the melt.

Continuity equation reads:

$$\nabla\cdot\vec{\mathrm{v}}=\frac{\partial u}{\partial r}+\frac{\partial \mathrm{v}}{\partial z}+\frac{u}{r}=0 \quad (4)$$

3. Numerical method

The calculations were performed using Volume of Fluid method where the description of multiphase fluxes is carried out by introducing volume fraction of each fluid; volume fractions represent the space, occupied by each phase and conservation of mass and momentum laws are fulfilled for each phase.

The qth fluid volume fraction in the cell is denoted as α_q, then the following three conditions are possible:

$\alpha_q = 0$: The cell is empty (of the qth fluid).

$\alpha_q = 1$: The cell is full (of the qth fluid).

$0 < \alpha_q < 1$: The cell contains the interface between the qth fluid and one or more other fluids.

The properties appearing in the transport equations are determined by the presence of the component phases in each control volume. In a two-phase system, for example, if the phases are denoted by the subscripts 1 and 2, and if the volume fraction of the second of them is being tracked, the density in each cell is given by $\rho = \alpha_2\rho_2 + (1-\alpha_2)\rho_1$.

The volume fraction equation will not be solved for the primary phase; the primary-phase volume fraction will be computed based on the following constraint: $\sum_{q=1}^{n}\alpha_q = 1$.

The tracking of the interface(s) between the phases is carried out by the solution of a continuity equation for the volume fraction of one (or more) of the phases. For the qth phase, this equation has the following form:

$$\frac{\partial}{\partial t}(\alpha_q\rho_q) + \nabla\cdot(\alpha_q\rho_q\vec{v}_q) = \sum_{p=1}^{n}(\dot{m}_{pq} - \dot{m}_{qp}) \tag{5}$$

where $\dot{m}_{pq}$ is the mass transfer from phase q to phase p and $\dot{m}_{qp}$ is the mass transfer from phase p to phase q.

The no-slip condition is imposed for the at the rigid boundaries:

$$r = R_{out},\quad z = 0, L:\quad u = \mathrm{v} = 0$$

For volume fractions on the rigid boundaries a fixed value for fluid and for air, conditions of fixed contact line are imposed:

$$\begin{aligned} z = 0, L:\quad 0 < r < R_c:\ & \alpha_{fluid} = 1,\ \alpha_{air} = 0 \\ r > R_c:\ & \alpha_{fluid} = 0,\ \alpha_{air} = 1 \end{aligned} \tag{6}$$

The calculations were carried out using the commercial package ANSYS Fluent. The analysis was limited by axisymmetric regimes and so 2ddp version of package was used. Implicit scheme of the second order of discretization by time was used. For better resolution the computational grid was condensed near free surface and rigid boundaries. In order to determine the parameters of computational grid, optimal from the view point of computation precision and computational time expenses the test runs with the use of different grids were carried out. On the basis of these computations the grid 100x100 was chosen for main computations.

The influence of vibrations was taken into account by introducing a periodically varying volumetric force in Source Terms. The fields of average velocity and other average quantities were calculated by averaging the functions over the five periods of vibrations in User Defined Function.

4. Numerical results

Calculations were carried out for parameters of fluid, corresponding to growth of silicon crystals: density was $\rho = 2.53$ g/cm³, dynamic viscosity was $\mu = 8.855\times10^{-3}$ g/cm/s, surface tension was $\gamma = 720$ dyn/cm. Bridge radius was $R_c = 0.5$ cm, height $L = 1$ cm. Vibration parameters were varied.

In Figs. 2 - 4 the average velocity vector fields and average shapes of free surface at the vibration amplitude 0.15 mm and different vibration frequencies are presented.

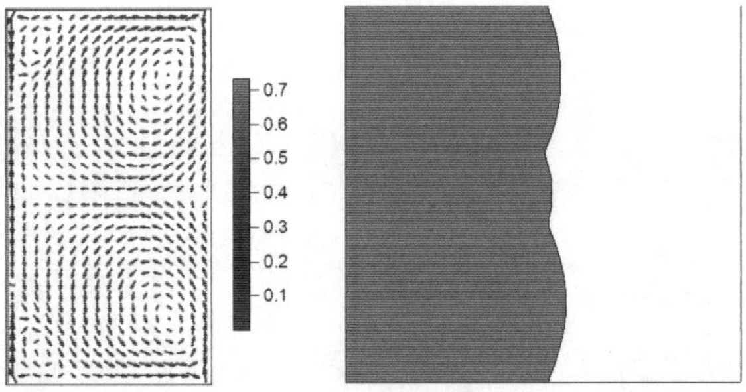

Figure 2 Average velocity vector fields and meniscus deformations for the vibration amplitude equal to 0.15 mm and frequency 10 Hz

The calculations show that the vibrations generate the waves on free surface, propagating from the oscillating rods toward zone center, the amplitude and number of waves depend on the amplitude and frequency of vibrations. The average flow, generated by the surface waves, is directed near free surface from the rigid rods to zone center. At frequency 10 Hz (Fig. 2) the average flow in the bridge consists of two toroidal surface-wave-induced vortices of large size and intensity with the flow direction near free surface from the rigid rods toward zone center and two small toroidal vortices, located near the rigid rods close to the axis, these vortices are generated by the Schlichting mechanism of generation of average flow.

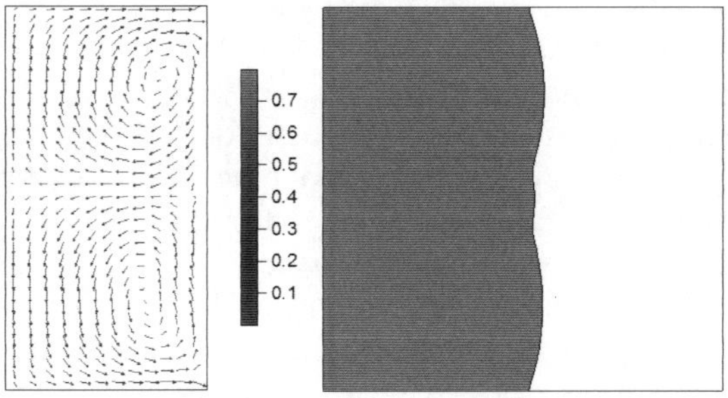

Figure 3 Average velocity vector fields and meniscus deformations for the vibration amplitude equal to 0.15 mm and frequency 50 Hz

With the increase of vibration frequency, the number of waves on the free surface increases, the flow becomes more intensive, the Schlichting vortices are suppressed by the vortices induced by surface waves.

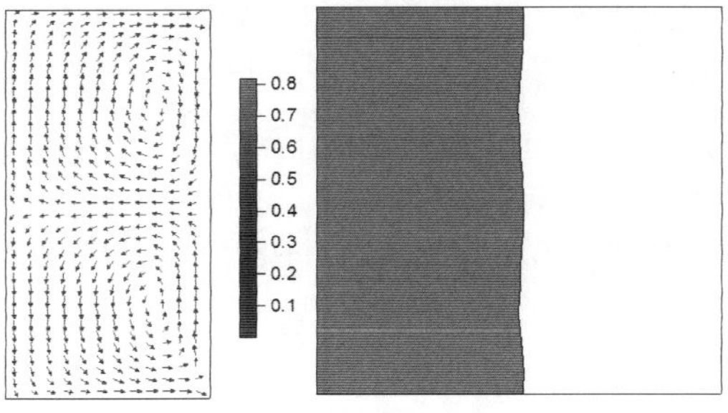

Figure 4 Average velocity vector fields and meniscus deformations for the vibration amplitude equal to 0.15 mm and frequency 100 Hz

In Fig.5 the evolution of instantaneous shape of free surface during the vibration period is shown. As one can see, the waves propagate along the free surface.

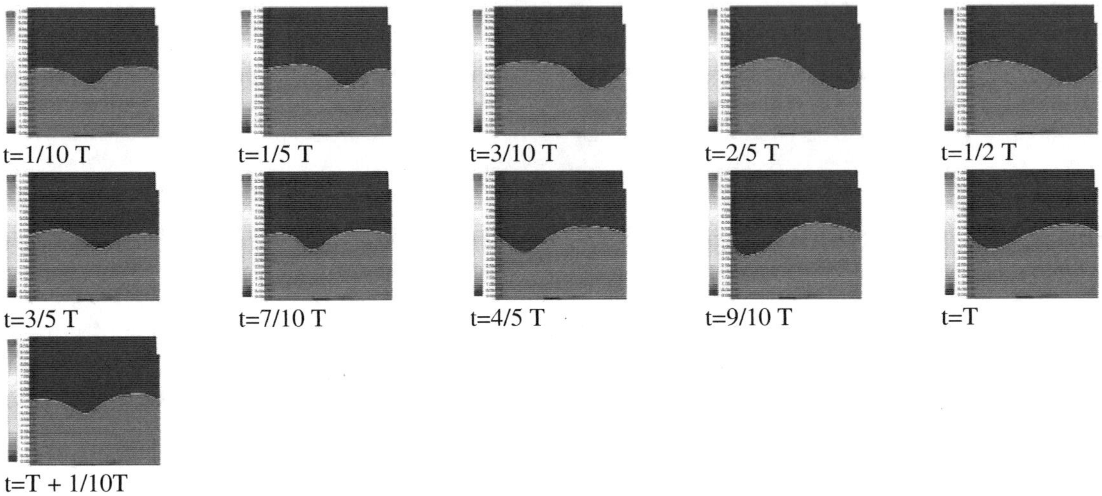

Figure 5 Evolution of instantaneous shape of free surface during the vibration period

5. Conclusion

Meniscus deformations and flows in isothermal liquid bridge subjected to axial vibrations of finite amplitude and frequency in conditions of microgravity have been studied numerically. The calculations were performed in non-averaged approach for fluid parameters which correspond to silicon melt. Numerical data on instantaneous and average velocity fields and instantaneous and average shapes of the meniscus are obtained for different frequencies and amplitudes of vibrations. It is found that under the influence of vibrations the average flow in the form of toroidal vortices of Schlichting origin with the direction near rigid rods from the free surface toward the bridge axis is generated near the rigid rods. Additionally, the rods vibrations generate the waves on the free surface, propagating from the rods toward the bridge center. The average flow induced by surface waves is directed near free surface from oscillating rods toward the cavity center. With the increase of vibration frequency the relative contribution of the Schlichting mechanism of average flow generation decreases, and that of the surface waves mechanism increases. During crystals growth by the floating zone method with ring heater located in the zone center, the thermo-capillary flow develops in the zone; it has the form of two toroidal vortices with the direction of flow near free surface from the heated center of zone toward the cold rigid rods. The average flow, induced by surface waves mechanism of generation, represent two toroidal vortices with the direction, opposite to thermocapillary flow. This makes promising the use of axial vibrations for the improvement of quality of grown crystals.

Acknowledgments

The work was carried out with financial support of Russian Foundation for Basic Research (grant 15-01-09069).

References

[1] Strutt J. W. (Baron Rayleigh). The Theory of Sound 1894 **1** *Macmillan*
[2] Schlichting H., Gersten K., Boundary Layer Theory, 8th ed. 2004 *Springer-Verlag*
[3] Longuet-Higgins M.S. Mass transport in water waves 1953 *Philosophical Transactions A* **245** 535
[4] Gershuni G.Z., Lyubimov D.V. Thermal Vibrational Convection 1998 *Wiley*

[5] Avdonin N.A., Zharikov E.V., Kalis Kh.E., Storozhev N.R. Analiz vtorichnykh techenij v zhidkosti vblizi vibriruyshchej poverkhnosti [Analysis of secondary flow in a liquid near a vibrating surface]. Preprint No. **90**, *General Physics Institute, Academy of Sciences USSR, Moscow,* 1989. 17 p.
[6] Zharikov E.V., L.V.Pribod'ko, Storozhev N.R. Fluid flow formation resulting from forced vibration of a growing crystal 1990 *J. Crystal Growth* **99**, 910-914
[7] Anilkumar A.V., Grugel R.N., Shen X.F., Lee C.P., Wang T.G. Control of thermocapillary convection in a liquid bridge by vibration 1993 *J. Appl. Phys*. **73**, **9** 41654170
[8] Lyubimov D., Lyubimova T., Roux B. Mechanisms of vibrational control of heat transfer in a liquid bridge 1997 *Int. J. Heat Mass Transfer* **40 17** 4031
[9] Lyubimov D.V., Lyubimova T.P., Skuridin R.V., Chen G., Roux B. Numerical investigation of meniscus deformation and flow in an isothermal liquid bridge subject to high-frequency vibrations under zero gravity conditions 2002 *Computers & Fluids* **31**, **4-7** 663
[10] Lyubimova T.P., Scuridin R.V., Croell A., Dold P. Influence of high frequency vibrations on fluid flow and heat transfer in a floating zone 2003 *Crystal Research and Technology* **38 7-8** 635

IOP Conf. Series: Materials Science and Engineering **208** (2017) 012025 doi:10.1088/1757-899X/208/1/012025

Structure of critical perturbations in a horizontal layer of melted water with the prescribed heat flux at the boundaries

V A Sharifulin[1] and T P Lyubimova[2]

[1] Perm National Research Polytechnic University, 29, Komsomolsky ave., Perm, 614990, Russia
[2] Institute of Continuous Media Mechanics UB RAS, 1, Korolev St., Perm, 614013, Russia

E-mail: vadsharif@bk.ru

Abstract. The paper deals with the investigation of critical perturbations structure in a horizontal layer of fluid with temperature inversion of density. The lower boundary of the layer is rigid and upper boundary is free and undeformable. The constant vertical heat flux is imposed at both boundaries. The thermocapillarity, evaporation and radiation are neglected. The temperature dependence of the density is assumed to be quadratic. The coordinate of the density inversion point which shows the position of the plane of maximal density in the fluid layer in the conductive state is used for the description of density inversion effect. The influence of the location of density inversion point on structure of critical perturbations is studied. It is shown that, depending on the location of the inversion point, velocity profile of longwave critical perturbations can have two-floor or three-floor structure. Stability to finite-wavelength perturbations is studied for the entire range of their existence. The asymptotic formulas for critical values of the Rayleigh number and wave number are obtained, the structure of finite-wavelength critical perturbations .is determined.

1. Introduction

Non-monotonic, i.e. non-linear dependence of thermal expansion of water was mentioned in the Book [1] by G. Ostroumov on thermal convection. Veronis G. in [2] suggested to approximate this non-linear temperature dependence of water density by simple quadratic formula. In 1963 it was shown independently by Eklund H. [3] that in normal conditions in the range between zero and eight degrees the specific volume of water is well described by quadratic dependence on temperature. Independently of [2,3], the quadratic temperature dependence of density was obtained by Goren S.L. in [4]. In modern notations it has the form:

$$\rho = \rho_m (1 - \alpha (T - T_i)^2) \tag{1}$$

where $\alpha = 0.8 \cdot 10^{-5} ({}^{\circ}C)^{-2}$, $T_i = 4.00^{\circ}C$.

The onset of convection in a horizontal fluid layer heated from below in the presence of density inversion described by (1) was studied for the first time in [2]. In [2] the case of isothermal boundaries was considered and the subcritical excitation of convection was detected. The mathematical equivalency of the conductive state stability problem for a fluid layer with rigid isothermal boundaries in the presence of density inversion and the Couette flow between rotating cylinders was noted in this paper. In the latter case, the ratio of angular velocities of inner and outer cylinders plays the same role as the ratio of the height of stably stratified part of the layer to its full thickness. A simple algebraic formula relating the Taylor and Rayleigh numbers was derived.

In [5] the important cases of fixed temperature and fixed heat flux (directed vertically from the bottom to the top, see figure 1) at the boundaries were firstly investigated; however, an error was made in deriving the linearized equations. Correct formulation of the problem of the conductive state stability for the case of rigid lower and free and undeformable upper boundary was given in [6]. The dimensionless inversion point coordinate z_i was used for the description of density inversion effect. The parameter z_i has simple physical meaning: this parameter describes the position of a maximum density plane within a fluid layer in the conductive state.

At $z_i < 0$ the density stratification in layer is stable throughout the entire layer and the conductive state is absolutely stable; at $z_i > 1$ (the dimensionless thickness of the layer equals to 1) stratification is unstable throughout the entire fluid layer and at $0 < z_i < 1$ the density stratification is unstable only in a part of the layer, namely, at $0 < z < z_i$.

The thermal conditions corresponding to a fixed heat flux are favorable for the development of longwave perturbations which, in the case of linear temperature dependence of the density (the Boussinesq fluid) are the most dangerous [7]. In [6] it was shown that differ from the Boussinesq fluid, in the case of fluid with density inversion this is not always the case. The longwave perturbations remain the most dangerous if potentially unstable stratification exists throughout the entire fluid layer, or, at least, if the layer with stable stratification is thin. However, when the layer with unstable stratification is thin, the longwave perturbations are strongly stabilized and the finite-wavelength perturbations become the most dangerous.

In this paper we investigate the structure of both longwave and finite-wavelength critical perturbations.

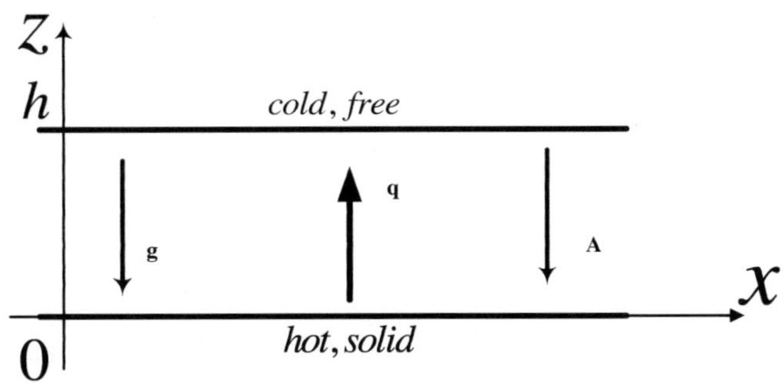

Figure 1. Horizontal layer with fixed heat flux at the boundaries.

2. Method of solution

In conditions of stationary heat flux A from bottom to top possible motionless equilibrium state. Let us consider normal monotonic perturbations of this state. In [6] it is shown that the form of normal critical perturbations and the values of critical Rayleigh number for fixed values of wave number k and parameter z_i can be obtained from the solution of the boundary-value problem:

$$\Delta^2 W + \mathrm{Ra}(z - z_i)k^2 \vartheta = 0, \quad \Delta\vartheta + W = 0,$$
$$z = 0: \vartheta' = W = W' = 0, \quad z = 1: \vartheta' = W = W'' = 0. \tag{2}$$

IOP Conf. Series: Materials Science and Engineering **208** (2017) 012025 doi:10.1088/1757-899X/208/1/012025

Here W and ϑ – dimensionless amplitudes of vertical velocity and temperature perturbations, $\mathrm{Ra} = 2\alpha g A^2 h^5 / \nu\chi$ -Rayleigh number. Analytical method for the long wave limit $k \to 0$ and numerical method for finite k are described in [6]. In the present work the form of critical perturbations is defined from the solution of the problem (2), such study was not performed earlier.

3. Results

Longwave instability occurs at $z_i = 5/9 \approx 0.55$ [6]. In the range $0.61 < z_i < \infty$ the longwave perturbations are more dangerous than perturbations with finite wavelength. The influence of z_i is presented in figures 2, 3 (the boundary between unstably and stably stratified layers is marked by the dotted line). The stable layer in these figures is large enough and comparable in its size with the unstable one.

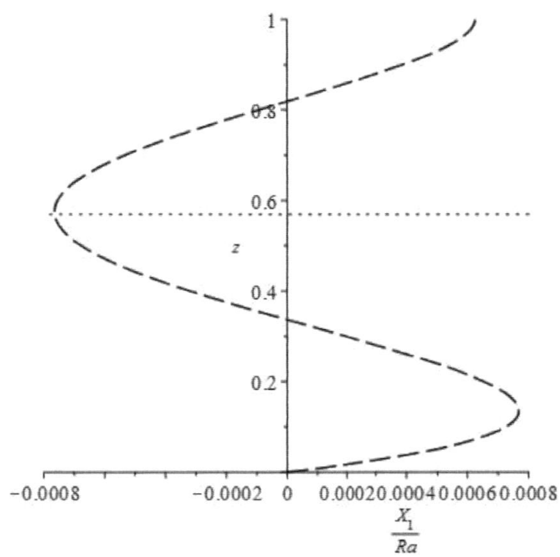

Figure 2. At the moment of the onset of longwave instability there are 3 layers with different form of critical perturbations. For this value of z_i the perturbations with finite wavelength are more dangerous. The profile of horizontal velocity X_1 of neutral longwave perturbations at $z_i = 0.57$ is presented. Velocity values are normalized by the value of Rayleigh number. The dotted line describes the upper boundary of the layer with unstable stratification.

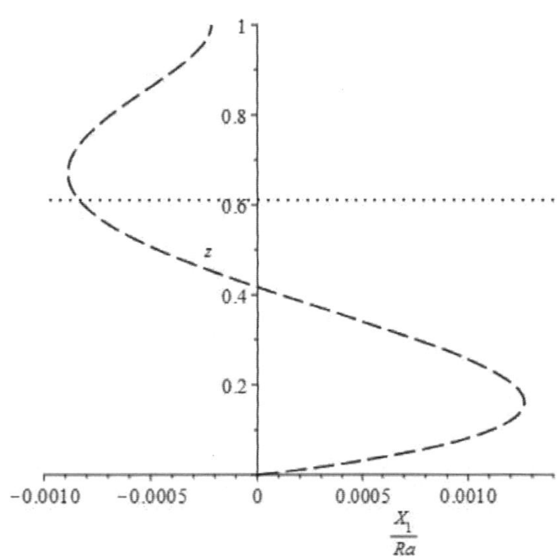

Figure 3. Profile of horizontal velocity X_1 of neutral longwave perturbations for $z_i = 0.6104$. For this value of z_i the perturbations with finite wavelength are not the most dangerous. The dotted line marks the upper boundary of the layer with unstable stratification.

Surprisingly, the perturbations presented in figure 3 has the velocity values on the free surface close to zero. Maximal velocity of fluid is replaced deep in unstable layer.

The growth of the thickness of unstable layer significantly changes the form of the profile of critical perturbations. The maximum value of velocity is reached now on the free surface (see figure 4 for $z_i = 0.8$). Further growth of z_i does not change significantly the form of critical longwave perturbations of the velocity.

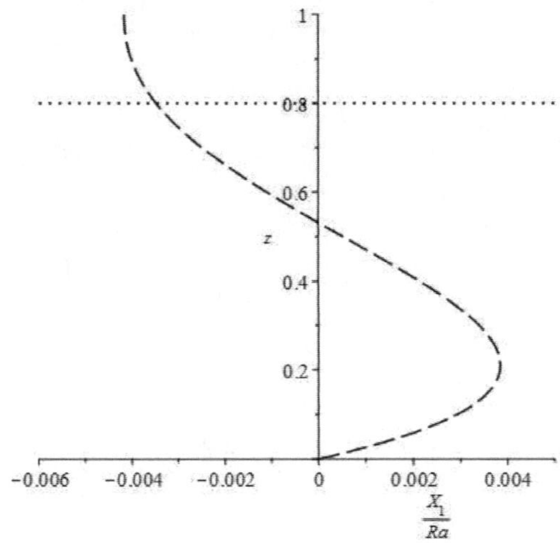

Figure 4. Profile of horizontal velocity X_1 of neutral longwave perturbations for $z_i = 0.8$. For this value of z_i the perturbations with finite wavelength are not the most dangerous. The dotted line marks the upper boundary of the layer with unstable stratification.

If the thickness of unstably stratified layer is less than 0.61 the finite-wavelength perturbations become more dangerous than the longwave ones. The neutral curves obtained by the solution of linear stability problem are presented in figure 5 for three values of z_i.

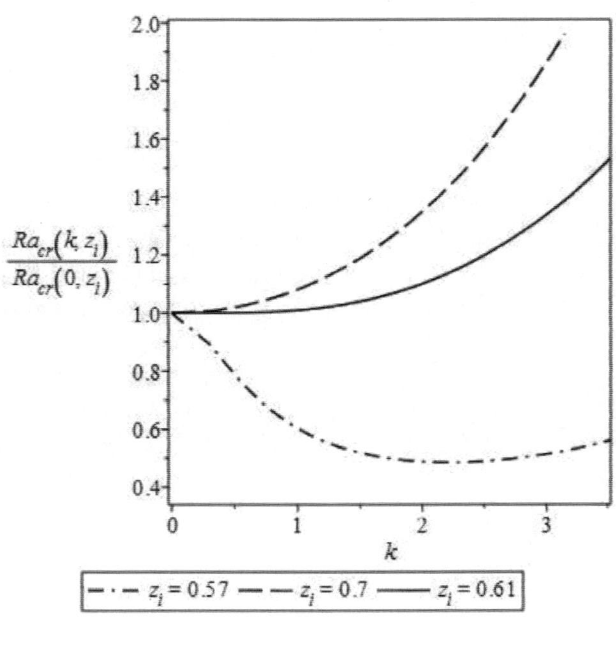

Figure 5. Neutral curves for three values of z_i

IOP Conf. Series: Materials Science and Engineering **208** (2017) 012025 doi:10.1088/1757-899X/208/1/012025

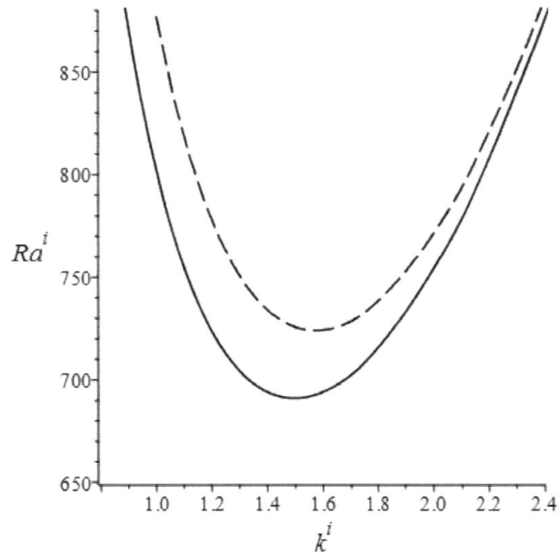

Figure 6. Neutral curves of finite-wavelength instability for $z_i = 1/2$ (dashed line) and $z_i = 1/3$ (solid line).

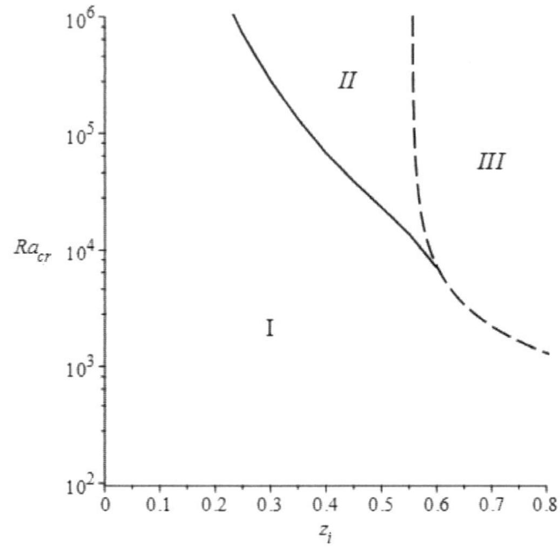

Figure 7. The conductive state stability boundaries to the longwave (dashed line) and finite-wavelength (solid line) perturbations.

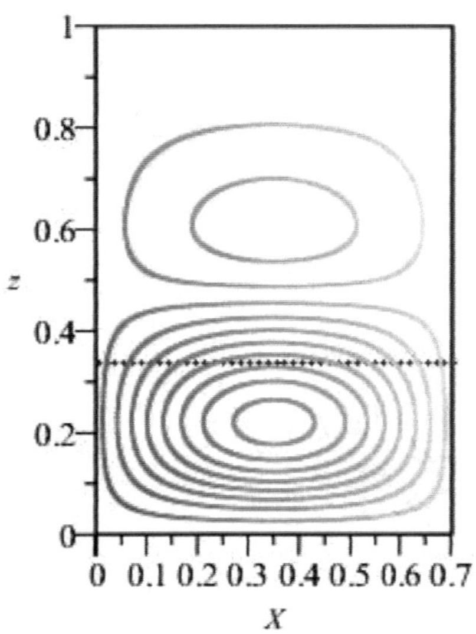

Figure 8. Streamlines of critical finite-wavelength perturbations for $z_i = 1/3$. Dotted line marks upper boundary of layer with unstable stratification. One can see that cells penetrate deeply in stably stratified layer of fluid.

We introduce the modified Rayleigh number Ra^i and the modified wave number k^i, determined using the thickness of unstably stratified layer, i.e. z_i. They are related to the Rayleigh number determined by the thickness of the layer as

$$Ra^i = z_i^5 Ra, \quad k^i = k / z_i . \tag{3}$$

The calculations show that in these variables the neutral curves are almost independent for $z_i \le 1/3$. This means that the instability boundary with respect to finite-wavelength perturbations (separating the regions I and II in figure 8) for $z_i \le 1/3$ can be described by the asymptotic formula:

$$Ra_{cr} = 591/z_i^5 \tag{4}$$

Conclusions

The structure of both longwave and finite-wavelength critical perturbations has been investigated для всех значений координаты точки инверсии $0 < z_i < \infty$.

In the case when unstable stratification occupies the whole layer or its major part ($0.61 < z_i < \infty$), the critical longwave perturbations have simple two-floor structure, beside at $z_i = 0.61$ the velocity of the perturbations at the free surface is close to zero. In the case $0.55 < z_i < 0.61$ when the longwave perturbations are less dangerous than the wavelength perturbations the longwave perturbations have exotic three-floor structure.

Finite-wavelength perturbations at $z_i < 0.55$ have two-floor structure and occupy the whole layer. The asymptotic formula is obtained for the dependence of critical Rayleigh number on z_i from the analysis of the obtained results.

Acknowledgments
The work was supported by the Ministry of Education and Science of Russian Federation (project No. 3.6990.2017/BCh).

References

[1] Ostroumov G A 1958 Free convection under the conditions of the internal problem. NASA Tech. Memo No. 1407
[2] Veronis G 1963 *Astrophys. J.* **137** 641
[3] Eklund H 1963 *Science* **142** 1457-8
[4] Goren S 1966 *Chemical Engineering Science* **21** 515
[5] Merker G, Waas P, Straub J and Grigull U 1976 *Warme und Stoffubertrag* **9** 99
[6] Lyubimov D, Lyubimova T and Sharifulin V 2012 *Fluid Dynamics* **47** 448
[7] Gershuni G Z and Zhukhovitskii E M 1972 *Convective Stability of Incompressible Fluids* (Moscow: Nauka) [in Russian]

IOP Conf. Series: Materials Science and Engineering **208** (2017) 012026 doi:10.1088/1757-899X/208/1/012026

Transport of dissolved gases through unsaturated porous media

B S Maryshev

Institute of Cotinous Media Mechanics UB RAS, 614013, 1 Akad. Korolev street, Perm, Russia

E-mail: bmaryshev@mail.ru

Abstract. The natural porous media (e.g. soil, sand, peat etc.) usually are partially saturated by groundwater. The saturation of soil depends on hydrostatic pressure which is linearly increased with depth. Often some gases (e.g. nitrogen, oxygen, carbon dioxide, methane etc.) are dissolved into the groundwater. The solubility of gases is very small because of that two assumptions is applied: I. The concentration of gas is equal to solubility, II. Solubility depends only on pressure (for isothermal systems). In this way some part of dissolved gas transfers from the solution to the bubble phase. The gas bubbles are immovably trapped in a porous matrix by surface-tension forces and the dominant mechanism of transport of gas mass becomes the diffusion of gas molecules through the liquid. If the value of water content is small then the transport of gas becomes slow and gas accumulates into bubble phase. The presence of bubble phase additionally decreases the water content and slows down the transport. As result the significant mass of gas should be accumulated into the massif of porous media. We derive the transport equations and find the solution which is demonstrated the accumulation of gases. The influence of saturation, porosity and filtration velocity to accumulation process is investigated and discussed.

1. Introduction

The fluid flow into the unsaturated porous media is investigated during last seventy years. The interest to unsaturated media is linked to the applications because usually the media contains two phases: air and liquid. This situation is observed for any natural porous media (groundwater flow, mining etc.) and for many filtration systems. The pioneer theory of two-phase filtration is published in [1] where the idea that saturation mainly depends on phase permeabilities is proposed. After that the authors of [2] assumed that permeability of fluid depends on capillary pressure into the pores and this theory is confirmed in [3] by the analysis of experimental data. Later, the solution for direct description of the dependence of water content (saturation) on pressure is proposed in [4]. After that many particular models [5, 6, 7, 8] are published, but most of them propose the "S"–shaped curve for the dependence of saturation on pore pressure. One of simplest solution is obtained in [9] and confirmed by experimental observations. In this work the pressure is assumed as linear function of depth and the saturation becomes the power function of depth. In this way the water content is increased with depth wherein the three zones of saturation is usually distinguished: I. Dry zone - zone near ground surface where water contains only in very small pores, water content is constant and equal to residual value into this zone. II. Partially saturated zone where the water content is nonlinearly increased with depth from residual value to the porosity of media by "S"–shaped curve. III. Saturated zone where

IOP Conf. Series: Materials Science and Engineering **208** (2017) 012026 doi:10.1088/1757-899X/208/1/012026

the water content is equal to porosity of media [3, 9]. This structure of saturation is confirmed by many experimental data.

The diffusion of solute gases in liquids is well-studied problem see e.g. [10, 11]. Usually the gas-liquid interface is presented as surface and the exchange processes between gas and liquid is described by the boundary conditions. But the flow of liquid with dissolved gas through porous media is a specific process. The free gas is presented as bubbles into the massive of porous media and the interface is distributed into the volume of media. Such media is named "bubbly" media [12, 13]. For moderate values of pressure (the value commensurate with the atmospheric pressure) the solubility of gases is small and the concentration of dissolved gas everywhere is equal to its solubility. The solubility depends only on pressure (for isothermal problem). In the case of mechanical equilibrium the pressure linearly depends on depth it means that the gas bubbles is formed in upper layer of media where the solubility is small. The bubbles of gas are trapped in porous media by surface-tension forces. They are immovable and the main mechanism of gas transport is diffusion of gas molecules into the liquid phase. It means that the decreasing of the proportion of liquid phase slows down the diffusion transport of gases in upper layer. In result the mass of gas accumulates into the upper layer of media. This effect can be very strong for unsaturated media.

The unsaturated porous media also saturated by gas (usually it is air for natural media) and liquid. The proportion of gas is great and it is also presented by bubbles. The dry zone and the top of partially saturated zone contain weak mass of liquid because of small values of hydrostatic pressure. The gas from the solute transits here to the bubble phase and this process additionally reduces the water content. All bubbles are immovable and they occupied most part of media pores as result the diffusive transport is blocked and all significant mass of gas accumulates near the boundary of dry zone. The main purpose of paper is description of the accumulation effect especially on the example of peat bogs where the methane is emitted at decomposition of organic substances.

The paper consists of introduction, three sections and conclusion. Into the introduction the overview of literature and our motivation is presented. The first section is devoted to the derivation of equations for gas transport description in unsaturated media. Into the second section the problem of one dimensional gas transport is formulated and the methods of solution are described. The third section presents the results of solution. The distribution of gas in bubble phase is shown and discussed the influence of governing parameters to the transport and accumulation processes. In the conclusion we summarize all results and discuss them.

2. Diffusion of gases in unsaturated porous media

In the present paper we consider the diffusion of gases in isothermal case at moderate values of pressure (less than 10 bar). In this case we assume that solubility of gas in liquid is small (e.g. it is 10^{-4} for methane) and concentration of gas is equal to the solubility everywhere. In this case the diffusion of gas can be described by the following equation [14]:

$$\frac{\partial}{\partial t}(Q) = \nabla\left(D\nabla C + UQ\right), \tag{1}$$

where D is the effective diffusivity, U is speed of media sedimentation, $Q = C + C_b$ is full volumic concentration of the gas while C and C_b is the parts of media volume which are occupied by the dissolved gas and the gas in the bubble phase (bubble phase consist of air and considered type of gas) respectively. The part of volume which is occupied by liquid phase in porous media is named water content and denotes by θ whereas bubble phase occupies in the part $\varphi - \theta$ where φ is porosity of media. In this case the solubility of gas and its volumic concentration in bubble phase is defined as $X = C/\theta$ and $X_b = C_b/(\varphi - \theta)$. For the considerable values of pressure the solubility depends linearly on pressure by the Henry law $X = P_g/K$ where K is Henry constant

[15]. The partial pressure of gas in bubble phase is determined by expression $P_g = X_b P/(\varphi - \theta)$ where P is full pressure. The substitution of all expressions for concentration to the equation (1) gives the transport equation in terms of X_b, saturation and pressure:

$$\frac{\partial}{\partial t}\left(\frac{\theta X_b P}{K(\varphi-\theta)} + (\varphi-\theta)X_b\right) = \nabla\left(D\nabla\frac{\theta X_b P}{K(\varphi-\theta)} + U\left(\frac{\theta X_b P}{K(\varphi-\theta)} + (\varphi-\theta)X_b\right)\right). \qquad (2)$$

For the closure of transport equation we need to use some model for media saturation. It is known [2] that for heterogeneous permeability of media the saturation depends only from capillary presure. We use the power dependence which is proposed within the model [9].

$$\begin{gathered} \theta = \theta_r, \;\; P < P_w, \\ \theta = \theta_r + (\varphi - \theta_r)\left[1 - (1+\alpha P^n)^{1/n-n}\right], \;\; P_w < P < P_s, \\ \theta = \theta_s = \varphi, \;\; P > P_s, \end{gathered} \qquad (3)$$

where the θ_r is the residual water content of dry porous media, P_w is characteristic pressure which needed for wetting of media, $\theta_s = \varphi$ is water content of full saturated media and P_s is pressure of media saturation. The empiric parameters n and α is described the speed of media saturation under the variation of pressure. The value of full water content of media is smaller than θ to the volumic concentration of gas in bubble phase, because of that the full water content $\theta_f = \theta - X_b$. We should use the θ_f instead of θ into the transport equations. For obtaining the close system of equation we should introduce the equation for pressure. In the present paper the quasistatic problem is considered when the moving into the media is linked only to the slow sedimentation of soil. In this case the pressure of fluid depends only on depth by the hydrostatic distribution $P = \rho g z$ where ρ is the density of liquid, g is gravity acceleration, z is depth. Finally the system of transport equations can be written as

$$\begin{gathered} \frac{\partial}{\partial t}\left(\frac{\theta_f \rho g z X_b}{K(\varphi-\theta_f)} + (\varphi-\theta_f)X_b\right) = \nabla\left(D\nabla\frac{\theta_f \rho g z X_b}{K(\varphi-\theta_f)} + U\left(\frac{\theta_f \rho g z X_b}{K(\varphi-\theta_f)} + (\varphi-\theta_f)X_b\right)\right), \\ \theta = \theta_r, \;\; z < z_w, \\ \theta = \theta_r + (\varphi-\theta_r)\left[1 - \left(1 + \alpha(\rho g)^n z^n\right)^{1/n-n}\right], \;\; z_w < z < z_s, \\ \theta = \theta_s = \varphi, \;\; z > z_s, \\ \theta_f = \theta - X_b, \end{gathered} \qquad (4)$$

where z_w is the depth of dry zone and z_s is depth of media saturation. The system (4) describes the transport of gas into the porous media with partial saturation and the forming of gas bubbles. The solution of (4) for one dimensional problem is presented in next section.

3. Problem statement

We restrict to consider only one dimensional problem about transport of gases in vertical direction (parallel to the gravity field). In this case the massif of unsaturated porous media can be divided to three zones by the saturation: dry, partially saturated and saturated zone. The main interest is the transport in partially saturated zone, because of that we will use its depth as spatial scale $L = z_s - z_w$. The considered porous media is a layer of depth equal to $3L$ where top part with depth L corresponds to dry zone, bottom part to saturated zone. Such choice of problem configuration is based on decay of perturbation outside the partially saturated zone. The configuration of problem is presented on Fig. 1.

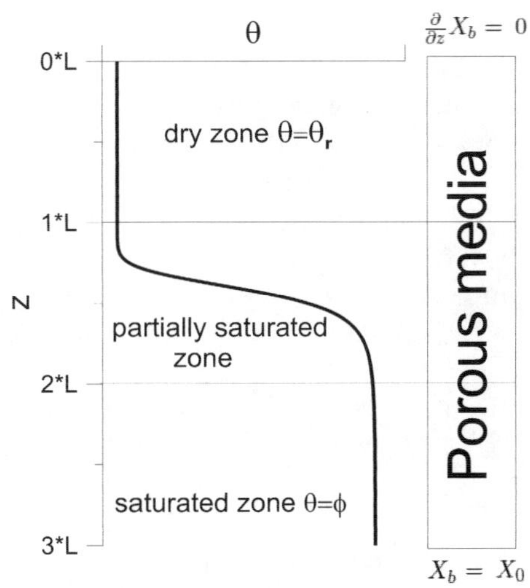

Figure 1. Problem configuration.

We assume that diffusivity and sedimentation speed are constant. Under this assumption the dimensionless form of equations (4) can be written in form

$$\begin{gathered}\frac{\partial}{\partial t}\left(\frac{\theta_f z C_0 X_b}{\varphi-\theta_f}+(\varphi-\theta_f)\,X_b\right)=\frac{\partial^2}{\partial z^2}\frac{\theta_f z C_0 X_b}{\varphi-\theta_f}+Pe\frac{\partial}{\partial z}\left[\frac{\theta_f z C_0 X_b}{\varphi-\theta_f}+(\varphi-\theta_f)\,X_b\right],\\ \theta=\theta_r-X_b,\ \ z<L,\\ \theta=\theta_r+(\varphi-\theta_r)\left[1-(1+Az^n)^{1/n-n}\right]-X_b,\ \ L<z<2L,\\ \theta=\theta_s=\varphi-X_b,\ \ z>3L,\\ \theta_f=\theta-X_b,\end{gathered} \tag{5}$$

where the units for time and depth are L^2/D and L. Equations (5) contain seven parameters $Pe = UL/D$ is the Peclet number (dimensionless speed of sedimentation), $C_0 = \rho g L/K$ specific solubility of gas at the scale of problem, φ is porosity of media and $A = C_0{}^n$. The meanings of parameters n, θ_s and θ_r were explained early. The boundary conditions for (5) is chosen in following form: $\frac{\partial}{\partial z} X_b|_{z=0} = 0$ and $X_b|_{z=3L} = X_0$. The first condition corresponds to the absent of gas flux on upper boundary because the bubble of gas is immovable. The second condition means that into the deep of saturated zone the air is absent and the concentration of gas in bubble phase depends only on full concentration and pressure. The value X_0 is defined by the expression $X_0 = Q_0 - 3C_0$ where Q_0 is full concentration of gas and $3C_0$ is the gas solubility at the depth $3L$.

One of the most interesting applications of considered problem is the accumulation of methane into the peat-bogs. Because of that the considered gas is methane and the porous media is peat soil. The estimation of governing parameters for this system gives $X_0 \sim C_0 \sim 10^{-4}$, $Pe \sim 0.1..10$, $A \sim 10^4$, $n \sim 5..6$ and $\varphi \sim 0.6..0.9$ this values can be found in [3, 9, 16].

The equations (5) with boundary conditions is solved numerically [17] by the semi-implicit finite difference scheme of second order accuracy in space and first order in time. The results of solution and the influence of governing parameters to accumulation are discussed in following section.

4. Results

The profiles of concentration of gas in bubble phase are presented into this section. All profiles demonstrate the effect of gas accumulation into partially saturated zone. The parameters of saturation and the concentration of gas in bubble phase at the depth $3L$ for all calculations are $A = 10^4$, $n = 6$ and $X_0 = 10^{-4}$. The concentration of accumulated gas reached the value $X_b = 0.09 \sim 100X_0$. The effect is strong into the unsaturated media because the small saturation of media and transition of gas into the bubble phase both slow down the diffusive transport. The sedimentation process speeds up the transportation and the mass of accumulated gas decreases with increasing of Peclet number. This effect is demonstrated in figure Fig. 2.

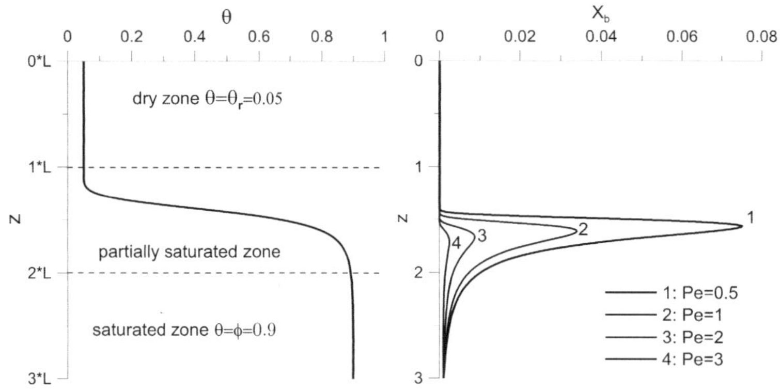

Figure 2. The profiles of gas concentration in bubble phase. The effect of Peclet number variation. The values residual water content and porosity are $\theta_r = 0.05$ and $\varphi = 0.9$.

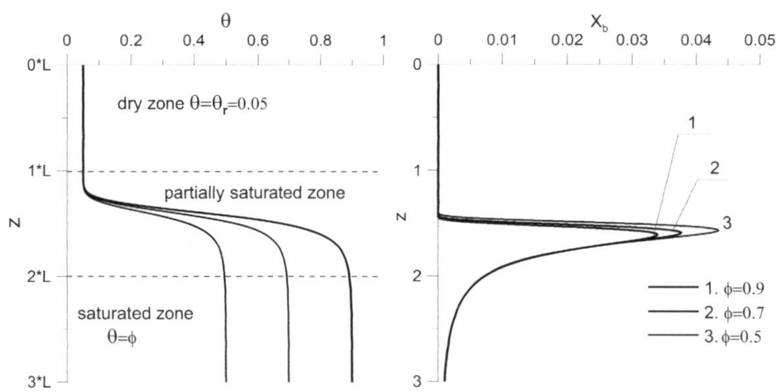

Figure 3. The profiles of gas concentration in bubble phase. The effect of porosity variation. The values of residual water content and Peclet number are $\Theta_r = 0.05$, $Pe = 1$.

The figure Fig. 3 demonstrates the influence of porosity variation. The decreasing of porosity leads to increasing the proportion of gas into the pores because the volumic concentration is the same. In result the concentration of gas in bubble phase increases. The effect of residual concentration variation is shown in Fig. 4. It is seen that the increasing of residual water content produced the increasing of mass of accumulated gas and the maximum value of concentration is moved to the upper boundary. This effect can be explained by the expansion of zone where gas can be transferred by the diffusion. This expansion can be illustrated by the Fig. 5. The profile of gas concentration Fig. 5 is obtained for the full saturated porous media. The diffusion does not block because the concentration of gas in bubble phase is not reach the critical value. The profile corresponds the problem of porous media saturation and we obtain the well known [3, 4, 2, 9, 16] "S"–shaped curve for full water content $\theta_f = \varphi - X_b$.

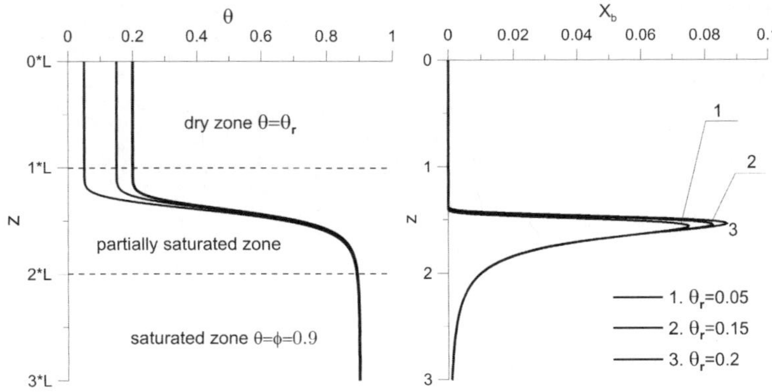

Figure 4. The profiles of gas concentration in bubble phase. The effect of residual water content θ_r variation. The values of porosity and Peclet number are $\varphi = 0.9$, $Pe = 0.5$.

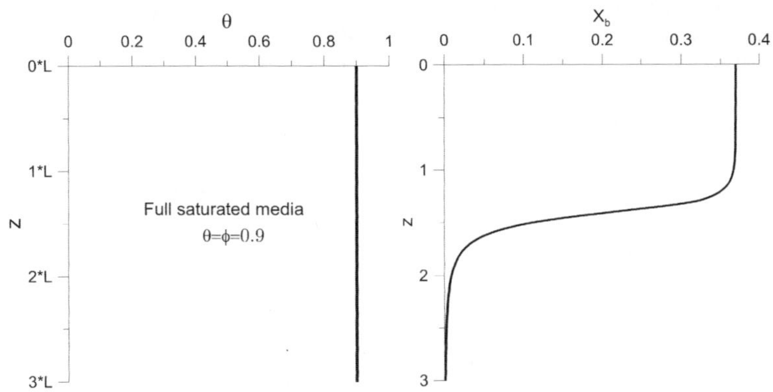

Figure 5. The concentration of gas in bubble phase under the constant water content $\theta = 0.9 = \varphi$. The value of Peclet number is $Pe = 0.5$.

5. Conclusion

The transport of dissolved in liquid gas into the unsaturated porous media is investigated. The gas concentration into the solute is equal to solubility which is dependent only on pressure. When the value of pressure is low the part of gas transits from the solute to the bubble phase. The main mechanism of gas transport is diffusion of gas molecules into the liquid. The bubbles are immovable because of that the transport slows by the bubbles generation. It is shown that the presence of air in unsaturated porous media additionally slows down the diffusion in upper zone of media. The diffusion is blocked and gas accumulates in thin layer. It is obtained that the concentration of accumulated gas may exceed the concentration in the deep layers of media to the hundred times. The equations for gas transport description are derived and the one dimensional problem is solved numerically. It is shown that the increasing of sedimentation speed leads to the reducing of accumulated gas mass due to speed up of transport. The variation of porosity weakly affects to the accumulation. The increasing of water content leads to expansion of zone where diffusive transport of gas is not blocked. Finally it is show that for full saturated media the one dimensional problem of gas transport becomes the problem of media saturation. The resulted profile of gas concentration has the well known "S"–shaped form.

Acknowledgments

This work was partially supported by a Grant of the President of Russian Federation (Grant No. MK-1447.2017.5).

References

[1] Buckley S E and Leverett M 1942 *Trans. AIME* **146** 107
[2] Burdine N 1953 *SPE JPT* **5** 71
[3] Brooks R H and Corey A T 1964 *Trans. ASAE* **7** 26
[4] Mualem Y 1976 *Water Resour. Res.* **12** 513
[5] Padilla I Y, Yeh T J and Conklin M H 1999 *Water Resour. Res.* **35** 3303
[6] Khaleel R, Yeh T C J and Lu Z 2002 *Water Resour. Res.* **38** WR000072
[7] Hilfer R 2006 *Physica* A **371** 209
[8] Gerke H H and Van Genuchten M T 1993 *Water Resour. Res.* **29** 305
[9] Van Genuchten M T 1980 *Soil Sci. Soc. Am. J.* **44** 892
[10] Bird R B, Hirschfelder J O and Curtiss C F 1954 *Molecular theory of gases and liquids* (New York: Wiley)
[11] Bird R B, Stewart W E and Lightfoot E 2007 *Transport Phenomena* (New York: Wiley)
[12] Yurkovetsky Y and Brady J F 1996 *Phys. Fluids* **8** 881
[13] Goldobin D S and Brilliantov N V 2011 *Phys. Rev.* E **84** 056328
[14] Davie M K and Buffett B A 2001 *J. Geophys. Res.* **106** 497
[15] Sander R 2015 *Atmos. Chem. Phys.* **15** 4399
[16] Latrille C and Neel M C 2013 *EPJ Web Conf.* **50** 04002
[17] Fort T 1948 *Finite differences and difference equations in the real domain* (Gloucestershire UK: Clarendon Press)

IOP Conf. Series: Materials Science and Engineering **208** (2017) 012027 doi:10.1088/1757-899X/208/1/012027

Phase coherence of 0.1 Hz microvascular tone oscillations during the local heating

I A Mizeva

Institute of Continuous Media Mechanics UB RAS, 1, Acad. Korolev St. 1, Perm, 614013 Russia

E-mail: mizeva@icmm.ru

Abstract.

The origin of the mechanisms of blood flow oscillations at low frequencies is discussed. It is known that even isolated arteriole demonstrates oscillations with the frequency close to 0.1 Hz, which is caused by the synchronous activity of myocyte cells. On the other hand, oscillations with close frequency are found in the heart rate, which are associated with quite different mechanism. The main purpose of this work is to study phase coherence of the blood flow oscillations in the peripheral vessels under basal and perturbed conditions. Local heating which locally influences the microvascular tone, as one of currently elucidated in sufficient detail physiological test, was chosen. During such provocation blood flow though the small vessels significantly increases because of vasodilation induced by the local synthesis of nitric oxide. In the first part of the paper microvascular response to the local test is quantified in healthy and pathological conditions of diabetes mellitus type 1. It is obtained that regardless of the pathology, subjects with high basal perfusion had lower reserve for vasodilation, which can be caused by the low elasticity of microvascular structure. Further synchronization of pulsations of the heated and undisturbed skin was evaluated on the base of wavelet phase coherency analysis. Being highly synchronised in basal conditions 0.1 Hz pulsations became more independent during heating, especially during NO-mediated vasodilation.

1. Introduction

Spontaneous blood flow pulsations in microvessels have recently attracted significant interest in the context of biomedical research [1], [2], [3] and provided a basis for new diagnostic methods for clinical practice [4]. It is known that even isolated arteriole demonstrates spontaneous oscillations of the lumen diameter. The frequency of such oscillations is close to 0.1 Hz, and they arise due to the synchronous activity of myocyte cells [5]. In-vivo measurements of the vessel lumen is quite difficult that is why blood flow oscillations, which carry an information about vessel vasomotion, are usually monitored.

Skin, as one of the most accessible organs, is a model for the investigation of microvascular dysfunction [6]. There are a variety methods of indirect measurements of blood flow oscillations, such as photopletismography [7], laser speckle [8], temperature variations [32]. Among them laser Doppler flowmetry (LDF) is widely applied [12]. This method has become an attractive technique for the intravital studies of cutaneous blood flow. It enables simple real-time monitoring of the relative changes of red blood cell flux in a cutaneous microvascular bed, which consists of nutritive capillaries and deeper elements of the skin vascular tree. Individual differences in skin thickness and capillary density have a significant impact on the person-to-person LDF signal variability. The mechanism of LDF signal formation is rather complicated and the problems concerning interpretation of such measurements are currently under discussion [9] [10].

In-vivo blood microcirculation system is affected by nerve, humoral and local factors, which influence both the average and oscillating components of the blood flow [11]. This leads to the presence of the fulfilled spectra of blood flow oscillations. Five assigned frequency bands fall in the range from 0.0095 to 1.5 Hz: cardiac [12], respiratory [13], myogenic [14], neurogenic [15] and endothelial [16]. Recently these mechanisms have been intensively studied [17], [18]. At the moment origin of blood flow variations at low frequencies are under discussion. In [17] it is assumed that the microvascular blood flow possesses local mechanisms [19] of generating low-frequency blood flow oscillations. On the other hand, it is found that central baroreflex mechanism [20] likely synchronizes low-frequency oscillations through the whole cardiovascular system. In the paper oscillations with the frequency close to 0.1 Hz are discussed. On the one hand these pulsations are associated with the synchronous action of myocyte cells and they can be not in phase in different locations. On the other hand, such oscillations are associated with central mechanisms [21]. In this case local action will not influence the oscillation syncronization. To answer this question the local action, which can influence the synchronization of different oscillating system in the human body [22], is provided. In this study well studied local heating test is used.

The issue of changing skin perfusion in response to local heating is currently elucidated in sufficient detail [23], [24], [25], and it depends on the heating regime (rate, maximum value, duration of heating). The perfusion – time curve during this physiological test demonstrates several phases. First, fast increase of the blood flow for the first few minutes after heater turning on takes place and it is associated with the axon-reflex. During the prolonged heating with the same temperature the LDF signal has a local minimum that is accompanied by an increase of perfusion. This slow vasodilation is driven by another physiological mechanism – local synthesis of a vasodilator agent NO. The appearance of this perfusion – time curve differs in health and pathological subjects [26], [27], [28]. In this studies the initial LDF curve is smoothed and LDF pulsations are not taken into account.

In the study it is included both average and pulsatile components of perfusion. In the first part of the paper we quantify the response of the microvascular blood flow measured by LDF to local heating in the zones under the heater and in its neighbourhood in two groups: control and patients with diabetes mellitus. In the second part phase coherence of 0.1 Hz pulsations of LDF signals collected at two points is discussed.

2. Materials and methods

2.1. Measurements

The study was carried out in control group of 12 healthy nonsmoking volunteers (CG) and 8 subjects with diabetes mellitus of type 1 without macrovascular complications (DM). All subjects voluntary signed an informed consent of participation in the study, which was proved by the local ethical committee of Perm Medical University. Subjects were all caffeine free for 12 hours before the test and under the controlled laboratory conditions for 15 minutes. Blood flow variations in the skin area were monitored by a two-channel laser Doppler flowmeter (Moor Instruments, UK). A skin heater unit (heat control accuracy $\pm$ 1°C) was mounted on the volar site of the forearm of the left hand of each volunteer using a double sided adhesive disk. A laser Doppler flowmeter sensor was installed into the heater through the hole. Measurements were recorded continuously. During the first 10 minutes the heating element was turned off, and the basal blood flow was recorded. In the eleventh minute, the heating element was turned on (heating up to 40° C during one minute was performed gradually - 1° C per 15 seconds). Then at 40° C the perfusion parameters were recorded during 40 minutes. 50 minutes after the start of the experiment the heating element was turned off, and the restore process was recorded during ten minutes.

The second LDF probe was placed in the neighbourhood of the heater element at a distance of 3 cm from the geometric center of the heater. The imaginary line connected LDF probes was perpendicular to the blood flow in main arteries so that the probes were not positioned one after another in relation to the blood flow.

2.2. Data preprocessing and analysis

The first stage of the statistical analysis of average LDF variations involves a 5 second moving average. On the smooth curve illustrating the LDF signal, the maximum value lies within the axon-reflex phase and the mean value within the plateau phase.

Analysis of oscillating components and signal synchronization was carried out by the original algorithm based on wavelet decomposition. Firstly, data preprocessing was done: large scale trends and random peaks were removed from the signal. Secondly, the signal was decomposed using a wavelet transform [30]:

$$W(\nu,\tau) = \nu \int_{-\infty}^{\infty} f(t)\psi^*(\nu(t-\tau))dt, \tag{1}$$

where * means complex conjugation. The Morlet wavelet written in the form

$$\psi(t) = e^{2\pi it}e^{-t^2/\sigma} \tag{2}$$

was used for the decomposition. Integrating the power over time gives the global wavelet spectrum

$$M(\nu) = \frac{1}{T}\int_0^T |W(\nu,t)|^2 dt. \tag{3}$$

Wavelet phase coherence is defined as

$$C(\nu) = \sqrt{\langle\cos\Delta\phi(\nu,\tau)\rangle^2 + \langle\sin\Delta\phi(\nu,\tau)\rangle^2}. \tag{4}$$

Function $C(\nu)$ belongs to the interval [0,1] and characterises the phase coherence of oscillations in two signals $f_i(t)$ ($i = 1,2$) at the given frequency ν. Phase shift $\Delta\phi(\nu,\tau)$ is defined for every frequency ν as $\Delta\phi(\nu,\tau) = \arg W_1(\nu,\tau)W_2^*(\nu,\tau)$, where W_i are wavelet coefficients for the f_i.

The value of $C(\nu)$ depends on the duration of the signal, short signals will demonstrate higher phase coherence, that's why level of significance should be estimated prior to the research. In this study surrogate data obtained from the initial by amplitude-adjusted Fourier transform are used. We conserve spectral property of the real data and destroy temporal information. In current work 5000 pairs of surrogate data were produced, the duration of the surrogate signals was 15 minutes and significance level was defined as 2 standard deviations above the surrogate mean. The level of significance of the coherency depends on the frequency of pulsations, and higher at lower frequencies.

The Mann-Whitney test was used to compare the intragroup results and the Wilcoxon statistical test for intergroup variations. All data processing was implemented on Mathmatica 8.0, Wolfram research.

3. Results

3.1. Average perfusion analsys

The characteristic LDF records collected in two regions of the hand of healthy volunteer are presented in figure 1. The LDF-time curve has a pronounced peak after the heater is switched

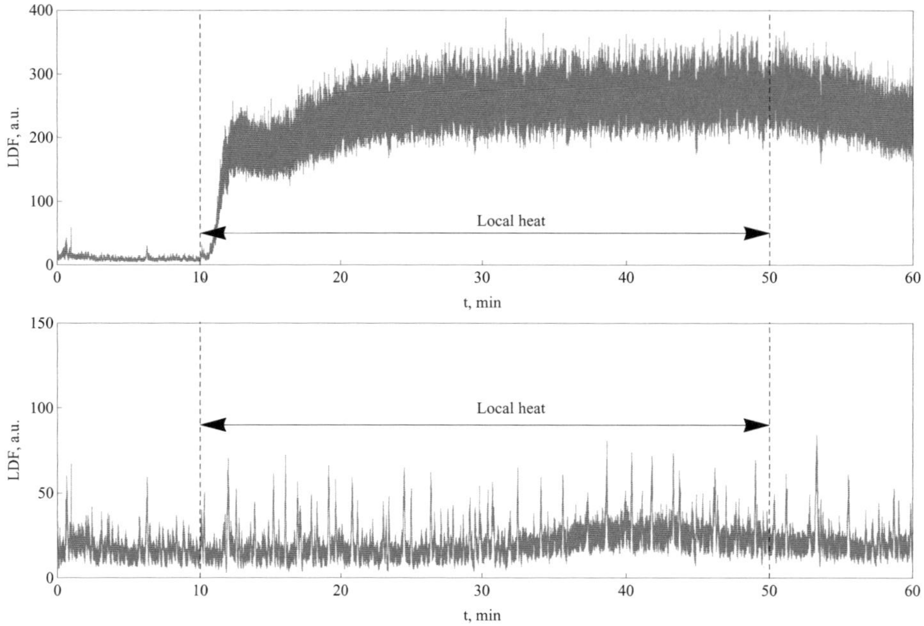

Figure 1. An example of LDF records under the heater (upper panel) and at a distance of 3 cm from the heater (lower panel)

on, which corresponds to axon-reflex vasodilation. After that, perfusion decreases and then slowly increases up to the NO mediated plateau stage. The form of the curve conforms to that described in [23]. In this example, the local heating induces blood flow variation not only directly under the heater but also in the neighboring region, which coincides with the results obtained in [22]. Note that the behaviour of this kind was not observed in all experiments.

Local vasodilation during the prolonged heating is caused by local synthesis of NO, and this effect is slightly observed at the distance of 3 cm, that's why it is assumed that NO concentration is smaller in the undisturbed skin in comparison with heating and input of NO-related mechanisms of vascular tone regulation is smaller in the undisterbed skin.

Averaged over all experiments values of the perfusion are summarized in the Table 1. The difference in the baseline perfusion ((9.5 ± 0.75) p.u. and (7.5 ± 1.4) p.u) is not statistically meaningful. An increase in the blood flow due to the axon reflex is lower in the DM (110 ± 50) p.u. than in CG (190 ± 90) p.u., but the difference is not significant (p=0.06). The mean value of the blood flow caused by the NO dependent vasodilation for CG (250 ± 100)p.u. is much higher that that of the DM (140 ± 70) p.u., p=0.03. The relative variations of perfusion caused by the axon-reflex response to the baseline and perfusion during the NO dependent vasodilation obtained in two groups differed only slightly. A linear correlation analysis ($R_{m,n}$) was carried out to study the mean values of perfusion within three time intervals, which are marked as $m, n =$"Basal", "Axon-reflex", "NO", 4-6 lines of the Table 1. The results indicated a moderate anticorrelation of perfusion under basal conditions and at time of axon-reflex mediated vasodilation (-0.44 for CG and -0.40 for DM). In addition, perfusion at the stage of NO mediated vasodilation is anti-correlated with perfusion observed under basal conditions (-0.65 for CG and -0.46 for DM subjects). In both groups, there was a significant correlation (0.96) in vasodilation between the axon-peak and NO-mediated plateau mechanisms.

A negative correlation of the blood flow during vasodilation and under basal conditions with

Table 1. Statistics of LDF variations

Parameter	*CG*	*DM*	p
Baseline perfusion, p.u.	9.5 ± 0.75	7.5 ± 1.4	0.3
Axon reflex perfusion, p.u.	190 ± 90	110 ± 50	0.06
NO dependent vasodilation perfusion, p.u.	250 ± 100	140 ± 70	0.03
$R_{Basal,Axon-reflex}$	-0.44	-0.40	
$R_{Basal,NO}$	-0.65	-0.46	
$R_{Axon-reflex,NO}$	0.96	0.96	

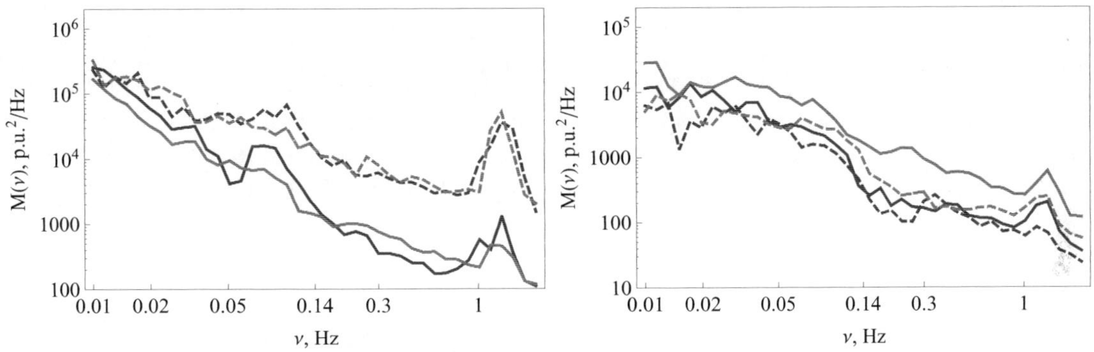

Figure 2. Averaged spectra of LDF signals in *CG* (blue lines) and *DM* (gray lines) in basal state (solid) and during axon reflex vasodilation stage (dashed). Spectra of LDF signals collected under the heater (left panel) and in the neighborhood (right panel).

close coefficients at both stages (axon-reflex mediated and NO vasodilation) demonstrates a lower reserve for the vasodilation of microvessels in subjects with high basal perfusion. This is independent of the mechanism responsible for vasodilation ($R_{A,NO}$ is high in both groups). In general, the second stage of microvascular system response to heating leads to a higher vasodilation difference ($p < 0.05$) between the control and *DM* groups.

3.2. Blood flow oscillations

In the figures 2, 3 averaged spectra in two time intervals corresponds to different stages of vasodilation are presented. In every plot spectra in basal conditions for reference are shown. Spectra of LDF pulsation in basal conditions had no significant difference in *CG* and *DM*. During the axon-mediated vasodilation amplitudes of pulsations increased in both groups, spectral slope changed, thus high frequency pulsations had higher impact in the LDF signals. No significant difference between *CG* and *DM* was founded.

Spectra of LDF pulsations for NO-mediated vasodilation plateau are presented in the figure 3. Energy of pulsations in all frequency bands were significantly higher in the heated tissue in comparison with basal state. High frequency oscillations demonstrated higher increment then lower ones. Significant difference between *CG* and *DM* was found in the 0.07-0.12 Hz frequency band. Note, that controls demonstrated higher energy of pulsations. Perfusion of the unheated skin also had higher pulsations in both groups, which indicates indirect influence of heating. Significant difference between groups in this time interval was found in the frequency band, which corresponds to the respiratory activity.

To study oscillations phase coherency, first, surrogate data were tested. Note that for reliable statistic of slow oscillations longer records duration is needed, and significant coherence of 0.01

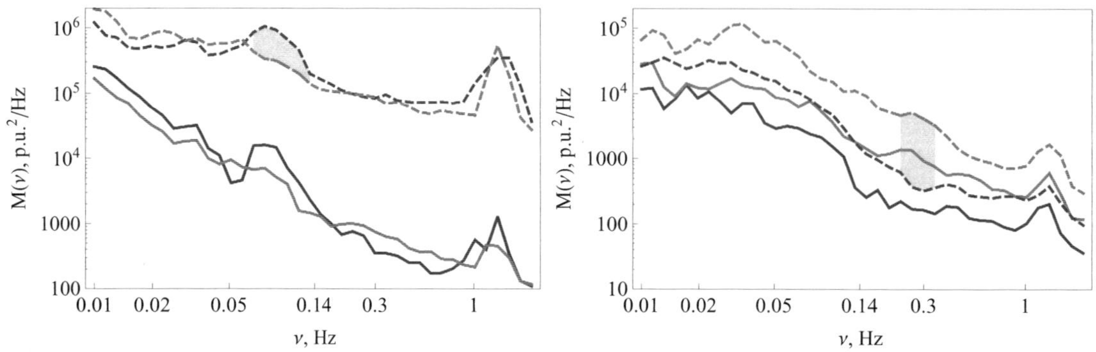

Figure 3. Averaged spectra of LDF signals in *CG* (blue lines) and *DM* (gray lines) in basal state (solid) and during NO-mediated vasodilation stage (dashed). Spectra of LDF signals ,collected under the heater (left panel) and in the neighborhood (right panel). Filling show the frequency band, where difference between *CG* and *DM*, estimated by Mann-Whitney test, was significant.

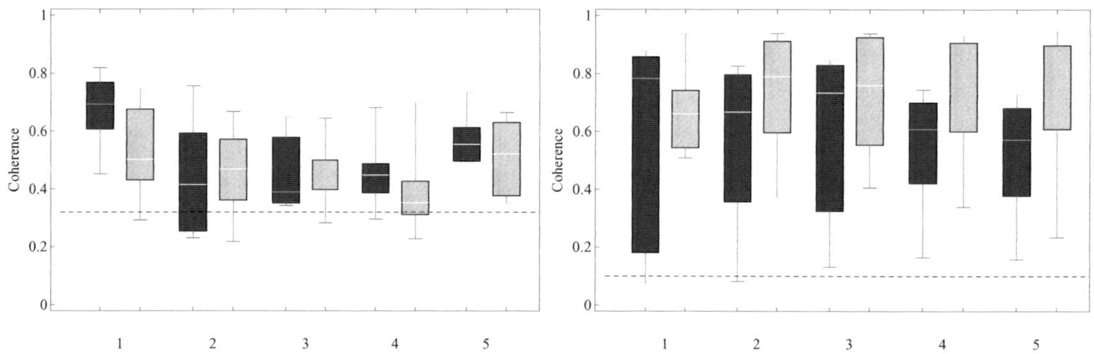

Figure 4. Box-whisker diagram for the coherence between two LDF records in different points for *CG* (blue rectangles) and *DM* (gray rectangles in five time intervals 1- basal state, 2- axon reflex-mediated vasodilation stage, transition process to the NO-mediated plateau, 4- NO-mediated plateau, 5 - restore processes after the heating switching off). Level of significance is shown by dashed lines. In the left panel synchronization of 0.1 Hz pulsations and 1 Hz - in the right panel are demonstrated.

Hz pulsations for the 15 minutes record is higher as 0.7. In this study we didn't get such high values of coherence, that's why only the frequency band 0.05-2 Hz was studied. Oscillations associated with cardiac activity and close to 0.1 Hz oscillations were analysed. The dynamics of phase coherence of pulsations in perfusion in two points of measurements on this two frequencies was traced (figure 4).

In the basal conditions 0.1 Hz LDF pulsations in two measurement points were coherent ($C = 0.7$). With the heating, phase coherence coefficient decreased, difference between basal conditions and heating conditions became significant in the NO-mediated vasodilation stage ($p < 0.05$). After the heater switched off, C restored up to the level in basal state. The dynamic is quite the same for healthy and diabetic subjects.

Using the same technique 1 Hz pulsation associated with cardiac activity were analysed (right panel of the figure 4). The level of significance for such pulsations are lower than for 0.1 Hz, and it is shown by the dashed line on the right panel of the figure 4. 1 Hz pulsations are statistically meaningful during the whole measurement interval, no significantly variation were obtained. Intergroup statistics also did not reveal any significant difference. Thus we don't see rapture of synchronization of oscillations associated with cardiac activity, this support the technique adequateness.

4. Conclusion

In the paper, the variations of the blood flow in the microvascular system caused by local heating were studied. It is found, that NO-mediated vasodilation is impaired in diabetic patients, which indicates an endothelial dysfunction [32] or weaker vasodilation response in pathological conditions. These results coincides with founded earlier impaired vasodilation caused by decreased production of (NO) [29], [31].

It has been established that the high basal perfusion measured by means of LDF leads to a lower reserve for microvessel vasodilation, and this effect does not depend on both the mechanism of vasodilation and on the group under consideration. The limit of tissue saturation is supposed to exist, and therefore the higher saturated tissue is able to get the lower volume of blood than the less saturated one. Another possible reason is the variation of the dynamic permeability range from subject to subject. The lower dynamic permeability range means that the system is less elastic. Note that this conclusion is correct for both groups, but mostly diabetes patients revealed lower dynamic range for vasodilation. Summing up, these characteristics are correct for both vasodilation mechanisms, and the NO mediated vasodilation is found to be more effective than one induced by the axon reflex.

The synchronisation of pulsations at two measurement points spaced 3 cm apart was studied at frequencies 1 Hz and 0.1 Hz. It has been found that the phase coherence of blood flow oscillations, associated with cardiac activity was not influenced by the local heating. At the same time 0.1 Hz were synchronous in the basal state and remained synchronous during the axon reflex mediated vasodilation. The situation was different over the time period of NO mediated vasodilation, namely, the index of phase coherence C significantly decreases in CG, as well as in DM. This fact indicates local NO associated vasodilation influence on the phase of 0.1 Hz vascular tone pulsations. The 0.1 Hz LDF pulsations associated with the myogenic vascular tone regulation were assumed to be locally originated, and therefore the local disturbance can change their characteristics.

Acknowledgment

The work is supported by the Russian foundation for basic research RFBR-ra under projects 17-41-590560 17-44-590755. The author kindly thank prof.E.Smirnova and E.Loran for the clinical base.

References

[1] Stefanovska A, Bracic M and Kvernmo H D 1999 *IEEE transactions on bio-medical engineering* **46** 1230–1239
[2] Rossi M, Carpi A, Galetta F, Franzoni F and Santoro G 2008 *Biomed. and Pharmacother.* **62** 541–545
[3] Schmidt J, Breit G, Borgstrm P and Intaglietta M 1995 *Inter. J. Microcirc.* **15** 28–36
[4] Martini R, Ticcinelli V and Bagno A 2014 *Clin. Hemorheol. Micro.* **4** 347–358
[5] Aalkjer C, Boedtkjer D and Matchkov V. 2011 *Acta Physiologica* **2**02 253–269
[6] Stirban A 2014 *Cur. Diabet. Rep.* **14** 1–9
[7] Allen J 2007 *Phys. Meas.* **28** R1-39
[8] Ansari M, Kang E, Manole M, Dreier J, Humeau-Heurtier A 2017 *Microvasc. Res.* **111** 49-59
[9] Obeid A N 1993 *Med. Biol. Eng. Comput.* **31** 3–52
[10] Mizeva I, Frick P and Podtaev S 2016 *J. Biomed. Opt.* **21** 085002
[11] Krupatkin A 2007 *Human Physiology* **33** 595–602
[12] Stefanovska A 1999 *Contemp. Phys.* **40** 31–55
[13] Bollinger A, Yanar A, Hoffmann U and Franzeck U K 1993 *Prog. Appl. Microcirc.* **20** 52–58
[14] Bertuglia S, Colantuoni A and Intaglietta M 1994 *Microvasc. Res.* **48** 68–84

IOP Conf. Series: Materials Science and Engineering **208** (2017) 012027 doi:10.1088/1757-899X/208/1/012027

[15] Kvandal P, Landsverk S A, Bernjak A, Stefanovska A, Kvernmo H D and Kirkeben K A 2006 *Microvasc. Res.* **72** 120-127

[16] Bernjak A and Stefanovska A 2007 *Conference proceedings : ... Annual International Conference of the IEEE Engineering in Medicine and Biology Society. IEEE Engineering in Medicine and Biology Society. Conference* **2007** 4064–4067

[17] Tankanag A, Grinevich A, Kirilina T, Krasnikov G, Piskunova G and Chemeris N 2014 *Microvasc. Res.* **95** 53–59

[18] Sheppard L W , Stefanovska A, McClintock P V E 2011 *Phys. Rev. E.* **85** 046205

[19] Salerud E, Tenland T, Nilsson G and Oberg P 1983 *Inter. J. Microcirc.* **2** 91–102

[20] Heistad D, Abboud F, Mark A and Schmid P 1973 *J. Appl. Physiol.* **35** 581–586

[21] Shvartz V, Karavaev A, Borovkova E, Mironov S, Ponomarenko V, Prokhorov M, Ishbulatov Y, Lapsheva E, Gridnev V and Kiselev A 2016 *Russian Open Medical Journal* **5**

[22] Liao F and Jan Y-K 2012 *Med. Bio. Eng. Comput.* **50** 1059–1070

[23] Johnson J and Kellogg D 2010 *J. Appl. Physiol* **109** 1229–1238

[24] Sorelli M, Stoyneva Z, Mizeva I and Bocci L 2017 *Phys. Meas.* *in* press

[25] Parshakov A, Zubareva N, Podtaev S and Frick P 2016 *Microcirc.* **23** 406–415

[26] Bandini A, Orlandi S, Manfredi C, Evangelisti A, Barrella M, Bevilacqua M and Bocchi L 2013 *J. of healthcare engeneering* **4** 541–554

[27] Fredriksson I, Larsson M, Nyström F H, Länne T, Östgren C J and Strömberg T 2010 *Diabetes* **59** 1578–1584

[28] Arora S, Smakowski P, Frykberg R G, Simeone L R, Freeman R, LoGerfo F W and Veves A 1998 *Diabet. Care.* **21** 1339–1344

[29] Stevens M, Dananberg J, Feldman E, Lattimer S, Kamijo M, Thomas T, Shindo H, Sima A and Greene D 1994 *J. Clin. Invest.* **94** 853–859

[30] Frick P, Mizeva I and Podtaev S 2015 *Bio. Sign. Proc. and Cont.* **21** 1–7

[31] Stevens M, Feldman E and Greene D 1995 *Diabet. Med.* **12** 566–579

[32] Smirnova E and Podtaev S and Mizeva I and Loran E 2013 *Diabet. Vasc. Dis. Res* **10** 489–497

IOP Conf. Series: Materials Science and Engineering **208** (2017) 012028 doi:10.1088/1757-899X/208/1/012028

Modeling of deformation of ground media on the basis of the particle method in a two-dimensional formulation

D V Berezhnoi, N F Gabsalikova, V G Izotov and V V Miheev[a]

Kazan Federal University, 18 Kremlyovskaya Street, Kazan, 420008, Russian Federation

E-mail: [a] miheevvova@mail.ru

Abstract. A two-dimensional version of the soil deformation method based on the particle method is implemented in this paper. For describing the behavior of a continuous medium, the previously used two-parameter "modified" Lennard-Jones interaction potential was chosen. A number of model problems describing the loss of stability of the soil embankment in the field of gravity and the destruction of the soil under the action of the liquid phase are solved.

1. Introduction

At present, most of the computational methods are based on the discretization of the region, most often the finite-element and finite-difference techniques [1-12]. Such an approach has perfectly proved itself in solving problems of deformation of complex heterogeneous media of various nature. However, there is a wide class of problems in which the loading of the medium is accompanied by multiple destruction and slippage of fragments, intensive mass transfer, including the effects of mixing masses, etc. When solving such problems, the use of the grid approach is very difficult. A promising class of numerical methods for the mechanics of a deformable solid body, adapted for modeling fracture, are particle methods [13-18].

One of the promising applications of the particle method is hydraulic fracturing, which is currently considered one of the most effective technologies for intensifying oil and gas production, both in traditional reservoirs and in low-permeability reservoirs known as shale hydrocarbons. The theoretical basis for calculating the resulting crack was laid by Christianovich and Zheltov [19]. Somewhat later Perkins and Kern [20] proposed another model for calculating a fissure. Nowadays, these models are currently used in applied fracture design systems. The models of Khristianovich-Zheltov and Perkins-Kern give a rather simplified performance development of the fissure of fracturing; in particular, they do not take into account the vertical and lateral heterogeneity of the reservoir according to mechanical characteristics, as well as the presence of natural fracturing. In addition, in real conditions we know the geological environment with some, sometimes rather large, uncertainty. Insufficient reliability of the design in conditions of stochastic character of the geological environment requires the control of the fracturing process to determine the actual characteristics of the fracture HF and the planning of further activities for the development of the well. Nowadays, the most commonly used monitoring method is the recording of microseismic activity accompanying the formation of a fracture by recording microseismic waves in neighboring wells [21] or from the earth's surface [22]. However, known analytical models do not relate the intensity of microseisms to the parameters of the crack, which makes interpretation difficult. More realistic models for calculating fracture propagation require the use of numerical multi-model numerical models that take into account the elastic interaction, fracture mechanics, fluid flow in the fracture and filtration processes in the formation.

An approach based on the Lagrange method is known [23], which allows numerical modeling of both hydraulic fracturing (HF) of a formation and generated microseismic waves within the framework of one numerical model. This makes it possible to increase the realistic modeling of the hydraulic fracturing process, in a complex geological environment. This also makes it possible to determine by calculation the dependence of the resulting microseismic waves on the characteristics of the fracture HF and thus increase the adequacy of the interpretation of microseismic methods of fracturing control of HF. Methods of numerical simulation of fracturing processes require the use of high-performance computing systems. At present, high-performance computing systems are increasingly equipped with computational accelerators based on GPGPUs, which are systems with mass parallelism (SIMD). SIMD architecture effectively implements simple linear algorithms without branching. Thus, to effectively model the fracturing process of HF, it is desirable to implement it as simply as possible. The method of discrete elements is used in [23]. According to this paper, the method of discrete elements gives quite realistic models of the behavior of the destruction of continuous media and the propagation of elastic waves. The method of discrete elements has a relatively complex algorithm, which includes, at each step of the simulation, the determination of the contacting elements, the calculation of the contact areas between them, and the calculation of the interaction forces.

We know a simpler approach based on the Lagrange method, based on the representation of the particles of the medium by nonstructural points interacting with each other on the basis of a potential that depends only on the distance between the points. This approach is called molecular dynamics and is widely used in computational chemistry, nanoparticle mechanics, and astrophysics. The attractiveness of this method lies in the computational simplicity. The method of molecular dynamics is based on the same type of calculations of potential forces between points. These calculations are well implemented on systems with mass parallelism. Modeling of the fracturing process should include modeling of fluid flow, including in pore channels, modeling of fracture and formation of seismic waves, including shear waves. There are known works on the molecular dynamics modeling of fluid flow of different scale, from macro-scale modeling by the well known method of large particles to nanopores [24]. The work on modeling the destruction of solids by the molecular dynamics method using the Lennard-Jones interaction potential has been conducted for a long time [17], however, high-energy supersonic processes are investigated in them, where brittle fracture plays no appreciable role. In this paper, we investigate the possibility of simulating less energy processes of brittle fracture and the formation of seismic waves in modeling the fracturing process.

The modern development of computer technology, including parallel computing systems, provides ample opportunities for modeling continuous media at the micro level. In recent years, such modeling is carried out on the basis of the particle method. The particle method is that a continuous medium is represented as a set of interacting material particles. As such particles can act as atoms and, molecules of matter, and mineral particles of ground materials. At present, the potentials of interatomic interaction for most materials are known, which can not be said about the potentials for describing the behavior of deformed continuous media, especially shear ones. One of the important advantages of the particle method is that much less information about the properties of the material is needed when using it. Complex mechanical processes can be modeled in some way using the simple Lennard-Jones potential. To describe each of these effects, a separate theory is required, while in particle modeling, these effects are obtained by integrating the equations of motion.

2. Formulation of the problem

The motion of the aggregate of material particles interacting in a potential field can be described on the basis of the following equation of motion

$$m\ddot{\mathbf{r}}_k = \sum_{n=1}^{N} \frac{1}{r_{kn}} f\left(r_{kn}\right)\mathbf{r}_{kn} + \boldsymbol{\varphi}\left(\mathbf{r}_{kn}\right), \quad (1)$$

IOP Conf. Series: Materials Science and Engineering **208** (2017) 012028 doi:10.1088/1757-899X/208/1/012028

where $\mathbf{r}_k$ – radius vector k -th particle

$$\mathbf{r}_{kn} \overset{def}{=} \mathbf{r}_k - \mathbf{r}_n, \quad (2)$$

m – mass of particle, $\varphi(r)$ describes an external conservative force field, energy dissipation was not taken into account. We consider the pair potential $\Pi(r)$, force of interaction $f(r)$ corresponding to it is defined as

$$f(r) \overset{def}{=} -\dot{\Pi}(r). \quad (3)$$

For describing the interaction of material particles of a deformed medium (soil), we used the "modified" Lennard-Jones potential [25, 26], the parameters and interaction force for which were taken in the form

$$D = 10^5, a = 0.4*1.28r, f(r) \equiv \frac{D}{r}\left[12(a/r)^{12} - 36(a/r)^9 + 14.4(a/r)^8 + 6(a/r)^6\right]. \quad (4)$$

To describe the deformation of the liquid phase, we used the Lennard-Jones potential, the parameters and the interaction force for which were taken in the form:

$$D = 4*10^5, a = 0.4r, f(r) \equiv \frac{12D}{r}\left[(a/r)^{12} - (a/r)^6\right]. \quad (5)$$

To describe the motion of a set of material points, a system of equations of motion was used. Its integration was carried out using the Verlet method [27]. For calculations, heterogeneous computing clusters using graphic accelerators (GPUs) were used. To implement calculations on graphics accelerators, an open source VexCL library was used [28]. This library simplifies the development of applications using OpenCL technology and allows using intuitive notation for recording basic operations of linear algebra [29, 30].

3. Numerical example

To illustrate the possibilities of the implemented methodology, two model problems were solved, for which the qualitative character of the deformation of the soil massif, the possibility of its destruction and loss of stability of slopes, is determined, so a certain arbitrariness was allowed in choosing the parameters of the Lennard-Jones potential. The parameters of the "modified" Lennard-Jones potential were selected on the basis of the parameters of the ordinary Lennard-Jones potential, starting from the qualitative condition of the appearance of a shear wave when considering the problem of propagation of seismic waves in a geological environment [26]. In the first task, the soil layer modeled by 800 particles is initially formed between two absolutely rigid barriers. After the "packing" of material particles in the form of a rectangle, the lateral obstacles instantly disappear and the layer of soil under the influence of its own weight creeps, resulting in its partial destruction. Figures 1-4 show the final configurations, which acquire a soil massif in the process of loss of stability and partial destruction, depending on the magnitude of acceleration of gravity g $0.5g$, $0.8g$, $0.85g$ and g respectively) using the Lennard-Jones potential and the "modified" Lennard-Jones potential.

In the second task, a soil layer defined by 1000-th material points with a rectangular notch formed by absolutely rigid walls is formed (Fig. 5). In the cavity rectangular area consisting of 200 particles a system of material particles is formed, representing the liquid phase failure. At some point in time, the lower wall that bounds the liquid phase disappears instantaneously, and the fluid under pressure pours into the ground. The so-called hydraulic fracturing of the formation is being realized. Figures 6-9 show the different cases of the soil breaking out with a liquid phase, when the magnitude of acceleration of gravity - $0.9g$, g and $1.1g$ respectively.

IOP Conf. Series: Materials Science and Engineering **208** (2017) 012028 doi:10.1088/1757-899X/208/1/012028

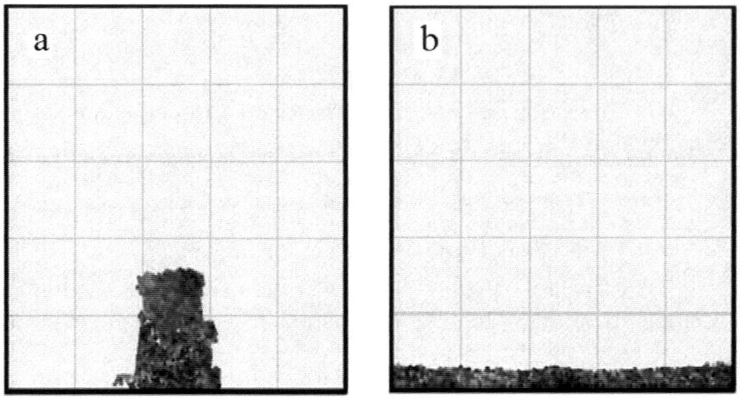

Figure 1. The final configuration of the soil layer after the process of sample destruction for the Lennard-Jones potential (b) and its modified version (a) when the magnitude of acceleration of gravity - $0.5g$.

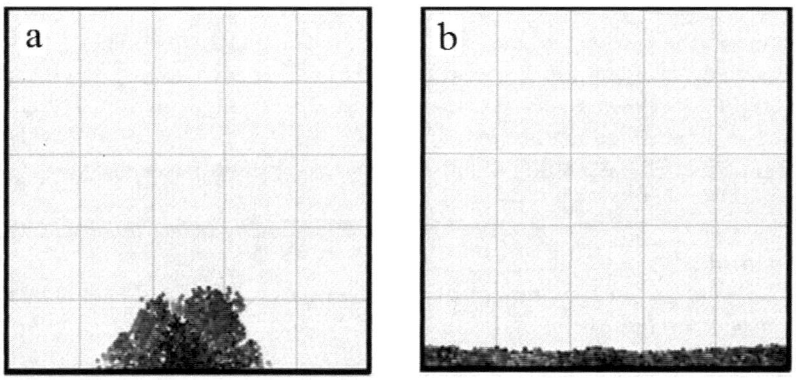

Figure 2. The final configuration of the soil layer after the process of sample destruction for the Lennard-Jones potential (b) and its modified version (a) when the magnitude of acceleration of gravity - $0.8g$.

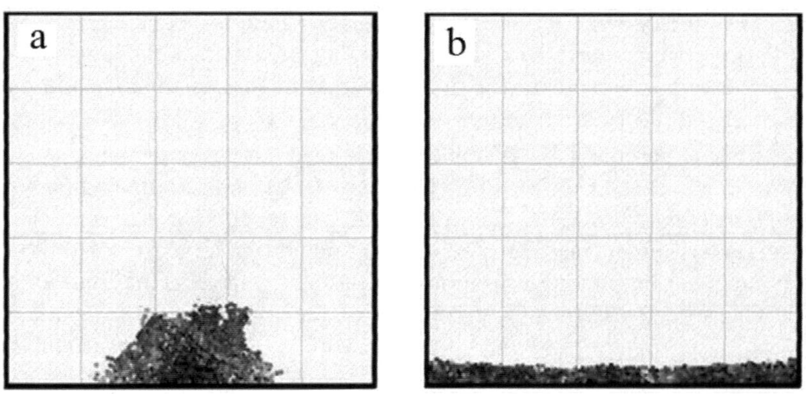

Figure 3. The final configuration of the soil layer after the process of sample destruction for the Lennard-Jones potential (b) and its modified version (a) when the magnitude of acceleration of gravity - $0.85g$.

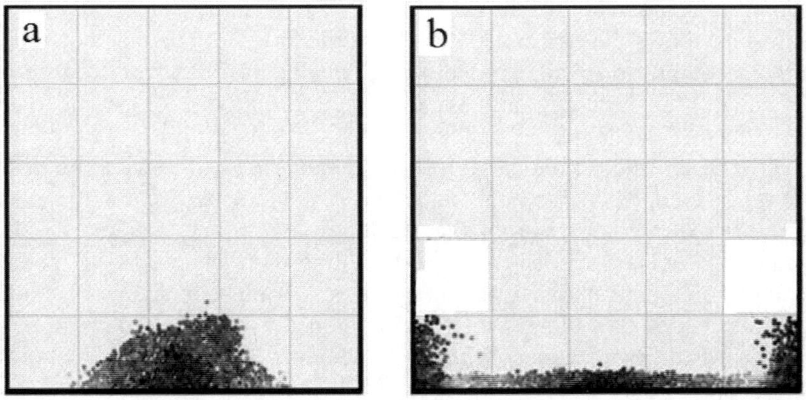

Figure 4. The final configuration of the soil layer after the process of sample destruction for the Lennard-Jones potential (b) and its modified version (a) when the magnitude of acceleration of gravity - g .

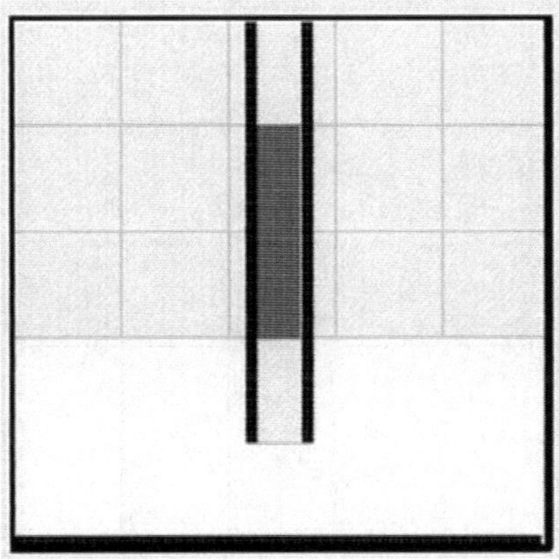

Figure 5. Initial position of the solid and liquid phases.

Figure 6. Fracturing process and soil degradation when the magnitude of acceleration of gravity - $0.9g$.

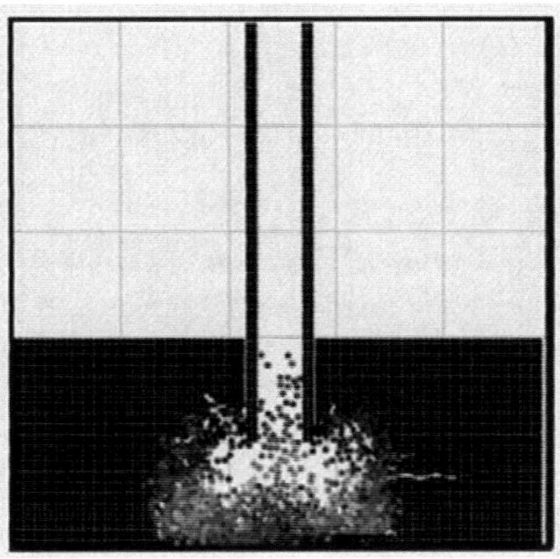

Figure 7. Fracturing process and soil degradation when the magnitude of acceleration of gravity - g .

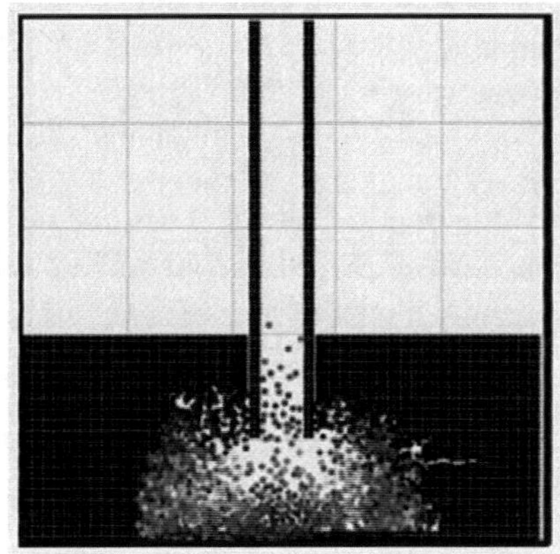

Figure 8. Fracturing process and soil degradation when the magnitude of acceleration of gravity - $1.1g$.

4. Conclusion

Analysis of the results shows that the use of the "modified" Lennard-Jones potential to describe the ground phase allows the brittle fracture process to be realized in the ground, whereas the Lennard-Jones potential will allow reproducing only viscous fracture. In the work, studies were carried out to identify the parameters of the "modified" Lennard-Jones potential, but it should be noted that these parameters are difficult to find numerically for an adequate description of the deformation of real earth. In the future, a comprehensive analysis of the "modified terms" of the potential should be carried out on the basis of a comparison of the results of solving the simplest problems for different chilas of the material particles used. It is also worth noting that the "modified" potential shows itself well in the modeling of a deformable solid medium. It can be seen from the experiments that the process of cracks formation takes place, which shows that the modified potential is really capable of describing the brittle properties of the medium.

Acknowledgments

The reported study was supported by Government of the Republic of Tatarstan and RFBR research projects No. 15-41-02555 and 15-07-05380.

References

[1] Badriev I B, Makarov M V and Paimushin V N 2016 *Procedia Engineering* **150** 1050–1055
[2] Badriev I B, Makarov M V and Paimushin V N 2016 *Procedia Engineering* **150** 1056–1062
[3] Shigapova F A, Mustakimova R F, Saleeva G T and Sachenkov O A *International Journal of Applied Engineering Research* **10 (24)** 44711–44714
[4] Sachenkov O A, Mitryaikin V I, Zaitseva T A and Konoplev Yu G 2014 *Applied Mathematical Sciences* **8 (159)** 7889–7897
[5] Sultanov L U 2015 *Procedia Earth and Planetary Science* **15** 119–124
[6] Golovanov A I and Sultanov L U 2005 *International Applied Mechanics* **41 (6)** 614–620
[7] Sultanov L U 2014 *Applied Mathematical Sciences* **8 (143)** 7117–7124
[8] Zienkiewicz O 1977 *The finite element method* (Mc Graw-Hill) 787
[9] Wilkins M L 1967 *Vychislitelnye metody v gidrodinamice* (Mir) 212-263
[10] Evans M W and Harlow F H 1957 *Los Alamos Scientific Laboratory Report No. LA-2139* (Los Alamos)
[11] Berezhnoi D V, Balafendieva I S, Sachenkov A A and Sekaeva L R 2016 *IOP Conference Series: Materials Science and Engineering* **158** 012018
[12] Mokshin E V and Berezhnoi D V 2016 *IOP Conference Series: Materials Science and Engineering* **158** 012069
[13] Hockney R W and Eastwood J W 1981 *Computer simulation using particles* (McGraw-Hill)
[14] Grigorjev Yu N and Vshivkov V A 2000 *Numerical methods "particle-in-cell"* (Nauka)
[15] Bogomolov S V, Zamaraeva A A, Karabelly H and Kuznetsov K V 1998 *Computational Mathematics and Mathematical Physics* **38 (9)** 1602
[16] Monaghan J J 2005 *Reports on Progress in Physics* **68 (8)** 1703-1759
[17] Krivtsov A M and Krivtsova N V 2002 *Far Eastern Mathematical Journal* **3 (2)** 254-276
[18] Psahje S G, Smolin A Yu, Stefanov Yu P, Makarov P V and Chertov M A 2004 *Letters Technical Physics* **30 (17)** 7-13
[19] Zheltov Y P and Hristianovich S A 1955 *Izvestia AN USSR* **5** 3-41
[20] Perkins T K and Kern L R 1961 *J. Petrol. Technol.* **9** 937–949
[21] Maxwell S C, Rutledge J, Jones R and Fehler M 2010 *Geophysics* **75 (5)** 75A129–75A137
[22] Duncan P and Eisner L 2010 *Geophysics* **75 (5)** 75A139–75A146.
[23] Qi Zhao, VitaeAndrea Lisjak, Omid Mahabadi, Qinya Liu and Giovanni Grasselli 2014 *Journal of Rock Mechanics and Geotechnical Engineering* **6 (6)** 574–581
[24] Soong C Y, Yen T H and Tzeng P Y 2007 *Phys. Rev. E.* **76** 036303.
[25] Berezhnoy D V, Gabsalikova N F and Balafendieva I S 2015 *International Journal of Applied*

Engineering Research **10 (24)** 44677-44681
[26] Berezhnoi D V, Gabsalikova N F and Miheev V V 2016 *IOP Conference Series: Materials Science and Engineering* **158** 012016
[27] Verlet L 1967 *Phys. Rev.***159 (98)** 103
[28] Galimov M R and Birjaltsev E V 2010 *Numerical methods and programming* **11** 77-93
[29] Demidov D *VexCL: a C++ vector expression template library for OpenCL/CUDA* https://github.com/ddemidov/vexcl
[30] Demidov D, Karsten A, Rupp K and Gottschling P 2013 *SIAM Journal on Scientific Computing* **35 (5)** 453-472
[31] Karsten A, Demidov D and Mulansky M 2014 *Numerical Computations with GPUs* (Springer International Publishing) 125-157

Effect of the universal acid-base indicator on the formation of the concentration-dependent diffusion instability

E A Mosheva and A V Shmyrov

Institute of Continuous Media Mechanics UB RAS, 1, Acad. Korolev St., Perm, 614013, Russia

E-mail: mosheva@icmm.ru

Abstract. The effect of the universal acid-base indicator on the pattern formation and mass transfer in a two-layer system composed of two reactive miscible liquids in a vertical Hele-Shaw cell is studied experimentally. The reaction we study is a neutralization one. It turns out that the presence of the indicator leads to a change in the spatio-temporal characteristics of the system and even in the mass transfer mechanism near the reaction front—from diffusive to convective. The conditions, where the universal indicator does not affect the reaction and can be used as a visualizing mean, are reported.

1. Introduction

Different types of indicators are used to visualize chemical composition of the reactive system in different fields of chemohydrodynamics increasingly often [1-4]. Particularly, various acid-base indicators are utilized in studies on the neutralization reaction in a two-layer system in order to visualize spatial distribution of reagents. Usually, physical parameters (viscosity, density, diffusivity) of an indicator and reagents are different. Hence, the use of an indicator may results in appearance of additional driving forces of convection and even gives rise to the pattern formation in the system.

There are several studies on how the indicator affects buoyancy-driven instability of the acid-base front. In [5] it was demonstrated both numerically and experimentally, that an indicator can affect the instability scenarios by changing the density profile of the system. An attempt to determine the conditions, under which the indicators do not have a significant impact on the system and can be used as a visualizing mean, was made in [6]. The authors examined the neutralization reaction between aqueous solutions of hydrochloric acid and bromocresol green, which acts as a base. They reported the change in the scenario of instability development due to concentration variation of reactants. In [7] the neutralization reaction between the aqueous solutions of hydrochloric acid and sodium hydroxide with a dissolved color indicator was studied. In some cases the employed indicator was bromothymol blue, in other it was phenolphthalein. The authors reported that chemically reacting flows are drastically changed depending on the type of *p*H indicator. In [8] the same group of researchers repeated the experiment, but they added aqueous solution of ethanol to the base containing bromothymol blue. They showed that the predominant mechanism of mass transfer at the reaction front changes from diffusive to convective depending on the mass fraction of ethanol. There are also interesting series of studies where the reaction is caused by a droplet containing the *p*H indicator bromothymol blue that falls dropwise into an aqueous alkaline solution [9-10]. Authors investigated the short-term transport processes across the interface, including the fingering instability of the reaction front.

In our previous work [11], we showed that the dependence of the diffusion coefficients of the reactants on their concentration in a two-layer reactive system leads to the formation of a periodic cellular convective structure localized within the reactive zone; we called it the concentration-dependent diffusion instability (henceforth CDD instability). For visualization of emerging convective structures we used the acid-base universal indicator. We supposed that the indicator concentrations we used were sufficiently small and did not influence the reaction. However, the above-mentioned investigations raise the question whether the indicator we used could lead us to misleading results. In this regard, here we focus on the detailed study of the effect of the universal indicator on the onset of CDD instability and mass transfer processes near the reaction front in a miscible two-layer system composed of aqueous solutions of NaOH and HCl.

2. Experiment

The experiments were performed in a vertically oriented Hele-Shaw cell. The cell was made of two glass plates separated by a thin gap of 0.12 cm (figure 1). The reaction under consideration was the neutralization of hydrochloric acid by sodium hydroxide. The standard enthalpy change for this reaction is $\Delta H = -57$ kJ/mol. Molar concentrations of acid and base aqueous solutions were $C_a = 0.7$ mol/l and $C_b = 0.5$ mol/l, respectively. Such concentrations were chosen due to the fact that CDD instability in HCl/NaOH system is observed at certain ratio of initial concentrations, specifically $C_a/C_b = 1.4$ [11].

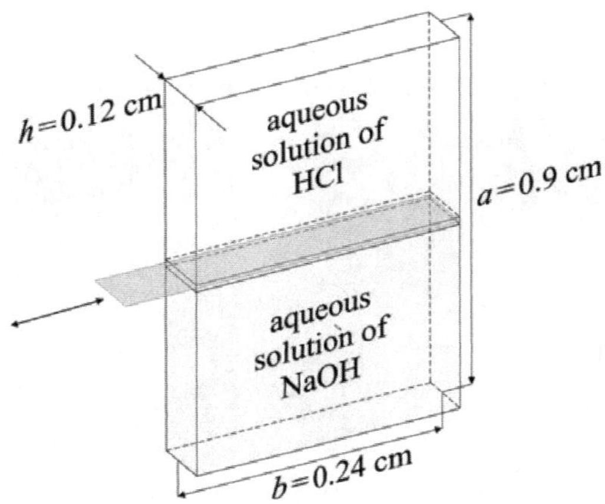

Figure 1. Sketch of the two-layer miscible system filling a vertical Hele-Shaw cell.

In order to provide an initially flat horizontal interface between the reagents, we used special thin plastic shutter that was separating solutions before their contact. At first, a lower layer was filled with NaOH solution, then the solution was tightly covered by the shutter, and then an upper layer was filled with HCl solution. After that, the shutter was carefully taken out of the cell, the reactants were brought into contact, and the reaction started.

Fizeau interferometer was used to visualize the refractive index field. In order to visualize the space distribution of the reagents and the reaction product, we used the universal indicator. An advantage of the universal indicator is that it provides an image with a distinct boundary between the reagents and the reaction product. Note that the universal indicator is a combination of several indicators (thymol blue, methyl orange, bromothymol blue, phenolphthalein, tropaeolin OO, bromocresol green, and bromocresol purple), that display colors at *p*H values inside the certain transition range. That is why the universal indicator displays color changes over a wide *p*H value range from 1 to 10. Colors from yellow to red indicate an acidic solution, colors from light to dark blue indicate bases, and green color indicates that a solution is neutral. The indicator was prepared on the basis of the aqueous solution of ethanol with mass concentration $w = 80\%$. The initial concentration and density of the indicator were

IOP Conf. Series: Materials Science and Engineering **208** (2017) 012029 doi:10.1088/1757-899X/208/1/012029

2×10^{-4} M and 0.849 g/cm^3, respectively. To provide high-contrast images of *p*H distribution, the Hele-Shaw cell was illuminated by white-light toroidal lamp. Experiments were recorded by digital camera with high spatio-temporal resolution. All experiments were performed at room temperature (24±1°C).

3. Results and discussions

At first, we would like to describe the typical experiment without any indicator. When reactants come into contact, the transition zone originates between them where the reagents are supplied towards the reaction front only due to diffusion mechanism, which substantially increases the duration of the reaction and leads to low rate of the reaction front propagation, which is approximately $\upsilon = 10^{-4}$ cm/s (see figure 4). Then, the occurrence of a depleted layer located above the diffusion zone gives rise to the formation of plumes which results in the development of weak buoyancy-driven convection in the entire upper layer, while the area below the diffusion zone remains motionless. After certain time (in our experiment, it was approximately 480 seconds), the fluid flow in the form of a horizontal array of convective cells is formed within the diffusion zone just above the reaction front. The convective cells structure exists between two parts of immobile fluid with stable density stratification, which indicates the occurrence of inflection on the density profile and consequently the formation of a localized density "pocket" with unstable density stratification. The refractive index distribution which reflects the density distribution in this experiment is presented in figure 2. A more detailed description of the formation mechanism of CDD instability can be found in [11].

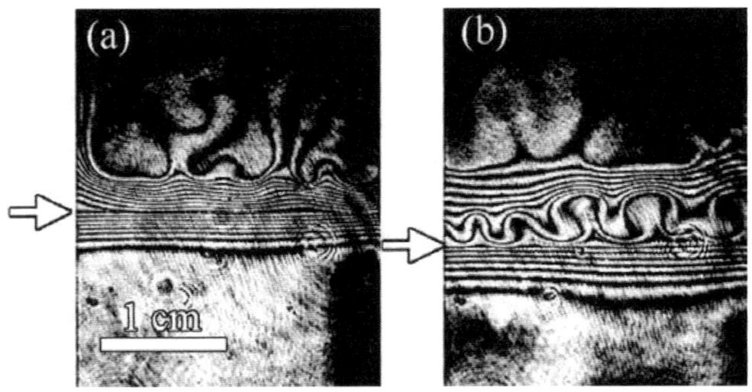

Figure 2. Interference images showing the development of CDD instability in experiment without indicator. The arrows indicate the reaction front position, x_f observed at (a) 400 s, (b) 1300 s after the aqueous solutions were brought into contact.

In order to investigate the influence of the universal indicator, three different situations were considered: (i) the universal indicator was dissolved in both layers, (ii) the indicator was dissolved only in the upper layer with acid, (iii) the indicator was dissolved only in the lower layer, containing the base. The indicator was dissolved in the solutions of reagents prior to the beginning of the experiments. The range of the volume concentration of the dissolved indicator is $V = (0.15 \div 3.3)\%$. Convective patterns observed at 2000 second for the experiments with different conditions are presented in figure 3. All images clearly demonstrate the appearance of the localized cellular structure. The purple downstreams indicates the acid entrained by the flow. The yellow upstreams enriching with salt points out on that the reaction occurs at the lower edge of the cellular structure, where the reaction front is.

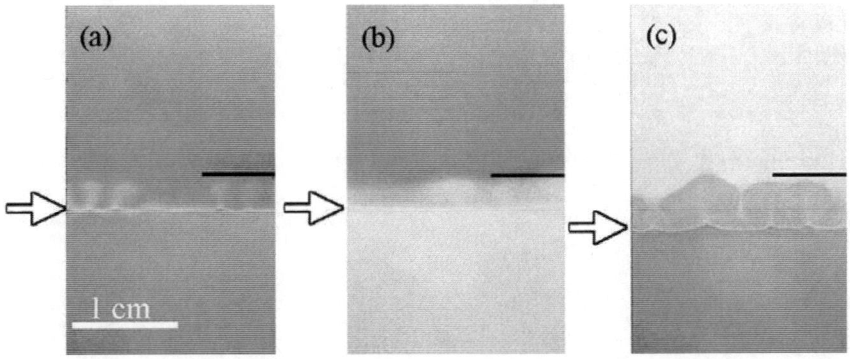

Figure 3. Chemoconvective patterns observed in the system with the universal indicator dissolved (a) in each layer; (b) in the upper layer, (c) in the lower layer. Volume concentration of the indicator is $V = 1.5\%$. The arrows indicate the reaction front position, x_f. Black line: initial contact line. (Colored online).

Comparison of the temporal evolution of the reaction front in the system with and without the indicator is shown in figure 4. It is interesting to note that in the case, where the volume concentration of indicator is $V = 0.15\%$, the reaction front rate is the same as in the experiment without the indicator, no matter in which layer the indicator was dissolved. An increase in the indicator concentration (up to $V = 2\%$) does not essentially affect the front rate when it is dissolved in both layers or only in the layer with acid (in the figure they are indicated as half shaded and white symbols, respectively), but, as it will be shown later, it drastically changed the onset time of the instability. However, if the indicator is added only to the lower layer (black symbols), the front propagation rate noticeably increases. The addition of the indicator with concentration $V > 2\%$ to both layers results in completely different reaction scenario. The CDD instability observed previously is not forming. Instead, in the upper layer of the system containing acid, an intense convective motion develops, which magnifies the mass transfer near the reaction front, leading to a significant increase in the front propagation rate. Thus, the addition of even such low indicator concentrations leads to a change in the mass transfer mechanism from diffusive to convective.

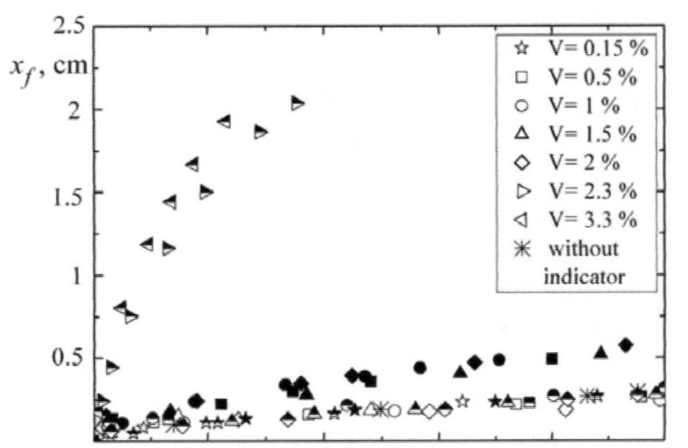

Figure 4. Temporal evolution of the reaction front position. Half shaded, shaded, and open symbols indicate experiments when the indicator is dissolved in both layers, in the lower layer, and in the upper layer, respectively.

In figure 5 the dependence of the onset time of CDD instability on the volume concentration of the dissolved indicator is presented. The experiment showed that the presence of the indicator in both or only in the upper layer always leads to delay of the instability development, whereas its addition only to the lower layer always reduces the instability onset time, t_{onset}. By varying the indicator concentrations and selecting the layer in which it will be dissolved, it is possible to approach the results obtained in the experiment in the absence of the indicator.

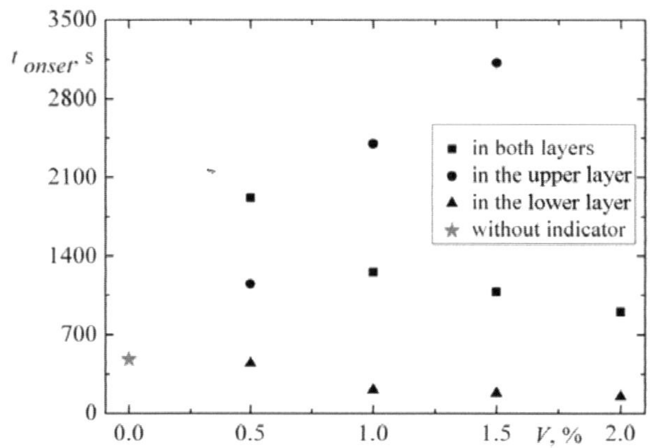

Figure 5. CDD instability onset time versus volume concentration of the dissolved indicator.

Such a behavior may be explained as follows. Due to low density of the indicator solution, the density of the layer in which the indicator is dissolved decreases. Thus, in experiments with indicator on the upper layer the system becomes more stable. Specifically, the density difference, which triggers the formation of the density pocket inside the reaction area, is smaller than the density difference forming in the experiment without indicator. Therefore, the time required for the density pocket formation, due to which the CDD instability develops, increases. It explains the observed delay in the formation of convective structure. Moreover, the greater the concentration of the indicator the more stable the system becomes. Experimental results show that the predominant mechanism of mass transfer near the reaction front is diffusion; therefore, the front propagation rate as shown above is almost the same as in the experiment without indicator.

In the reverse case, where the indicator is dissolved only in the lower layer, the value of the density difference that triggers the formation of the density pocket inside the reaction area becomes bigger. Thus, the onset time of CDD instability is almost the same as in the experiment without indicator. An increase of the indicator concentration further destabilizes the reaction area. Hence, the time required for the formation of the density pocket decreases. The main mechanism of mass transfer near the reaction front is also diffusion. However, due to the fact that CDD instability onsets earlier, the mass transfer near the reaction front is more intense than in the experiments without indicator; therefore, the front propagation rate is noticeably higher.

The presence of the indicator in both layers leads to a symmetrical change in their physical parameters, i.e., the value of the density pocket should not differ from its value in the experiment without indicator. That is why in this case, the front propagation rate coincides with the one observed without indicator. However, an increase of indicator concentration ($V > 2\%$) triggers completely different scenario of the reaction. When reagents come into contact, a convective motion in the form of rising plumes develops above the reaction front. Formed plumes quite quickly reach the upper boundary of the cell. As a result, buoyancy-driven convection develops in the entire upper layer, but in contrast to the experiment without indicator, it is more intense. In this way, an intense convective motion that continuously supplies the fresh reagent to the reaction area and performs a removal of the reaction product is formed above the reaction front. Constant refreshment of the reactive layer is fast enough, which leads to the rapid propagation of the reaction front until it reaches the lower boundary of the cell. The rate of the reaction front plotted in figure 4 is approximately $\upsilon = 10^{-3}$ cm/s.

4. Conclusions

In the experimental study of the effect of the universal indicator on the process of instability formation, the conditions, where the universal indicator does not make a significant impact on the system and can be used as a visualizing mean, are found. It is shown that the volume concentrations of the indicator $V < 0.2\%$ does not affect the onset time of CDD instability, nor the mass transfer near the reaction front. Despite the fact that such indicator concentrations are negligible, it is enough to have a good visualization.

5. Acknowledgments

Authors want to thank A. I. Shmyrova and S. N. Lysenko for help with experiments. Appreciation goes to A. I. Mizev for fruitful discussions. The reported study was supported by RFBR, research project No. 15-01-04842_a.

6. References

[1] Cardoso S S S and Andres J T H 2014 Geochemistry of silicate-rich rocks can curtail spreading of carbon dioxide in subsurface aquifers *Nat. Commun.* **5** 5743

[2] Outeda R, Hasi C El, D'Onofrio A and Zalts 2014 An Experimental study of linear and nonlinear regimes of density-driven instabilities induced by CO_2 dissolution in water *Chaos* **24** 013135

[3] Thomasa C, Lemaigrea L, Zaltsb A, D'Onofrio A and de Wit A 2015 Experimental study of CO_2 convective dissolution: the effect of color indicators *Int. J. of Greenhouse Gas Control* **42** 525–533

[4] Eckert K, Acker M and Shi Y 2004 Chemical pattern formation driven be a neutralization reaction. Mechanism and basic features *Phys. Fluid* **16**(2) 385–399

[5] Almarcha C, Trevelyan P M J, Riolfo L A, Zalts A, Hasi C El, D'Onofrio A and de Wit A 2010 Active role of a color indicator in buoyancy-driven instabilities of chemical fronts *J. Phys. Chem. Lett.* **1** 752–757

[6] Kuster S, Riolfo L A, Zalts A, Hasi C El, Almarcha C, Trevelyan P M J and de Wit A, D'Onofrio A 2011 Differential diffusion effects on buoyancy-driven instabilities of acid-base fronts: the case of a color indicator *Phys. Chem. Chem. Phys.* **13** 17295–303

[7] Nisimura T, Tanoue K, Watanabe T, Itoh Y and Kunitsugu K 2006 Instabilized Fluid Flow at interface of chemical reaction in liquid phase *Trans. Japan Soc. Mech. Eng.* **72**(719) B 1773–80

[8] Tanoue K, Ikemoto H, Yoshitomi M and Nisimura T 2009 Instabilized heat and mass transfer in exothermic chemically reacting flows. *Thermal Science and Engineering* **17**(3) 121–129

[9] Tsuji K and Müller S C 2012 Chemical reaction evolving on a droplet *J. Phys. Chem. Lett.* **3** 977–980

[10] Tsuji K and Müller S C 2013 Observation of a falling droplet with pH indicator *Interfac. Phenomena and Heat Transfer* **1**(3) 289–299

[11] Bratsun D, Kostarev K Mizev A and Mosheva E 2015 Concentration-dependent diffusion instability in reactive miscible fluids *Phys. Rev.* E **92** 011003(R)

IOP Conf. Series: Materials Science and Engineering **208** (2017) 012030 doi:10.1088/1757-899X/208/1/012030

A search for optimal parameters of resonance circuits ensuring damping of electroelastic structure vibrations based on the solution of natural vibration problem

D Oshmarin, N Sevodina, M Iurlov and N Iurlova

Institute of Continuous Media Mechanics UB RAS, 1, Acad. Korolev St., Perm, 614013, Russia

E-mail: yurlova@icmm.ru

Abstract. In this paper, with the aim of providing passive control of structure vibrations a new approach has been proposed for selecting optimal parameters of external electric shunt circuits connected to piezoelectric elements located on the surface of the structure. The approach is based on the mathematical formulation of the natural vibration problem. The results of solution of this problem are the complex eigenfrequencies, the real part of which represents the vibration frequency and the imaginary part corresponds to the damping ratio, characterizing the rate of damping. A criterion of search for optimal parameters of the external passive shunt circuits, which can provide the system with desired dissipative properties, has been derived based on the analysis of responses of the real and imaginary parts of different complex eigenfrequencies to changes in the values of the parameters of the electric circuit. The efficiency of this approach has been verified in the context of natural vibration problem of rigidly clamped plate and semi-cylindrical shell,, which is solved for series-connected and parallel –connected external resonance (consisting of resistive and inductive elements) *R-L* circuits. It has been shown that at lower (more energy-intensive) frequencies, a series-connected external circuit has the advantage of providing lower values of the circuit parameters, which renders it more attractive in terms of practical applications.

1. Introduction

Providing the engineering structures with piezoelectric elements and their shunting by external electric circuits (shunt circuits) to suppress system vibrations is the technique, which is in great request in such engineering applications, for which the small size and weight requirements are of critical importance and access to a power source is not available. Shunting allows a straightforward and low cost control of structure vibrations by damping unwanted oscillations at the prescribed frequencies.

The above technique of designing smart structures with piezoelectric elements, which are connected to various shunt electric circuits to provide passive damping, has a number of advantages over the traditional methods of damping the mechanical vibrations (for example, with the aid of mechanical damper). Among these is the ease of fitting the structure with external electric circuits which, because of their simple configuration, include only a few standard electronic components. In some situations there is no need for additional electronic devices or power sources. In some other cases one can dispense with the feedback devices. Moreover, such an approach to vibration damping compares favorably with the other techniques due to its reliability and stability- it does not require complicated digital processors and in many cases does not require the development of parametric models in designing the system for damping structure vibrations [1].

The first work, in which the above approach is considered more fully, is the paper by Hagood N.W. and Von Flotow A. [2]. The resonance series shunt circuit was proposed in works [3-6], and parallel circuit – in [7].

In paper [2], the authors developed an approach for selecting optimal parameters of the external circuit, which is based on the analysis of the transfer function constructed for the system with piezoelectric element and external electric circuit. The analysis of the literature showed that in terms of efficiency of finding the appropriate parameters for external circuits providing passive vibration damping, this approach is still most popular when analyzing the dynamic behavior of electroelastic systems with external electric circuits. In this case, the optimal values of the external circuit parameters estimated on the basis of the transfer functions can be obtained by different ways, for example, by optimizing the transfer functions with respect to $\mathcal{H}_2$- or $\mathcal{H}_\infty$- norms [8,17], assuming the equality of amplitudes of transfer functions at the characteristic intersection points of curves [2, 6, 10], using the technique for placing the poles of the transfer function (pole placement technique) [9-10, 18], etc. The result of all these peculiarities is that, depending on the ways of constructing the transfer function and selecting the optimization criterion, the optimal parameters of the external circuit parameters can take different values [2, 10, 17].

In work [19-21], it was shown that both, the series and parallel resonance *RL*- circuits can provide equally effective damping of structure vibrations at the prescribed frequency.

Among the parameters, characterizing the dynamic behavior of structures, the most important are the resonance frequencies and parameters that specify the damping properties of the system. As a rule, the damping properties of smart structures are estimated by the magnitude of the amplitude in the resonance mode or by the rate of transition processes. In the first case it is necessary to solve the problem of forced steady-state vibrations and in the second case – a dynamic problem with initial conditions.

The investigation made in work [22] supports the view that the natural vibration problem is most convenient and informative in terms of efficiency of identification and optimization of the system dissipative properties. The complex eigenvalues found by solving this problem will determine the eigenfrequencies of vibrations (real parts) and damping ratios (imaginary parts). The efficiency of the above problem in finding the optimal dissipative characteristics of electromechanical systems can be attributed to their independence on the external loads and initial conditions.

In this paper, to solve the problem of passive damping of vibrations in electroelastic systems with external electric circuits, we use as a framework the natural vibration problem, since it allows us to directly construct the dependence of the damping ratio on the parameters of the external electric circuit for any types of electric circuits and eventually to determine their optimal values, at which a considerable decrease in the amplitudes of system vibrations can be reached [23-24].

2. Mathematical statement of the problem and methods of its solution

Consider a piece-wise homogeneous body (smart-structure) of volume $V = V_1 + V_2$, where V_1 consists of homogeneous elastic parts and volume V_2 of piezoelectric elements. A part of the surface of the body volume V_2 is covered by electrodes. One of the electrode-covered surfaces is connected to the external passive electric circuit of arbitrary configuration consisting of resistive elements (resistors) of resistance *R*, inductive elements (induction coils) of inductance *L* and capacitive elements (capacitors) of capacitance *C*.

The variational equation of natural vibrations for a body consisting of elastic and piezoelectric elements is formulated using the relations of the linear elasticity theory and quasi-static Maxwell equations [25-27]:

$$\int_{V_1} \sigma_{ij}\delta\varepsilon_{ij} + \rho\ddot{u}_i\delta u_i \, dV + \int_{V_2} \sigma_{ij}\delta\varepsilon_{ij} - D_i\delta E_i + \rho\ddot{u}_i\delta u_i \, dV +$$
$$+\sum_{p=1}^{n_L}\frac{1}{L_p}\iint \varphi_1^{L_p} - \varphi_2^{L_p} \, \delta\varphi dt dt + \sum_{q=1}^{n_R}\frac{1}{R_q}\int \varphi_1^{R_q} - \varphi_2^{R_q} \, \delta\varphi dt + \sum_{r=1}^{n_C} C_r \, \varphi_1^{C_r} - \varphi_2^{C_r} \, \delta\varphi = 0 \quad (1)$$

Here D, E are the vectors of the electric inductance and electric field strength; σ_{ij} are the components of the symmetric Cauchy stress tensor; ε_{ij} are the components of the tensor of linear deformations; u_i are the components of the displacement vector; φ is the electric potential; $\varphi_1 - \varphi_2$ - the potential difference in the external circuit element, n_L, n_R, n_C are the number of inductive and resistive elements, respectively, L_p, R_q, C_r - values of inductance, resistance and capacitance of the corresponding circuit element.

For electric field, the condition of equipotentiality is fulfilled:

$$\varphi_{,i} = -E_i \quad (2)$$

For elastic parts (V_1), the physical equations of the linear elasticity theory hold true

$$\sigma_{ij} = C_{ijkl}\varepsilon_{kl} \quad (3)$$

For isothermal processes the electroelastic parts (V_2) are governed by the following physical relations:

$$\left.\begin{aligned} \sigma_{ij} &= \bar{C}_{ijkl}\varepsilon_{kl} - \beta_{ijk}E_k \\ D_k &= \beta_{ijk}\varepsilon_{ij} + e_{ki}E_i \end{aligned}\right\} \quad (4)$$

where C_{ijkl}, $\bar{C}_{ijkl}$ are the tensors of elastic constants, β_{ijk} and e_{ki} are the tensors of piezoelectric and dielctric coefficients.

The constraints imposed on the examined system are given in the form of mechanical and electrical boundary conditions. The mechanical boundary conditions are defined as

$$u_i\big|_{\Gamma_u} = u_{i0}, \quad (5)$$

where Γ_u is the part of the surface of the body V, on which the displacements are prescribed.
The expressions for electrical boundary conditions on the piezoelectric parts of volume V_2 are written as

$$\varphi\big|_{\Omega_{el}^{\varphi}} = \varphi_0, \quad (6)$$

here Ω_{el}^{φ} is the part of the bounding electrode-covered surface of body volume V_2, on which the electric potential is prescribed. In the absence of electrode covering all electrical boundary conditions take zero values.

The eigenvectors of vibrations are determined in the following form:

$$u_i \; x,t \; = \bar{u}_i \; \vec{x} \; e^{i\omega t}, \; \varphi \; x,t \; = \bar{\varphi} \; \vec{x} \; e^{i\omega t} \quad (7)$$

where $\omega = \omega_{\mathrm{Re}} + i\omega_{\mathrm{Im}}$ is the complex eigenfrequency of vibrations. In this relation ω_{Re} corresponds to the eigenfrequency and ω_{Im} corresponds to the rate of vibration damping (damping coefficient), $\bar{u}_i \; \vec{x}$, $\bar{\varphi} \; \vec{x}$ are the vibration eigenmodes.

A complete mathematical formulation of the problem is presented in work [28].

The stated problem was solved using the finite element method (FEM). Its numerical implementationwas performed based on the new algorithm developed by the authors of the paper, in which, to derive benefits from the commercial ANSYS software package (license agreement ANSYS Academic Research Mechanical and CFD № 1064623), the APDL macros and FORTRAN programs were incorporated. The obtained algebraic problem of complex eigenvalues was solved by the Muller method with a search through a variety of initial approximations.

3. Computational schemes for different system configurations

3.1. 2D system

Consider a rigidly clamped elastic plate with a fastened piezoelectric element (figure 1(a)). The plate has the following dimensions: $l_1 = 0.21$ m, $b_1 = 0.026$ m, $h_1 = 0.0006$ m. The piezoelectric element made in shape of rectangle with the dimensions $l_p = 0.05$ m, $b_p = 0.02$ m, $h_p = 0.00036$ m. The plate is made from material with the following mechanical characteristics: $E = 2 \cdot 10^{11}$ Pa, ν=0.3, ρ=7800 kg/m^3. The piezoelectric element located 12 mm from the fixed plate edge symmetrically to the plate axis.

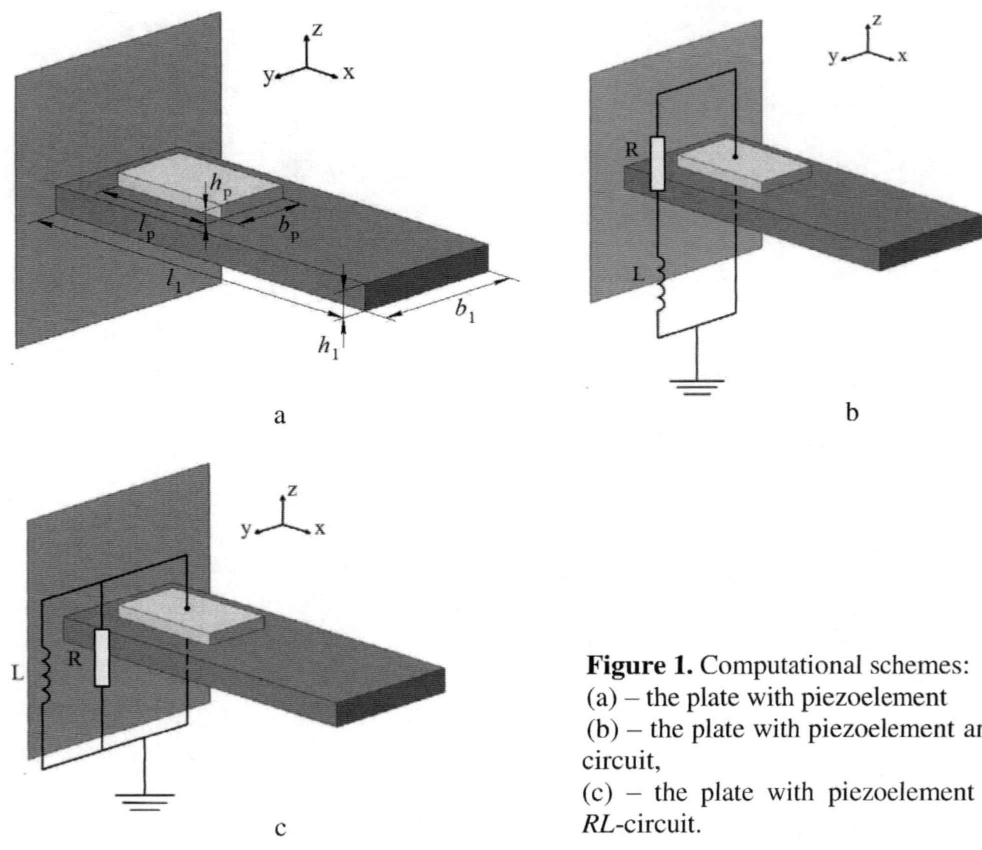

Figure 1. Computational schemes:
(a) – the plate with piezoelement
(b) – the plate with piezoelement and series *RL*-circuit,
(c) – the plate with piezoelement and parallel *RL*-circuit.

3.2. 3D system

As an example of a thin-walled structure, we consider a shell in the form of a semi-cylinder, which is rigidly clamped at the edges and is simply supported along the generatrix (figure 1(b)). The shell with $r_1 = 0.076$ m, $l_1 = 0.3$ m, $h_1 = 0.00025$ m is made from the elastic isotropic material showing the following mechanical properties: $E=1.96\cdot10^{11}$ Pa, $\nu=0.3$, $\rho=7700$ kg/m^3. The shell structure comprises a piezoelectric element in the form of the ring segment, which is attached to the shell surface. The element has the following dimensions: $r_p=0.07625$ m, $\varphi_p=15.08°$, $h_p=0.00036$ m. The center of mass of the element is located at a distance of 15mm from the clamped edges of the shell and is shifted by 90° with respect to the angular coordinate from the simply supported generatrixes.

In both variants of the structure the piezoelectric element is made of the material PZT-4 with the following mechanical characteristics (which for a shell are set in the cylindrical coordinates): $C_{11}=C_{22}=13.9\cdot10^{10}$ N/m^2, $C_{12}=7.78\cdot10^{10}$ N/m^2, $C_{13}=C_{23}=7.43\cdot10^{10}$ N/m^2, $C_{33}=11.5\cdot10^{10}$ N/m^2, $C_{44}=3.06\cdot10^{10}$ N/m^2, $C_{55}=C_{66}=2.56\cdot10^{10}$ N/m^2, $\beta_{31}=\beta_{32}=-5.2$ C/m^2, $\beta_{33}=15.1$ C/m^2, $\beta_{52}=\beta_{61}=12.7$ C/m^2, $e_{11}=e_{22}=6.45\cdot10^{-9}$ F/m, $e_{33}=5.62\cdot10^{-9}$ F/m, $\rho=7700$ kg/m^3. In the case of a plate, piezoceramics has polarization in the z direction, and in the case of a shell the direction of polarization coincides with the direction of the r–axis. The upper and lower surfaces of the element are fitted with electrodes.

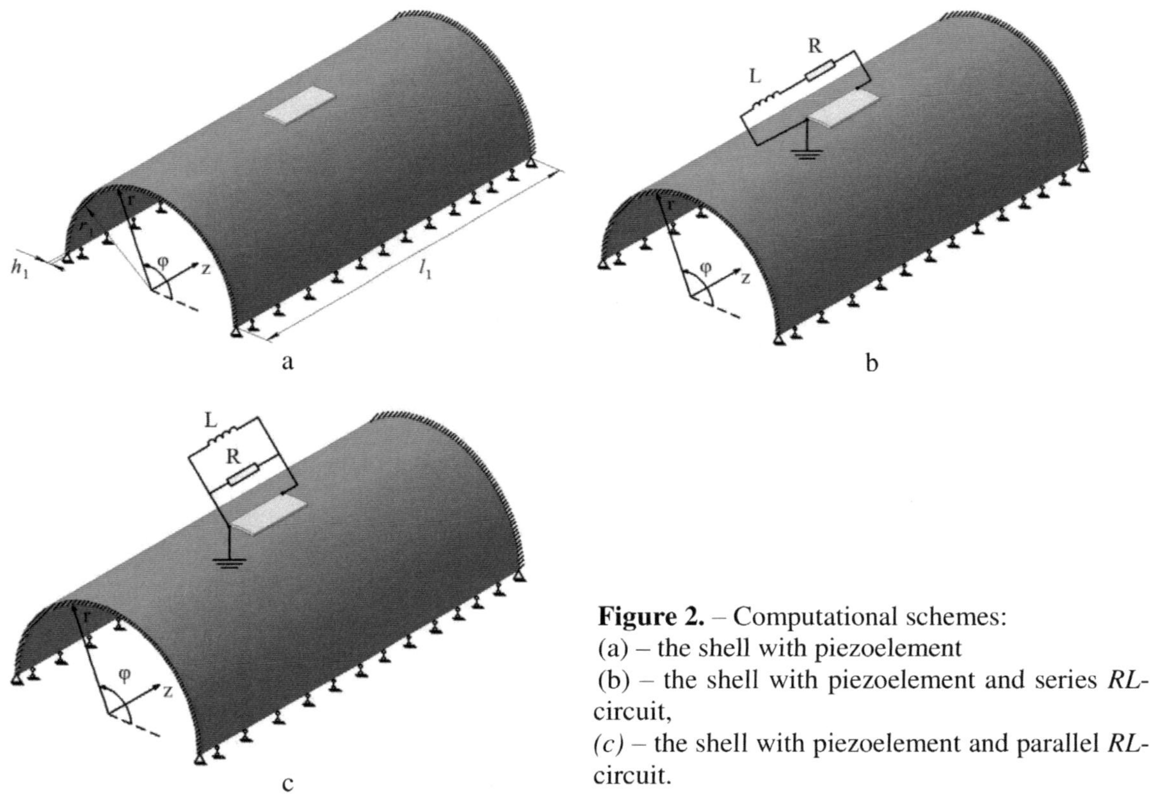

Figure 2. – Computational schemes:
(a) – the shell with piezoelement
(b) – the shell with piezoelement and series *RL*-circuit,
(c) – the shell with piezoelement and parallel *RL*-circuit.

4. Criterion of a search for optimal parameters of the external electric circuit

A search for optimal parameters of the external electric circuits is accomplished with the method of scanning in the range of parameters *R* and *L* using complex eigenfrequencies $\omega=\omega_{\rm Re}+{\rm i}\omega_{\rm Im}$. obtained by solving the natural vibration problem. Here, the real part $\omega_{\rm Re}$ is the vibration frequency and the imaginary part $\omega_{\rm Im}$ – is the damping ratio.

IOP Conf. Series: Materials Science and Engineering **208** (2017) 012030 doi:10.1088/1757-899X/208/1/012030

According to the results of work [23], the optimal damping characteristics of a structure with piezoelectric element and external electric circuit in the region of the *j-th* resonance can be achieved in the case when the natural *j-th* frequency of vibrations $\omega^s_{\mathrm{Re}\ j}$ and the vibration frequency ω^e_{Re} of the electric loop, composed of the intrinsic capacitance of piezoelectric element and external electric circuit, coincide.

$$\omega^s_{\mathrm{Re}\ j} = \omega^e_{\mathrm{Re}} \tag{8}$$

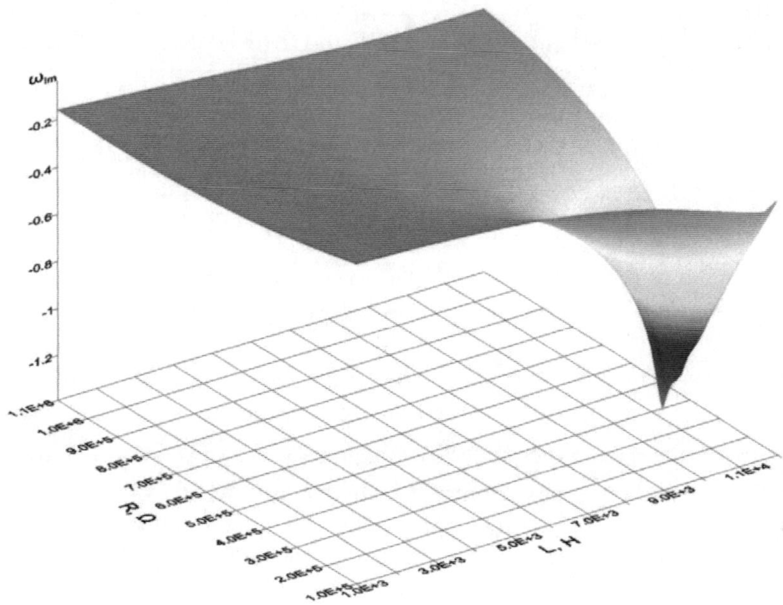

Figure 3. The imaginary part ω_{Im} of the complex eigenfrequency of the structure as a function of the values of parameters, *R* and *L*.

In work [23], it was also shown that along with this peculiarity, the imaginary part of the eigen frequency of the structure exhibits a pronounced extremum (fig. 3).

Thus, as an optimality criterion for selection of external circuit parameters we can use the condition of maximum absolute value of the imaginary part of the complex eigenfrequency of the structure (8). Physically, it means that at these values of parameters *R* and *L* the rate of vibration damping is a maximum. Mathematically, this criterion of optimality can be written as

$$\left|\omega^s_{\mathrm{Im}}\right|_j \rightarrow \max \tag{9}$$

5. Numerical results

At a certain location of the piezoelelctric element on the structure surface a situation may arise, wherein the vibrations of the structure at the examined frequency does not cause the deformation of the element or lead to its symmetric deformation, which implies the absence of the potential normally generated on its electrode –covered surface. Therefore, such an arrangement of the piezoelectric element does not allow damping of vibrations at this frequency by means of transforming a part of the mechanical energy into the electric energy and its subsequent dissipation in the external electric circuit. In what follows, we exclude such frequencies form our consideration

Table 1 presents the values of eigenfrequencies of the plate and the shell with piezoelectric element in the short circuit (s/c) and open circuit (o/c) regimes Here, the first and the fifth eigenfrequencies were subject to damping.

The results of computations with account of condition (9), which were done to find the optimal parameters of the external electric circuits ensuring maximal damping effect for the first and fifth eigenfrequencies of the structure with piezoelectric element, were presented in table 2 for a series and parallel *RL*-circuits.

Table 1. The eigenfrequencies of the plate

Resonant frequency number	Plate		Shell	
	$\omega_{s/c}$ (Hz)	$\omega_{o/c}$ (Hz)	$\omega_{s/c}$ (Hz)	$\omega_{o/c}$ (Hz)
1	13.56	13.86	554.25	557.41
2	72.15	72.33	755.77	759.57
3	197.90	198.32	800.13	803.79
4	399.79	404.15	1287.62	1293.42
5	665.45	674.96	1471.64	1482.21

As is evident from the results obtained for the plate, for the preset accuracy of computations, $\varepsilon = 10^{-5}$, the values of the imaginary part of the first eigenfrequency ($\omega_{\text{Im}}^{1ser}$)of the system connected to a series external circuit differ from the values of imaginary parts of the first eigenfrequency ($\omega_{\text{Im}}^{1par}$)in the case of a parallel external circuit by about 1 %, whereas for the fifth eigenfrequency the difference in the imaginary parts is ~ 4%. Furthermore, the values of the optimal parameters , *R* and *L*, for the series- connected circuit are lower than for the parallel- connected circuit, which makes the former circuit more promising for practical realization at low frequencies.

Table 2. The eigenfrequencies of the plate and shell with external electrical circuits.

Structure without external circuit		Structure with series *RL*-circuit			Structure with parallel *RL*-circuit		
eigenfrequency		Optimum circuit parameters		eigenfrequency	Optimum circuit parameters		eigenfrequency
s/c (Hz)	o/c (Hz)	*L* (H)	*R* (kOhm)		*L* (H)	*R* (kOhm)	
Plate							
13.56	13.86	11200	401	13.64 – 1.42*i*	12172	2515	13.43 – 1.40*i*
665.45	674.96	5.1	7.2	687.68 – 52.16*i*	5.4	66.1	673.41 – 54.34*i*
Shell							
554.25	557.41	6.9	5.1	553.60 – 28.30*i*	7.12	110.3	554.70 – 32.78*i*
1471.64	1482.21	1.1	2.6	1493.29 – 83.48*i*	1.08	37.9	1453.23 – 74.40*i*

From the results of computation for shell structures it follows that for the above accuracy of computation, selection of the optimal parameters on the basis of condition (9) provides in the case of a parallel external circuit a greater damping ratio (by 14-15%) compared to the case of a series circuit. However, as in the case of a plate, the optimal resistance and inductance, ensuring vibration damping at the prescribed frequency, take much higher values at low frequencies compared to the case of a series-connected circuit.

Conclusions

In this paper, in the framework of the mathematical formulation of the natural vibration problem for viscoelastic bodies with external electric circuits, a new approach has been proposed for determining the optimal parameters of the external electric circuit. The idea behind this approach is to find in the range of R and L parameters the damping ratio being maximum in absolute magnitude - i.e., an imaginary part of the complex eigenfrequency, at which the structure vibrations are undesirable.

It has been shown that the proposed approach is equally effective for systems of different types (both for plane and spatial structures) and can be applied for determining the optimal parameters of the external resonance *RL*-circuits connected in series or in parallel.

From the viewpoint of practical applications dealing with low (most energy intensive) frequencies, the preference in the examined structures should be given to series-connected circuits, in which the optimal resistance and inductance determined on the condition of maximal value of the imaginary part of the system eigenfrequency take lower values than in the case of a parallel circuit.

Acknowledgments

The work is supported by the RFBR (projects No 15-01-03976-a, No 17-01-590152-r-ural-a).

References

[1] Niederberger D 2005 *Smart damping Materials using Shunt Control* (Dissertation for the degree of Dr. of Sci., Zurich) p 210

[2] Hagood N W and von Flotow A 1991 Damping of structural vibrations with piezoelectric materials and passive electrical networks *Journal of Sound and Vibration* **146:2** 243-268

[3] Lesieutre G A 1998 Vibration damping and control using shunted piezoelectric materials *The Shock and Vibration Digest May* **30** pp.187-195

[4] Caruso G 2001 A critical analysis of electric shunt circuits employed in piezoelectric passive vibration damping *Smart Mater. Struct.* **10** 1059-1068

[5] Moheimani S O R, Fleming A J and Behrens S 2002 On the feedback structure of wideband piezoelectric shunt damping systems *Smart Mater. Struct.* **12** 49-56

[6] Park C H, Inman D J 2003 Enhanced Piezoelectric Shunt Design *Shock and Vibration* **10:2** 127-133

[7] Wu S Y 1996 Piezoelectric Shunts with Parallel R-L Circuit for Structural Damping and Vibration Control *Proc. SPIE Smart Structures and Materials, Passive Damping and Isolation; SPIE* **2720** 259-269

[8] Fleming A J, Moheimani S O R 2005 Control Orientated Synthesis of High-Performance Piezoelectric Shunt Impedances for Structural Vibration Control *IEEE Transactions on control systems technology* **13:1** 97-113

[9] Wu D, Yang Z and Sun H 2009 Vibration control efficiency of piezoelectric shunt damping system *Front. Mech. Eng. China* **4:4** 441-446

[10] Thomas O, Ducarne J and Deü J-F 2012 Performance of piezoelectric shunts for vibration reduction *Smart Materials and Structures* **21:1** 0150082012

[11] Berardengo M, Cigada A, Manzoni S and Vanali M 2015 Vibration Control by Means of Piezoelectric Actuators Shunted with *LR* Impedances: Performance and Robustness Analysis *Shock and Vibration* 704265

[12] Høgsberg J and Le Cöent A 2014 Explicit solution format for complex-valued natural frequency

of beam with R-shunted piezoelectric laminate transducer *Proc IMechE Part C: J Mechanical Engineering Science* **228:1** 31-44

[13] Preumont A 2011 *Vibration control of active structures* 3rd ed. (Heidelberg: Springer)

[14] Moheimani S O R and Fleming A J 2006 *Piezoelectric transducers for vibration control and damping* (London, Springer)

[15] Delpero T, Bergamini A E and Ermanni P 2012 Identification of electromechanical parameters in piezoelectric shunt damping and loss factor prediction *Journal of Intelligent Material Systems and Structures* **11** 287-298

[16] Porfiri M, Maurini C and Pouget J 2007 Identification of electromechanical modal parameters of linear piezoelectric structures *Smart Mater Struct.* **16** 323-331

[17] Soltani P, Kerschen G, Tondreau G and Deraemaeker A 2014 Piezoelectric vibration damping using resonant shunt circuits: an exact solution *Smart Materials and Structures* **23** 125014

[18] Heuss O, Salloum R, Mayer D and Melz T 2016 Tuning of a vibration absorber with shunted piezoelectric transducers *Arch. Appl. Mech.* **86:10** 1715-1732

[19] Park C H 2003 Dynamics modeling of beams with shunted piezoelectric elements *Journal of Sound and Vibration* **268:1** 115-129

[20] Wu S Y 1996 Piezoelectric shunts with a parallel R-L circuit for structural damping and vibration control *Proceedings of the SPIE 2720* (San Diego, SPIE) 259-269

[21] Park C H and Inman D J 1999 A uniform model for series R-L and parallel R-L shunt circuits and power consumption *Proceedings of the SPIE 3668* (California, SPIE) 797-804

[22] Kligman E P and Matveenko V P 1997 Natural Vibration Problem of Viscoelastic Solids as Applied to Optimization of Dissipative Properties of Constructions *Journal of Vibration and Control* **3:1** 87-102

[23] Ivanov A S, Matvienko V P, Oshmarin D A, Sevodina N V, Yurlov M A and Yurlova N A 2016 Justification of equivalent substitution circuits used to optimize the dissipative properties of electroelastic bodies with external electric circuits *Mechanics of Solids* **51:3** 273–283

[24] Matveenko V P, Kligman E P, Yurlov M A and Yurlova N A 2012 Simulation and optimization of dynamic characteristics of piezoelectric smart structures *Physical Mesomechanics* **15:3-4** 190-199

[25] Washizu K 1982 *Variational Methods in Elasticity and Plasticity* (Pergamon Press, London)

[26] Parton V Z, Kudryavtsev B A 1988 *Electromagnetoelasticity of piezoelectric and electroconductive bodies* (Nauka, Moscow)

[27] Karnaukhov V G, Kirichok I F 1988 *Electrothermal Viscoelasticity* (Naukova Dumka, Kiev)

[28] Matvvenko V P, Oshmarin D A, Sevodina N V and Iurlova N A 2016 Natural vibration problem for electroviscoelstic body with external electric circuits and finite-element relations for its numerical implementation *Computational Continuum Mechanics* **9:4** 476-485

IOP Conf. Series: Materials Science and Engineering **208** (2017) 012031 doi:10.1088/1757-899X/208/1/012031

Eddy current flowmeter for sodium flow

A Pavlinov[1], R Khalilov[1], A Mamikyn[1], I Kolesnichenko[1,2]

[1] Institute of Continuous Media Mechanics, Perm, Russia
[2] Perm National Research Polytechnic University, Perm, Russia

E-mail: pam@icmm.ru

Abstract. The eddy current flowmeter for liquid metal passing through the cylindrical channel is studied experimentally and numerically. The alternating magnetic field is created by an annular-shape coil and resulting field is measured by two symmetric coils. The dependence of the signal on the magnetic field frequency was obtained for a solid duralumin cylinder. The phase shift and amplitude ratio linearly depends on the flowrate in a wide range of parameters. The workability of the method was tested on a liquid sodium, flowing along a cylindrical channel. The dependencies for liquid sodium and solid duralumin cylinder differ due to the existence of a non-uniform velocity profile along the radius of the fluid. This necessitates an additional calibration procedure on the liquid metal in addition to research on the solid conductor. The developed flowmeter is used to measure the flowrate in the sodium loop.

1. Introduction

There are a large number of liquid metal velocity sensors designs [1]. Each of them has it's own advantages and disadvantages. Potential probes [2, 3] and flowmeters [1] are widespread. Their advantage is the simplicity of implementation and the wide possibilities of the velocity pulsations characteristics measurement. Disadvantages are the need for a good electrical contact with liquid metal and the vulnerability of permanent magnets to high temperatures. There are techniques based on the measurement of cross correlations between the signals of thermocouples installed along the flow [4, 5]. The advantage of the method is a direct measurement of velocity without the need of a calibration. The disadvantages are the need to immerse the thermocouples into the flow, as well as the reliable applicability of this technique only for the flows with a sustainable direction. It is also possible to note modern methods based on measuring the Lorentz force [6] and the induced motion [7], which are in the stage of research and experimental application.

At the same time, there are induction velocity measuring methods based on eddy currents generation [8]. They were actively studied and used for many years (e.g. [1, 9–13]). Some devices are immersed into the flow and represent a streamlined capsule combining both the generating and measuring part [14]. There are also more compact designs, which are actually the local velocity sensors [15]. They use a permanent magnet to generate the initial magnetic field and Hall sensors to measure its perturbations. With all the advantages of these designs, their main drawback is their vulnerability to high temperatures due to the complexity of the cooling systems. Therefore, they are used at relatively low temperatures.

The flowmeters with external placement of generating and measuring parts received the main development. The main advantage of such systems is the possibility of implementing the measuring coils thermostabilization system and the cooling of the entire system. This allows them to be used even at high temperatures. Additionally, such flowmeters provide the ability to implement the liquid metal electrical conductivity measurement [9], or the registration of a gas phase in the liquid metal [16]. In most designs, the coils are put on a cylindrical pipe, although there are designs in which they are located on the side of the pipe [17]. Also, there are designs that use a single measuring coil, and an alternating magnetic field is generated by a traveling magnetic field pump [18]. More often, it is necessary to implement a more universal scheme, which can be used independently of other parts of a main experimental setup. Such a scheme was actively investigated theoretically (e.g. [11, 19–21]) and experimentally (e.g. [22–24]).

One of the main advantages of this scheme, in our opinion, is the ability to process the acquired signal so as to exclude the influence of temperature on the measurements. Electrical conductivity of the liquid metal, which significantly determines the characteristics of the process, depends on the temperature. In most measurement schemes, it is necessary to calibrate the device not only at different velocities or flowrates, but also at different temperatures of the liquid metal. In this scheme, when measuring signals S_1 and S_2 from two symmetrical coils can be recorded simultaneously, the following manipulations can be performed. The value $A_1 = S_1 - S_2$ represents the component of voltage due to motion induced voltage and the value $A_2 = S_1 + S_2$ represents the component of voltage due to transformer action. Thus, their ratio $A = A_1/A_2$ is independent on the electrical conductivity of the medium [13, 25]. This essentially lightens the calibration process of the device and improves the accuracy of the results. The usage of this method was simplified relatively recently with the active introduction of computers into the experiment. Computers provide real-time calculations and output of measuring flowrate value allowing to control the process. Nevertheless the possibility of eliminating the electrical conductivity dependence was known for a relatively long time (e.g. [9]). But in the absence of computers, this technique was implemented using complex electronic circuits. Thus, for example, two pairs of measuring coils and compensation circuit were used in [9]. In addition to using the described technique, it is possible to determine the flowrate by a conventional method by means of a phase shift (e.g. [12]).

These studies show the possibility of successful implementation of the methodology in practice. However, the operation of this method is essentially determined by geometric factors, the frequency of the alternating magnetic field, and the characteristics of the medium. The set of these characteristics is unique for each implementation, that reduces the universality of the previously obtained results. This was the motivation to study the method of flowrate measurement applicable for our sodium loop [26]. We chose a scheme with a generating coil and two symmetrical measurement coils of the annular shape located on a cylindrical channel (figure 1(a)). The aim of the work was to find out the dependence of the the measuring module sensitivity on the supply current frequency and to check its operation on the liquid sodium flow. The existence of the skin effect should, on the one hand, dictate the choice of the lowest available frequency as the best. On the other hand, it limits the range of measured flowrates [21]. Therefore, the main effort in our work is directed to experimental study of the process and to test method's sensitivity.

2. Methods

The main element of the experimental setup is a measuring module that registrates the interaction between the moving electrically conductive medium and a magnetic field (figure 1). The module contains a generating coil 1 and two measuring coils 2 (figure 1(a)). The generating coil has an average diameter of $D_s = 100mm$, a thickness of $H_s = 20mm$ and a width of $L_s = 20mm$. The coil has $N_s = 260$ turns of a copper wire having diameter of $d_s = 1mm$ and insulation. The generating coil is connected to the low-frequency generator "G3-109" which provides high voltage stability for a wide range of frequences. The coil was fed by harmonic signal of varied frequency f_s and a fixed voltage $U_s = 1.6V$ and current $I_s = 0.39A$ during the experiments. The measuring coils have an average diameter of $D_m = 112mm$, a thickness of $H_m = 20mm$ and a width of $L_m = 10mm$ and were located symmetrically with respect to the center of the generating coil. The number of turns of measuring coils is $N_m = 500$. Copper wire of $d_m = 0.4mm$ in diameter was used for both measuring coils. The distance $L_b = 20mm$ between the center of the generating coil and centers of the measuring coils was chosen based on the results of the calculations. The generating and measuring coils were placed on a heat-resistant fiberglass frame.

EMF generated in two measuring coils S_1 and S_2 were recorded using the National Instruments NI 9239 data acquisition board. Signal analysis gave, firstly, the phase shift P, which was determined using a fast Fourier transform. Secondly, ratios between the amplitudes of the signals $A_1 = S_1 - S_2$, $A_2 = S_1 + S_2$ and $A = A_1/A_2$ were calculated.

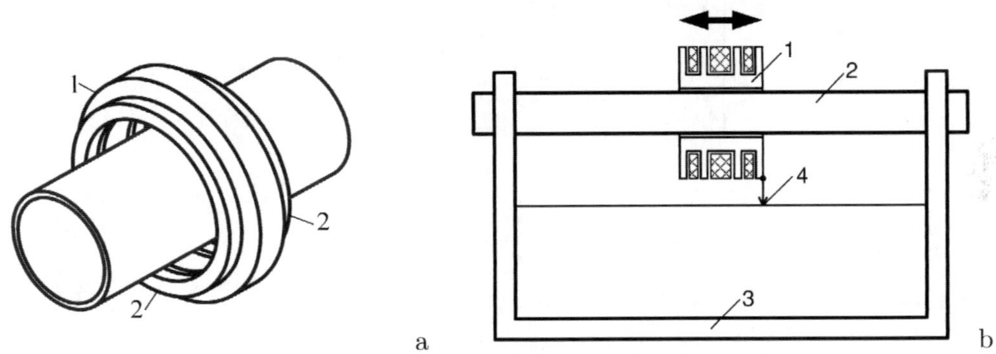

Figure 1. (a) – Sketch of the module assembly, (b) – motion of the measuring module relative to solid duralumin cylinder (not to scale).

In the first series of experiments the solid cylindrical volume of duralumin 2 (figure 1(b)) was used as an electrically conductive medium. The metal has the electrical conductivity $\sigma_d = 11.6 \cdot 10^6 Sm/m$ at temperature $T_d = 25^oC$. The temperature T_d was varied during this series of experiments . The diameter of the cylindrical volume is $D_d = 70mm$ and the distance at which measurements were taken is $L_d = 890mm$. The cylindrical volume was placed on the frame 3. The measuring unit 1 was placed on a cylindrical volume 2 and could be moved freely along the cylinder axis. The sliding contact 4 was used to get reference values of the relative movement. It consisted of a nichrome wire stretched parallel to the cylinder and a copper contact mounted on the measuring module. The ends of the nichrome wire were connected to a DC voltage source. When the measuring module was moved along the cylinder, the voltage between one of the ends of the nichrome wire and the contact varied proportionally to the distance. Analysis of the voltage signal gave the value of the movement velocity of the conductive cylinder V_d relative to the measuring module. In the first series of experiments, the dependencies of the signals $A_d(Q)$ and $P_d(Q)$ on the flowrate of the moving medium $Q = V_d\pi(D_d/2)^2$ were determined for different temperatures T_d of the medium. The frequency f_s varied from 20 to $500Hz$. In this case, the value of the skin layer $\delta_d = \sqrt{1/\pi\mu_0 f_s \sigma_d}$ ranged from $33mm$ to $6.6mm$ at $T_d = 25^oC$.

In the second series of experiments a liquid sodium passing through the cylindrical pipe 2 (figure 2) was used as an electrically conductive medium. The experiments were carried out at a fixed temperature $T_n = 150^oC$ and a fixed frequency $f_s = 50Hz$. The liquid sodium has the electrical conductivity $\sigma_n = 8.86 \cdot 10^6 Sm/m$ at this temperature. In this case the value of the skin layer has a fixed value equal to $\delta_n = \sqrt{1/\pi\mu_0 f_s \sigma_n} = 24mm$. The pipe 2 with an internal diameter of a $D_n = 68mm$ was made of stainless steel, the thickness of its wall is $d_w = 4mm$, the electrical conductivity is $\sigma_{ss} = 1.38 \cdot 10^6 Sm/m$ at a temperature T_n. Measuring module 1 has been mounted and fixed on the pipe 2. The pipe was built into the system 3, consisting of two tanks. A pumping of liquid sodium between these tanks was carried out by a pressure difference using argon. Contact probes 4 mounted in the tanks recorded liquid sodium volume levels. It was possible to determine the flowrate of liquid sodium Q through the pipe by analyzing the signals from these probes. The experiment is described in more detail in [27]. In the second series of experiments, the dependence of the signals $A_n(Q)$ on the flowrate of the moving medium Q was determined.

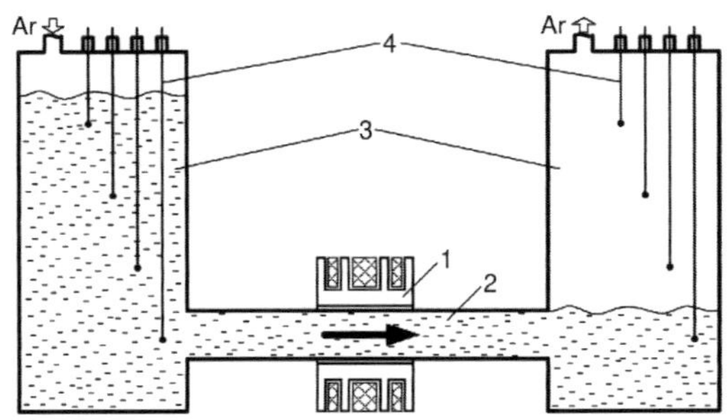

Figure 2. Measuring module installed on a pipe connecting two tanks, filled with a liquid sodium: scheme of a calibration experiment (not to scale).

Preliminary numerical simulations of the magnetic field created by the generating coil 1 of the measuring module (figure 1(a)) were carried out. The measuring coils 2 are connected to high-impedance inputs of DAQ system, so they do not disturb the resulting magnetic field and are excluded from the simulation. The simulations are based on solving the Maxwell equations system where generating coil is a magnetic field source which is being disturbed by a liquid metal flow. The pipe and surrounding area are taken into account. The simulations are carried out for the stationary isothermal case.

3. Results and discussion

Calculations showed that the magnetic field generated by an annular-shape coil is highly inhomogeneous in the region of the electrically conductive medium. Increasing the frequency of the magnetic field f_s does not change the profile, but it significantly changes the maximum value of the axial distribution (figure 3(a)). This result indicates that increasing the frequency reduces the overall sensitivity of the technique. Also, the axial dependence helps to choose a distance L_b between measuring coils and the center of the generating coil. As mentioned earlier, it is not possible to place measuring coils close to the generating coil, because it is necessary to ensure the presence of a structural gap for installation and thermostabilization. A large distance from the generating coil will reduce the sensitivity of the method. The strongest magnetic field gradient is located on the first third of the dependence. It is best to place the measuring coils in this area. Also increasing the magnetic field frequency f_s significantly reduces the maximum value in the radial distribution (figure 3(b)) due to the skin effect. Thus, at higher frequencies the near-wall flow will give the greatest contribution to the resulting signal, which may distort the result. However, up to $50Hz$ this effect appears to be insignificant.

Increasing the electrical conductivity of the wall σ_w relative to the electrical conductivity of the liquid metal σ_n, the maximum value of the magnetic field also decreases, due to the skin effect (figure 4). Consequently, the use of a poorly conductive liquid metal (like lead) and a good conductive material for the wall (like copper) will significantly weaken the signal level. This is one of the reasons for choosing sodium as an electrically conductive medium and stainless steel for the wall. The electrical conductivity of sodium is about six times higher than the electrical

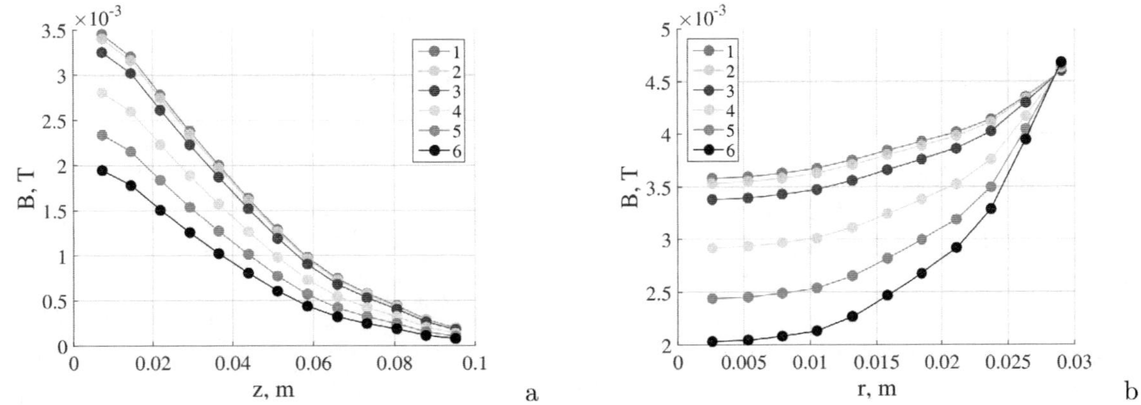

Figure 3. Magnetic field component B_z, calculated for different current frequences f_s (a) – as a function of distance along z-axis, (b) – as a function of distance along radial direction r. Curves No. 1,2..6 are obtained at $f_s = 10Hz$, $25Hz$, $50Hz$, $100Hz$, $150Hz$, $200Hz$ correspondingly.

conductivity of stainless steel, so in our case there is an insignificant weakening of the useful signal.

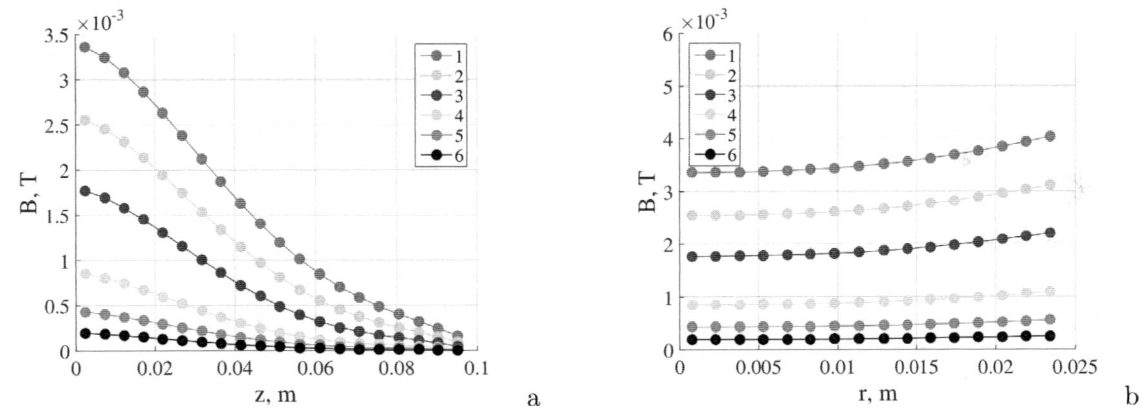

Figure 4. Magnetic field component B_z profiles calculated for different conductivities relations $k = \sigma_n/\sigma_w$ (a) – along z-axis, (b) – along radial direction r. Curves No. 1,2..6 are obtained at $k = 1,\ 1/5,\ 1/10,\ 1/25,\ 1/50,\ 1/100$ correspondingly.

The experiment on a solid conductor in the isothermal case showed that both the phase shift P (figure 5(a)) and the amplitude ratio A (figure 5(b)) give a linear dependence on the flowrate Q for different frequencies of the magnetic field f_s. The linear dependencies make it possible to calculate the slope coefficients k_P and k_A for these signals. Increasing the frequency leads to a weakening of the signal level. The part of electrically conductive medium that contributes to the measured signal decreases as the frequency increases due to the skin effect. This also forces to use small frequencies f_s for high conducting media.

Dependencies of the slope coefficients k_P and k_A on the frequency f_s show that these quantities behave differently (figure 6(a)). An increase to the maximum value is observed followed by a decrease for the phase shift. A decrease is observed for the amplitude ratio in the entire frequency range (figure 6(a)).

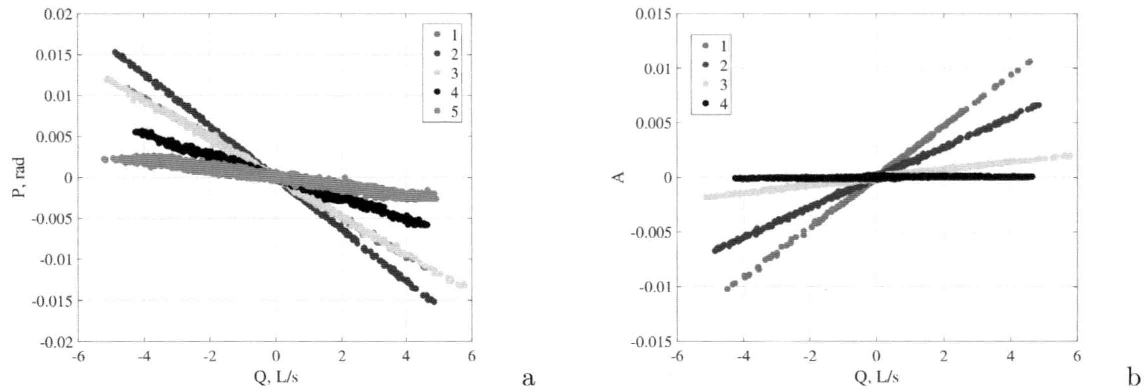

Figure 5. Phase shift P (a) and amplitudes relation A (b) measured at different generating coil current frequences *vs.* flowrate Q. Curves No. 1,2..5 are obtained at $f_s = 20Hz$, $40Hz$, $80Hz$, $150Hz$, $300Hz$ correspondingly.

Experiments carried out on a liquid electrically conductive medium have also shown a linear dependence of amplitude ratio A on a flowrate Q (figure 6(b)). The dependencies for liquid sodium and solid duralumin cylinder differ. This is due to the existence of a non-uniform velocity profile along the radius of the fluid that flows along the cylindrical channel. In a solid conductor, such a profile is uniform that causes the difference. Perhaps at very high flowrates when the profile is strongly turbulized and tends to be uniform, the difference will become small.

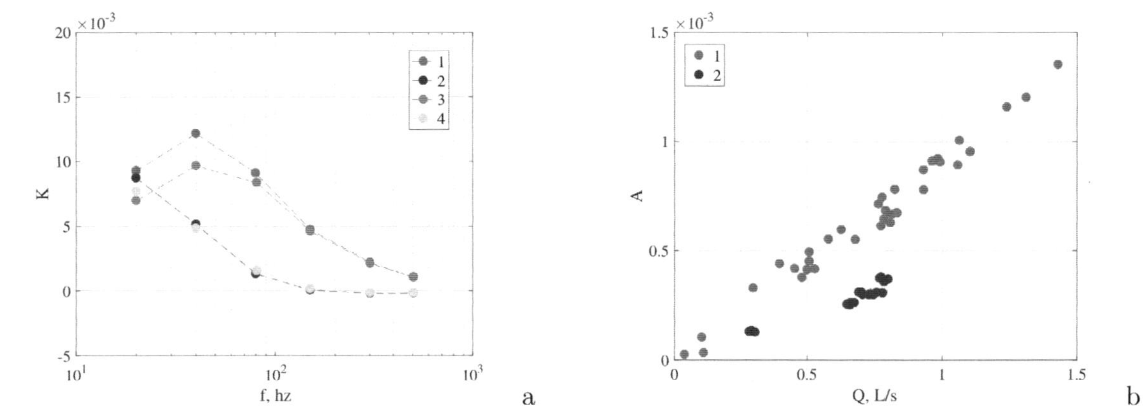

Figure 6. (a) – Slope coefficients k_P (1,3) and k_A (2,4) as a function of generating coil current frequency f_s at two different temperatures: $T = 23°C$ (1,2) and $T = 93°C$ (3,4) , (b) – amplitudes relation A measured during experiments on the solid duralumin cylinder (1) and on the liquid sodium flow (2).

4. Conclusions

We consider an implementation of eddy current flow measurement technique for different feeding current frequencies. This technique for measuring the motion of an electrically conductive medium has a number of features. Annular-shape generating coil creates a magnetic field that decrease quite fast along the axis. This limits the distance at which the measuring coils can be placed without losing module's sensitivity. They should be placed so close to the generating coil, as far as the design and the requirement for their thermal stabilization permit.

Experiments on a solid conductor made it possible to reveal a number of features of the technique that were difficult to obtain on a liquid metal because of limitations of the facility and large time costs when working with liquid sodium. The phase shift and amplitude ratio linearly depend on the flowrate in a wide range of parameters. The presence of a maximum in the slope-to-frequency dependence for the phase shift indicates the possibility of selecting such a frequency of the current of the generating coil, which will ensure the best sensitivity of the method. However, this parameter will depend essentially on the electrical conductivity of the medium. Therefore, its value should be determined during experiments on the medium for which the methodology is intended to be applied. The dependence of the slope coefficient on the frequency for the amplitude ratio did not reveal the presence of an extremum. Consequently, the greatest sensitivity of the technique should be expected at the lowest possible frequency. These data were obtained for a solid conductor of the aluminum alloy, which has a high electrical conductivity. Conductivity is lower for liquid conductors and therefore, such a maximum is likely to be absent for the available frequency range.

The chosen parameters allowed one to determine the flowrate of the conductive medium with sufficient accuracy. The signal level is well above the noise level. Using the results of experiments on liquid sodium we obtained the dependencies of the amplitude ratio on the flowrate which are close to linear. Since the difference between the results for solid and liquid conductors, which is due to the presence of a velocity profile for a liquid metal, it is necessary to perform additional calibration procedure on liquid metal in addition to studies on solid conductor. The amplitude ratio is preferable to the phase shift because of the lack of temperature dependence. So, the technique described in this study can provide reliable measurements of liquid metal flowrate.

References

[1] Shercliff J A 1962 *Theory of electromagnetic flow measurement* (UK: Cambridge University Press)
[2] von Weissenfluh T 1985 *Int. J. Heat Mass Transf.* **28** 1563–1574
[3] Ricou R and Vives C 1982 *Int. J. Heat Mass Transf.* **25** 1579–1588
[4] Belyaev I A, Razuvanov N G, Sviridov V G and Zagorsky V S 2015 *Magnetohydrodynamics* **51** 673–684
[5] Frick P, Khalilov R, Kolesnichenko I, Mamykin A, Pakholkov V, Pavlinov A and Rogozhkin S 2015 *EPL* **109** 14002
[6] Thess A, Votyakov E and Kolesnikov Y 2006 *Phys. Rev. Lett.* **96** 164501(4)
[7] Priede J, Buchenau D and Gerbeth G 2009 *Magnetohydrodynamics* **45** 451–458
[8] Lehde H and Lang W 1948 AC electromagnetic induction flowmeter. US Patent 2.435.043
[9] Poberejsky L 1963 *Tech. Phys. (in Russian)* 1464–1469
[10] Tzirkunov V E, Jeigur B D, Sermons G Y and Kalnin R K 1973 *Contactless control of liquid metal flux* (Riga: "Zinatne" (in Russian))
[11] Convert D, Marthy P, Alemany A and Poinsot S 1993 *Magnitnaya gidrodinamika (in Russian)* **29** 91–96
[12] Schulenberg T and Stieglitz R 2010 *Nucl. Eng. Des.* **240** 2077–2087
[13] Poornapushpakala S, Gomathy C, Sylvia J and Babu B 2014 *Flow Meas. Instrum.* **38** 98–107
[14] Cowley M 1965 *J. Sci. Instrum.* **42** 406–409
[15] Miralles S, Verhille G, Plihon N and Pinton J F 2011 *Rev. Sci. Instrum.* **82** 095112
[16] Nakamoto K, Tamura S, Ishii K, Kuwahara H, Ohyama N and Muramatsu T 1984 *Nucl. Eng. Des.* **82** 393–404
[17] Priede J, Buchenau D and Gerbeth G 2011 *Meas. Sci. Technol.* **22** 055402
[18] Araseki H, Kirillov I R and Preslitsky G V 2012 *Nucl. Eng. Des.* **243** 111–119
[19] Wiegand D E 1968 *IEEE Trans. Nucl. Sci.* **15** 28–36
[20] Baker R C 1977 *Prog. Nucl. Energy* **1** 41–61
[21] Kawabe R 1982 *J. Nucl. Sci. Technol.* **19** 169–172
[22] Wiegand D E and Michels C W 1969 *IEEE Trans. Nucl. Sci.* **16** 192–195
[23] Dementev S, H Glasbrenner H, Kirchner T, Heinrich F, Bucenieks I, Platacis E, Pozdnjaks A, Krisko S, Kirshtein G and Mikoelov A 2001 *Magnetohydrodynamics* **37** 417–426

[24] Dementjev S, Barbogallo F, Groeschel F, Bucenieks I, Krysko S and Poznyaks A 2002 *Magnetohydrodynamics* **38** 413–417
[25] Sharma P, Kumar S S, Nashine B, Veerasamy R, Krishnakumar B, Kalyanasundaram P and Vaidyanathan G 2010 *Ann. Nucl. Energy* **37** 332–338
[26] Kolesnichenko I, Khalilov R, Shestakov A and Frick P 2016 *Magnetohydrodynamics* **52** 87–94
[27] Khalilov R, Kolesnichenko I, Mamykin A and Pavlinov A 2016 *Magnetohydrodynamics* **52** 53–60

IOP Conf. Series: Materials Science and Engineering **208** (2017) 012032 doi:10.1088/1757-899X/208/1/012032

Aeroacoustic investigations of subsonic jet in PNRPU anechoic chamber

V V Palchikovskiy[a], I V Khramtsov, V V Ershov, D A Gornova, A A Selivanova

Perm National Research Polytechnic University, Perm, 614990, Komsomolsky prospect, 29, Russia

E-mail: [a] vvpal@bk.ru

Abstract. A study of the aeroacoustic characteristics of the PNRPU jet rig is considered. The design of the jet rig is described. The velocity of the jet along the axis is measured at a certain distance from the nozzle exit section. The comparison of the velocity with a semi-empirical model is performed. A good agreement of the velocities is observed. The noise of the jet is measured for different velocities in the direction of 30 and 90 degrees. The obtained spectra are compared with the results of measurements of the jet rig in the anechoic chamber AC-2 TsAGI. Localization of the noise sources of the jet with the Bruel & Kjaer microphone array is also carried out. The investigations are carried out in a new anechoic chamber at PNRPU.

1. Introduction

The noise of an aircraft engine jet makes a dominant contribution to the overall noise of the aircraft at take-off [1]. The development of effective measures to reduce this negative impact requires a more thorough study of the noise generation by turbulent flows. The basic mechanisms of these processes are studied with model jets. In order to carry out such studies in laboratory conditions, the availability of an anechoic chamber is primarily required, where it is possible to carry out acoustic measurements in free field conditions. Then, it is required experimental rig that supplies air turbulent flow into the anechoic chamber. In 2014-2016, these facilities were created at the Perm National Research Polytechnic University (PNRPU). The qualification tests have shown that the anechoic chamber allows the acoustic measurements to be performed to obtain quantitative results [2]. However, it is also required to investigate the aeroacoustic quality of a subsonic turbulent jet.

2. Jet rig

The availability of a jet rig in the anechoic chamber gives a wide field for aeroacoustic research. However, chose of the jet type is largely related to the financial possibilities, as well as the availability of the necessary rooms for the installation of the entire supply-exhaust system of the jet rig. For example, a supersonic jet with a cocurrent flow requires the powerful compressed air sources and a developed network of air ducts to deliver the air flow into the chamber and exhaust it [3]. The hot jet complicates the design of the anechoic chamber as a whole [4]. To implement jet noise study in the research university, a cold submerged jet can be chosen [5].

Also the key issue is choosing the type of compressed air source. As a rule, these are compressor units that pre-pump special vessels with high pressure and carry out pumping air into the vessel during the experiment. A simpler, but rare, variant is to use a fan to supply the jet into the anechoic chamber [6]. This method does not require the use of pressure vessels, it allows continuous supply of air for a long time (several hours) at the maximum flow rate, while the preparation time for testing is minimal (1-2 minutes). However, there are also disadvantages; in particular, it is necessary to provide

IOP Conf. Series: Materials Science and Engineering **208** (2017) 012032 doi:10.1088/1757-899X/208/1/012032

protection from the penetration of the fan unit noise into the anechoic chamber through the air ducts. In connection with the factors listed above, a variant of the submerged cold jet was chosen in PNRPU, the supply of which into the anechoic chamber is carried out by fans.

The diagram of the supply-exhaust system of the jet rig is shown in figure 1. Two fans in tandem with a power of 45 kW each allow to accelerate the jet to 0.7 M. The fans are controlled programmatically via frequency converters, which allows smoothly adjusting velocity of the jet. To prevent penetration of the fan noise into the anechoic chamber, mufflers are provided in the air duct net. In the anechoic chamber, a plenum chamber with circular muffler (there are 3 layers of sound-absorbing material inside it) is mounted before the nozzle. This gives an additional reduction in fan noise. The measurements of the jet velocity with and without the plenum chamber have shown that it does not affect the magnitude of the velocity.

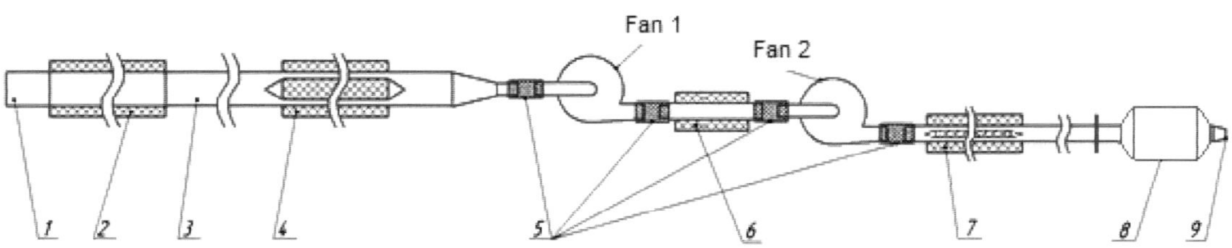

Figure 1. Diagram of jet supply-exhaust system.
1 – inlet via collector; 2 – muffler; 3 – air duct; 4 – circular muffler; 5 – flexible joint; 6 – muffler; 7 – circular muffler; 8 – plenum chamber with circular muffler; 9 – nozzle.

3. Measurements of the jet velocity

The study of aeroacoustic characteristics of the jet is primarily required to begin with a study of jet velocity, since its correct determination significantly affects the accuracy of the results in estimating the levels of the sound pressure generated by the turbulent jet. In this study the jet velocity was measured with Pitot-Prandtl tube. The tube was installed at a different distance from 0 to 20 nozzle-exit diameters. At the first 0.4 m, measurements were taken every 0.05 m, then every 0.1 m.

The evaluation of the results was carried out by comparing the measured velocity with the semiempirical model, which makes it possible to estimate the velocity on the axis of the jet at Mach number M = 0.28 [7]:

$$\frac{U}{U_j} = 1 - \exp\left(\frac{6.5D}{4.1D - X}\right) \tag{1}$$

where U is jet velocity at measuring point; U_j is velocity at nozzle exit section; $D = 0.04$ m is nozzle-exit diameter; X is distance from nozzle exit section.

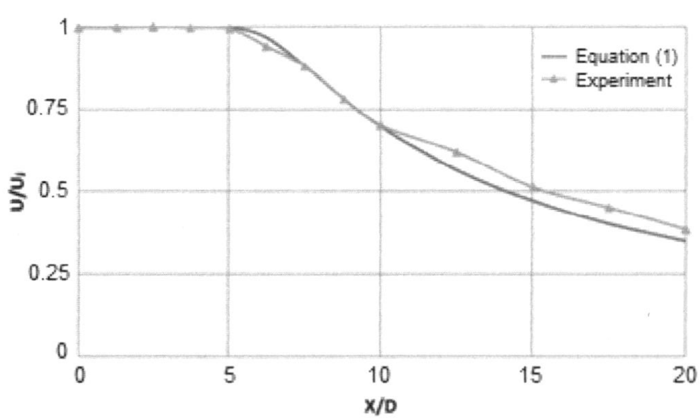

Figure 2. Comparison of the experimental results and the semiempirical model.

As can be seen in figure 2, the velocity distribution on the axis of the jet is in good agreement with the semi-empirical model, especially in the region of 5-10 nozzle-exit diameters, where the location of the jet noise source is expected. The initial region of the jet where the velocity is constant is approximately 5 nozzle-exit diameter. Discrepancy in the region of 10-20 nozzle-exit diameters does not exceed 5% and can be caused, both by the measurement errors, and by the flow conditions in the anechoic chamber.

4. Measurements of the jet noise

The next important study in the estimation of the aeroacoustic quality of the turbulent jet was the measurement of its noise in the far field. The measurements were performed with a microphone ½" Bruel & Kjaer 4192-C-001. The microphone was installed at a height of 2.5 m from the surface of the hard floor, which corresponds to the height of the jet rig axis. The floor of the anechoic chamber was covered with sound-absorbing wedges when measuring the noise. Measurements were made in the direction of 30 and 90 degrees relative to the axis of the jet, at a distance of 4.6 m and 2.3 m from the nozzle exit section. Position of the microphone in anechoic chamber when measuring the noise of the turbulent jet in the direction of 30 degrees is shown if figure 3.

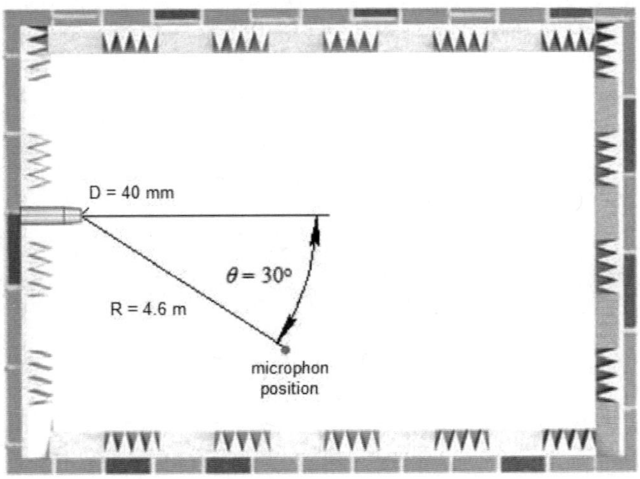

Figure 3. Position of the microphone in anechoic chamber when measuring the noise of turbulent jet.

In order to evaluate the possibility of measuring the noise of the turbulent jet in the anechoic chamber, the obtained measurement results had to be compared with the known results of similar tests in another anechoic chamber. For this purpose, it was used the results of measurements in anechoic chamber AC-2 TsAGI, which were performed for the jet rig with nozzle diameter D = 40 mm with a 6-microphones array of 0.85 m radius (figure 4) within the investigation of the azimuthal decomposition of turbulent jet noise [8]. From the data set, the points with coordinates $\theta = 30°$, R = 1.7 m and $\theta = 90°$, R = 0.85 m were taken. To perform the comparison, the sound levels obtained in the PNRPU were calculated to the points corresponding to the measurements in the AC-2 TsAGI. The power spectral density (PSD) for noise of the jets with the similar velocities is shown in figure 5.

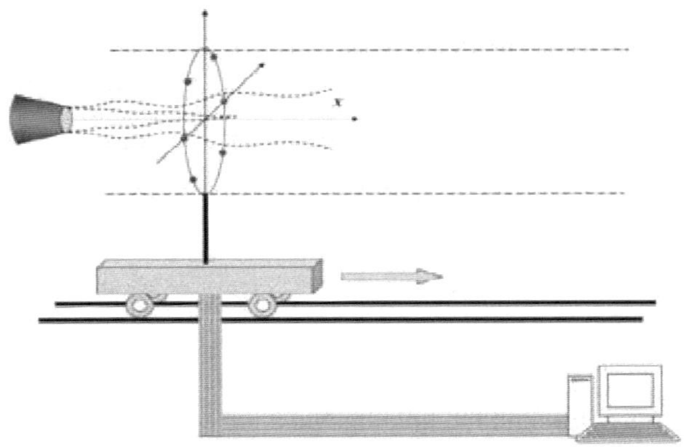

Figure 4. Diagram of jet noise measurements in anechoic chamber AC-2 TsAGI.

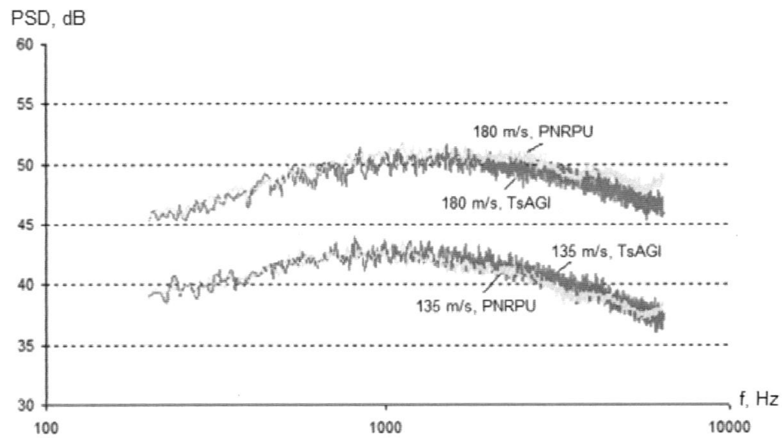

a)

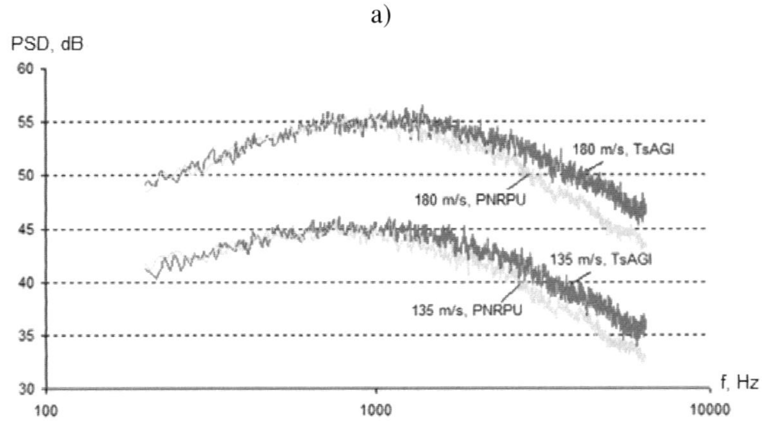

b)

Figure 5. Power spectral density for the similar velocities of PNRPU and TsAGI jet rigs a) at angle 30°, b) at angle 90°.

It can be seen that, in general, the noise spectra of the jet obtained in the anechoic chamber of the PNRPU are qualitatively similar to the spectra of AC-2 TsAGI. Oscillations in the noise spectrum of the jet rig at the frequencies above 3500 Hz may be because the microphone was mounted on a stand without a long holder, so that at high frequencies the reflections from the microphone stand have a significant effect on the quality of measurements.

IOP Conf. Series: Materials Science and Engineering **208** (2017) 012032 doi:10.1088/1757-899X/208/1/012032

5. Localization of the jet noise source

Additional information on the quality of the noise generated by the jet rig can be obtained by localizing the noise source: in good quality subsonic jet the noise source is located on the axis of the jet at a distance of 5-6 diameters from the nozzle-exit section. For the localization of the noise sources a 9-armed 54-microphone array of Bruel & Kjaer was used. The microphones arrangement in the array are optimized for the dynamic range, the level of phantom sources is less than 9-15 dB the maximum noise source level at this frequency (frequency band). The measurements use Bruel & Kjaer 4944-W-008 microphones. The received signals are transferred from microphones to analyzers Bruel & Kjaer 3055-B-120 with a frequency range up to 25.6 kHz, and then to the specialized software PULSE.

This software allows performing calculations to localize acoustic sources in a given frequency range using the beamforming method [9]. The main idea of the beamforming method consists in coherent summation of the measurement results from various microphones to improve the signal emitted from the focal point and minimize the contribution of signals from all other points. Thus, the beamforming method is a method of summation of data measured by microphones to determine the spatial distribution of noise sources.

Figure 6 shows a photo of the experiment (left) and some localization results of the noise sources in turbulent jet (right) obtained by beamforming method. The localization was carried out with a dynamic range of 6 dB, i.e. the sources with an acoustic pressure of 2 or more times less than the maximum are not shown. The obtained results demonstrate that the source is located on the axis of the turbulent jet at a distance of 5-6 nozzle-exit diameters. This result also indicates a good quality of the jet rig.

Figure 6. Localization of noise sources at frequency of 800 and 2000 Hz at jet velocity of M=0.5.

6. Conclusion

The following conclusions can be drawn as a result from the current study.

1. Velocity distribution on the axis of the jet is in good agreement with the semiempirical models, which is important for the correct quality of jet noise generation.
2. Source of noise in the jet is located at the initial region of the jet at a distance of 5-6 nozzle-exit diameters, which is typical for the noise of turbulent jets.

3. Noise of the jet in the far field in the direction of 30 and 90 degrees is in good agreement with the noise of the jet rig in AC-2 TsAGI.

Thus, the created jet rig in the anechoic chamber of PNRPU has a good aeroacoustic quality and makes it possible to carry out studies of the noise of turbulent flows, the jet-obstacles interaction noise, and testing passive and active methods for controlling the noise of turbulent jets.

Acknowledgments
The authors are grateful to Prof. V.F. Kopiev for providing experimental data obtained on the anechoic jet rig AC-2 TsAGI.

Results have been obtained in the framework of the government task "Researchers working in scientific laboratories created by the government program "Megagrants", contract No. 9.8130.2017/P220.

Experiments were carried out on unique scientific installation "Acoustic anechoic chamber with aerodynamic noise sources".

References

[1] Leylekian L, Lebrun M and Lempereur P *Journal Aerospace Lab.* 2014 **7**
[2] Kopiev V F, Palchikovskiy V V, Belyaev I V, Bersenev Yu V, Makashov S Yu, Khramtsov I V, Korin I A and Sorokin E V *Acoustical Physics.* 2017 **63** 114-126
[3] Anechoic jet rig AC-2 TsAGI, http://www.tsagi.ru/experimental_base/akusticheskaya-kamera-ak-2/
[4] Ahuja K K *AIAA Paper.* 2003 **706**
[5] Bridges J E and Hussain A K M F *Journal of Sound and Vibration.* 1987 **117** 289-311
[6] Guj G, Camussi R, Carley M, Ragni A, D'Alessio T and Bosi M *AIAA Paper.* 1998 **2358**
[7] Lau J C, Morris P J and Fisher M J *Journal of Fluid Mechanics.* 1979 **93** 1-27
[8] Kopiev V F, Zaitsev M Yu, Velichko S A, Kotova A N and Belyaev I V *AIAA Paper.* 2008 **2887**
[9] Johnson D H and Dudgeon D E Array signal processing: concepts and techniques (Prentice - Hall, Upper Saddle River, NJ) 1993 p 537

IOP Conf. Series: Materials Science and Engineering **208** (2017) 012033 doi:10.1088/1757-899X/208/1/012033

Specific interface area in a thin layer system of two immiscible liquids with vapour generation at the contact interface

Anastasiya V Pimenova[1], **Ilias M Gazdaliev**[2] **and Denis S Goldobin**[1,2]

[1]Institute of Continuous Media Mechanics UB RAS, 1 Akademika Koroleva street, 614013 Perm, Russia

[2]Department of Theoretical Physics, Perm State University, 15 Bukireva street, 614990 Perm, Russia

E-mail: Anastasiya.Pimenova@gmail.com, Denis.Goldobin@gmail.com

Abstract. For well-stirred multiphase fluid systems the mean interface area per unit volume, *or* "specific interface area" S_V, is a significant characteristic of the system state. In particular, it is important for the dynamics of systems of immiscible liquids experiencing interfacial boiling. We estimate the value of parameter S_V as a function of the heat influx $\dot{Q}_V$ to the system or the average system overheat $\langle\Theta\rangle$ above the interfacial boiling point. The derived results can be reformulated for the case of an endothermic chemical reaction between two liquid reagents with the gaseous form of one of the reaction products. The final results are restricted to the case of thin layers, where the potential gravitational energy of bubbles leaving the contact interface is small compared to their surface tension energy.

1. Introduction

For multiphase liquid systems the mean contact interface area per unit volume, *or* the "specific interface area" $S_V \equiv (\delta S/\delta V)$, is an important characteristic of the state. It is especially significant for the systems where this interface is active chemically or in some other way. The systems of immiscible liquids experiencing interfacial boiling [1, 2, 3, 4, 5, 6, 7, 8, 9, 10, 11, 12, 13, 14] are an example of the systems where parameter S_V becomes especially important. For the interfacial boiling the vapour phase grows if the sum of saturated vapour pressures of the liquids brought in contact excesses the atmospheric pressure, which means that this boiling can occur under conditions where no boiling in the bulk of both liquids is possible. The problem of calculation of S_V cannot be addressed rigourously and any direct numerical simulation, being extremely challenging and CPU-time consuming, will provide results pertaining to specific system set-ups. Thus, some general assessments on S_V can be highly beneficial.

In this paper we provide the assessments for the process of gas generation at the direct contact interface for a thin layer. The generated vapour volume in the system will be proportional to the heat influx to the interface. Therefore, our results will be relevant both for the case of endothermic chemical reaction between two liquids with the gaseous form of one of the reaction products and for the case of interfacial boiling [14, 15]. At the direct contact interface, a vapour layer grows and produces bubbles which breakaway of the interface and rise. The presence of vapour bubbles changes the fluid buoyancy and performs a "stirring" of the system. Additional

IOP Conf. Series: Materials Science and Engineering **208** (2017) 012033 doi:10.1088/1757-899X/208/1/012033

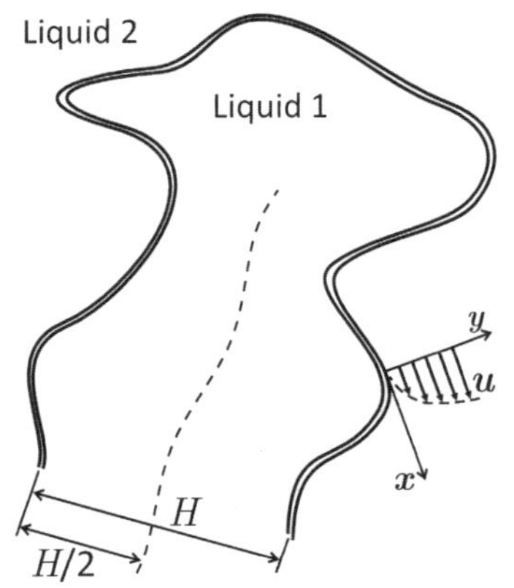

Figure 1. Sketch of the vapour layer between two immiscible liquids and the average liquid flow u within the turbulent boundary layer

kinetic energy is brought into the system by bubble poppings at the liquid–atmosphere interface. In [15], the theory has been developed for the case where one can neglect the latter contribution into the balance of mechanical energy. In this work we consider an opposite limiting case; the liquid layer is thin in a sense that the potential gravitational energy of rising bubbles is small compared to their surface tension energy.

For a two-liquid system experiencing direct contact boiling, the quantity of our interest depends on parameters of liquids and characteristics of the evaporation process, which are controlled by the mean overheating and the bubble production rate [9]. In this work the volumes of both components are assumed to be commensurable, no phase can be considered as a medium hosting dilute inclusions of the other phase. The characteristic width of the neighborhood of the vapour layer, beyond which the neighborhood of another vapour layer lies, is (see figure 1)

$$H \sim S_V^{-1}. \tag{1}$$

The process of boiling of a mixture above the bulk boiling temperature of the more volatile liquid is well-addressed in the literature [3, 4, 5, 6, 7, 8, 9]. Hydrodynamic aspects of the process of boiling below the bulk boiling temperature [13, 14] have to be essentially similar at the macroscopic level; rising and popping vapour bubbles drive the stirring of system, working against the gravitational stratification into two layers with a flat horizontal interface, the surface tension forces tending to minimize the interface area, and viscous dissipation of the flow kinetic energy. Specifically, the behaviour of parameter S_V depending on macroscopic characteristics of processes in the system should be the same as for systems with superheating of the more volatile component. In what follows, we perform an analytical assessment of the dependence of S_V on the evaporation rate for a well-stirred system. The evaporation rate can be controlled either by the average overheat Θ of the system above the interfacial boiling point T_* (section 4) or heat influx (section 5).

2. Energy flux balance in a well-stirred system

Let us derive the rough relationships between the macroscopic parameter S_V of the system state and the heat influx rate per unit volume $\dot{Q}_V = \delta Q/(\delta V\,\delta t)$ for a statistically stationary process of interfacial boiling.

The flow and consequent stirring in the system are enforced by the buoyancy and poppings of the vapour bubbles, while other mechanisms counteract the stirring of the system. These other mechanisms are gravitational stratification of two liquids, surface tension tending to minimise the interface area and viscous dissipation of the flow energy. Since the latent heat of phase transitions and heat of temperature inhomogeneities are enormously large compared to the realistic values of the kinetic energy of microscopic motion and gravitational potential energy

IOP Conf. Series: Materials Science and Engineering **208** (2017) 012033 doi:10.1088/1757-899X/208/1/012033

(*e.g.*, see [15] for quantitative estimates), the latter can be neglected in consideration of the heat balance. Hence, all the heat inflow into the system can be considered to be spent for the vapour generation; $\dot{Q}_V V \to (\Lambda_1 n_1^{(0)} + \Lambda_2 n_2^{(0)})\dot{V}_\mathrm{v}$, where V is the system volume, $\dot{V}_\mathrm{v}$ is the volume of the vapour produced in the system per unit time, Λ_j is the enthalpy of vaporization per one molecule of liquid j, and $n_j^{(0)}$ is the saturated vapour pressure of liquid j. Thus,

$$\dot{V}_\mathrm{v} = \frac{\dot{Q}_V V}{\Lambda_1 n_1^{(0)} + \Lambda_2 n_2^{(0)}} \,. \tag{2}$$

The mechanical energy income from one rising and popping vapour bubble E_bub is converted into the kinetic energy of liquid flow, the potential energy of a stirred state of the two-liquid system, the surface tension energy and dissipated by viscosity forces. In a statistically stationary state, the mechanical kinetic and potential energies do not change averagely over time and all the energy influx is to be dissipated by viscosity;

$$E_\mathrm{bub} N \longrightarrow \dot{W}_\mathrm{l,k}\tau \,, \tag{3}$$

where $\dot{W}_\mathrm{l,k}$ is the rate of viscous dissipation of energy, τ is the time of generation of the vapour volume V_v, $V_\mathrm{v} = \dot{V}_\mathrm{v}\tau$, N is the number of bubbles which reach the atmosphere–liquid interface per time τ. Hence,

$$E_\mathrm{bub}\frac{\dot{V}_\mathrm{v}}{V_\mathrm{bub}} \longrightarrow \dot{W}_\mathrm{l,k} \,, \tag{4}$$

where V_bub is the characteristic volume of a bubble.

2.1. Energy income E_bub from one bubble

The average potential gravitational energy of the rising bubble is $\rho g V_\mathrm{bub} h/2$, where we assume a nearly uniform distribution of the bubble formation sites over height, h is the liquid layer thickness, ρ is the characteristic density of two liquids.

The surface tension energy of the bubble is $4\pi r_\mathrm{bub}^2\sigma$, where σ is the characteristic surface tension for two liquids. Here, one should also take into account the energy of the vapour compression by the surface tension forces, which will be released as the bubble pops. The pressure increase by the surface tension forces is $\delta P = 2\sigma/r_\mathrm{bub}$. As the bubble pops, the pressure excess over the atmospheric pressure drops to zero, and the work made by the expanding vapour is $A_{\delta V} = \int P\,\mathrm{d}V$. For a reference bubble radius $\sim 1\,\mathrm{mm}$ [14, 16], the reference pressure increase $\sim 10^2$Pa, which is small compared to the atmospheric pressure, and the integrand of the latter integral remains nearly constant; therefore, $A_{\delta V} \approx P\delta V$. For a quick process of the bubble popping, one can neglect the energy dissipation due to molecular heat conduction and consider the expansion process to be adiabatic; $PV^{c_P/c_V} = const$, where c_V and c_P are the isochoric and isobaric specific vapour heat capacities, respectively, and thus $P^{-1}(-\delta P) + (c_P/c_V)V^{-1}\delta V = 0$ (here we take into account, that in our terms the pressure change is $(-\delta P)$). Hence, $A_{\delta V} = (c_V/c_P)V\delta P$.

While the surface tension forces act immediately on the liquid and one can assume the surface tension energy to be released into the kinetic flow energy of the liquid, the expanding vapour enforces both the motion of the surrounding liquid and gas flow. Thus, for the vapour expansion work, one should introduce a coefficient ϑ of the efficiency of the energy conversion into the mechanical energy of the liquid flow. With this conversion coefficient,

$$E_\mathrm{bub} = \rho g V_\mathrm{bub}\frac{h}{2} + 4\pi r_\mathrm{bub}\sigma + \vartheta A_{\delta V} = \rho g V_\mathrm{bub}\frac{h}{2} + \frac{3\sigma}{r_\mathrm{bub}}V_\mathrm{bub}\vartheta_\Sigma \,. \tag{5}$$

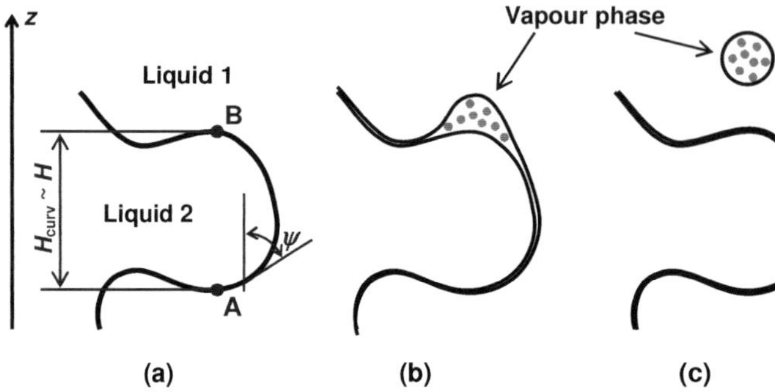

Figure 2. Process of bubble formation from vapour layer in a well-stirred system

where we introduce notation

$$\vartheta_\Sigma = 1 + \frac{2}{3}\vartheta\frac{c_V}{c_P}. \tag{6}$$

Let us compare the gravitational and surface tension terms in $E_{\rm bub}$. They are equal for

$$h_* = \frac{6\sigma\vartheta_\Sigma}{\rho g r_{\rm bub}}. \tag{7}$$

For the n-heptane–water system, the characteristic observed value of $r_{\rm bub}$ was 0.5–1 mm [14, 16], which yields $h_* \sim$ 5–10 cm. The case of $h \gg h_*$ has been considered in [15] and in this paper we assume $h \ll h_*$. Henceforth in this paper,

$$E_{\rm bub} = \frac{3\sigma\vartheta_\Sigma}{r_{\rm bub}}V_{\rm bub}\,. \tag{8}$$

Now one has to evaluate $r_{\rm bub}$. In [13] for a well-stirred system, the vapour layer thickness reached before the vapour breakaway, which is sketched in figure 2, has been evaluated

$$L_* = \left(\frac{48\eta_{\rm v}}{\rho g}\right)^{1/3} S_V^{-2/3}\left(\frac{\dot{V}_{\rm v}}{V}\right)^{1/3}, \tag{9}$$

where $\eta_{\rm v}$ is the dynamic viscosity of the vapour. This vapour breakaway occurs over the interface area $S_{\rm bub}$ between points **A** and **B** in figure 2a, which has the second horizontal scale (perpendicular to the sketch plane) $\sim H$. Thus, $S_{\rm bub} \sim H \times (H/\langle\cos\psi\rangle)$, where the averaging of $\cos\psi$ is performed along the interface and, therefore, $\langle\cos\psi\rangle \approx 1/2$. The vapour volume forming a bubble is

$$V_{\rm bub} = S_{\rm bub}L_* = 2S_V^{-2}L_* = 4\left(\frac{6\eta_{\rm v}}{\rho g}\right)^{1/3} S_V^{-8/3}\left(\frac{\dot{V}_{\rm v}}{V}\right)^{1/3}, \tag{10}$$

and equation (8) can be finally rewritten as

$$\frac{E_{\rm bub}}{V_{\rm bub}} = 3\sigma\vartheta_\Sigma\left(\frac{\pi}{3}\right)^{1/3}\left(\frac{\rho g}{6\eta_{\rm v}}\right)^{1/9} S_V^{8/9}\left(\frac{\dot{V}_{\rm v}}{V}\right)^{-1/9}. \tag{11}$$

2.2. Viscous dissipation and heat transfer within the turbulent boundary layer near the interface
In [15], the flow in the system has been shown to be essentially turbulent. With this turbulent flow momentum and heat transfer in the system is governed by the laws derived for the turbulent boundary layer [17, 18]. According to these laws, as derived in [15], the viscous energy loss rate

$$\frac{\delta\dot{W}_{\rm 1,k}}{\delta V} \approx S_V\frac{2\rho u_*^3}{\varkappa}\ln\frac{Hu_*}{2\xi_0\nu} \tag{12}$$

and the space-average kinetic energy of liquid flow

$$\frac{\delta W_{1,\mathrm{k}}}{\delta V} \approx \frac{\rho u_*^2}{2\varkappa^2} \ln^2 \frac{H u_*}{2e\xi_0 \nu}\,, \tag{13}$$

where empiric constants

$$\varkappa \approx 0.4 \qquad \text{and} \qquad \xi_0 \approx 0.13 \tag{14}$$

are known from experiments with turbulent boundary layers, ν is the characteristic kinematic viscosity of two liquids, e is the Euler's number, and u_* measures the characteristic turbulent pulsations of the flow velocity. Physically, u_* is determined by flow kinetic energy $W_{1,\mathrm{k}}$ (equation (13) provides the relation between them) and governs viscous energy dissipation $\dot{W}_{1,\mathrm{k}}$ (see equation (12)).

The the heat flux to the interface from one liquid phase can be determined from the temperature boundary layer theory [19, 20] (see [15] for details on the implementation of the theory to our case);

$$q_T \approx \frac{\varkappa \rho c_P u_* \langle \Theta \rangle}{\beta \ln \dfrac{H u_*}{2\nu}}\,, \tag{15}$$

where turbulent Prandtl number $\beta \approx 0.9$ for the turbulent boundary layer is known from experiments, c_P is the liquid specific heat capacity and $\langle \Theta \rangle$ is the average temperature excess above the interfacial boiling point.

2.3. Approximation of mechanical energy equipartition

We have to establish the relationship between the flow kinetic energy and the mechanical potential energy in the system. Rising vapour bubbles pump the mechanical energy into the system, while its stochastic dynamics is governed by interplay of its flow momentum and the forces of the gravity and the surface tension on the interface. In thermodynamic equilibrium, the total energy is strictly equally distributed between potential and kinetic energies related to quadratic terms in Hamiltonian (this statement is frequently simplified to a less accurate statement, that energy is equally distributed between kinetic and potential energies associated with each degree of freedom). Being not exactly in the case where one can rigorously speak of thermalization of the stochastic Hamiltonian system dynamics, we still may estimate the kinetic energy of flow to be of the same order of magnitude as the mechanical potential energy of the system. Thus,

$$W_{1,\mathrm{k}} \sim W_{1,\mathrm{pg}} + W_{1,\mathrm{p}\sigma}\,, \tag{16}$$

where $W_{1,\mathrm{pg}}$ and $W_{1,\mathrm{p}\sigma}$ are the gravitational potential energy and the surface tension energy, respectively. In [14] it has been shown, that for physically plausible regimes of interfacial boiling of a well-stirred system, the surface tension energy is large compared to the gravitational potential energy and the latter can be neglected (as in [15]);

$$W_{1,\mathrm{k}} \sim W_{1,\mathrm{p}\sigma}\,. \tag{17}$$

The surface tension energy is

$$W_{1,\mathrm{p}\sigma} \sim (\sigma_1 + \sigma_2) V S_V\,, \tag{18}$$

where we neglected the interface area of the stratified state compared to the area $V S_V$ in the well-stirred state. Due to the presence of the vapour layer between liquids the effective surface tension coefficient of the interface is $(\sigma_1 + \sigma_2)$ but not σ_{12} as it would be in the absence of the vapour layer.

3. Specific interface area S_V providing energy transfer balance

3.1. S_V as a function of average overheat $\langle\Theta\rangle$

From equations (2), (4), (11) and (12), one finds for $\dot{Q}_V$:

$$\dot{Q}_V \approx (\Lambda_1 n_1^{(0)} + \Lambda_2 n_2^{(0)}) S_V^{1/8} \frac{(3/\pi)^{3/8}}{(3\vartheta_\Sigma)^{9/8}} \left(\frac{6\eta_\mathrm{v}}{\rho g}\right)^{1/8} \left(\frac{2\rho u_*^3}{\varkappa} \ln \frac{u_*}{2\xi_0 \nu S_V}\right)^{9/8} . \tag{19}$$

The relation between S_V and u_* follows from equations (17) and (13);

$$2\sigma S_V \approx \frac{\rho u_*^2}{2\varkappa^2} \ln^2 \frac{u_*}{2e\xi_0 \nu S_V} , \tag{20}$$

which can be recast as

$$2\varkappa \sqrt{\sigma/\rho}\, S_V^{1/2} \approx u_* \ln \frac{u_*}{2e\xi_0 \nu S_V} , \tag{21}$$

On the other hand, the heat inflow $\dot{Q}_V = 2q_T S_V$ for a turbulent boundary layer is given by equation (15);

$$\dot{Q}_V \approx S_V \frac{2\varkappa \rho c_P u_* \langle\Theta\rangle}{\beta \ln \dfrac{u_*}{2\nu S_V}} . \tag{22}$$

Equations (19)–(22) form a self-content equation system for S_V, u_* and $\dot{Q}_V$. In these equations one can neglect discrepancies in the logarithm arguments as the argument is large $u_*/(\nu S_V) \gg 1$.

Excluding $\dot{Q}_V$ and logarithms from equations (19)–(22), one can find

$$u_* = \left(\frac{3}{4}\vartheta_\Sigma\right)^{\frac{9}{2}} \left(\frac{\pi}{3}\right)^{\frac{3}{2}} \left(\frac{\rho g}{6\eta_\mathrm{v}}\right)^{\frac{1}{2}} \left(\frac{\rho^2 c_P}{\beta(\Lambda_1 n_1^{(0)} + \Lambda_2 n_2^{(0)})}\right)^4 (\rho\sigma)^{-\frac{17}{4}} \langle\Theta\rangle^4 S_V^{-\frac{3}{4}} . \tag{23}$$

Substitution of u_* into equation (21) yields

$$S_V = \alpha_{\Theta 1}^{4/5} \langle\Theta\rangle^{16/5} \ln^{4/5} \frac{\alpha_{\Theta 2} \langle\Theta\rangle^4}{S_V^{7/4}} , \tag{24}$$

where

$$\alpha_{\Theta 1} \equiv \frac{1}{2\varkappa} \left(\frac{3}{4}\vartheta_\Sigma\right)^{\frac{9}{2}} \left(\frac{\pi}{3}\right)^{\frac{3}{2}} \left(\frac{\rho^2 g}{6\eta_\mathrm{v}\sigma}\right)^{\frac{1}{2}} \left(\frac{\rho^2 c_P}{\beta(\Lambda_1 n_1^{(0)} + \Lambda_2 n_2^{(0)})}\right)^4 (\rho\sigma)^{-\frac{17}{4}} , \tag{25}$$

$$\alpha_{\Theta 2} \equiv \frac{1}{2e\xi_0\nu} \left(\frac{3}{4}\vartheta_\Sigma\right)^{\frac{9}{2}} \left(\frac{\pi}{3}\right)^{\frac{3}{2}} \left(\frac{\rho g}{6\eta_\mathrm{v}}\right)^{\frac{1}{2}} \left(\frac{\rho^2 c_P}{\beta(\Lambda_1 n_1^{(0)} + \Lambda_2 n_2^{(0)})}\right)^4 (\rho\sigma)^{-\frac{17}{4}} . \tag{26}$$

One can express the solution to transcendent equation (24) in the form of a continuous logarithm-fraction [15];

$$S_V = \alpha_{\Theta 1}^{4/5} \langle\Theta\rangle^{16/5} \ln^{4/5} \cfrac{\alpha_\Theta \langle\Theta\rangle^{-8/5}}{\ln^{7/5} \cfrac{\alpha_\Theta \langle\Theta\rangle^{-8/5}}{\ln^{7/5} \cfrac{\alpha_\Theta \langle\Theta\rangle^{-8/5}}{\cdots}}} , \tag{27}$$

where $\alpha_\Theta \equiv \alpha_{\Theta 2}/\alpha_{\Theta 1}^{7/5}$. This result can be rewritten as

$$S_V = \alpha_{\Theta 1}^{\frac{4}{5}} \langle\Theta\rangle^{\frac{16}{5}} F_{\frac{7}{5},\infty}\left(\alpha_\Theta \langle\Theta\rangle^{-\frac{8}{5}}\right) , \tag{28}$$

IOP Conf. Series: Materials Science and Engineering **208** (2017) 012033 doi:10.1088/1757-899X/208/1/012033

where

$$F_{\frac{7}{5},n}(Z) \equiv \ln^{\frac{4}{5}}\Big(\underbrace{Z\ln^{-\frac{7}{5}}\Big(\dots Z\ln^{-\frac{7}{5}}}_{(n-1)\text{ times}}\Big(Z\Big)\dots\Big)\Big) \tag{29}$$

for $n = 1, 2, 3, \dots$. Due to a large value of Z the convergence of this function with respect to n is fast; for physically relevant values of Z, function $F_{7/5,2}(Z)$ deviates from $F_{7/5,\infty}(Z)$ by less that 1%.

3.2. Systems driven by heat inflow $\dot{Q}_V$

If the system state is controlled by a given heat inflow $\dot{Q}_V$ (or, equivalently, vapour production rate $\dot{V}_\mathrm{v}/V$, which is related to $\dot{Q}_V$ via equation (2)), one can express from equations (19), (21):

$$u_* = \left(\frac{3\vartheta_\Sigma}{4(\Lambda_1 n_1^{(0)} + \Lambda_2 n_2^{(0)})}\right)^{\frac{1}{2}} \left(\frac{\pi}{3}\right)^{\frac{1}{6}} \left(\frac{\rho g}{6\eta_\mathrm{v}}\right)^{\frac{1}{18}} (\rho\sigma)^{-\frac{1}{4}} \dot{Q}_V^{\frac{4}{9}} S_V^{-\frac{11}{36}}. \tag{30}$$

Then equation (21) can be rewritten as

$$S_V = \alpha_{Q1}^{\frac{36}{29}} \dot{Q}_V^{\frac{16}{29}} \ln^{\frac{36}{29}}\left(\alpha_{Q2} \dot{Q}_V^{\frac{4}{9}} S_V^{-\frac{47}{36}}\right), \tag{31}$$

where

$$\alpha_{Q1} \equiv \frac{\sqrt{\rho/\sigma}}{2\varkappa} \left(\frac{3\vartheta_\Sigma}{4(\Lambda_1 n_1^{(0)} + \Lambda_2 n_2^{(0)})}\right)^{\frac{1}{2}} \left(\frac{\pi}{3}\right)^{\frac{1}{6}} \left(\frac{\rho g}{6\eta_\mathrm{v}}\right)^{\frac{1}{18}} (\rho\sigma)^{-\frac{1}{4}}, \tag{32}$$

$$\alpha_{Q2} \equiv \frac{1}{2e\xi_0\nu} \left(\frac{3\vartheta_\Sigma}{4(\Lambda_1 n_1^{(0)} + \Lambda_2 n_2^{(0)})}\right)^{\frac{1}{2}} \left(\frac{\pi}{3}\right)^{\frac{1}{6}} \left(\frac{\rho g}{6\eta_\mathrm{v}}\right)^{\frac{1}{18}} (\rho\sigma)^{-\frac{1}{4}}. \tag{33}$$

Solution to equation (31) can be found in the form of a continuous logarithm-fraction;

$$S_V = \alpha_{Q1}^{\frac{36}{29}} \dot{Q}_V^{\frac{16}{29}} F_{\frac{47}{29},\infty}\left(\alpha_Q \dot{Q}_V^{-\frac{8}{29}}\right), \tag{34}$$

where $\alpha_Q \equiv \alpha_{Q2}/\alpha_{Q1}^{47/29}$ and

$$F_{\frac{47}{29},n}(Z) \equiv \ln^{\frac{36}{29}}\Big(\underbrace{Z\ln^{-\frac{47}{29}}\Big(\dots Z\ln^{-\frac{47}{29}}}_{(n-1)\text{ times}}\Big(Z\Big)\dots\Big)\Big) \tag{35}$$

for $n = 1, 2, 3, \dots$.

4. Conclusion

On the basis of the energy flux balance in the system, the assumption of the system stochastization and transport laws for turbulent boundary layers, we have evaluated the specific interface area S_V as a function of parameters which can control the system state: average overheat $\langle\Theta\rangle$ and heat inflow $\dot{Q}_V$ (or, equivalently, vapour generation rate $\dot{V}_\mathrm{v}/V$). The results are derived in the form of continuous logarithm-fractions (28) and (34), which possess fast convergence properties (with respect to the number of iterations) for physically relevant value of parameters.

Acknowledgments
The work has been financially supported by the Russian Foundation for Basic Research (grant no. 15-01-04842).

References
[1] Krell E 1982 *Handbook of Laboratory Distillation* (Amsterdam: Elsevier) chapter 4.3
[2] Geankoplis C J 2003 *Transport Processes and Separation Process Principles* (New Jersey: Prentice Hall)
[3] Simpson H C, Beggs G C and Nazir M 1974 *Desalination* **15** 11–23
[4] Celata G P, Cumo M, D'Annibale F, Gugliermetti F and Ingui' G 1995 *Int. J. Heat Mass Transfer* **38** 1495–504
[5] Roesle M L and Kulacki F A 2012 *Int. J. Heat Mass Transfer* **55** 2160–5
[6] Roesle M L and Kulacki F A 2012 *Int. J. Heat Mass Transfer* **55** 2166–72
[7] Sideman S and Isenberg J 1967 *Desalination* **2** 207–14
[8] Kendoush A A 2004 *Desalination* **169** 33–41
[9] Filipczak G, Troniewski L and Witczak S 2011 *Evaporation, Condensation and Heat transfer* ed Ahsan A (Rijeka: InTech) p 123–50
[10] Gordon K F, Singh T and Weissman E Y 1961 *Int. J. Heat and Mass Transfer* **3** 90–3
[11] Prakash C B and Pinder K L 1967 *Can. J. Chem. Engineering* **45** 210–4
[12] Prakash C B and Pinder K L 1967 *Can. J. Chem. Engineering* **45** 215–20
[13] Pimenova A V and Goldobin D S 2014 *JETP* **119** 91–100
[14] Pimenova A V and Goldobin D S 2014 *Eur. Phys. J.* E **37** 108
[15] Goldobin D S and Pimenova A V 2017 *Eur. Phys. J.* ST **226** 1155
[16] Pimenova A V and Goldobin D S 2016 *J. Appl. Mech. Tech. Phys.* **57** 1182–9
[17] von Karman Th 1930 Mechanische Ähnlichkeit und Turbulenz *Nachrichten von der Gesellschaft der Wissenschaften zu Göttingen, Fachgruppe 1 (Mathematik)* **5** 58–76 (*English translation*: von Karman Th 1931 Mechanical Similitude and Turbulence *Tech. Mem. NACA* **611**)
[18] Prandtl L 1933 Neuere Ergebnisse der Turbulenzforschung *Z. Ver. dtsch. Ing.* **77** 105–14 (*English translation*: Prandtl L 1933 Recent results of turbulence research *Tech. Mem. NACA* **720**)
[19] Landau L D and Lifshitz E M 1986 *Fluid Mechanics* (Moscow: Nauka)
[20] Schlichting H and Gersten K 2000 *Boundary-Layer Theory* (Springer)

IOP Conf. Series: Materials Science and Engineering **208** (2017) 012034 doi:10.1088/1757-899X/208/1/012034

The adhesive and antiadhesive non-local interaction of solids

V Presnetsova, S Romashin, L Frolenkova, V Shorkin[a] and S Yakushina

Orel State University named after I.S. Turgenev, 95, Komsomolskaya, 302026, Orel, Russia

E-mail: [a] VShorkin@yandex.ru

Abstract. The inequalities which must be satisfied the characteristics of elastic state of the materials of contacting bodies at their adhesion (coalescence) and its absence (antiadhesion) were obtained. These are the result of the analysis of adhesion phenomena and its absence. The analysis is made on the basis of a special variant of a nonlocal theory of elasticity. Its main hypothesis is infinitely small particles of the continuous elastic medium interact with each other at finite distances with the help of many-particle potential forces. The results of using criterial inequalities were confirmed by known experimental data.

1. Introduction

When creating high-tech and science-intensive products, depending on the functions performed, there are situations when adhesion is required between contacting bodies (layered structures) or it should be absent. In the first case a composition of materials of the layered structure elements is selected, which provides its greatest strength. In the second case, antiadhesive layer or coating is used. The material of this layer (coating) should have antiadhesive properties in relation to the protected material (it should not stick together).

At the present time the industry uses a variety of experimentally tested materials with appropriate properties. At the same time, there are no criteria by which to predict theoretically the adhesive or antiadhesive properties of a particular pair of materials. This work is devoted to a partial solution of the problem of constructing such a criterion for linearly elastic materials. It is based on non-local model, which takes into account the paired and triple potential interactions of particles interacting materials at finite distances. At the same time, we use a number of assumptions about the character of these interactions of the individual particles and the contacting bodies in general.

2. Nonlocal potential interaction of particles of a continuous elastic medium

Adhesion of solids $B_{(1)}$ and $B_{(2)}$ is their coalescence [1]. Therefore, the antiadhesion can be called a property of a material, which is resistant to sticking of other material to it. The phenomenon of adhesion is caused by the action of Van der Waals forces between neutral particles or Coulomb forces between charged particles [2, 3] of the interacting bodies. Because of this antiadhesion (no adhesion) can be explained either by a lack of attractive forces between particles of different materials or their groups, or the occurrence of repulsive forces between them.

The continuous elastic medium is a mathematical model of real materials with their discrete atomic-molecular structure. This structure is characterized by a non-local potential interaction of composing them particles. In the works [4 – 7] this property is axiomatically transferred to the interaction of particles of a continuous elastic medium.

The body $B = B_{(1)} \cup B_{(2)}$ ($B_{(1)} \cap B_{(2)} = \varnothing$) corresponds to a discrete structure $N = N_{(1)} + N_{(2)}$ of material points with masses of $m_{i(1)}$ and $m_{j(2)}$, and the radius vectors $\vec{r}_{i(1)}$ and $\vec{r}_{j(2)}$. Its potential interaction energy of all particles is determined by expression [8]:

$$U\left(\vec{r}_1,...,\vec{r}_N\right) \equiv U\left(\vec{r}_{1(1)},...,\vec{r}_{N(1)};\vec{r}_{1(2)},...,\vec{r}_{N(2)}\right) = U_{(11)}\left(\vec{r}_{1(1)},...,\vec{r}_{N(1)}\right) + U_{(22)}\left(\vec{r}_{1(2)},...,\vec{r}_{N(2)}\right)$$
$$+U_{(12)}\left(\vec{r}_{1(1)},...,\vec{r}_{N(1)};\vec{r}_{1(2)},...,\vec{r}_{N(2)}\right) \quad (1)$$

The first and second terms of expression determine the potential energy of the interaction of particles in the bodies $B_{(1)}$ and $B_{(2)}$. The third term is the potential energy of interaction between the particles, which belong to different bodies. At the same time, there are works [9] in which the adhesive interactions are not considered as volumetric (three-dimensional), but as surface (two-dimensional) interactions. To carry out the necessary calculations, we use representation of these potential energies in the form of sums of all pairs of interacting particles, their triplets, etc.

$$U\left(\vec{r}_1,...,\vec{r}_i,...,\vec{r}_N\right) = \sum_{\alpha=1}^{2}\left(\sum_{i}\sum_{j>1} U^{(2)}_{(\alpha)}\left(\vec{r}_{i(\alpha)},\vec{r}_{j(\alpha)}\right) + \sum_{i}\sum_{j>i}\sum_{k>j>i} U^{(3)}_{(\alpha)}\left(\vec{r}_{i(\alpha)},\vec{r}_{j(\alpha)},\vec{r}_{k(\alpha)}\right) + ...\right)$$
$$+\left(\sum_{i(\alpha),j(\beta)\geq 1} U^{(2)}_{(\alpha\beta)}\left(\vec{r}_{i(\alpha)},\vec{r}_{j(\beta)}\right) + \sum_{i(\alpha),j(\beta)\geq 1}\sum_{k(\gamma)>j(\beta),i(\alpha)} U^{(3)}_{(\alpha\beta\gamma)}\left(\vec{r}_{i(\alpha)},\vec{r}_{j(\beta)},\vec{r}_{k(\gamma)}\right) + ...\right) \quad (2)$$

Here $\alpha, \beta, \gamma = 1, 2$ is the number of the material (body), $\beta \neq \gamma$. Generally only pair interactions of particles are considered (the first double sum in parenthesis of expression (2)) [10] with potentials such as the potential of the Lennard-Jones, London, Morse, etc, which provide the stability of the lattice structures as each of the bodies $B_{(1)}$ and $B_{(2)}$, as well as the combined body $B = B_{(1)} \cup B_{(1)}$. In papers [11, 12] it was shown that the triple interactions can lead not only to the attraction of the particles, but also their repulsion. Therefore, this may contribute to the appearance of antiadhesion.

In the proposed for further consideration model, the role of particles is played by the infinitely small particles $\mathrm{d}B_{(\alpha)}$ of each of the bodies $B_{(\alpha)}$ ($\alpha = 1, 2$). They are obtained by a mental partition

$$B_{(\alpha)} = \lim_{N_{(\alpha)}\to\infty} \bigcup_{n=1}^{N_{(\alpha)}} \Delta B_{(\alpha)n} = \int \mathrm{d}B_{(\alpha)} \qquad (n \neq m) \Rightarrow \Delta B_{(\alpha)n} \cap \Delta B_{(\alpha)m} = \varnothing \quad (3)$$

Partitions (3) are used in the construction of integral sums in the calculation of additive characteristics of state of the solid body material $B_{(\alpha)}$.

Also as for the discrete point systems [8], the potentials of pair and triple interactions are described by empirical functions with the parameters characteristic of the test material, or a pair of materials. These functions correspond to the conditions of thermo-dynamic system. To perform the first requirement it is assumed that partial potentials $\mathrm{d}^n\Psi^{(n)}$ of interaction n particles (both one and different materials) are infinitely small quantities of n-th order with respect to volumes $\Delta V_{j(\alpha)} = \mathrm{d}V_{j(\alpha)} \to 0$ ($j = 1, 2, ..., n$):

$$\mathrm{d}^n\Psi^{(n)} = \Phi^{(n)}\mathrm{d}V_{1t(\alpha)}\mathrm{d}V_{2t(\beta)}...\mathrm{d}V_{nt(\gamma)} \quad (4)$$

The coefficient of proportionality $\Phi^{(n)}$ (further the potential) depends only on the relative distances $l_{ij(\alpha\beta)} = \left|\vec{l}_{ij(\alpha\beta)}\right|$ between the interacting particles, $\vec{l}_{ij(\alpha\beta)} = \vec{r}_{j(\beta)} - \vec{r}_{i(\alpha)}$ – the radius vector of the particle $\mathrm{d}B_{(\beta)}$ relative to the particle $\mathrm{d}B_{(\alpha)}$ in the current configuration of the body $B = B_{(\alpha)} \cup B_{(\beta)}$. If between any pair of particles the distance is $l_{i\,j(\alpha\beta)} \to \infty$, then $\Phi^{(n)} \to 0$. This happens with the speed required for the convergence of all integrals.

Also, as for the discrete point systems [8], the potentials of pair and triple systems interactions of infinitely small particles of the same material, which form a continuous solid, satisfy the stability condition of specific configuration of the particle system.

For particles of the same material $\alpha = \beta$) dependence of $\Phi^{(n)}$ on $l_{i\,j(\alpha\beta)}$ is such that the internal forces, which are generated by the deformation of the material, always keen to return it to its original state.

The interaction potentials of groups of particles of different materials, as well as the bodies themselves, can satisfy and not satisfy the condition of stability. In the first case, this corresponds to the presence of adhesion, in the second – to its absence.

As a result, the analogue of representations (1) and (2) is an expression

$$W = \frac{1}{2!}\sum_{\alpha=1}^{2} \int_{V_{t(\alpha)}} \int_{V_{t(\alpha)}} \Phi^{(2)}_{(\alpha\alpha)}\, \mathrm{d}V_{1t(\alpha)}\, \mathrm{d}V_{2t(\alpha)} + \frac{1}{3!}\sum_{\alpha=1}^{2} \int_{V_{t(\alpha)}} \int_{V_{t(\alpha)}} \int_{V_{t(\alpha)}} \Phi^{(3)}_{(\alpha\alpha\alpha)}\, \mathrm{d}V_{1t(\alpha)}\, \mathrm{d}V_{2t(\alpha)}\, \mathrm{d}V_{3t(\alpha)} + \ldots$$

$$+\frac{1}{2!}\sum_{\substack{\alpha,\beta=1\\ \alpha\neq\beta}}^{2} \int_{V_{t(\alpha)}} \int_{V_{t(\beta)}} \Phi^{(2)}_{(\alpha\beta)}\, \mathrm{d}V_{1t(\alpha)}\, \mathrm{d}V_{2t(\beta)} + \frac{1}{3!}\sum_{\substack{\alpha,\beta,\gamma=1\\ \alpha\neq\beta}}^{2} \int_{V_{t(\alpha)}} \int_{V_{t(\beta)}} \int_{V_{t(\gamma)}} \Phi^{(3)}_{(\alpha\beta\gamma)}\, \mathrm{d}V_{1t(\alpha)}\, \mathrm{d}V_{2t(\beta)}\, \mathrm{d}V_{3t(\gamma)} \quad (5)$$

Here $\Phi^{(2)}_{(\alpha\alpha)}, \Phi^{(2)}_{(\alpha\beta)} \equiv \Phi^{(2)}$, $\Phi^{(3)}_{(\alpha\alpha\alpha)}, \Phi^{(3)}_{(\alpha\beta\gamma)} = \Phi^{(3)}$ – are the potentials of pair and triple interactions of particles from materials with corresponding numbers; $V_{t(\alpha)}$ – the area (or its volume) occupied at the time t by body $B_{(\alpha)}$, $\mathrm{d}V_{it(\alpha)}$ – volume of particle $\mathrm{d}B_{i(\alpha)}$ at this time (it is assumed that the deformation is small, so the time index t is further discarded). In this expression we restricted ourselves to taking into account no more than three-particle interactions.

In condition of absence external mechanical influences on the body $B = B_{(1)} \cup B_{(2)}$ you can obtain expressions for the energy $\mathrm{d}W_{(1)}(\vec{r}_{(1)})$ of infinitely small particle, for example, $\mathrm{d}B_{(1)}$:

$$\mathrm{d}W_{(1)}(\vec{r}_{(1)}) = w_{(1)}(\vec{r}_{(1)})\, \mathrm{d}V_{(1)} = (w_{(11)} + w_{(12)})\, \mathrm{d}V_{(1)}$$

$$= \left[\int_{V_{(1)}} \Phi^{(2)}_{12(11)}\, \mathrm{d}V_{(1)} + \frac{1}{2!} \int_{V_{(1)}} \int_{V_{(1)}} \Phi^{(3)}_{123(111)}\, \mathrm{d}V_{1(1)}\, \mathrm{d}V_{2(1)} \right] \mathrm{d}V_{3(1)}$$

$$+ \left[\int_{V_{(2)}} \Phi^{(2)}_{12(12)}\, \mathrm{d}V_{(2)} + \frac{1}{2!} \sum_{\beta,\gamma=1}^{2} \int_{V_{(\beta)}} \int_{V_{(\gamma)}} \Phi^{(3)}_{123(1\beta\gamma)}\, \mathrm{d}V_{1(\beta)}\, \mathrm{d}V_{2(\gamma)} \right] \mathrm{d}V_{3(1)} \quad (6)$$

In this equation $w_{(11)}$ – is the bulk density of the potential energy, which arose due to the interaction between particles of the body $B_{(1)}$; $w_{(12)}$ – the addition to the value $w_{(11)}$, which arose due to interaction of particles of body $B_{(1)}$ with particles of body $B_{(2)}$.

The force $\vec{f}_{(1)}\, \mathrm{d}V_{(1)}$ of action on the particle $\mathrm{d}B_{(1)}$ from the environment is defined by the equality:

$$\vec{f}_{(1)} = -\nabla_{\vec{r}}\, w_{(1)} = -\nabla_{\vec{r}}\, w_{(11)} - \nabla_{\vec{r}}\, w_{(12)} = \vec{f}_{(11)} + \vec{f}_{(12)} \quad (7)$$

The force $\vec{f}_{(11)}\,\mathrm{d}V_{(1)}$ acts on the particle $\mathrm{d}B_{(1)}$ only from the side of the body $B_{(1)}$, while the force $\vec{f}_{(12)}\,\mathrm{d}V_{(1)}$ is caused by the action on a particle $\mathrm{d}B_{(1)}$ of all particles $\mathrm{d}B_{(2)} \subset B_{(2)}$. The first term in the expression (7) characterizes the cohesive interaction within the body $B_{(1)}$. The second term is the adhesive interaction between the bodies $B_{(1)}$ and $B_{(2)}$. It is seen, that both types of bulk density of the potential energy of an elementary particle and also the forces acting on it, are the sums of two terms. One of them is determined by pair interactions, the second by triple ones.
In this paper, the empirical dependences $\Phi^{(2)}$ and $\Phi^{(3)}$ of potentials of pair and triple particle interactions are chosen as a Morse potential.

$$\Phi^{(2)}_{(\alpha\beta)} = \Phi^{(2)}_{0(\alpha\beta)}\,\varphi_{ij(\alpha\beta)} \tag{8}$$

$$\Phi^{(3)}_{(\alpha\beta\gamma)} = \Phi^{(3)}_{0(\alpha\beta\gamma)}\varphi_{12(\alpha\beta)}\,\varphi_{31(\alpha\gamma)} \tag{9}$$

$$\varphi_{ij(\alpha\beta)} = e^{-2\eta_{ij(\alpha\beta)}} - 2e^{-\eta_{ij(\alpha\beta)}} \qquad \eta_{ij(\alpha\beta)} = \beta_{(\alpha\beta)}\,l_{ij(\alpha\beta)} \tag{10}$$

In equations (8) – (10) it is acceptable the coincidence $\alpha, \beta, \gamma = 1, 2$. The parameters $\beta_{(\alpha\beta)}$, $\Phi^{(2)}_{0(\alpha\beta)}$ and $\Phi^{(3)}_{0(\alpha\beta\gamma)}$ characterize the properties of interacting materials and are determined by the methods described in papers [4 – 7].

3. The phenomena of adhesion, antiadhesion and their modeling

To answer the question whether or not there is the adhesion between two solid materials, the interaction of two semi-infinite bodies $B_{(1)}:\ 0 < x \equiv x_{(1)} < +\infty$, $B_{(2)}:\ -\infty < x_{(2)} < 0$ is considered. To simplify the mathematical reasoning and increase the visibility of their results, the bodies are mentally divided into cylindrical parts of the same infinitely small cross-section, which are perpendicular to the plane of contact $A_{(12)}:\ x = 0$. So we have taken into account only those interaction forces that occur between particles of different materials, which belong to the coaxial cylinders.

The distribution of the bulk density of the potential energy of interaction of the cylinder from the body $B_{(2)}$ to a particle $dB_{(1)}$, located at a distance $x_{(1)}$ from the contact surface, is given by

$$w(\eta) = \left(\frac{\Phi^{(2)}_{0(12)}}{\beta^3_{(12)}}\right)\left(\frac{1}{2}e^{-2\eta} - 2e^{-\eta}\right) - \left(\frac{\Phi^{(3)}_{0(122)}}{\beta^6_{(12)}}\right)\left(\frac{1}{2}e^{-2\eta} - 2e^{-\eta}\right)^2 \tag{11}$$

where $\eta = \beta_{(12)}x_{(1)}$ is a dimensionless distance.
When

$$\eta = \beta_{(12)}x_{(1)} = 0:\ \ w(0) = -\frac{3}{2}\left(\frac{\Phi^{(2)}_{0(12)}}{\beta^3_{(12)}} + \frac{3}{2}\frac{\Phi^{(3)}_{0(122)}}{\beta^6_{(12)}}\right) \tag{12}$$

The distribution of the bulk density of the interaction force is determined:

$$f_{(12)} = -\frac{\mathrm{d}w}{\mathrm{d}x_{(1)}} = \beta_{(12)}\left(e^{-2\eta} - e^{-\eta}\right)\left[\left(\frac{\Phi^{(2)}_{0(12)}}{\beta^3_{(12)}}\right) + 2\left(\frac{\Phi^{(3)}_{0(122)}}{\beta^6_{(12)}}\right)\left(2e^{-\eta} - \frac{1}{2}e^{-2\eta}\right)\right] \tag{13}$$

When

IOP Conf. Series: Materials Science and Engineering **208** (2017) 012034 doi:10.1088/1757-899X/208/1/012034

$$x_{(1)} \to 0: \; f_{(12)} \to = (-2\eta)\left[\left(\frac{\Phi^{(2)}_{0(12)}}{\beta^3_{(12)}}\right) + 3\left(\frac{\Phi^{(3)}_{0(122)}}{\beta^6_{(12)}}\right)\right] \tag{14}$$

When

$$\left(\frac{\Phi^{(2)}_{0(12)}}{\beta^3_{(12)}}\right) > 0 \qquad \left(\frac{\Phi^{(3)}_{0(122)}}{\beta^6_{(12)}}\right) > 0 \tag{15}$$

it turns out that $w(0) < 0$.

At the same time, if $\mathrm{d}B_{(1)}$ is displaced from the contact surface by a infinitesimal distance $x_{(1)} > 0$, then a restoring force $f_{(12)} \approx (-x_{(1)})$ appears. The system $\mathrm{d}B_{(1)} \cup B_{(2)}$ is stable. There is adhesion.

When

$$\left(\frac{\Phi^{(2)}_{0(12)}}{\beta^3_{(12)}}\right) > 0 \qquad \left(\frac{\Phi^{(3)}_{0(122)}}{\beta^6_{(12)}}\right) < 0 \tag{16}$$

we get that

$$W(0) = -\left(\frac{3}{2\beta^3_{(12)}}\right)\left(\frac{\Phi^{(2)}_{0(12)}}{\beta^3_{(12)}} - \frac{3}{2}\left|\frac{\Phi^{(3)}_{0(122)}}{\beta^6_{(12)}}\right|\right) \tag{17}$$

If

$$2\left(\frac{\Phi^{(2)}_{0(12)}}{\beta^3_{(12)}}\right) < 3\left|\frac{\Phi^{(3)}_{0(122)}}{\beta^6_{(12)}}\right| \tag{18}$$

then $W(0) > 0$, and when $x_{(1)} \to 0: \; \eta \to 0$ the force $\Delta P_{(12)} \approx (+x_{(1)}) > 0$ is repulsive. The system $\Delta B_{(1)} \cup B_{(2)}$ is unstable. There is no adhesion.

If

$$\left(\frac{\Phi^{(2)}_{0(12)}}{\beta^3_{(12)}}\right) < 3\left|\frac{\Phi^{(3)}_{0(122)}}{\beta^6_{(12)}}\right| < 2\left(\frac{\Phi^{(2)}_{0(12)}}{\beta^3_{(12)}}\right) \tag{19}$$

then $W(0) < 0$, but when $x_{(1)} \to 0: \eta \to 0$ the force $\Delta P_{(12)} \approx (+x_{(1)}) > 0$ is repulsive. As $\int_0^{+\infty} \Delta P_{(12)}\, \mathrm{d}x_{(1)} = \int_0^{+\infty} -\frac{\mathrm{d}W}{\mathrm{d}x_{(1)}} \mathrm{d}x_{(1)} = W(0)$, that is, the energy of $W(0)$ is the result of the interaction of all particles from the cylinder in $B_{(1)}$, then the stability of the system ($W(0) < 0$) is explained by the attraction of distant particles which prevails over repulsion of neighbors.

Expressions (12) – (19) are criterial. They can be used in the theoretical verification of the presence or absence of adhesion between a given pair of materials. They can be used for selection of a material, which has good adhesion properties with respect to the specified value. For example, the selection of elements of a strong layered structure or not at all entering into a state of adhesion with it (for example, when selecting a solid lubricant material or a protective release layer).

The parameters $\left(\Phi^{(2)}_{0(12)} / \beta^3_{(12)}\right)$ of potentials of interparticle interactions are expressed through the Young's modulus and the shift of the contacting materials. For this, the methods described in [4 – 7]

were used. As a result, conditions that provide both good adhesion and its absence are formed. Young's and shear modules of the contacting materials are selected according to these conditions.
Now let $B_{(1)}$ and $B_{(2)}$ be two parts of the same material. We consider their cohesion. Then

$$\beta_{(12)} = \beta_{(11)} = \beta_{(22)} = \beta > 0 \qquad \Phi^{(2)}_{0(12)} = \Phi^{(2)}_{0(11)} = \Phi^{(2)}_{0(22)} = \Phi^{(2)}_{0} \tag{20}$$

$$\Phi^{(3)}_{0(112)} = \Phi^{(3)}_{0(122)} = \Phi^{(3)}_{0(111)} = \Phi^{(3)}_{0(222)} = \Phi^{(3)}_{0} \tag{21}$$

Also known [4 – 7]:

$$\left(\frac{\Phi_0^{(2)}}{\beta^3}\right) = \frac{E}{294\pi}\frac{2+41\nu}{(1+\nu)(1-2\nu)} \tag{22}$$

$$\left(\frac{\Phi_0^{(3)}}{\beta^6}\right) = \frac{27E}{1764\pi^2}\frac{4\nu-1}{2(1+\nu)(1-2\nu)} \tag{23}$$

This indicates that $\left(\Phi_0^{(2)}/\beta^3\right) > 0$, if $\nu > -2/41$. While $\nu < 0.25$: $\left(\Phi_0^{(3)}/\beta^6\right) < 0$. Hence, even for homogeneous materials situation with the lack of adhesion between their parts – theoretically, and therefore with weak adhesion – in actual conditions, is possible.

For example, when

$$\nu = 0: \quad 0.0068\,E = \left(\frac{\Phi_0^{(2)}}{\beta^3}\right) < 3\left|\frac{\Phi_0^{(3)}}{\beta^6}\right| = 0.0073\,E < 2\left(\frac{\Phi_0^{(2)}}{\beta^3}\right) = 0.0136\,E \tag{24}$$

Hence, a material with $\nu = 0$ is kept in a consistent state by the attraction of particles remote from the contact surface.

There are a number of anisotropic materials, which have in one direction $\nu \le 0$ (determined by the methods described in [13]), that have anti-adhesive properties: talc, graphite, molybdenum disulfides (MoS_2) and tungsten (WS_2), etc. This is taken into account in calculating the repulsive force $\sigma_{(12)}$ for pairs: Talc – Cu, Talc – Fe, Cu – Fe. For the first pair the following is obtained: $\sigma_{(12)} = 65 \cdot 10^9$ N/m$^2 > 0$. Copper repels talc, there is no adhesion between them. For the second pair: $\sigma_{(12)} = 523 \cdot 10^9$ N/m$^2 > 0$. Iron repels talc, there is no adhesion. For the third pair: $\sigma_{(12)} = -53 \cdot 10^9$ N/m$^2 < 0$. Copper attracts steel, adhesion exists. In this case, the value $\left|\sigma_{(12)}\right|$ can be considered as a theoretical tensile strength of adhesive compound. As the initial characteristics of the elastic state of the contacting materials the following values of Young's modulus E and Poisson's ratio ν were adopted: talc [14, 15]: $E = 0.16 \cdot 10^{11}$ N/m^2, $\nu = 0$; copper [16]: $E = 1.10 \cdot 10^{11}$ N/m^2, $\nu = 0.31$; iron [16]: $E = 2.12 \cdot 10^{11}$ N/m^2, $\nu = 0.26$.

The obtained conclusions correspond to the conclusions known from technical and technological practice on the adhesion interaction of the considered pairs of materials.

4. Conclusion

The paper offers a description of the phenomenon of adhesion in the framework of the mechanics of a continuous deformable solid, namely a special variant of a nonlocal theory of elasticity. The proposed variant has a certain advantage in describing the phenomenon of adhesion, identifying criteria for its presence or absence, despite the fact that the analogy with solid-state physics, statistical mechanics of the discrete media in the construction of the environment model is used.

A possible presence of the adhesion between two solid materials, its energy and strength, or its absence can be judged by the values of characteristics of their elastic state, for example, Young's modulus and shear. Thus, there is no need for information about the internal atomic structure and chemical composition, internal physical properties of materials. The results of the work can be applied in the calculation of the adhesive compounds and the selection of pairs of materials with good adhesion or its absence.

Acknowledgment
In the Russian Federation, this work was performed within the framework of the governmental assignment for 2017–2019, project code 1.5265.2017/БЧ

References

[1] Kendall K 1994 *Science* **263** (5154) 1720–26
[2] Deryagin B V and Krotova N A 1948 *Sov. Phys. Usp.* **34** (3) 387–406
[3] Johnson K L 1998 *Tribology International* **31** (8) 413–18
[4] Frolenkova L Y and Shorkin V S 2013 *PNRPU Mechanics Bulletin* **1** 235–259.
[5] Romashin S N, Presnetsova V Yu, Frolenkova L Yu and Shorkin V S 2016 *Advanced Structured Materials* **42** (1) 363–76
[6] Shorkin V S 2011 *Mechanics of Solids* **6** 898–912
[7] Romashin S N and Shorkin V S 2016 *Proc. of the 10th All-Russian Scientific Conf. named after Yu I Neimarka* September 26–29 Nizhny Novgorod (Nizhny Novgorod: The Publishing House "Our Home") pp. 646–54
[8] Ruelle D 1969 *Statistical mechanics: Rigorous results* (New York, Amsterdam: Benjamin)
[9] Belov P A and Lurie S A 2007 *Mekhanika kompozitsionnykh materialov i konstruktsii* **13** (4) 519–34
[10] Maugis D 2000 *Contact, adhesion and rupture of elastic solids* (Springer-Verlag, Berlin Heidelberg New York)
[11] Barash Yu S 1988 *Sily Van-Der-Vaal'sa* [*Van der Waals Forces*] (Moscow: Nauka)
[12] Vekilov Yu Kh, Verner V D and Samsonova M B 1987 *Sov. Phys. Usp.* **30** 172–192
[13] Goldstein R V, Gorodtsov V A and Lisovenko D S 2009 *Doklady Physics* **54** (12) 546–48
[14] Vakhromeev G S, Yerofeev L Ya, Kanaikin V S and Nomokonova G G 1997 *Petrofizika* [*Petrophysics*] (Tomsk: TPU Publ. House)
[15] Erofeev L Ya, Vakhromeev G S, Zinchenko V S and Nomokonova G G 2006 *Fizika gornykh porod* [*Physics of rock properties*] (Tomsk: TPU Publ. House)
[16] Andreeva T V, Bolgar A S, Vlasova M V et al, ed. G V Samsonov 1976 *Svojstva jelementov. Chast' 1: Fizicheskie svojstva* [*Properties of elements. Chapter 1: Physical properties*] (Moscow: Metallurgy)

IOP Conf. Series: Materials Science and Engineering **208** (2017) 012035 doi:10.1088/1757-899X/208/1/012035

Exact Solutions for Stationary and Unsteady Layered Convection of a Viscous Incompressible Fluid with the Specified Velocities at the Bottom

E Yu Prosviryakov[1,2,a] and L F Spevak[1]

[1]Institute of Engineering Science, Ural Branch of the Russian Academy of Sciences, 34 Komsomolskaya St., Ekaterinburg, 620049, Russia
[2]Federal State Budgetary Educational Institution of Higher Education Kazan National Research Technical University named after A.N.Tupolev-KAI, 10, K.Marx Street, Kazan, Republic of Tatarstan 420111, Russia
E-mail: [a] evgen_pros@mail.ru

Abstract. The layered convective flow of a viscous incompressible fluid is considered with the specified velocities at the bottom of an infinite layer. A new exact stationary and nonstationary solution of the Oberbeck-Boussinesq system is presented. The account of fluid velocity at the bottom is characterized by the presence of two stagnant points, this being indicative of the nonmonotonic kinetic energy profile with two local extrema.

1. Introduction

The layered convective flows of a viscous incompressible fluid in the standard nomenclature in the Cartesian coordinate system are described by the Oberbeck-Boussinesq equations [1–6]:

$$\frac{\partial V_x}{\partial t}+V_x\frac{\partial V_x}{\partial x}+V_y\frac{\partial V_x}{\partial y}=-\frac{\partial P}{\partial x}+\nu\left(\frac{\partial^2 V_x}{\partial x^2}+\frac{\partial^2 V_x}{\partial y^2}+\frac{\partial^2 V_x}{\partial z^2}\right),$$

$$\frac{\partial V_y}{\partial t}+V_x\frac{\partial V_y}{\partial x}+V_y\frac{\partial V_y}{\partial y}=-\frac{\partial P}{\partial y}+\nu\left(\frac{\partial^2 V_y}{\partial x^2}+\frac{\partial^2 V_y}{\partial y^2}+\frac{\partial^2 V_y}{\partial z^2}\right),$$

$$\frac{\partial P}{\partial z}=g\beta T,$$

$$\frac{\partial T}{\partial t}+V_x\frac{\partial T}{\partial x}+V_y\frac{\partial T}{\partial y}=\chi\left(\frac{\partial^2 T}{\partial x^2}+\frac{\partial^2 T}{\partial y^2}+\frac{\partial^2 T}{\partial z^2}\right),$$

$$\frac{\partial V_x}{\partial x}+\frac{\partial V_y}{\partial y}=0. \qquad (1)$$

The overdetermined system of hydrodynamics equations (1) is solvable with the use of the exact solution [1–4]

$$V_x = U(z,t),\ V_y = V(z,t),$$

$$P = P_0(z,t) + xP_1(z,t) + yP_2(z,t),\ T = T_0(z,t) + xT_1(z,t) + yT_2(z,t). \tag{2}$$

The velocity field (2) identically satisfies the incompressibility equation (1). In this case, the unknown coefficients of the class of exact solutions (2) are calculated from the momentum conservation equation and the heat equation. The information on the use of exact solutions of form (2) can be found in [1–16].

Substituting the hydrodynamic fields (2), we obtain a loosely coupled system of equations. This system consists of heat-type evolution equations and stationary gradient relationships,

$$\frac{\partial T_1}{\partial t} = \chi \frac{\partial^2 T_1}{\partial z^2},\ \frac{\partial T_2}{\partial t} = \chi \frac{\partial^2 T_2}{\partial z^2},$$

$$\frac{\partial P_1}{\partial z} = g\beta T_1,\ \frac{\partial P_2}{\partial z} = g\beta T_2,$$

$$\frac{\partial U}{\partial t} = -P_1 + \nu \frac{\partial^2 U}{\partial z^2},\ \frac{\partial V}{\partial t} = -P_2 + \nu \frac{\partial^2 V}{\partial z^2},$$

$$\frac{\partial T_0}{\partial t} + UT_1 + VT_2 = \chi \frac{\partial^2 T_0}{\partial z^2},\ \frac{\partial P_0}{\partial z} = g\beta T_0. \tag{3}$$

System (3) describes the nonlinear properties of the flow of a viscous incompressible fluid. The nonlinearity of system (3) is due to the conservation of the convective derivative in the caloric equation of system (1).

We formulate the initial-boundary value problem for finding a particular solution of system (3). Consider that, at the initial time $t = 0$, the velocities and the reference temperature are zero,

$$U = V = 0,\ T_0 = 0, T_1 = 0,\ T_2 = 0. \tag{4}$$

At the lower fluid layer boundary described by the plane equation $z = 0$, the following boundary conditions are specified for $t \geq 0$:

$$U = W\cos\alpha,\ V = W\sin\alpha,\ T_0 = 0,\ T_1 = 0,\ T_2 = B. \tag{5}$$

The boundary conditions (5) determine the perturbation of the velocity field at the lower boundary. This condition can be interpreted as the non-satisfaction of the no-slip conditions, as well as the specification of the law of motion of the lower boundary in experimental hydromechanical studies.
On the free boundary ($z = h$), the boundary conditions of the form found in [8, 10] are valid,

$$P_0 = 0,\ P_1 = P_2 = 0,\ T_0 = 0,\ T_1 = A,\ T_2 = 0,\ \eta \frac{\partial U}{\partial z} = -\sigma T_1 = -\sigma A,\ \eta \frac{\partial V}{\partial z} = -\sigma T_2 = 0. \tag{6}$$

Here, σ is the temperature surface tension coefficient, η is the dynamic (shear) viscosity coefficient.

The curvature of the free boundary is negligible when large-scale processes are considered [8, 10]. Equations (1) describe large-scale flows through the use of a mathematical model in which the effect of vertical velocity is neglected for the motion of a viscous incompressible fluid.

IOP Conf. Series: Materials Science and Engineering **208** (2017) 012035 doi:10.1088/1757-899X/208/1/012035

2. Steady-state layered thermocapillary convection

To solve the initial-boundary value problems of layered thermocapillary convection, we reduce equation (3) and conditions (4)–(6) to the non-dimensional form. We select h, l, $\frac{h^2}{\nu}$, $\frac{gT\beta h^2}{\nu}$, $T = Al$ and $g\beta Th$ as the measures of the lateral dimension, horizontal scale, time, velocity, temperature and reduced pressure, respectively. To make the analysis more convenient, we introduce the following non-dimensional complexes: $\Delta = \frac{B}{A}$, $\delta = \frac{h}{l}$ (a characteristic scale ratio), $\Pr = \frac{\chi}{\nu}$, $\mathrm{We} = \frac{\rho h^2 g\beta}{\sigma}$, $\mathrm{Gr} = \frac{g\beta\Theta h^3}{\nu^2}$ and $\mathrm{Re} = \frac{Wl}{\nu}$ (the Prandtl, Weber, Grashof and Reynolds numbers) [8, 10]. Further dimensionless variables and dimensionless functions will be denoted by the same symbols as dimensional ones. The only exception is the dimensionless coordinate $Z = \frac{z}{h}$.

System (3) describing steady convective flows of a viscous incompressible fluid has the following nondimensional form:

$$\frac{d^2T_1}{dZ^2} = 0, \quad \frac{d^2T_2}{dZ^2} = 0,$$

$$\frac{dP_1}{dZ} = T_1, \quad \frac{dP_2}{dZ} = T_2,$$

$$\frac{d^2U}{dz^2} = \delta P_1, \quad \frac{d^2V}{dz^2} = \delta P_2,$$

$$\frac{d^2T_0}{dZ^2} = \frac{\delta \mathrm{Gr}}{\Pr}\left(UT_1 + VT_2\right), \quad \frac{dP_0}{dZ} = T_0. \tag{7}$$

The boundary conditions (5) and (6) in the dimensionless form are

$$U = \frac{\delta \mathrm{Re}}{\mathrm{Gr}}\cos\alpha, \; V = \frac{\delta \mathrm{Re}}{\mathrm{Gr}}\sin\alpha, \; T_0 = 0, \; T_1 = 0, \; T_2 = \Delta. \tag{8}$$

$$P_0 = 0, \; P_1 = P_2 = 0, \; T_0 = 0, \; T_1 = 1, \; T_2 = 0, \; \frac{\partial U}{\partial Z} = -\frac{1}{\mathrm{We}}, \; \frac{\partial V}{\partial z} = 0. \tag{9}$$

The boundary value problem (7)–(9) in the dimensionless form has the exact stationary solution

$$T_1 = Z, \; T_2 = \Delta(-Z+1), \; P_1 = \frac{1}{2}\left(Z^2 - 1\right), \; P_2 = -\frac{\Delta}{2}(Z-1)^2,$$

$$U = \delta\left(\frac{Z^4}{24} - \frac{Z^2}{4} + \frac{Z}{3}\right) - \frac{1}{\mathrm{We}}Z + \frac{\delta \mathrm{Re}}{\mathrm{Gr}}\cos\alpha,$$

$$V = \Delta\delta\left(-\frac{Z^4}{24} + \frac{Z^3}{6} - \frac{Z^2}{4} + \frac{Z}{6}\right) + \frac{\delta \mathrm{Re}}{\mathrm{Gr}}\sin\alpha,$$

$$T_0 = \mathrm{GrPr}\left[\frac{Z\delta}{12\mathrm{We}} - \frac{Z^4\delta}{12\mathrm{We}} - \frac{Z^4}{12\mathrm{We}} - \frac{41Z\delta^2}{2520} + \frac{Z^4\delta^2}{36} - \frac{Z^5\delta^2}{80} + \frac{Z^7\delta^2}{1008} - \frac{Z\delta^2\Delta^2}{126} + \frac{Z^3\delta^2\Delta^2}{36} - \right.$$

$$-\frac{5Z^4\delta^2\Delta^2}{144} + \frac{Z^5\delta^2\Delta^2}{48} - \frac{Z^6\delta^2\Delta^2}{144} + \frac{Z^7\delta^2\Delta^2}{1008} + \frac{Z^3\delta\mathrm{Re}}{6\mathrm{Gr}}\cos\alpha + \frac{Z^2\Delta\delta\mathrm{Re}}{2\mathrm{Gr}}\sin\alpha -$$

$$\left. -\frac{Z^3\Delta\delta\mathrm{Re}}{6\mathrm{Gr}}\sin\alpha + \frac{1}{12\mathrm{We}} - \frac{\delta\mathrm{Re}}{6\mathrm{Gr}}\cos\alpha - \frac{\Delta\delta\mathrm{Re}}{3\mathrm{Gr}}\sin\alpha\right],$$

$$P_0 = -\mathrm{GrPr}\left[-\frac{\delta}{40\mathrm{We}} + \frac{Z^2\delta}{24\mathrm{We}} - \frac{Z^5\delta}{60\mathrm{We}} - \frac{Z^4}{60\mathrm{We}} + \frac{61\delta^2}{13440} - \frac{41Z^2\delta^2}{5040} + \frac{Z^5\delta^2}{180} - \frac{Z^6\delta^2}{480} + \right.$$

$$+\frac{Z^8\delta^2}{8064} + + \frac{11\delta^2\Delta^2}{8064} - \frac{1}{252}Z^2\delta^2\Delta^2 + \frac{1}{144}Z^4\delta^2\Delta^2 - \frac{1}{144}Z^5\delta^2\Delta^2 + \frac{1}{288}Z^6\delta^2\Delta^2 -$$

$$-\frac{Z^7\delta^2\Delta^2}{1008} + \frac{Z^8\delta^2\Delta^2}{8064} + \frac{Z^4\delta\mathrm{Re}}{12\mathrm{Gr}}\cos\alpha + \frac{Z^3\Delta\delta\mathrm{Re}}{6\mathrm{Gr}}\sin\alpha - \frac{Z^4\Delta\delta\mathrm{Re}}{24\mathrm{Gr}}\sin\alpha +$$

$$\left. +\left(\frac{1}{12\mathrm{We}} - \frac{\delta\mathrm{Re}}{6\mathrm{Gr}}\cos\alpha - \frac{\Delta\delta\mathrm{Re}}{3\mathrm{Gr}}\sin\alpha\right)Z - \frac{1}{15\mathrm{We}} + \frac{\delta\mathrm{Re}}{12\mathrm{Gr}}\cos\alpha + \frac{5\Delta\delta\mathrm{Re}}{24\mathrm{Gr}}\sin\alpha\right]. \quad (10)$$

The study of the temperature and pressure gradients is trivial; therefore, we will analyze the velocity field. The velocity V_x assumes one zero value (figure 1) when the inequality

$$U(0)U(1) = \frac{\mathrm{Re}}{\mathrm{Gr}}\cos\alpha\left(\frac{1}{8} - \frac{1}{\delta\mathrm{We}} + \frac{\mathrm{Re}}{\mathrm{Gr}}\cos\alpha\right) < 0 \quad (11)$$

is satisfied.

For the velocity V_y, the condition for the existence of a critical value has the form

$$0 < \frac{\delta\mathrm{Re}}{\Delta\mathrm{Gr}}\sin\alpha < \frac{1}{24} \quad (12)$$

Counterflows at $\mathrm{Re} = 0$ (no slip at the bottom) were studied for thermal and concentration convection in [1, 2, 4, 16]. When the no-slip condition at the bottom is satisfied, the stagnant point is formed only with respect to the velocity V_y. The velocity V_x is of a constant sign. If there is a slip on the bottom of the fluid layer, the velocity V_x can have one stagnant point or two stagnant points (see figures 1, 2).

3. Nonstationary layered thermocapillary convection

On a specified time interval, the parabolic equations of system (3) are sequentially solved in time steps with the corresponding boundary conditions (4) and (5) by the boundary element method, similarly to the way it was done in [14].

Figure 1 shows velocity graphs at different moments of time for the following parameter values: $\mathrm{Pr} = 6.7$, $\mathrm{Gr} = 70.5$, $\mathrm{Re} = 3.6$, $\delta = 0.1$, $\mathrm{We} = 50$, $A = 1$, $\Delta = 0.5$, $\alpha = 1.3737$. In this case, at each instant of time, one stagnant point is observed, similarly to stationary flow.

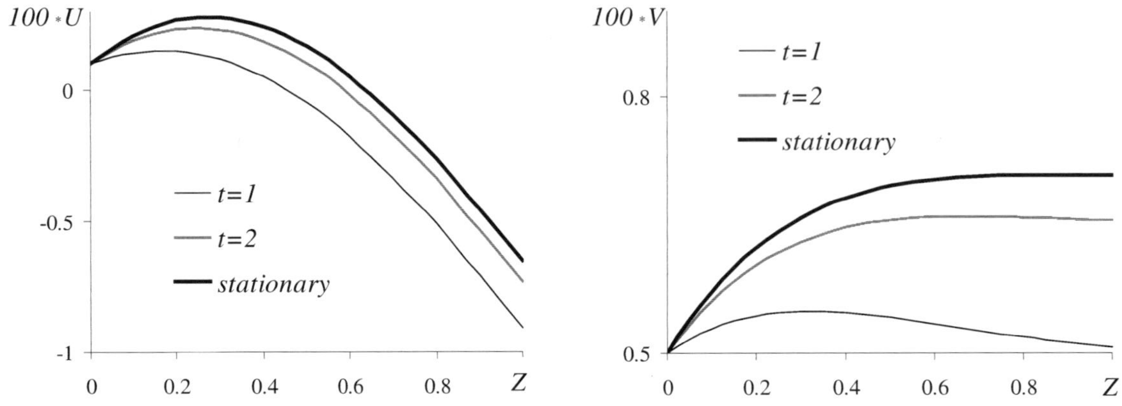

(a) (b)

Figure 1. Velocity components at different moments of time, one stagnant point: (a) – V_x, (b) – V_y ($\mathrm{Pr}=6.7$, $\mathrm{Gr}=70.5$, $\mathrm{Re}=3.6, \delta=0.1$, $\mathrm{We}=50$, $A=1$, $\Delta=0.5$, $\alpha=1.3737$).

The velocity graphs are shown in figure 2 at different times for $\alpha=1.7679$. The other parameters take their previous values. Here, two stagnant points are observed starting from a certain instant of time.

The velocity profiles shown in figures. 1, 2 testify to a considerable effect of dissipative mechanisms on the stress-strain state of a layered medium. The stratification of hydrodynamic fields results from the allowance made for diffusion effects in the motion equations and convective terms in the equation for heat conducting incompressible materials.

Figure 3 shows velocity hodographs for the solutions corresponding to figure 2.

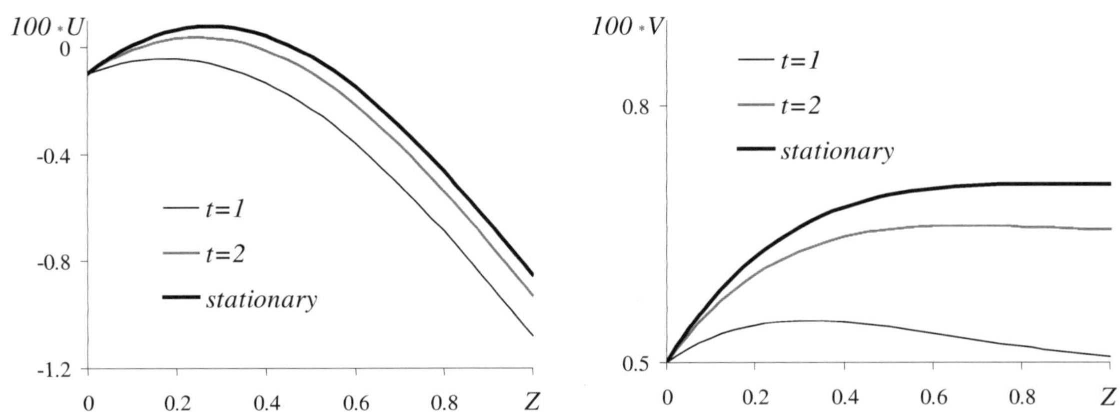

(a) (b)

Figure 2. Velocity components at different moments of time, two stagnant points: (a) – V_x, (b) – V_y $\mathrm{Pr}=6.7$, $\mathrm{Gr}=70.5$, $\mathrm{Re}=3.6, \delta=0.1$, $\mathrm{We}=50$, $A=1$, $\Delta=0.5$, $\alpha=1.7679$.

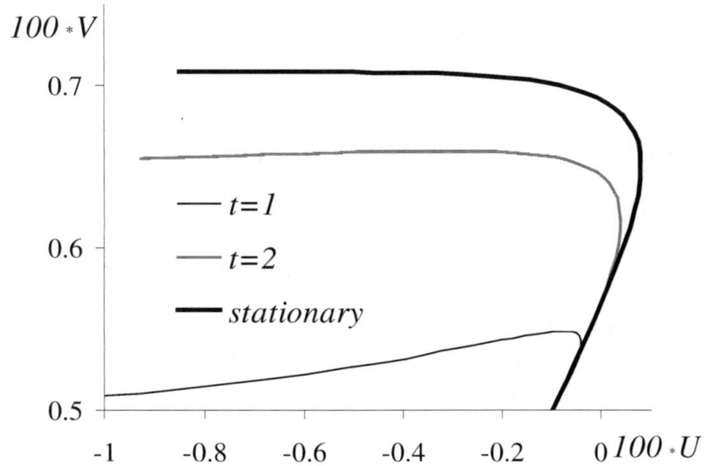

Figure 3. Velocity hodographs at different moments of time.

4. Conclusion

The convective flow of a viscous incompressible fluid has been studied. The velocity is set at the lower boundary of the fluid layer. An exact solution of the overdetermined Oberbeck-Boussinesq equation system has been obtained, and this solution can be used with any dimensionless Reynolds and Grashof numbers. This exact solution describes incompressible fluid flows with a nonmonotonic kinetic energy profile. In this case, the velocity field can take a zero value at two points.

Acknowledgment

This work was supported by the grant "Numerical and physical modelling of aerodynamic and aeroacoustic characteristics of rotor systems of future concept aircarft" (No. 9.1577.2017/ P Ch) of the Ministry of Education and Science of the Russian Federation.

References

[1] Aristov S N and Prosviryakov E Yu 2013 *Rus. J. Nonlin. Dyn.* **9** 651
[2] Aristov S N, Prosviryakov E Y and Spevak L F 2015 Computational Continuum Mechanics **8** 445
[3] Aristov S N, Prosviryakov E Yu 2016 *Theor. Found. Chem.* Eng. **50** 286
[4] Aristov S N, Prosviryakov E Y and Spevak L F 2016 *Theor. Found. Chem. Eng.* **50** 132
[5] Sidorov A F 1989 J. Appl. Mech. Tech. Phy. **30** 197
[6] Shvarz K G 2014 *Fluid Dynamics* **4** 438
[7] Birich R.V. 1966 J. Appl. Mech. Tech. Phy. **3** 69
[8] Gershuni G Z and Zhukhovitskii E M 1976 *Convective stability of incompressible liquid* (Jerusalem: Wiley)
[9] Napolitano L G 1980 *Acta Astronaut.* **7** 461
[10] Landau L D and Lifshitz E M 1987 *Fluid Mechanics* (PergamonPress, Oxford)
[11] Goncharova O N and Kabov O A. 2009 *Microgravity Sci. Tec* **21** 129
[12] Aristov S N, Knyazev D V and Polyanin A D 2009 *Theor. Found. Chem. Eng.* **43** 642
[13] Pukhnachev V V 2011 The News of Altai State University **1-2** 62
[14] Andreev V K and Bekezhanova V B 2013 *J. Appl. Mech. Tech. Phy.* **54** 171
[15] Aristov S N and Knyazev D.V. 2014 *Fluid Dynamics* **49** 565
[16] Prosviryakov E Yu and Spevak L F *AIP Conference Proceedings* **1785** 040048

IOP Conf. Series: Materials Science and Engineering **208** (2017) 012036 doi:10.1088/1757-899X/208/1/012036

Modeling the behaviour of shape memory materials under large deformations

A A Rogovoy and O S Stolbova

Institute of Continuous Media Mechanics, Ural Branch, Russian Academy of Sciences, Perm, Russia

E-mail: rogovoy@icmm.ru

Abstract. In this study, the models describing the behavior of shape memory alloys, ferromagnetic materials and polymers have been constructed, using a formalized approach to develop the constitutive equations for complex media under large deformations. The kinematic and constitutive equations, satisfying the principles of thermodynamics and objectivity, have been derived. The application of the Galerkin procedure to the systems of equations of solid mechanics allowed us to obtain the Lagrange variational equation and variational formulation of the magnetostatics problems. These relations have been tested in the context of the problems of finite deformation in shape memory alloys and ferromagnetic materials during forward and reverse martensitic transformations and in shape memory polymers during forward and reverse relaxation transitions from a highly elastic to a glassy state.

1. Introduction

Shape memory alloys (SMA), ferromagnetic shape memory alloys (FSMA) and shape memory polymers (SMP) refer to the class of functional (smart) materials. Constructions made of such materials may significantly change their configuration under the action of external thermal, magnetic or electric fields. SMAs have two phases, each with a different crystal structure and therefore different properties [1]. The phase transformation occurs by shear distortion of the lattice structure. On cooling, the crystal structure changes from austenite (the high symmetry parent phase) to martensite (the low symmetry product phase) and the material experiences inelastic deformation. When the material is heated being in the martensitic phase, the crystal structure transforms back to austenite and inelastic strain is recovered (the shape memory effect).

The combination of ferromagnetic properties and the structural (martensitic) phase transition makes FSMA a rather promising material in search for reliable ways of controlling the shape of a sample by varying temperature, pressure, and external magnetic field [2]. With the help of the magnetic field the phase transition process can be effectively controlled, as was shown in experimental studies [3]. If the direct or reverse phase transitions in the ferromagnetic material are realized under the action of the magnetic field, the specific temperatures of the process change according to the generalized Clausius-Clapeyron law and depend on the magnetic and stress fields. Such effect of the magnetic and stress fields allows us to vary temperatures of direct and reverse phase transformations and thereby to control the process of the austenite-to-martensite phase transition.

When SMP is cooled over the transition temperature range from the highly elastic to the glassy state, the deformation is "stored" in the material due to the formation of additional polymer chains and their interactions, which severely redstricts the large-scale conformational motion of the polymer as long as low temperature is maintained [4, 5, 6]. Such process of deformation storage does not occur in the material when heated over the transition temperature range. In this case, the material exhibits the so-called one-way shape memory effect when the material in the temporary shape at low temperatures remembers its permanent shape at high temperatures.

2. Basic relations

2.1. Kinematic relations

The behavior of media under finite deformations can be described using, according to [7, 8], the following process characteristics: the initial undeformed κ_0 and current (actual) κ configurations, the deformation gradient $\mathbf{F}$, converting the initial configuration to the actual one, the Cauchy-Green deformation measure $\mathbf{C}=\mathbf{F}^T\cdot\mathbf{F}$ and tensor $\mathbf{E}=(\mathbf{C}-\mathbf{g})/2$ ($\mathbf{g}$ is the identity tensor). Following [9], the history of any deformation process can be traced by introducing the intermediate κ_* configuration closely approaching the current one. To formalize this closeness, a small parameter ε is used. For the intermediate configuration the deformation gradient is written as $\mathbf{F}=(\mathbf{g}+\varepsilon\mathbf{h})\cdot\mathbf{F}_*=(\mathbf{g}+\varepsilon(\mathbf{e}+\mathbf{d}))\cdot\mathbf{F}_*$, where $\mathbf{F}_*$ is the deformation gradient converting the initial configuration to the intermediate one, $\mathbf{h}$ is the gradient of the small displacement vector, which connects the configuration κ_* and κ, $\mathbf{e}$ and $\mathbf{d}$ are the tensors of small strains and rotations for the intermediate configuration (symmetric and skew-symmetric parts of the tensor $\mathbf{h}$). Thus, we obtain the following relations:

$$\mathbf{C}=\mathbf{C}_*+2\varepsilon\mathbf{F}_*^T\cdot\mathbf{e}\cdot\mathbf{F}_*,\qquad \mathbf{E}=\mathbf{E}_*+\varepsilon\mathbf{F}_*^T\cdot\mathbf{e}\cdot\mathbf{F}_*=\mathbf{E}_*+\Delta\mathbf{E}. \tag{1}$$

According to these relations, in the limit of going from the intermediate κ_* to the current configuration κ, the rates of changes in $\mathbf{C}$ and $\mathbf{E}$ are given by: $\dot{\mathbf{C}}=2\,\mathbf{F}^T\cdot\mathbf{D}\cdot\mathbf{F}$, $\dot{\mathbf{E}}=\mathbf{F}^T\cdot\mathbf{D}\cdot\mathbf{F}$, where $\mathbf{D}=\dot{\mathbf{e}}$ is the strain rate tensor.

2.2. State equation

Out of the available equivalent forms of writing the constitutive equations for simple materials, which satisfy the objectivity principle [7], we choose the following form: $\mathbf{T}=J^{-1}\mathbf{F}\cdot\widetilde{\mathbf{g}}(\mathbf{C}_E,\Theta)\cdot\mathbf{F}^T$, where $\mathbf{T}$ is the true stress tensor, $\widetilde{\mathbf{g}}(\mathbf{C}_E,\Theta)$ is the function of the material response (second-order tensor), $\mathbf{C}_E$ is the Cauchy-Green measure of elastic strains, $J=I_3(\mathbf{F})$ is the third invariant of $\mathbf{F}$ (Jacobian, determining a relative variation in the volume element) and Θ is the absolute temperature. The tensor $\widetilde{\mathbf{g}}$ in this equation is known as the second (symmetric) Piola-Kirchhoff stress tensor $\mathbf{P}_{II}\equiv\widetilde{\mathbf{g}}$, which in the theory of finite elastic deformations is represented in terms of elastic potential $W(\mathbf{C}_E)$: $\mathbf{P}_{II}=2(\partial W/\partial\mathbf{C}_E)$. For the intermediate configuration, this constitutive equation is written as

$$\mathbf{P}_{II}=\mathbf{P}_{II*}+\varepsilon\,\mathbf{L}_*^{IV}\cdot\cdot\,\mathbf{e}_E=\widetilde{\mathbf{P}}_{II*}+\widetilde{\mathbf{L}}_*^{IV}\cdot\cdot\,\Delta\mathbf{E}_E. \tag{2}$$

Here $\mathbf{L}_*^{IV}$ and $\widetilde{\mathbf{L}}_*^{IV}$ are the fourth-rank tensors specifying the material responses to the small elastic strains $\varepsilon\mathbf{e}_E$ and to the increment of elastic strain $\Delta\mathbf{E}_E$ pertinent to the intermediate configuration Passing to the limit yields $\dot{\mathbf{P}}_{II}=\mathbf{L}^{IV}\cdot\cdot\,\mathbf{D}_E=\widetilde{\mathbf{L}}^{IV}\cdot\cdot\,\dot{\mathbf{E}}_E$. Here and below superscript “dot” differentiating with respect to time.

IOP Conf. Series: Materials Science and Engineering **208** (2017) 012036 doi:10.1088/1757-899X/208/1/012036

2.3. Phase transition in the shape memory alloys (SMA and FSMA)

Kinematics of the process is defined by the deformation gradients $\mathbf{F}_A$ in the austenitic state and $\mathbf{F}_M$ in the martensitic state and expressions for other kinematic tensors such as $\mathbf{C}_A$, $\mathbf{C}_M$, $\mathbf{E}_A$, $\mathbf{E}_M$ are similar to those in (1). We introduce the following kinematic equation [10]:

$$\dot{\mathbf{E}} = (\varphi_A \mathbf{E}_A)^{\bullet} + (\varphi_M \mathbf{E}_M)^{\bullet} + \dot{\mathbf{E}}_{Ph} + \dot{\mathbf{E}}_P + \dot{\mathbf{E}}_{\Theta}, \tag{3}$$

where φ_M and φ_A are the parameters that specify the volume fractions of the martensitic and austenitic phases in the material ($\varphi_M + \varphi_A = 1$), $\dot{\mathbf{E}}_{\Theta} = \alpha \dot{\Theta} \mathbf{C}$ is the rate of change in the temperature strains (α is the linear thermal expansion factor), $\dot{\mathbf{E}}_P = 3\dot{\sigma}_i/(2H\sigma_i)\,(\mathbf{F}^T \cdot \mathbf{S} \cdot \mathbf{F})$ is the rate of change in the plastic strains ($\mathbf{S}$ is the deviator of $\mathbf{T}$, σ_i is the stress intensity, H is the modulus of plastic strain hardening), $\dot{\mathbf{E}}_{Ph}$ is the rate of a change in the phase strains.

In order to approximate the dependence of φ_M on the temperature, we use the following relation:

$$\varphi_M(\xi) = \begin{cases} 0, & \xi \le 0; \\ 0.5\left(1 - \cos(\pi\xi)\right), & 0 < \xi < 1; \\ 1, & \xi \ge 0. \end{cases} \qquad \begin{aligned} \xi &= \frac{M_s - \Theta}{M_s - M_f}, \quad M_f \le \Theta \le M_s \ (d\varphi > 0); \\ \xi &= 1 + \frac{A_s - \Theta}{A_f - A_s}, \quad A_s \le \Theta \le A_f \ (d\varphi < 0). \end{aligned} \tag{4}$$

Here M_s, M_f, A_s, A_f are the characteristic temperatures at the beginning and end of the direct and reverse martensitic transformations in the absence of the stress and magnetic fields. In the case, when the thermal phase transitions occur in the presence of the stress and magnetic fields, there is a shift in $\Theta_c = M_c = (M_s + M_f)/2$ for direct martensitic transformation and $\Theta_c = A_c = (A_s + A_f)/2$ for reverse martensitic transformation according to the generalized Clausius-Clapeyron law [10]:

$$\Theta_c = \Theta_{c0} + \frac{\Theta_{c0}}{[q]}\left(\mathbf{P}_{II} \cdot\cdot [\mathbf{E}_{Ph}] + \mu_0 \mathbf{H} \cdot [\mathbf{M}]\right), \tag{5}$$

where Θ_c is a new value of the average temperature, Θ_{c0} is the value of the average temperature in the absence of the stress and magnetic fields, $\mathbf{H}$ is the magnetic field vector, $[q]$, $[\mathbf{E}_{Ph}]$ and $[\mathbf{M}]$ are changes in the released or absorbed heat, phase strain and magnetization during phase transition, and μ_0 is the magnetic permeability of vacuum.

To describe the rate of change in the phase strains (without regard to reversible shape memory effect of reverse transformation), we used the equations, which were proposed for small strains in [11]. In our study, these relations were generalized to finite strains: $\dot{\mathbf{E}}_{Ph} = \widetilde{\mathbf{E}}_{Ph}\,\dot{\varphi}_M$, where

$$\widetilde{\mathbf{E}}_{Ph} = (a\mathbf{g} + b\hat{\mathbf{P}}_{II} + c\,\mathbf{E}_{Ph}), \ \dot{\varphi}_M > 0; \qquad \widetilde{\mathbf{E}}_{Ph} = \left(\frac{c\,\mathbf{E}_{Ph}^{(0)}}{\exp(c\varphi_M^{(0)})} + c\,\mathbf{E}_{Ph}\right), \ \dot{\varphi}_M < 0. \tag{6}$$

In these expressions a, b, c are the material parameters, $\hat{\mathbf{P}}_{II}$ is the deviator of $\mathbf{P}_{II}$, $\varphi_M^{(0)}$, $\mathbf{E}_{Ph}^{(0)}$ are the values of the martensite phase parameter and phase strain at the starting point of the reverse transformation process.

2.4. Relaxation transition in the shape memory polymers

Kinematics of the process is defined by the deformation gradients $\mathbf{F}_r$ in the highly elastic (rubbery) state and $\mathbf{F}_g$ in the glassy state and the expressions for other kinematic tensors such as $\mathbf{C}_r$, $\mathbf{C}_g$, $\mathbf{E}_r$, $\mathbf{E}_g$ are similar to those in (1). We introduce the following kinematic equation [12]:

$$\dot{\mathbf{E}} = (\varphi_r \mathbf{E}_r)^{\bullet} + (\varphi_g \mathbf{E}_g)^{\bullet} + \dot{\mathbf{E}}_f + \varphi_g \dot{\mathbf{E}}_V + \dot{\mathbf{E}}_{\Theta}, \tag{7}$$

where φ_g and φ_r are the parameters that specify the volume fractions of the glassy and the highly elastic phases in the material ($\varphi_r + \varphi_g = 1$), $\dot{\mathbf{E}}_{\Theta}$ is the rate of change in the temperature strains and is defined by the relation same as in the previous subsection, $\dot{\mathbf{E}}_V = \mu^{-1}(\mathbf{F}^T \cdot \mathbf{T} \cdot \mathbf{F})$ is the rate of a change in the viscous strains (μ is the viscosity), $\dot{\mathbf{E}}_f$ is the rate of freezing of highly elastic strains in the cooling process.

In order to approximate the dependence of φ_g on temperature, we use the following relation [4]: $\varphi_g(\Theta) = 1 - 1/[1 + A(\Theta_h - \Theta)^4]$, where A is the material parameter, Θ_h is the characteristic temperature of the relaxation transition. According to [4]:

$$\dot{\mathbf{E}}_f = \begin{cases} \dot{\varphi}_g \mathbf{E}_r, & \dot{\Theta} < 0, \\ (\dot{\varphi}_g / \varphi_g)\, \mathbf{E}_f, & \dot{\Theta} > 0, \\ 0, & \dot{\Theta} = 0. \end{cases} \tag{8}$$

2.5. Description of the elastic behavior

To formulate the constitutive equation for SMA and FSMA, we use a simplified Signorini law [8] for describing both the austenitic and martensitic phase behavior:

$$\begin{gathered} \mathbf{P}_{II\gamma} = J(\mathbf{C}_\gamma)\left[(k_{1\gamma} + k_{2\gamma})\mathbf{C}_\gamma^{-1} - 2k_{2\gamma}\mathbf{C}_\gamma^{-2}\right], \quad \gamma = A, M, \\ k_{1\gamma} = \Lambda_\gamma\left[3 - I_1(\mathbf{C}_\gamma^{-1})\right]/2 + (\Lambda_\gamma + G_\gamma)\left[3 - I_1(\mathbf{C}_\gamma^{-1})\right]^2/8, \; k_{2\gamma} = G_\gamma - (\Lambda_\gamma + G_\gamma)\left[3 - I_1(\mathbf{C}_\gamma^{-1})\right]/2, \end{gathered} \tag{9}$$

where Λ_γ and G_γ are the material parameters, which coincide with the Lame parameter and the shear modulus of the linear theory of elasticity. Hence, in accordance with (2) we obtain for SMA and FSMA

$$\begin{gathered} \tilde{\mathbf{L}}_\gamma^{IV} = J(\mathbf{C}_\gamma)\Big\{(k_{1\gamma} + k_{2\gamma})\mathbf{C}_\gamma^{-1}\mathbf{C}_\gamma^{-1} - k_{2\gamma}\mathbf{C}_\gamma^{-1}\mathbf{C}_\gamma^{-2} + (\Lambda_\gamma + G_\gamma - 2k_{2\gamma})\mathbf{C}_\gamma^{-2}\mathbf{C}_\gamma^{-1} - \\ - 2(k_{1\gamma} + k_{2\gamma})\mathbf{C}_\gamma^{-1} \cdot \mathbf{C}_{II} \overset{2}{*} \mathbf{C}_\gamma^{-1} + 4k_{2\gamma}\left[\left(\mathbf{C}_\gamma^{-1} \cdot \mathbf{C}_{II} \overset{2}{*} \mathbf{C}_\gamma^{-1}\right) \overset{2}{*} \mathbf{C}_\gamma^{-1} + \mathbf{C}_\gamma^{-1} \cdot \left(\mathbf{C}_\gamma^{-1} \cdot \mathbf{C}_{II} \overset{2}{*} \mathbf{C}_\gamma^{-1}\right)\right]\Big\}. \end{gathered} \tag{10}$$

Here $\mathbf{C}_{II}$ is the second isotropic fourth-order tensor [8] and the operation $\mathbf{B}^{IV} \overset{2}{*} \mathbf{A}$ means a positional scalar multiplication (from the right) of the second-order tensor $\mathbf{A}$ by the second basic vector of the fourth-rank tensor $\mathbf{B}^{IV}$.

As the elastic potential for SMP, we use the Peng–Landel potential for describing both the highly elastic and the glassy state behavior:

$$W_\gamma = G_\gamma\left[I_1(\mathbf{C}_\gamma)\, J^{-2/3}(\mathbf{C}_\gamma) - 3\right]/2 + B_\gamma\left[J(\mathbf{C}_\gamma) - 1\right]^2/2, \quad \gamma = r, g, \tag{11}$$

where G_γ and B_γ are the material parameters, which coincide with the shear modulus and the bulk modulus of the linear theory of elasticity. Hence, in accordance with (2) we obtain for SMP

$$\begin{aligned}\tilde{\mathbf{L}}_\gamma^{IV} = -\frac{2}{3} J^{-2/3}(\mathbf{C}_\gamma)\mathbf{C}_\gamma \left[\mathbf{g}\mathbf{C}_\gamma^{-1} + \mathbf{C}_\gamma^{-1}\mathbf{g} - \frac{1}{3} I_1(\mathbf{C}_\gamma)\,\mathbf{C}_\gamma^{-1}\mathbf{C}_\gamma^{-1} - I_1(\mathbf{C}_\gamma)\,\mathbf{C}_\gamma^{-1}\cdot\mathbf{C}_{II} \overset{2}{*} \mathbf{C}_\gamma^{-1} \right] + \\ + 2J(\mathbf{C}_\gamma)B_\gamma \left[\left(J(\mathbf{C}_\gamma) - \frac{1}{2} \right)\mathbf{C}_\gamma^{-1}\mathbf{C}_\gamma^{-1} - \left(J(\mathbf{C}_\gamma) - 1 \right)\mathbf{C}_\gamma^{-1}\cdot\mathbf{C}_{II} \overset{2}{*} \mathbf{C}_\gamma^{-1} \right].\end{aligned} \tag{12}$$

3. Verification of the model

3.1. Boundary-value problem

The variational formulation of the boundary value problem in the initial configuration in the absence of a magnetic field (for SMA and SMP) can be written as

$$\int_{S_0} J\sqrt{\mathbf{n}\cdot\mathbf{C}^{-1}\cdot\mathbf{n}}\,\mathbf{Q}\cdot\delta\mathbf{U}\,dS_0 + \int_{V_0^{(in)}} \rho_0\,\mathbf{K}\cdot\delta\mathbf{U}\,dV_0 - \int_{V_0^{(in)}} \mathbf{P}_{II}\cdot\delta\mathbf{E}\,dV_0 = 0, \tag{13}$$

where $\mathbf{n}$ is the external unit normal to the surface S_0 of the body with mass density ρ_0 and with volume $V_0^{(in)}$ in the initial configuration, $\mathbf{U}$ is the displacement vector connecting the initial and current configurations, $\mathbf{Q}$ and $\mathbf{K}$ are the surface and mass forces, δ is the symbol of variation.

Under the action of the magnetic field (for FSMA) it is necessary to solve the connected magnetomechanical problem for the magnetizable body of finite dimensions located in the space. In this case, one more variational equation is considered:

$$\begin{aligned}\int_{V_0^{(in)}} J\,\chi\,\mathbf{H}_0\cdot\mathbf{F}^{-T}\cdot\delta(\nabla\psi)\,dV_0 = \int_{V_0^{(in)}} J(1+\chi)(\mathbf{F}^{-T}\cdot\nabla\psi)(\mathbf{F}^{-T}\cdot\delta(\nabla\psi))\,dV_0 + \\ + \int_{V_0^{(ex)}} J(\mathbf{F}^{-T}\cdot\nabla\psi)(\mathbf{F}^{-T}\cdot\delta(\nabla\psi))\,dV_0.\end{aligned} \tag{14}$$

Here $V_0^{(ex)}$ is the environmental space volume, $\mathbf{H}_0$ is the applied external magnetic field, ψ is the scalar magnetic potential $(\mathbf{H} = \mathbf{H}_0 - \tilde{\nabla}\psi)$, $\tilde{\nabla}$ is the Hamilton operator in the actual configuration while ∇ is the Hamilton operator relative to the initial configuration and $\chi = \varphi_M\chi_M + \varphi_A\chi_A$ is the magnetic susceptibility in the mixed austenitic-martensitic state.

We use the step-wise procedure for numerical realization of equations (13) and (14).

3.2. Shape memory alloys

To verify the model for SMA we consider the following problems on the deformation of a plate with one rigidly clamped end (the left end in figure 1 where the results of solving these problems are shown).

Problem 1. Two plates of the same length made of beryllium bronze and SMA (nickel titanium) fastened together along the length without tension. The sample is first cooled to the temperature Θ_M corresponding to the martensitic state of the SMA and then reheated to the temperature Θ_A corresponding to the austenitic state. Thus, the SMA first undergoes the forward martensitic transformation (this leads to bending of the two-layer plate) and then the reverse martensitic transformation (the plate takes the same shape as it has in the initial state).

Problem 2. Before the two plates are fastened together (see Problem 1), the SMA plate at the temperature Θ_A corresponding to the fully austenitic state of the material is subjected to uniaxial uniform tension along under the action of tensile stress 100 MPa. Then, the plate is cooled to the temperature Θ_M corresponding to a fully martensitic state of the SMA, after which the load is removed. The plates having the same length are fastened along the length without tension. This two-layer plate is heated to the temperature Θ_A (the bending of the plate) and then cooled to the temperature Θ_M (return to the original position).

Problem 3. The plate made of SMA is bended by a shear stress of 20 MPa applied to the free (right) end at the temperature Θ_A and then is cooled to the temperature Θ_M, after which the load is removed (the plate remains in the bent state). After that the plate is reheated to the temperature Θ_A (the plate takes the same shape as that in the initial state).

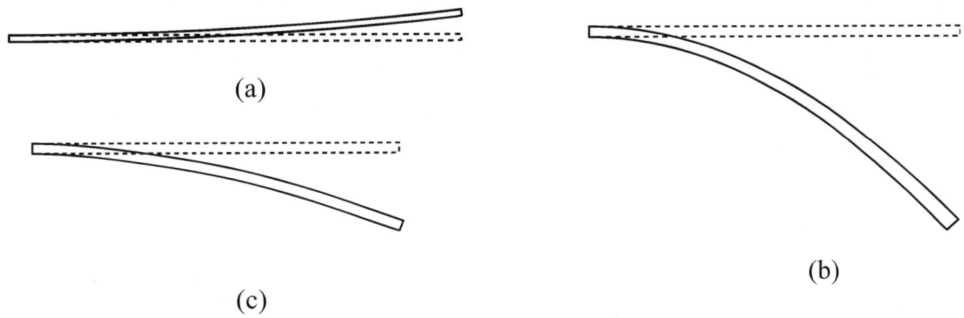

Figure 1. Bending of the plates for Problem 1 (a), Problem 2 (b) and Problem 3 (c): the initial and final states of the plates are denoted by the dashed and solid lines, respectively. In figures (a) and (b) the SMA plate is situated at the bottom.

3.3. Ferromagnetic shape memory alloys

To verify the model for FSMA we used the experimental data of work [3], where the possibility of magnetic control of the shape memory effect was analyzed. The austenite-to-martensite phase transition in the specimen made of polycrystalline Ni-Mn-Fe-Ga alloy was realized at a fixed external temperature due to a shift of the specific phase transition temperature induced by the magnetic field. An approximately linear dependence of the martensite and austenite start temperatures on the applied magnetic field (in the absence of stress) was obtained experimentally in paper by [3].

We consider the following boundary-value problem corresponding to the above mentioned experiment. At the temperature of the austenitic state Θ_0 the straight cantilevered bar-shaped specimen (figure 2 (a)) was subjected to a follower tangential force 5 MPa uniformly distributed along the surface of the free end of the bar (the specimen was bent). Then the bar was cooled down to the martensitic state temperature and unloaded. This process proceeded in the absence of the magnetic field. Then the bar was exposed to the external magnetic field $\mathbf{H}_0 = 80 \cdot 10^5$ A/m, which led to a significant increase in the characteristic temperature of the reverse phase transition. After this the bar was heated up to the temperature Θ_0, $A_s(\mathbf{H}=\mathbf{0}) < A_f(\mathbf{H}=\mathbf{0}) < \Theta_0 < A_s(\mathbf{H}=\mathbf{H}_0) < A_f(\mathbf{H}=\mathbf{H}_0)$, which was kept constant, the magnetic field was gradually removed and the specimen underwent the phase transition to the austenitic state at a fixed temperature Θ_0.

Figure 2 (b) demonstrates the specimen configuration and the distribution of the axial component of the phase strain calculated in compliance with the above relations at the beginning of the magnetic field removal. Red and blue lines correspond to stretching and compressing strains in the bar. The dashed line is the median surface of the bar. When the magnetic field vanishes the phase strains vanish, too, and the bar regains its initial configuration.

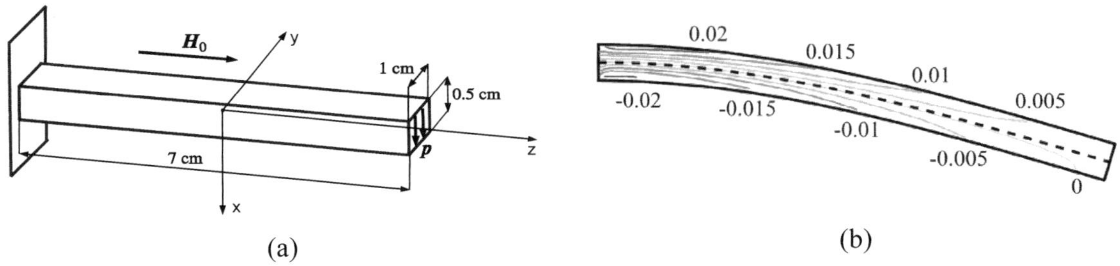

Figure 2. The specimen (a) and distribution of axial component of phase strain (b).

3.4. Shape memory polymers

The model for SMP was verified using the experimental data presented in [5, 6]. In experiments [5], the sample of a shape memory polymer (epoxy polymer) was initially in the unstrained and unstressed state at a temperature Θ_h (high elasticity). In the first stage, the specimen was subjected to uniaxial force applied along its axis at a constant temperature until certain axial strain E_* was reached. After that, the material was cooled to a temperature Θ_g (glassy state) at a fixed strain. This was accompanied by an increase in the stress. When the SMP in the glassy state was unloaded to zero stresses, the axial strains in the material remained frozen. Heating of the material to a temperature Θ_h led to unfreezing of the strains, and the polymer returned to the initial state. Figure 3 (a, b, c) shows curves of the nonzero component of the true stress tensor $\mathbf{T}$ versus relative temperature, obtained for elastic material behavior (solid curves), viscoelastic material behavior (dot-and-dashed curves) and in experiment [5] (dashed curves) at the initial axial strain $E_* = 9.1\%$, $E_* = 0\%$ and $E_* = -9.1\%$.

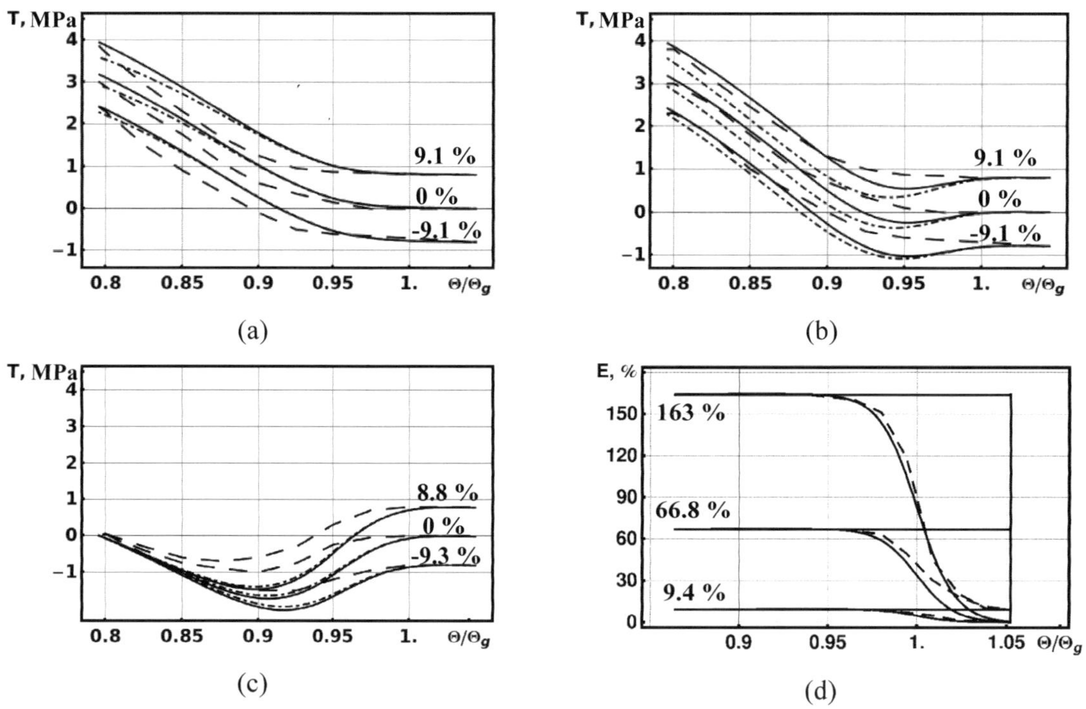

Figure 3. True stress in the sample [5] during its cooling (a), heating after cooling (b), heating after cooling and unloading (c) and Cauchy–Green axial strain in the sample [6] (d).

Figure 3 (a) shows the true stress in the sample during its cooling, figure 3 (b) shows true stresses in the sample when heated in the state, corresponding to completion of the cooling process and figure 3 (c) shows the true stresses in the sample when heated in the state, corresponding to completion of the cooling process and unloading (with the final axial strain $E = 8.8\%$, $E = 0\%$ and $E = -9.3\%$).

In [6], the uniaxial loading experiments were performed with a sample of Veriflex^{TM} shape memory polymer. Figure 3 (d) shows the Cauchy–Green axial strain versus relative temperature curves, which were plotted using the developed model for elastic material behavior (solid curves) and the data from experiments [6] (dashed curves). At a temperature Θ_h (high elasticity) the sample was stretched by the axial force to different axial strains $E_* = 163\%$, $E_* = 66.8\%$ and $E_* = 9.4\%$, and was cooled to a temperature Θ_g (glassy state). After that the sample was unloaded to zero axial stress and heated to a temperature Θ_h.

4. Conclusion

The models, describing the behavior of shape memory alloys, ferromagnetic materials and polymers have been constructed, using a formalized approach to the development of constitutive equations for complex media under large deformations. A comparison between the results obtained in the framework of the constructed models and the available experimental data has shown that the proposed models can describe with acceptable accuracy the behavior of shape memory materials.

Acknowledgements

Authors appreciate greatly the financial support of the Russian Fund of Fundamental Research (grants 14-01-00080 and 16-31-00161 mol_a) and are thenkful to Mrs L V Semukhina for her assistance in preparation of the paper English variant.

References

[1] Otsuka K and Wayman C M 1998 *Shape Memory Materials* (Cambridge: Cambridge University Press)
[2] Vasil'ev A N, Buchel'nikov V D, Takagi T, Khovailo V V and Estrin E I 2003 *Phys. Usp.* **46** 559–588
[3] Cherechukin A A, Dikshtein I E, Ermakov D I, Glebov A V, Koledov V V, Kosolapov D A, Shavrov V G, Tulaikova A A, Krasnoperov E P and Takagi T 2001 *Phys. Lett.* A **291** 175–183
[4] Baghani M, Naghdabadi R and Arghavani J 2013 *J. Intelligent Material Systems and Structures* **24** 21–32
[5] Liu Y, Gall K, Dunn M L, Greenberg A R and Diani J 2006 *Int. J. Plasticity* **22** 279–313
[6] Volk B L, Lagoudas D C, Chen Yi-Chao 2010 *Smart Materials and Structures* **19** 75006–16
[7] Truesdell C 1972 *A First Course in Rational Continuum Mechanics* (Baltimore: J. Hopkins University)
[8] Lur'e A I 1980 *Nonlinear Elasticity Theory* (Moscow: Nauka)
[9] Rogovoy A A 2012 *Contin. Mech. Thermodyn.* **24** 81–114
[10] Rogovoy A A and Stolbova O S 2016 *Int. J. Plasticity* **85** 130–155
[11] Movchan A A, Shelymagin P V and Kazarina S A 2001 *Appl. Mech. Tech. Phys.* **36** 864–871
[12] Rogovoy A A and Stolbova O S 2015 *Appl. Mech. Tech. Phys.* **56** 1059–70

IOP Conf. Series: Materials Science and Engineering **208** (2017) 012037 doi:10.1088/1757-899X/208/1/012037

Hierarchical modeling of deformation and damage of metal matrix composite under uniaxial loading conditions

S V Smirnov[1], **A V Konovalov**[1], **M V Myasnikova**[1], **Yu V Khalevitsky**[1] **and A S Smirnov**[1]

[1]Institute of Engineering Science, Ural Branch of the Russian Academy of Sciences, 34 Komsomolskaya St., Ekaterinburg, 620049, Russia

E-mail: marina@imach.uran.ru

Abstract. A three-dimensional model for deformation of metal matrix composite with aluminum matrix and silicon carbide reinforcement particles is developed. The model accounts for an internal structure of the composite, as well as rheology of its constituents. The model is further used in numerical simulations in order to study the evolution of stress-strain state parameters in a randomly-chosen composite microstructure fragment under uniaxial tension and compression loading on micro- and macroscale. The parameters include the stress stiffness coefficient, the Lode-Nadai coefficient and equivalent (von Mises) strain. It is found that local deformation regions and internal tensile stress concentration regions appear in the material of composite matrix. Adhering to a phenomenological damage theory, a damage development is computed in the matrix metal. We present damage fields and damage distributions for uniaxial tension and compression.

1. Introduction

Studies of mechanical behavior of composites under loading are crucial for the assessment of reliability, durability and usability of machine parts and structural elements produced from composites. Due to a hierarchical composite structure, these studies need to cover multiple scale levels. The established modern paradigm for such studies is the multilevel approach for describing structurally inhomogeneous materials [1–4]. Multilevel models of plastic deformation and damage are advantageous over the classical approach adopted in mechanics in a way that they allow one to observe stress concentration phenomenon leading to microcrack emergence at early deformation stages.

It is known that under conditions of a complexly changing stress-strain state, the microscopic crack stage (also known as the hidden or scattered damage accumulation stage) is described phenomenologically by damage criteria (e.g. Kolmogorov criterion, Lemaitre criterion, etc.) and it indirectly characterizes damage evolution in a microscopic volume of continuous medium [5–8]. The damage level is associated with local plastic strain and assessed by effective plastic strain before failure which is in its turn dependent on the evolution of dimensionless stress-strain state parameters: the stress stiffness coefficient and the Lode-Nadai stress state coefficient. Thus, in order to adequately describe damage process, one needs to acquire the data on stress-strain state history under severe plastic deformation conditions. To achieve this goal, numerical simulations

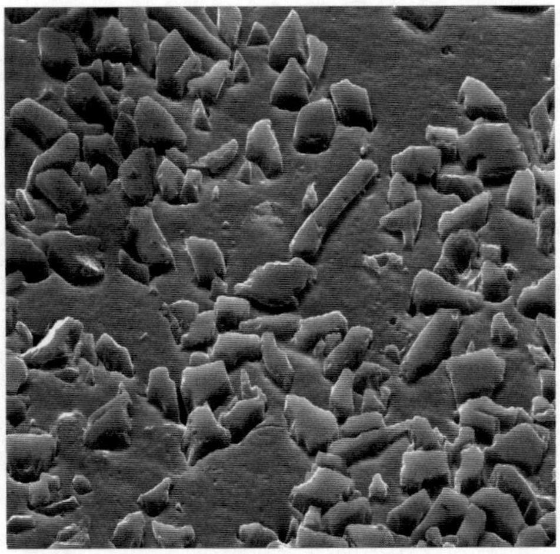

Figure 1: The model material matrix composite microstructure.

employing different software are widely used [2, 3, 9–12]. Simulation results allow one to obtain sufficient data for further computations.

This paper presents a numerical simulation of damage accumulation in a randomly-chosen microstructure fragment of an aluminum matrix composite under uniaxial loading conditions. The simulation is based on the multilevel approach for describing a material, a damage mechanics criterion and considers the rheology of composite constituents.

2. Material and Methodology

A specific metal matrix composite is considered to be a model material. The constituents of the composite are 99.8% commercially pure aluminum and silicon carbide reinfig:forcement particles. The dominant particle shape is considered to be irregularly prismatic; particle sizes are in ranges of 1 . . . 5 µm and 15 . . . 20 µm.

Particles make up 50 vol% of the composite. The composite microstructure is depicted in Fig. 1, which is obtained by means of scanning electron microscope. Experimental studies observe strong adhesive bonding between the matrix and reinforcement particles [13, 14].

The computational model has been implemented adhering to the two-level structural-phenomenological approach, which connects problem solutions on micro- and macroscale [15, 16]. According to the approach, the composite volume on microlevel is modeled by three-dimensional continuum which represents the aluminum matrix with embedded silicon carbide particles. The microstructural properties of the metal matrix composite have been chosen according to metallographic investigation [13, 14]. The composite microvolume has the shape of a cube with the edge size of 30 µm. The structurally inhomogeneous microvolume is surrounded with a buffer layer. The layer has smeared macroscopic mechanical properties of the composite and dilates evenly from microvolume borders. The volume thickness is equal to the microvolume linear size. Thus, the whole microvolume is a cube with edge size of 90 µm This problem statement allows us to connect solutions on micro- and macroscale and properly meet a challenge of atypical boundary behavior. An in-house software was developed by authors in previous work [17]. The software is capable of three-dimensional model generation for structurally inhomogeneous materials with a complex internal structure. It is compatible with ANSYS finite element suite input format. The three-dimensional metal matrix composite computational model is depicted in Fig. 2. Detailed information is published in [18, 19].

IOP Conf. Series: Materials Science and Engineering **208** (2017) 012037 doi:10.1088/1757-899X/208/1/012037

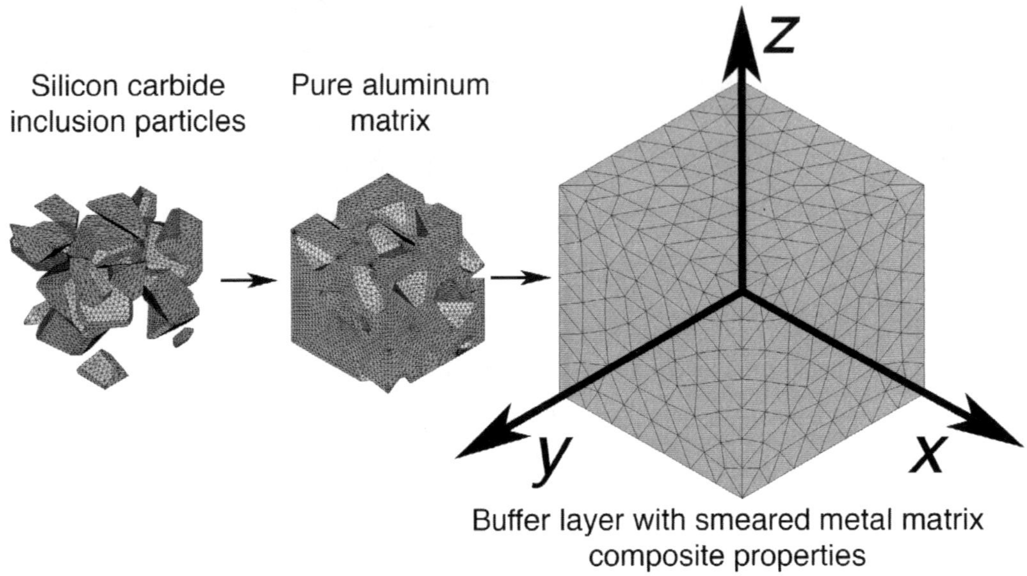

Figure 2: Three-dimensional metal matrix composite computational model.

Rheological properties of commercially pure aluminum were set as strain-hardening curves according to compression tests with cylindrical specimens on macro-level. Tests were conducted with a strain rate of 1 s^{-1} at 300 $°C$[1]. The materials of the matrix and the buffer layer have been considered to be isotropic plastically incompressible elasto-plastic medium. The silicon carbide particle material has been considered to be isotropic linear plastic. The elastic proprieties have been set as follows: Youngs modulus $E = 70$ GPa and the Poisson coefficient $\nu = 0.34$ for pure aluminum [20]; $E = 380$ GPa and $\nu = 0.19$ for silicon carbide [21]. The elastic proprieties of the buffer layer have been obtained by the rule of mixtures [22] using composite constituent volume fractions as follows: $E = 225$ GPa and $\nu = 0.265$.

The numerical simulation of metal matrix composite deformation has been conducted in the quasi-static statement with ANSYS finite element suite installed on a URAN GPU cluster of IMM UB RAS. Boundary conditions have been set in displacements of buffer layer facets in such a way that metal matrix composite could be considered to be 1/8 part of a body being subjected to uniaxial tension or compression along y-axis. Some facets have been placed on symmetry planes. The displacements have been set in such a way that equivalent macroscopic strain of tension or compression could reach $\varepsilon = 0.2$ on the final load step.

The simulation allowed us to obtain stress tensor σ_{ij} and strain increment $\Delta\varepsilon_{ij}$ tensor data in each node of the metal matrix composite computational model. The data have been further used to determine the stress stiffness coefficient k_n and the Lode-Nadai coefficient $\mu_{\sigma n}$ on each computation step:

$$k_n = \frac{\sigma}{T}, \tag{1}$$

where σ denotes mean normal (hydrostatic) stress and T denotes tangential stress intensity equal to shear yield stress in the plastic region.

$$\mu_{\sigma n} = 2\frac{\sigma_2 - \sigma_3}{\sigma_1 - \sigma_3} - 1, \tag{2}$$

[1] Experimental research was conducted in Collective centre "Plastometria" of IES UB RAS

where σ_1, σ_2, σ_3 denote principal stresses on n-th computational step.

The equivalent (von Mises) strain increment on n-th computation step has been expressed in terms of strain tensor component increments $\Delta\varepsilon_{ij}$ obtained on this step:

$$\Delta\varepsilon_n = \sqrt{\frac{2}{3}\Delta\varepsilon_{ij}\Delta\varepsilon_{ij}} \quad (3)$$

Thereafter, the whole accumulated equivalent deformation ε in every node is computed as follows:

$$\varepsilon = \sum_{n=1}^{N} \Delta\varepsilon_n, \quad (4)$$

where N denotes total computational step number for deformation.

In [13, 14] it is determined experimentally that in metal matrix composite specimen loading process, first cracks appear in the matrix. Considering this failure behavior, the matrix damage accumulation has been studied in scope of this work. The phenomenological theory authored by Kolmogorov V L [6, 7] has been used. This theory assumes that material damage ω lies in range $0 \dots 1$, where 0 means undeformed material and 1 implies material failure and crack emergence. The material damage on computational step of deformation is equal to the ratio of equivalent strain increment to equivalent plastic strain to fracture and damage accumulation considered to be linear. The fracture locus of commercially pure aluminum for 300 °C has been taken from the experimental investigation [23]. A fracture locus determine the dependence of ultimate shear strain Λ_f at fracture on stress state parameters k and μ_σ: $\Lambda_f = \Lambda_f(k, \mu_\sigma)$. Ultimate strain at fracture was calculated as follows: $\varepsilon_f = \Lambda_f/\sqrt{3}$

The damage model has been further used to compute damage and damage accumulation in every finite element node representing the metal matrix. A node has been considered to be destroyed if an accumulated damage in this node reached 1. Thus the condition of failure after N computational steps has the following notion in every finite element mesh node:

$$\omega = \sum_{n=1}^{N} \frac{\Delta\varepsilon_n}{\varepsilon_f(k_n, \mu_{\sigma n})} = 1 \quad (5)$$

For considered loading scenarios, damage distribution fields have been visualized for each computational step in whole volume of the metal matrix.

3. Result and Discussion

Finite element method simulations show that the stress-strain state in the matrix on microscale is significantly inhomogeneous on every computation step. Even on initial loading stages, the matrix material appears to have tensile stress concentration regions (characterized by $k > 0$), as well as substantial local plastic deformation regions. These peculiarities emerge in close proximity of reinforcement particles. This holds true both for tension and compression simulations. Volume fraction of such regions increases with equivalent macroscopic strain. For further information on the microvolume stress-strain state peculiarities and stress-strain state evolution during the deformation process, we refer to [18, 19].

It is known that severe tensile stresses contribute to intensive plastic dilatancy and accelerate fracture process [5, 6, 24]. This conclusion is also confirmed by numerical simulations of damage ω accumulation in the matrix metal within microvolume. It is found that the most possible regions of failure initiation (i.e. regions, where equation (5) holds true) are strain localization regions where adverse tensile stresses prevail. As an example, figures 3 and 4 depict accumulated damage distribution in central cross section xy of the metal matrix within the microvolume depending on equivalent macroscopic strain ε for uniaxial tension and compression.

IOP Conf. Series: Materials Science and Engineering **208** (2017) 012037 doi:10.1088/1757-899X/208/1/012037

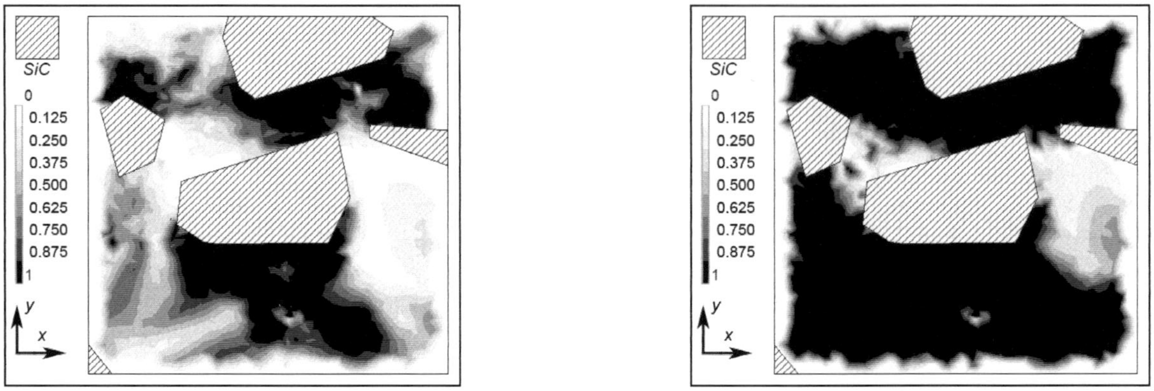

(a) (b)

Figure 3: Damage ω distribution in matrix. Central xy cross-section of metal matrix microvolume, tension simulation at equivalent macroscopic strain $\varepsilon = 0.04$ (a) and $\varepsilon = 0.2$ (b).

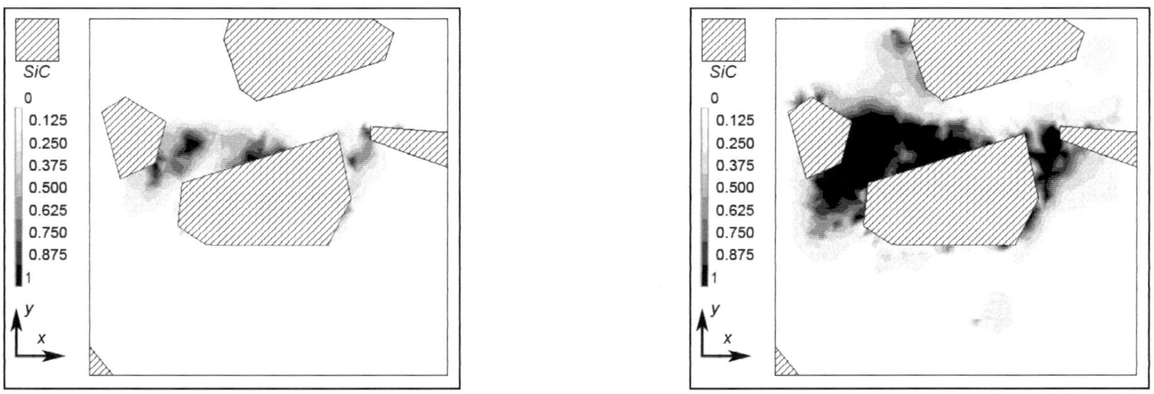

(a) (b)

Figure 4: Damage ω distribution in matrix. Central xy cross-section of metal matrix microvolume, compression simulation at equivalent macroscopic strain $\varepsilon = 0.04$ (a) and $\varepsilon = 0.2$ (b).

Obviously, a damaged region grows with increase of equivalent macroscopic strain. On the tension simulation step with equivalent macroscopic strain $\varepsilon = 0.04$ the failure criterion (5) holds true for order of 30% nodes, whereas the step with $\varepsilon = 0.2$ shows that more than 70% of nodes would experience failure. Compression loading scenario is less adverse in terms of internal damage. In case of compression for analogous equivalent macroscopic strain, the failure criterion (5) holds true for less than 3% and 10% of matrix nodes within the microvolume.

4. Conclusion

Adhering to the multilevel material description approach, the three dimensional computational model of metal matrix composite deformation has been developed. The model takes into account internal structure of the composite material, as well as the rheology of its constituents. The numerical simulations with randomly-chosen microvolume of Al/SiC composite have been conducted for uniaxial tension and compression.

Adhering to the phenomenological theory of damage, a simulation of damage accumulation in composite matrix has been conducted. The evolution of stress-strain state parameters (the stress stiffness coefficient and the Lode-Nadai coefficient) has been taken into account. It is found that for considered loading scenarios the most probable failure initiation regions are regions with plastic strain localization and adverse stress state with prevalence of tensile stresses. Initial cracks can emerge in these regions, spreading through the volume of the composite with deformation increase.

Acknowledgments
The authors gratefully acknowledge the help of A.S. Igumnov, the lead developer in N.N. Krasovskii Institute of Mathematics and Mechanics of the Ural Branch of the Russian Academy of Sciences (IMM UB RAS) for help with numerical simulations with ANSYS.

This work was partly supported by the Russian Scientific Foundation (project 14-19-01358) in the part of Al/SiC MMC numerical model development.

References

[1] Panin V E, Lihachev V A and Grinyev J V 1985 *Structural Levels of Deformation of Solids (Strukturnye urovni deformatsii tverdykh tel)* (Moskow: Nauka) (in Russian)
[2] Balokhonov R R, Romanova V A, Schmauder S and Martynov S A 2014 *Composites: Part B* **66** 276–286
[3] Smirnov S V, Myasnikova M V and Pugacheva N B 2016 *Int J of Damage Mech* **25**(2) 251–265 URL `http://dx.doi.org/10.1177/1056789515577401`
[4] Pyo S H and Lee H K 2009 *Intl J Solids Structures* **46**(17) 3138–3149
[5] Smirnov S V 2013 *Key Eng Mater* **528** 61–69 URL `https://www.scientific.net/KEM.528.61`
[6] Kolmogorov V L 2001 *Metal forming mechanics (Mechanika obrabotki metallov davleniem)* 2nd ed (Ekaterinburg: GOU VPO UGTU-UPI) (in Russian)
[7] Lemaitre J A and Lippmann H A 1996 *Course on Damage Mechanics* (Berlin: Springer-Verlag)
[8] Broek D 1984 *Elementary engineering fracture mechanics* (The Hague: Martinus Nijhoff Publishers)
[9] Balokhonov R R, Zinovyev A V, Romanova V A and Bakeev R A 2016 *Phys Mesomech* **19**(4) 430–440
[10] Pugacheva N B, Myasnikova M V and Michurov N S 2016 *Phys Met Metallogr+* **117**(2) 195–203 URL `http://dx.doi.org/10.1134/S0031918X15120078`
[11] Williams J J, Segurado J, LLorca J and Chawla N 2012 *Mater Sci Eng A.* **557** 113–118
[12] Romanova V A and Balokhonov R R 2012 *Comp Mater Sci* **61** 71–75
[13] Pugacheva N B, Michurov N S and Bykova T M 2015 *Diagnostics, Resource and Mechanics of materials and structures* (6) 6–18 URL `http://dx.doi.org/10.17804/2410-9908.2015.6.006-018`
[14] Pugacheva N B, Michurov N S and Bykova T M 2016 *Phys Met Metallogr+* **117**(6) 654–550 URL `http://dx.doi.org/10.1134/S0031918X16060119`
[15] Vildeman V E, Sokolkin Y V and Tashkinov A A 1997 *Mechanics of Elastic Deformation and Fracture of Composte Materials (Mekhanika neuprugogo deformirovanija i razrushenija kompozitsionnykh materialov)* (Moskow: Nauka) (in Russian)
[16] Haritos G K, Hager J W, Amos A K, Salkind M J and Wang A S D 1988 *Intl J Solids Structures* **24**(11) 1081–1096
[17] Khalevitsky Y V, Myasnikova M V and Konovalov A V 2014 *Mathematical modeling in natural science (Matematicheskoe modelirovanie v estestvennykh naukakh)* **1** 277–280 (in Russian)
[18] Smirnov S V, Konovalov A V, Myasnikova M V, Khalevitsky Y V, Smirnov A S and Igumnov A S 2016 A hierarchial modeling of stress-strain state of multiphase material subjected to uniaxial loading *AIP Conf Proc* vol 1785 URL `http://dx.doi.org/10.1063/1.4967123`
[19] Smirnov S V, Konovalov A V, Myasnikova M V, Khalevitsky Y V and Smirnov A S 2016 *PNRPU Mechanics Bulletin* **3** 175–187 URL `http://dx.doi.org/10.15593/perm.mech/2016.3.12`
[20] Davis J 1993 *Aluminum and Aluminum Alloys, ASM specialty handbook* (ASM International)
[21] Shackelford J and Alexander W 2000 *CRC Materials Science and Engineering Handbook* 3rd ed (CRC Press)
[22] Askeland D, Fulay P and Wright W 2010 *The Science and Engineering of Materials* (Cengage Learning)
[23] Smirnov S V, Vichuzhanin D I, Nesterenko A V and Igumnov A S 2016 A fracture locus for commercially pure aluminum at 300 °C *AIP Conf Proc* vol 1785 URL `http://dx.doi.org/10.1063/1.4967124`
[24] Bogatov A A, Mizhiritsky O I and Smirnov S V 1984 *Metal Plasticity Margin in Metal Forming (Resurs plastichnosti metallov pri obrabotke davleniem)* (Moskow: Metallurgiya) (in Russian)

IOP Conf. Series: Materials Science and Engineering **208** (2017) 012038 doi:10.1088/1757-899X/208/1/012038

Kinematic dynamo in a tetrahedron composed of helical Fourier modes

R. Stepanov[1] **and F. Plunian**[2]

[1]Institute of Continuous Media Mechanics, Korolyov 1, Perm, 614013, Russia

[2]Université Grenoble Alpes, Institut des Sciences de la Terre, 38058 Grenoble Cedex 9, France

E-mail: rodion@icmm.ru

Abstract. It is generally believed that helicity can play a significant role in turbulent systems, e.g. supporting the generation of large-scale magnetic fields, but its exact contribution is not clearly understood. For example there are well-known examples of large scale dynamos produced by a flow which is pointwise non-helical. In any case a break of mirror symmetry seems to be always at the heart of the dynamo mechanism. A fruitful framework to analyze such processes is the use of helical mode decomposition. In pure hydrodynamics such framework has proved its availability in study of the processes responsible for helicity cascades. It has also been used in the analysis of MHD helical mode interactions. The present work deals with the kinematic dynamo problem, solving the induction equation within the framework of helical Fourier modes decomposition. We show that the simplest modes configuration leading to an unstable solution has the form of a tetrahedron. Then the dynamo is produced by only two scales flow. We find necessary conditions for such dynamo action, not certainly related to flow helicity. The results help to understand generic dynamo flows like the one studied by G.O. Roberts (1972).

1. Introduction

It is generally believed that helicity can play a significant role in turbulent systems, e.g. supporting the generation of large-scale magnetic fields, but its exact contribution is not clearly understood. For example there are well-known examples of large scale dynamos produced by a flow which is pointwise non-helical but in any case a break of mirror symmetry seems to be always at the heart of the dynamo mechanism. Most of the dynamo studies are based on solutions of the mean field equations, direct numerical simulations or shell model of turbulence. Another framework to analyze such processes is the use of helical mode decomposition. In pure hydrodynamics such framework provide comprehensive studies the processes responsible for helicity cascades [1–4]. It has also been used in the analysis of MHD mode interactions [5].

The present work deals with the kinematic dynamo problem, solving the nondiffusive induction equation within the framework of helical Fourier modes decomposition. We look for the simplest modes configuration leading to an unstable solution with only two flow scales. We look for necessary conditions for the dynamo action in terms of helical mode amplitudes, not directly related to flow helicity. The results should help for better understanding generic dynamo flows like the one studied by G.O. Roberts [6].

2. Helical decomposition

The induction equation that governs the time evolution of the magnetic induction **b** for a given flow **u** is

$$\left(\partial_t - \eta\nabla^2\right)\mathbf{b}(\mathbf{x}) = \nabla\times\left(\mathbf{u}(\mathbf{x})\times\mathbf{b}(\mathbf{x})\right), \quad (1)$$

where η is a magnetic diffusivity. Assuming triply periodic boundary conditions in a cube of volume L^3, both fields, flow and magnetic induction, can be expanded into discrete Fourier series:

$$\mathbf{u}(\mathbf{x}) = \sum_{\mathbf{k}} \mathbf{u}(\mathbf{k}) \mathrm{e}^{\mathrm{i}\mathbf{k}\cdot\mathbf{x}}, \qquad \mathbf{b}(\mathbf{x}) = \sum_{\mathbf{k}} \mathbf{b}(\mathbf{k}) \mathrm{e}^{\mathrm{i}\mathbf{k}\cdot\mathbf{x}}, \tag{2}$$

where $\mathbf{u}(\mathbf{k})$ and $\mathbf{b}(\mathbf{k})$ are the complex Fourier coefficients. Then in Fourier space the induction equation takes the form

$$\left(\partial_t + \eta k^2\right) \mathbf{b}(\mathbf{k}) = \sum_{\substack{\mathbf{p},\mathbf{q} \\ \mathbf{k}+\mathbf{p}+\mathbf{q}=0}} \mathrm{i}\,\mathbf{k} \times \left(\mathbf{u}^*(\mathbf{p}) \times \mathbf{b}^*(\mathbf{q})\right). \tag{3}$$

Following the approach presented by [7] one can introduce, in Fourier space, a base of polarised helical waves $\mathbf{h}^{\pm}$ defined as the eigenvectors of the curl operator [8–10],

$$\mathbf{ik} \times \mathbf{h}^{\pm} = \pm \mathrm{k}\mathbf{h}^{\pm}. \tag{4}$$

Note that the helical vectors $\mathbf{h}^{\pm}(\mathbf{k})$ are defined up to an arbitrary rotation of axis $\mathbf{k}$. Taking

$$\mathbf{h}^{\pm}(\mathbf{k}) = \mathbf{u}_2(\mathbf{k}) \pm i\mathbf{u}_1(\mathbf{k}) \tag{5}$$

with $\mathbf{u}_1(\mathbf{k}) = (\mathbf{z}_\mathbf{k} \times \mathbf{k})/|(\mathbf{z}_\mathbf{k} \times \mathbf{k})|$ and $\mathbf{u}_2(\mathbf{k}) = \mathbf{u}_1(\mathbf{k}) \times \mathbf{k}/k$, where $\mathbf{z}_\mathbf{k}$ is an arbitrary vector that, in general, may depend on $\mathbf{k}$, though it is not proportional to $\mathbf{k}$. This approach has been extended to MHD with the following line of argument [11].

The Fourier modes of velocity and magnetic fields are expanded on that helical base

$$\mathbf{u}(\mathbf{k}) = u^+(\mathbf{k})\,\mathbf{h}^+(\mathbf{k}) + u^-(\mathbf{k})\,\mathbf{h}^-(\mathbf{k}), \tag{6}$$

$$\mathbf{b}(\mathbf{k}) = b^+(\mathbf{k})\,\mathbf{h}^+(\mathbf{k}) + b^-(\mathbf{k})\,\mathbf{h}^-(\mathbf{k}). \tag{7}$$

Replacing the expressions (6-7) for $\mathbf{u}$ and $\mathbf{b}$ in Eqs. (3), and projecting on to the helical base $\mathbf{h}^{s_k}(\mathbf{k})$ (with $s_k = \pm 1$) leads to the following equation

$$\left(d_t + \eta k^2\right) b^{s_k}(\mathbf{k}) = -s_k k \sum_{\Delta_{\mathbf{kpq}}} \sum_{s_p,s_q=\pm 1} g_{s_k,s_p,s_q}(\mathbf{k},\mathbf{p},\mathbf{q}) \left(u^{s_p}(\mathbf{p}) b^{s_q}(\mathbf{q}) - b^{s_p}(\mathbf{p}) u^{s_q}(\mathbf{q})\right)^*, \tag{8}$$

where $\Delta_{\mathbf{kpq}}$ means a sum over all possible triads ($\mathbf{k}+\mathbf{p}+\mathbf{q} = 0$) containing $\mathbf{k}$ and two arbitrary vectors $\mathbf{p}$ and $\mathbf{q}$, g is a function of $\mathbf{k}$, $\mathbf{p}$, $\mathbf{q}$, s_k,s_p,s_q is defined as

$$g_{s_k,s_p,s_q}(\mathbf{k},\mathbf{p},\mathbf{q}) \equiv -\frac{1}{\mathbf{h}^{s_k}(\mathbf{k})^* \cdot \mathbf{h}^{s_k}(\mathbf{k})} \left(\mathbf{h}^{s_k}(\mathbf{k})^* \times \mathbf{h}^{s_p}(\mathbf{p})^*\right) \cdot \mathbf{h}^{s_q}(\mathbf{q})^*. \tag{9}$$

Considering a single triadic interaction, it is not necessary to introduce an arbitrary unit vector $\mathbf{z}_\mathbf{k}$ to define the unit vectors $\mathbf{u}_1$ and $\mathbf{u}_2$. Indeed, there is a natural direction which is represented by the unit vector perpendicular to the plane of the triad:

$$\boldsymbol{\lambda} = (\mathbf{k} \times \mathbf{p})/|\mathbf{k} \times \mathbf{p}| = (\mathbf{p} \times \mathbf{q})/|\mathbf{p} \times \mathbf{q}| = (\mathbf{q} \times \mathbf{k})/|\mathbf{q} \times \mathbf{k}|. \tag{10}$$

A second unit vector $\boldsymbol{\mu}_\mathbf{k} = \mathbf{k} \times \boldsymbol{\lambda}/k$ is introduced and the helical vectors are then defined as

$$\mathbf{h}^{s_k}(\mathbf{k}) = \mathrm{e}^{i s_k \varphi_\mathbf{k}} \left(\boldsymbol{\lambda} + i\, s_k\, \boldsymbol{\mu}_\mathbf{k}\right). \tag{11}$$

The angle $\varphi_\mathbf{k}$ defines the rotation around $\mathbf{k}$ needed to transform the base $(\boldsymbol{\mu}_\mathbf{k}, \boldsymbol{\lambda})$ onto the base $(\mathbf{u}_1(\mathbf{k}), \mathbf{u}_2(\mathbf{k}))$. Since the base $(\boldsymbol{\mu}_\mathbf{k}, \boldsymbol{\lambda})$ depends on the triad, the angle $\varphi_\mathbf{k}$ is also a function of $(\mathbf{k}, \mathbf{p}, \mathbf{q})$. The coupling constant for this triad then simply reduces to

$$g_{s_k,s_p,s_q} = -\frac{1}{2} \mathrm{e}^{-i(s_k\varphi_k + s_p\varphi_p + s_q\varphi_q)}\, s_k\, s_p\, s_q\, \left(s_k \sin\alpha_k + s_p \sin\alpha_p + s_q \sin\alpha_q\right) \tag{12}$$

with

$$\sin\alpha_k = \frac{Q_{kpq}}{2\,p\,q} \qquad \sin\alpha_p = \frac{Q_{kpq}}{2\,k\,q} \qquad \sin\alpha_q = \frac{Q_{kpq}}{2\,k\,p}, \tag{13}$$

and $Q_{kpq} = \sqrt{2\,k^2\,p^2 + 2\,q^2\,p^2 + 2\,q^2\,k^2 - k^4 - q^4 - p^4}$.

3. Solution in one triad

For one triad $\Delta_{\mathbf{kpq}}$ where $\mathbf{k}$, $\mathbf{p}$ and $\mathbf{q}$ are fixed, and using the notation u^{s_k} for $u^{s_k}(\mathbf{k})$, $u_k^{\pm}$ for $u^{\pm}(\mathbf{k})$, and g_{s_k,s_p,s_q} for $g_{s_k,s_p,s_q}(\mathbf{k},\mathbf{p},\mathbf{q})$, the diffusionless induction equation is given by

$$\begin{aligned} d_t b^{s_k} &= -s_k k \sum_{t_p,t_q=\pm 1} g_{s_k,t_p,t_q}(u^{t_p}b^{t_q} - b^{t_p}u^{t_q})^* \\ d_t b^{s_p} &= -s_p p \sum_{t_q,t_k=\pm 1} g_{t_k,s_p,t_q}(u^{t_q}b^{t_k} - b^{t_q}u^{t_k})^* \\ d_t b^{s_q} &= -s_q q \sum_{t_k,t_p=\pm 1} g_{t_k,t_p,s_q}(u^{t_k}b^{t_p} - b^{t_k}u^{t_p})^*. \end{aligned} \tag{14}$$

which is equivalent to

$$\begin{aligned} d_t b_k^+ &= -k[g_{+,+,+}(u_p^+b_q^+ - b_p^+u_q^+)^* + g_{+,-,-}(u_p^-b_q^- - b_p^-u_q^-)^* \\ &\quad + g_{+,+,-}(u_p^+b_q^- - b_p^+u_q^-)^* + g_{+,-,+}(u_p^-b_q^+ - b_p^-u_q^+)^*] \\ d_t b_p^+ &= -p[g_{+,+,+}(u_q^+b_k^+ - b_q^+u_k^+)^* + g_{-,+,-}(u_q^-b_k^- - b_q^-u_k^-)^* \\ &\quad + g_{-,+,+}(u_q^+b_k^- - b_q^+u_k^-)^* + g_{+,+,-}(u_q^-b_k^+ - b_q^-u_k^+)^*] \\ d_t b_q^+ &= -q[g_{+,+,+}(u_k^+b_p^+ - b_k^+u_p^+)^* + g_{-,-,+}(u_k^-b_p^- - b_k^-u_p^-)^* \\ &\quad + g_{+,-,+}(u_k^+b_p^- - b_k^+u_p^-)^* + g_{-,+,+}(u_k^-b_p^+ - b_k^-u_p^+)^*] \\ d_t b_k^- &= +k[g_{-,+,+}(u_p^+b_q^+ - b_p^+u_q^+)^* + g_{-,-,-}(u_p^-b_q^- - b_p^-u_q^-)^* \\ &\quad + g_{-,+,-}(u_p^+b_q^- - b_p^+u_q^-)^* + g_{-,-,+}(u_p^-b_q^+ - b_p^-u_q^+)^*] \\ d_t b_p^- &= +p[g_{+,-,+}(u_q^+b_k^+ - b_q^+u_k^+)^* + g_{-,-,-}(u_q^-b_k^- - b_q^-u_k^-)^* \\ &\quad + g_{-,-,+}(u_q^+b_k^- - b_q^+u_k^-)^* + g_{+,-,-}(u_q^-b_k^+ - b_q^-u_k^+)^*] \\ d_t b_q^- &= +q[g_{+,+,-}(u_k^+b_p^+ - b_k^+u_p^+)^* + g_{-,-,-}(u_k^-b_p^- - b_k^-u_p^-)^* \\ &\quad + g_{+,-,-}(u_k^+b_p^- - b_k^+u_p^-)^* + g_{-,+,-}(u_k^-b_p^+ - b_k^-u_p^+)^*]. \end{aligned} \tag{15}$$

We look for a growing solution of the linear system of ODEs (15) in the form

$$b_k^{s_k} = \check{b}_k^{s_k} e^{\gamma t}, \quad \gamma = \pm\sqrt{a}. \tag{16}$$

Therefore a necessary condition for having a positive growthrate γ is to have a positive.

3.1. Single helical triad

In this case we assume $b^{-s_k} = b^{-s_p} = b^{-s_q} = u^{-s_k} = u^{-s_p} = u^{-s_q} = 0$. Then we solve an eigenvalue problem with three variables

$$\begin{aligned} d_t b^{s_k} &= -s_k k g_{s_k,s_p,s_q}(u^{s_p}b^{s_q} - b^{s_p}u^{s_q})^* \\ d_t b^{s_p} &= -s_p p g_{s_k,s_p,s_q}(u^{s_q}b^{s_k} - b^{s_q}u^{s_k})^* \\ d_t b^{s_q} &= -s_q q g_{s_k,s_p,s_q}(u^{s_k}b^{s_p} - b^{s_k}u^{s_p})^*. \end{aligned} \tag{17}$$

It leads to non trivial solutions

$$a = -s_k s_p s_q kpq |g_{s_k,s_p,s_q}|^2 F \tag{18}$$

where F is given by

$$F = s_k \frac{|u^{s_k}|^2}{k} + s_p \frac{|u^{s_p}|^2}{p} + s_q \frac{|u^{s_q}|^2}{q}. \tag{19}$$

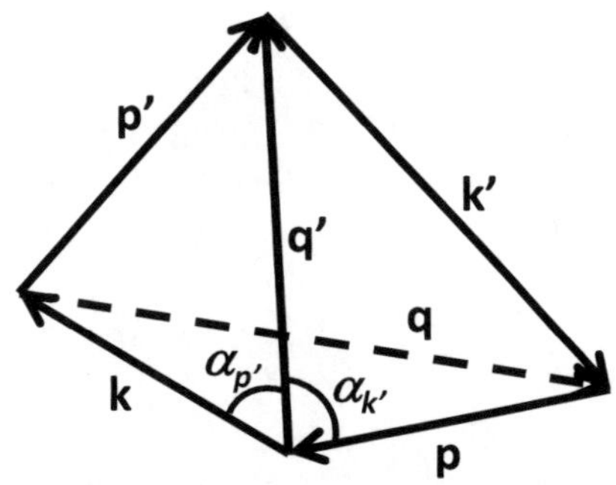

Figure 1. Tetrahedron configuration of interacting four triads $(\mathbf{k},\mathbf{p},\mathbf{q})$, $(\mathbf{k},\mathbf{p}',\mathbf{q}')$, $(\mathbf{p},\mathbf{q}',\mathbf{k}')$ and $(\mathbf{q},\mathbf{k}',\mathbf{p}')$.

Defining a potential vector $\mathbf{w}$ such that $\mathbf{u} = \nabla \times \mathbf{w}$, we have $\mathbf{w}(\mathbf{k}) = k^{-1}\left(u^{+}(\mathbf{k})\,\mathbf{h}^{+}(\mathbf{k}) - u^{-}(\mathbf{k})\,\mathbf{h}^{-}(\mathbf{k})\right)$, one can express a potential helicity in the form $F(\mathbf{k}) = \mathbf{u}\cdot\mathbf{w} = k^{-1}\left(|u^{+}(\mathbf{k})|^2 - |u^{-}(\mathbf{k})|^2\right)$. Therefore F above is the potential helicity for the whole triad $\Delta_{\mathbf{kpq}}$. We find that if $s_k = s_p = s_q$ then a is negative, suggesting that within a triad of only one type of helical modes it cannot lead to dynamo action. Taking $u^{s_k} = 0$ and $s_k = -s_p = -s_q$ leads to positive a. Then there is always a magnetic field scale k with helicity opposite to u such that dynamo action occurs. The importance of such potential helicity has already been put forward in the context of large scale dynamos [12] with indeed opposite helical modes between the large scale magnetic field and the flow.

3.2. Full triad

Considering all possible kinetic and magnetic helical modes within one triad $b_k^{\pm}$, $u_k^{\pm}$, $b_p^{\pm}$, $u_p^{\pm}$, $b_q^{\pm}$ and $u_q^{\pm}$ we find

$$a = -\frac{Q_{kpq}^2}{4}\left(\left|\frac{u_k^+ - u_k^-}{k}\right|^2 + \left|\frac{u_p^+ - u_p^-}{p}\right|^2 + \left|\frac{u_q^+ - u_q^-}{q}\right|^2\right), \tag{20}$$

provided we choose $\mathbf{z_k} = \mathbf{z_p} = \mathbf{z_q} = \boldsymbol{\lambda}$. Contrary to the previous cases a is always negative. It shows that one triad alone cannot lead to dynamo action (unless some helical magnetic modes are suppressed as previously seen). This result is consistent with the antidynamo theorem that says that no dynamo effect is possible for a magnetic field which depends only on two coordinates. Indeed for any triad $\Delta_{\mathbf{kpq}}$ there is always a rotational transformation which makes the problem two dimensional. So it is necessary to involve an additional magnetic field mode which does not belong to the plane $\Delta_{\mathbf{kpq}}$.

4. Solution in one tetrahedron

In addition to vectors $\mathbf{k}$, $\mathbf{p}$, $\mathbf{q}$ we introduce an arbitrary wave number vector $\mathbf{q}'$. Then the tetrahedron is formed by four triads: $(\mathbf{k},\mathbf{p},\mathbf{q})$, $(\mathbf{k},\mathbf{p}',\mathbf{q}')$, $(\mathbf{p},\mathbf{q}',\mathbf{k}')$ and $(\mathbf{q},\mathbf{k}',\mathbf{p}')$ as shown in figure 1. Vectors $\mathbf{k}'$ and $\mathbf{p}'$ have to be included for consistency.

We assume that the flow is given by the two vectors u_k and u_p only ($u_q^{\pm} = u_{k'}^{\pm} = u_{p'}^{\pm} = u_{q'}^{\pm} = 0$) and we solve the eigenvalue problem with all twelve magnetic field unknowns. We find a solution in the form

$$\begin{aligned} a = &-\frac{Q_{kp'q'}^2}{4}\left|\frac{u_k^+ - u_k^-}{k}\right|^2 - \frac{Q_{pq'k'}^2}{4}\left|\frac{u_p^+ - u_p^-}{p}\right|^2 \\ &\pm \frac{Q_{kp'q'}Q_{pq'k'}}{2}\frac{|\sin\psi_{q'}|}{q'}\sqrt{\left(\frac{|u_k^+|^2 - |u_k^-|^2}{k}\right)\left(\frac{|u_p^+|^2 - |u_p^-|^2}{p}\right)} \end{aligned} \tag{21}$$

where $\psi_{q'}$ is the angle between the planes defined by the triads $(\mathbf{k}, \mathbf{p}', \mathbf{q}')$ and $(\mathbf{p}, \mathbf{q}', \mathbf{k}')$. If the two triads are coplanar, corresponding to $\sin \psi_{q'} = 0$, then a is always negative, ruling out any possibility for dynamo action. If the flow is non helical ($u_k^+ = u_k^-$ and $u_p^+ = u_p^-$) then $a = 0$. Now if the two triads are non coplanar and the flow is helical then dynamo becomes possible. Oscillatory solutions are always possible provided $H(k) \cdot H(p) < 0$. If $H(k) \cdot H(p) > 0$ then only non local dynamo action is possible ($q' \ll k$ and $q' \ll p$).

5. Conclusion remarks

From solution (21) we can learn several things that are in favour of dynamo and provide highest growth rate: flow with maximum helicity, interaction of large magnetic field and small scale velocity field, orthogonal configuration of $\mathbf{u}(\mathbf{k})$, $\mathbf{u}(\mathbf{p})$ and $\mathbf{b}(\mathbf{q}')$. These are known since G.O. Roberts dynamo example [6]. He considered the flow $\mathbf{u} = (\sin y, \sin x, \cos x - \cos y)$ and found asymptotic unstable solution for $\mathbf{b}(x, y, z) = \mathbf{b}_0 e^{\mathrm{i}k_z z}$ for $k_z < 1$. If we consider this solution in terms of helical modes then $\mathbf{k} = (1, 0, 0)$, $\mathbf{p} = (0, 1, 0)$ and $\mathbf{q}' = (0, 0, k_z)$. Also $u_k^+ = 1/2$, $u_k^- = 0$ and $u_p^+ = -1/2$, $u_p^- = 0$. Extracted highlights form (21) are in full agreement. We show that the simplest modes configuration leading to an unstable solution has the form of a tetrahedron. Here the dynamo is produced by only two scales flow. We find necessary conditions for the dynamo action which is expressed by helicities of the individual helical modes and not directly related to flow helicity. For the case of one triad with single helical modes (section 3.1) we have found the growth rate depending on potential helicity [12]. Our general approach is similar the study of low-dimensional magnetohydrodynamic models containing three velocity and three magnetic modes [13]. These low-dimensional models exhibit a dynamo transition at a critical forcing amplitude that depends on the Prandtl number. In comparison we took an advantage of helical mode decomposition that prevents break of conservation laws automatically.

Acknowledgments

This work was supported by joint project of the Russian Science Foundation (grand RSF-16-41-02012) and the Department of Science Technology of the Ministry of Science and Technology of the Republic of India (grant INT/RUS/RSF/3).

References

[1] Biferale L, Musacchio S and Toschi F 2012 *Phys. Rev. Lett.* **108** 164501 (*Preprint* 1111.1412)
[2] Stepanov R, Golbraikh E, Frick P and Shestakov A 2015 *Phys. Rev. Lett.* **115** 234501 (*Preprint* 1508.07236)
[3] Kessar M, Plunian F, Stepanov R and Balarac G 2015 *Phys. Rev. E* **92**(3) 031004
[4] Alexakis A 2016 *ArXiv e-prints 1606.02540* (*Preprint* 1606.02540)
[5] Linkmann M, Berera A, McKay M and Jäger J 2016 *J. Fluid Mech.* **791** 61–96 (*Preprint* 1508.05528)
[6] Roberts G O 1972 *Philosoph. Transac. Royal Soc. London Ser. A* **271** 411–454
[7] Waleffe F 1992 *Phys. Fluids A: Fluid Dyn.* **4** 350–363
[8] Craya A 1958 *P.S.T. Ministère de l'Air (Paris)* **345**
[9] Herring J R 1974 *Phys. Fluids* **17** 859–872
[10] Cambon C and Jacquin J 1989 *J. Fluid Mech.* **202** 295–317
[11] Lessinnes T, Plunian F and Carati D 2009 *Theor. Comp. Fluid Dyn.* **23** 439–450
[12] Rädler K H and Brandenburg A 2008 *Phys. Rev. E* **77** 026405 (*Preprint* 0801.0602)
[13] Verma M K, Lessinnes T, Carati D, Sarris I, Kumar K and Singh M 2008 *Phys. Rev. E* **78**(3) 036409

Heat transfer in an infinite layer with fractal distribution of heating elements

V Titov[1] **and R Stepanov**[1,2]

[1] Institute of Continuous Media Mechanics UB RAS, 1, Acad. Korolev St., Perm, 614013, Russia

[2] Perm National Research Polytechnic University, 29, Komsomolsky ave., Perm, 614990, Russia

E-mail: titov.v@icmm.ru

Abstract. The effect of inhomogeneous temperature distribution at boundary on convection in an infinite horizontal plane layer is investigated. Direct numerical simulation of the compressible non-isothermal flow in a cubic cell with rigid horizontal walls is performed under periodic vertical boundary conditions. Laminar and weakly non-linear flow regimes are determined. The heated area on the cell bottom obeys regular or fractal distributions. The intensities of heat flux through the layer are compared for different heterogeneous distributions of heating elements at a specific temperature gradient in the case when the area of heated surface remains constant. Fractal geometry shows the appearance of the multiscale structure of the flow and the enhancement of heat transfer.

1. Introduction

At present, energy saving and increasing energy efficiency in various sectors of the economy is a priority of science, engineering and technology. The increase in energy efficiency is tightly linked to the improvement of the characteristics of heat exchangers. The complexity of a heat conduction problem is dependent upon the properties of interacting media and contact surface conditions. In fluid and gaseous media, two mechanisms are responsible for heat transfer: diffusion and convection. The value of diffusion heat flux is defined by the thermal conductive properties of the medium. Convective mechanism is determined by the pattern of medium motion. This gives researchers an opportunity to manage heat transfer so that its efficiency increases. The intensification of the vortex structure is one of the approaches used to increase the efficiency of heat transfer. This can be achieved by changing a surface heater shape and a heat-transfer agent [1, 2]. For the flat horizontal heated surface, inhomogeneous heating can be used. Therefore, even in the case of stable stratification, a vortex flow will arise and an increase in convective heat transfer will take place [3].

Large-scale circulation makes the greatest contribution to advective heat transfer throughout the whole region. However, in a thin boundary layer, the effective transfer can be ensured only through high turbulence. Turbulence implies the presence of a multiscale flow that arises as a result of instability at large Reynolds numbers [4]. A significant influence on the occurrence of such a flow can be provided by the inhomogeneity of temperature on the heated surface [5]. It is possible to imagine a heat-conducting surface consisting of heat-conducting and heat-insulating areas. Depending on the configuration (geometry) of the distribution of these regions, different convective flows over the surface and in the entire volume will form. Typically, a laminar flow over an inhomogeneously heated surface loses its stability and leads to the appearance of secondary structures [6]. These structures can play an important role and are therefore studied in detail experimentally and theoretically [7]. For the given temperature gradient and the area of the heat-conducting region, one can consider the problem of finding the optimal geometry for the formation of a developed turbulent flow, which in combination with large-scale circulation will provide the maximum heat flow. In this sense, fractal geometry may be of particular interest. There are also applied studies aimed at improving the efficiency of heat transfer, and the use of a fractal is one of the simplest and cheapest examples of their implementation [8].

The aim of this work is to numerically study the possibility of increasing the efficiency of heat transfer due to the inhomogeneity of the thermal conductive properties of the heater. We are focused on investigating the fractal distribution of heating elements. Depending on the dimension of a fractal, it is possible to obtain different energy distribution of pulsations across the spectrum, which in turn will enhance cascade processes and turbulent heat transfer. We consider several variants of the geometry of the heated region with a regular distribution of identical elements and a distribution that repeats the fractal - the "Sierpinsky carpet".

2. Mathematical model

Convection of the compressible non-isothermal viscous flow is described by the following equations:

$$\begin{aligned} \frac{D\ln\rho}{Dt} &= -\nabla\cdot\mathbf{U} \\ \frac{D\mathbf{U}}{Dt} &= -c_V(\gamma-1)(T\nabla\ln\rho+\nabla T)+\mathbf{g}+2\nu(\mathbf{S}\cdot\nabla\ln\rho+\nabla\cdot\mathbf{S}) \\ \frac{DT}{Dt} &= -T(\gamma-1)\nabla\cdot\mathbf{U}+\frac{K}{\rho c_V}\Delta T+\frac{2\nu}{c_V}\mathbf{S}^2 \end{aligned} \tag{1}$$

where $D/Dt = \partial/\partial t + \mathbf{U}\cdot\nabla$ is the total advective derivative, ν is the kinematic viscosity, K is the heat conductivity, ρ is the density, $\mathbf{U}$ is the velocity, $\mathbf{g}$ is the gravitational acceleration, $\gamma = c_P/c_V$ is the ratio of specific heats at constant pressure and volume, respectively. The traceless rate-of-strain tensor $\mathbf{S}$ is given by

$$S_{ij} = \frac{1}{2}\left(\frac{\partial U_i}{\partial x_j}+\frac{\partial U_j}{\partial x_i}\right)-\frac{1}{3}\delta_{ij}\nabla\cdot\mathbf{U}. \tag{2}$$

Application of Boussinesq equations is a common approach to describing turbulent flows. The system of equations (1) is more complex, but allows us to investigate convection both in liquids and gases with relatively large variations in density caused by a change of temperature.

To introduce non-dimensional variables, the following characteristic values are used: velocity U_0, layer height L, temperature of cold surface T_0 (at the top), time L/U_0, density ρ_0, and sound speed $a_0^2 = \gamma c_V T(\gamma-1)$. Then the system (1) takes the form:

$$\begin{aligned} \frac{D\ln\rho}{Dt} &= -\nabla\cdot\mathbf{U} \\ \frac{D\mathbf{U}}{Dt} &= -\left(\mathrm{M}^2\gamma\right)^{-1}(T\nabla\ln\rho+\nabla T)-\mathrm{Fr}^{-2}\mathbf{e}_z+2\mathrm{Re}^{-1}(\mathbf{S}\cdot\nabla\ln\rho+\nabla\cdot\mathbf{S}) \\ \frac{DT}{Dt} &= -T(\gamma-1)\nabla\cdot\mathbf{U}+\gamma\left(\mathrm{RePr}\right)^{-1}\rho^{-1}\Delta T+2\mathrm{M}^2\gamma(\gamma-1)\mathrm{Re}^{-1}\mathbf{S}^2, \end{aligned} \tag{3}$$

where Mach number $\mathrm{M} = U_0/a_0$, Froude number $\mathrm{Fr} = U_0/\sqrt{gL}$, Reynolds number $\mathrm{Re} = LU_0/\nu$, Prandtl number $\mathrm{Pr} = \nu/\chi$. The choice of these control parameters allows us to control a

IOP Conf. Series: Materials Science and Engineering **208** (2017) 012039 doi:10.1088/1757-899X/208/1/012039

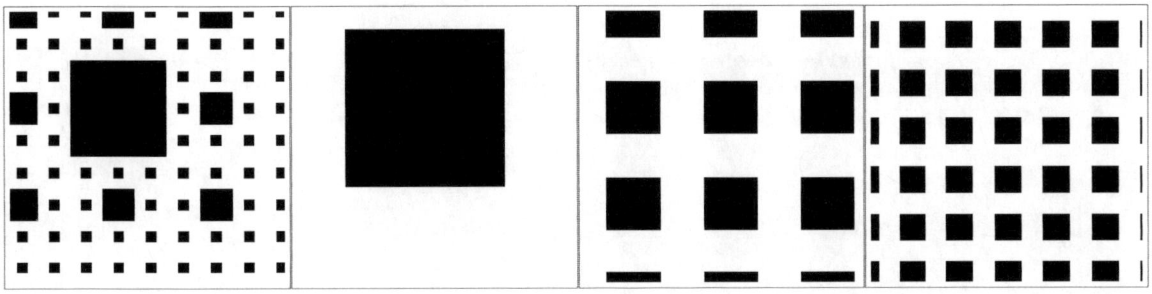

Figure 1. Four configurations of heat-conducting (black) and thermoisolated (white) areas: (from left to right) 3 levels of Sierpinsky carpet ($D = f$), one large square ($D = 1$), 9 middle size squares ($D = 9$), 36 small squares ($D = 36$).

transition from compressible (gas) to incompressible (fluid) medium by changing M, as well as a transition from laminar to turbulent flow by changing Re.

At the initial time, the medium is assumed to be stationary and homogeneous in temperature, so that $U_i = 0$, $T = 1$. The hydrostatic conditions allow us to formulate the initial density distribution $\ln \rho(x, y, z) = -\mathrm{Fr}^{-2}\gamma \mathrm{M}^2 z$. On horizontal rigid boundaries, the no-slip condition is satisfied, i.e. $U_i = 0$. At the upper boundary $T = 1$, and at the lower boundary $T = \theta$ in the heat-conducting region of the boundary and $\partial_z T = 0$ in the thermoisolated part, where θ is the relative temperature of a heater. The boundary condition for density $\partial_z \ln \rho = T^{-1}\left(\mathrm{Fr}^{-2}\gamma \mathrm{M}^2 + \partial_z T\right)$ comes from hydrostatic balance. Periodic boundary conditions are applied to the vertical boundaries of the studied domain.

The flow and heat transfer are calculated for 4 different heater configurations (see figure 1). The first heater (on the left) has a fractal configuration, which we are mostly interested in. For comparison, we consider three configurations with a regular distribution different in number and size of elements. The heaters have been selected in a such way that the total area of all the elements is equal for each configuration type. We introduce the parameter D to denote: $D = f$ - fractal heater, $D = 1$ - 1 solid square heater, $D = 9$ - 9 squares heater, $D = 36$ - 36 squares heater.

Equations (3) are numerically integrated using the sixth-order central differences for spatial derivatives and advance in time using a fifth-order Runge-Kutta-Felberga explicit scheme with adaptive time step. We control the time step so that the relative error doesn't exceed 10^{-4}. The code is parallelized for MPI clusters. Simulations are performed using 144 nodes. We choose the resolution of 216^3 meshpoints for more accurate discretization of the fractal distribution.

3. Results

In our simulations we used the following dimensionless parameters: $\theta = 3$, $\mathrm{M} = 0.1$, $\mathrm{Fr} = 0.088$, $\mathrm{Re} = 180$, $\gamma = 5/3$, $\mathrm{Pr} = 1$ and a variable parameter D. This approach is fully justified for implementing the basic idea – study of the characteristic structure of the flow and its effect on the heat transfer with different forms of heaters, which have an equal total surface area of heat exchange.

The main attention has focused on studying the four characteristics of the numerical solutions obtained in this work: the heat flux through the layer Q, the root-mean-square value of the velocity over the whole volume u_{rms}, the spectrum of kinetic energy in the horizontal plane, and the intensity of large-scale circulation.

Figure 2 shows $u_{\mathrm{rms}}(t)$ and $Q(t)$ for different values of D. For u_{rms} , only $D = f$ exhibits quasi-periodic nature. Also, $D = f$ demonstrates an essential increase of Q, compared to $D = 1$,

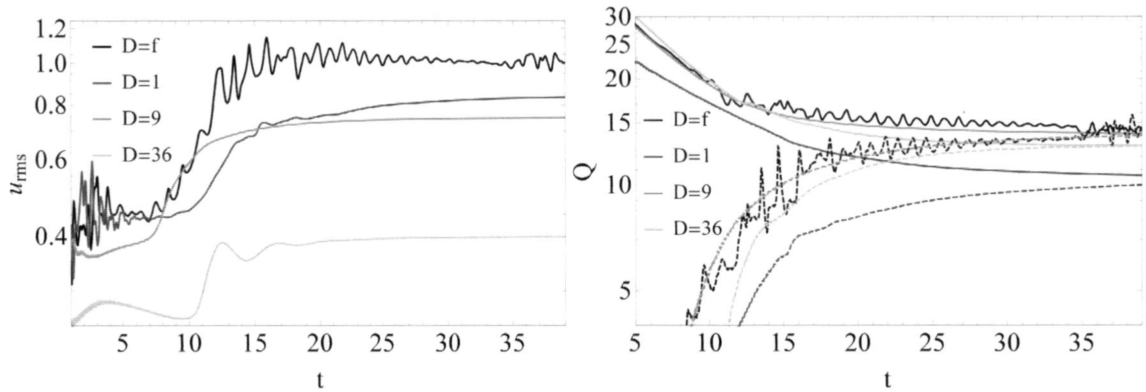

Figure 2. Evolution of u_{rms} (left) and $Q(t)$ (right) through upper (dashed line) and lower (solid line) boundaries for 4 values of D.

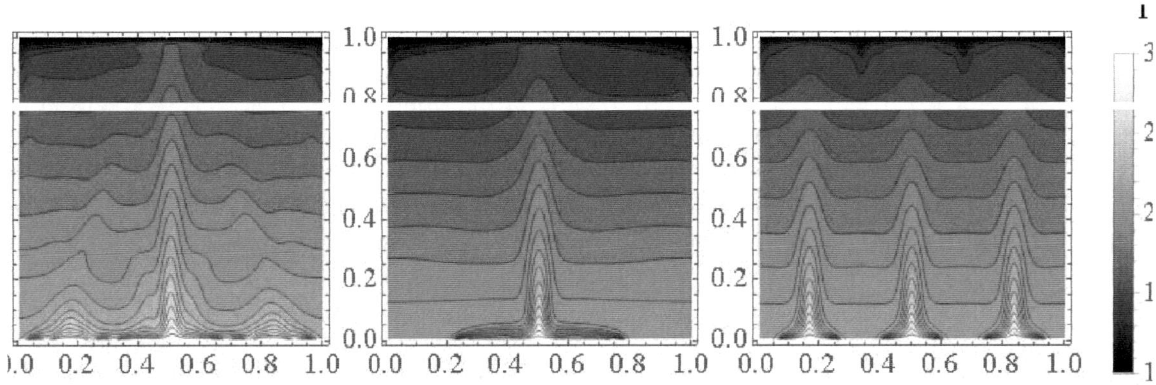

Figure 3. Vertical temperature distribution at $y = 0.5L$ for $D = f$, $D = 1$ and $D = 9$ (from left to right).

of about 50%. The same level of Q is produced by $D = 9$. Increasing number of squares to $D = 36$ gives the unexpected value of Q, which becomes lower by 10%. $D = f$ with the quasi-periodic behavior of u_{rms} and Q demonstrates the weak nonlinear nature of convective flow.

Temperature distributions are given in figure 3, where one can see a more complicated structure of these distributions when $D = f$ in comparison with the linear gradient. The more the number of squares used in configuration, the weaker the ascending fluid flow above the heaters. The vertical and horizontal slices of the velocity field are displayed in figure 4 at $y = 0.5L$ and $z = 0.5L$. Velocity fields demonstrate a pattern of the multiscale vortexes for $D = f$. The vertical distribution of velocity for $D = f$ suggests that the entire amount of the fluid at the cell bottom is included in the ascending stream, and the horizontal distribution of velocity for the same D represents at least five vortexes, of which the largest has a size of about $0.4L$. Velocity distributions for other values of D show similar large-scale circulations without any, even small, vortex. The spectra of total energy for each type of heater are shown in figure 5. The most developed spectrum is presented for $D = f$. For $D = 9$ and $D = 36$, we have obtained discrete spectra, where only a few scales are non zero.

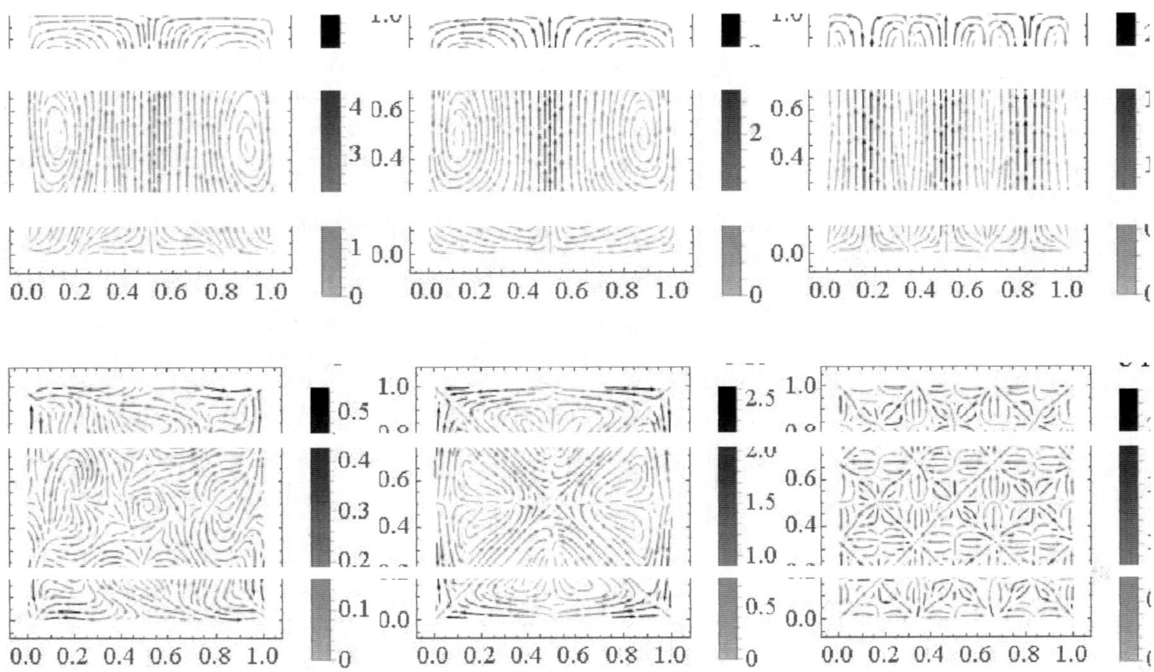

Figure 4. Vertical slice of the flow structure at $y = 0.5L$ (upper row) and its horizontal slice at $z = 0.5L$ (lower row) for $D = f$, $D = 1$ and $D = 9$ (from left to right).

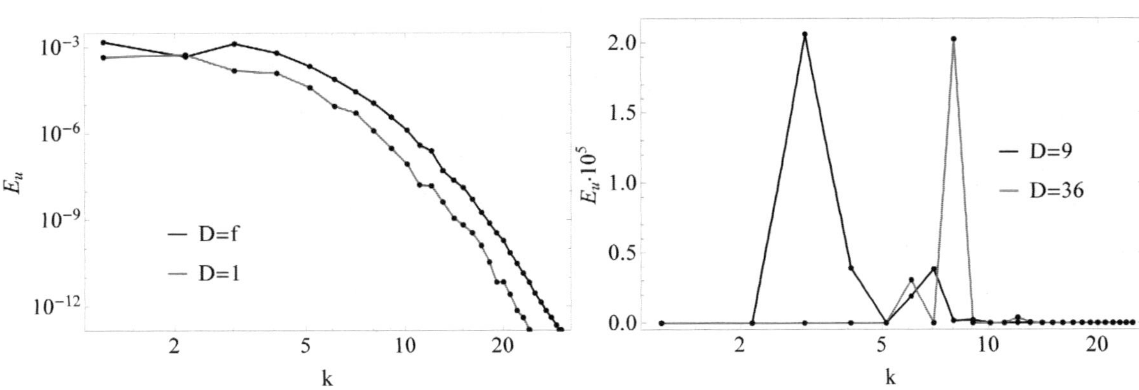

Figure 5. Kinetic energy spectrum: (left) for values $D = f$ and $D = 1$, in log-log scale, (right) $D = 9$ and $D = 36$, in log-linear scale. Energy of each scale for $D = 36$ is multiplied by a factor of 300 for comparison purposes.

4. Conclusions

We have investigated convection in an infinite plane layer over the inhomogeneous heating elements arranged in regular and fractal manner. Direct numerical simulations were performed for the compressible non-isothermal flow in the cubic cell with rigid horizontal walls under periodic vertical boundary conditions. Laminar and weak nonlinear flow regimes were considered at a given Reynolds number Re = 180. The comparative analysis of different heterogeneous heater distributions was carried out for the specific temperature gradient (heating at the bottom) in the case when the area of heated surface remains constant.

It is shown that this fractal heterogeneity, in comparison with all other cases of homogeneous heaters, provides a higher heat flux level throughout the entire layer, a more developed spatial (in horizontal plane) oscillation spectrum and the quasistochastic nature of the flow over the course of time. The results obtained at this stage are fundamental and do not pretend for practical application. We plan to investigate cascade process intensification and turbulent heat transfer when using a fractal heater thorough both numerical simulations and laboratory experimental studies.

5. Acknowledgements

Financial support from the Russian Foundation for Basic Research (project 16-41-590406-ural) is gratefully acknowledged. Computing resources of the supercomputers URAN and TRITON which were provided by the Institute of Mathematics and Mechanics UrB RAS and the Institute of Continuous Media Mechanics UrB RAS, respectively.

References

[1] Jensen M K, Bergles A E and Shome B 1997 *J. Enhan. Heat Trans.* **4** 1–6 ISSN 1065-5131
[2] Rainieri S, Bozzoli F and Cattani L 2014 *Journal of Physics: Conference Series* **547** 012002
[3] Verzicco R, Bakhuis D, Ostilla-Monico R, van der Poel E and Lohse D 2014 Inhomogeneous temperature boundary conditions in Rayleigh-Bénarrd convection *APS Meet. Abstr.* 16.004
[4] Kolesnichenko I V, Mamykin A D, Pavlinov A M, Pakholkov V V, Rogozhkin S A, Frick P G, Khalilov R I and Shepelev S F 2015 *Thermal Engin.* **62** 414–422
[5] Vasil'ev A Y, Kolesnichenko I V, Mamykin A D, Frick P G, Khalilov R I, Rogozhkin S A and Pakholkov V V 2015 *J. Tech. Phys.* **60** 1305–1309
[6] Sukhanovsky A, Batalov V, Teymurazov A and Frick P 2012 *Europ. Phys. J. B* **85** 9
[7] Sukhanovskii A, Evgrafova A and Popova E 2016 *Physica D Nonlin. Phen.* **316** 23–33
[8] Poltorak A 2016 Fractal heat transfer device US Patent US9228785 B2 URL `https://www.google.com/patents/US9228785`

IOP Conf. Series: Materials Science and Engineering **208** (2017) 012040 doi:10.1088/1757-899X/208/1/012040

Instability of cyclonic convective vortex

A Sukhanovskii[1,a], A Evgrafova[1,2] and E Popova[1]

[1]Institute of Continuous Media Mechanics UB RAS, 1, Acad. Korolev St., Perm, 614013, Russia

[2]Perm State University, 15, Bukireva St., Perm, 614990, Russia

E-mail: [a]san@icmm.ru

Abstract. Localized heating in the rotating layer of fluid leads to the formation of intensive cyclonic vortex. Cyclonic vortex becomes unstable at low values of viscosity and fast rotation of the experimental model. The instability of the vortex is tightly connected with a structure of the radial inflow. For moderate values of rotational Reynolds number *Re* the radial flows consist of several branches which transport angular momentum to the center of the model. When *Re* exceeds critical value (about 23) radial inflow changes its structure and appears as one wide branch which does not reach the center. As a result of strong anisotropy of radial inflow the cyclonic vortex is formed at some distance from the center. Further increase of *Re* leads to chaotic state with several vortices which appears at different locations near the periphery of the heating area. The map of regimes with stable and unstable vortices is presented.

1. Introduction

The present study is focused on the stability of laboratory vortex over localized source of heat in a rotating layer. There are only few laboratory realizations of cyclonic vortices driven by localized heating in rotating layers. The main motivation of these studies is a problem of formation of large-scale vortices in atmosphere. Laboratory hurricane model with an analog of latent heat release was proposed in [1] where resulting vortical flow was similar to the hurricane eye wall. Convection from isolated source at the bottom for low values of Rossby number was experimentally studied in [2,3]. These experiments were mainly directed to the study of small eddies formation in a buoyancy source region, their properties and evolution. Only short remark was made about formation of the single cyclonic vortex in the case of high values of convective Rossby number (more than 1). A qualitative experimental study of convective flow driven by a finite-sized circular heating plate at the bottom of a horizontal fluid layer, both with and without background rotation, was carried out in [4]. The studies [2-4] showed that in the case of water, for relatively low value of Prandtl number ($Pr \approx 7$) cyclonic motion in the lower layer exists only for slow rotation of the vessel.

Succesful laboratory model of hurricane-like vortex was proposed in [5,6]. The important differences of this experimental approach were more viscous fluids (in comparison with water) and using of a shallow layer. Later, series of the experiments [7] were done for the same configuration using PIV system for velocity measurements. The main focus of [7] was on integral characteristics of the azimuthal flows such as angular momentum and kinetic energy. Detailed study of different constraints of the steady-state hurricane-like vortex was carried out in [8]. PIV measurements showed that general structure of laboratory vortex is similar to the observational data [9].The three main dimensional parameters that define the vortex structure

IOP Conf. Series: Materials Science and Engineering **208** (2017) 012040 doi:10.1088/1757-899X/208/1/012040

for a fixed geometry – heating flux, rotation rate and viscosity were varied independently. It was found that viscosity is one of the main parameters that define the steady-state vortex structure. Increasing of kinematic viscosity may substantially suppress the cyclonic motion for fixed values of buoyancy flux and rotation rate. Strong competition between buoyancy and rotation provides the optimal ratio of the heating flux and rotation rate for achieving cyclonic vortex of maximal intensity.

The transitional stage when cyclonic vortex loses its stability and a system of smaller vortices appears instead was not studied before. The nature and description of this instability is our main goal. The paper is organized as follows. In section 2 we describe the experimental setup and measurement technique. Experimental results are presented in section 3 and conclusions are given in section 4.

2. Experimental setup

Experimental model is a cylindrical vessel of diameter $D = 300$ mm, and height $H = 40$ mm (figure 1(a)). The sides and bottom were made of plexiglass with a thickness 3 mm and 20 mm respectively. There was no cover or additional heat insulation at the sidewalls. The heater is a brass cylindrical plate mounted flush with the bottom. The diameter of the plate d is 104 mm, and its thickness is 10 mm. The brass plate is heated by an electrical coil placed on the lower side of the disc. Massive heater provides uniform heating which is optimal for vortex excitation. Cylindrical vessel was placed on a rotating horizontal table. Silicon oils with different values of kinematic viscosity, PMS-20, PMS-10 and PMS-5 (20, 10 and 5 cSt at at T = 25 ^{0}C) were used as working fluids. In all experiments, the depth of the fluid layer h was 30 mm and the surface of the fluid was open. The room temperature was kept constant by air-conditioning system, and cooling of the fluid was provided mainly by the heat exchange with surrounding air on the free surface and some heat losses through sidewalls. The velocity field measurements were made with a 2D particle image velocimetry (PIV) system Polis and the software package Actual Flow. The PIV velocity measurements were accurate to within 5%, estimated from calibration experiments in solid body rotation and long time series. Along with the dimensional parameters (heating flux, rotation rate and kinematic viscosity) we use the set of the non-dimensional parameters. These are the flux Grashof number Gr_f, rotational Reynolds number Re and Prandtl number Pr:

$$Gr_f = \frac{g\beta h^4 q}{c\rho\kappa\nu^2} \tag{1}$$

$$Re = \frac{\Omega h^2}{\nu} \tag{2}$$

$$Pr = \frac{\nu}{\kappa} \tag{3}$$

where g is the gravitational acceleration, h is the layer depth, β is the coefficient of thermal expansion, c is the thermal capacity, ρ is the density, ν is the coefficient of kinematic viscosity and κ is the thermal diffusivity, q is a heat flux ($q = P/S_h$, P is the power of the heater and S_h is the heater's surface area).

The parameters of experiments which are used for description of the flow are presented in Table 1.

3. Results

Detailed description of the basic flow structure can be found in [8] and here we present only brief description of the general structure of the large-scale flow. The heat flux in the central part of the bottom is a source of the intensive upward motion above the heater. Warm fluid cools at

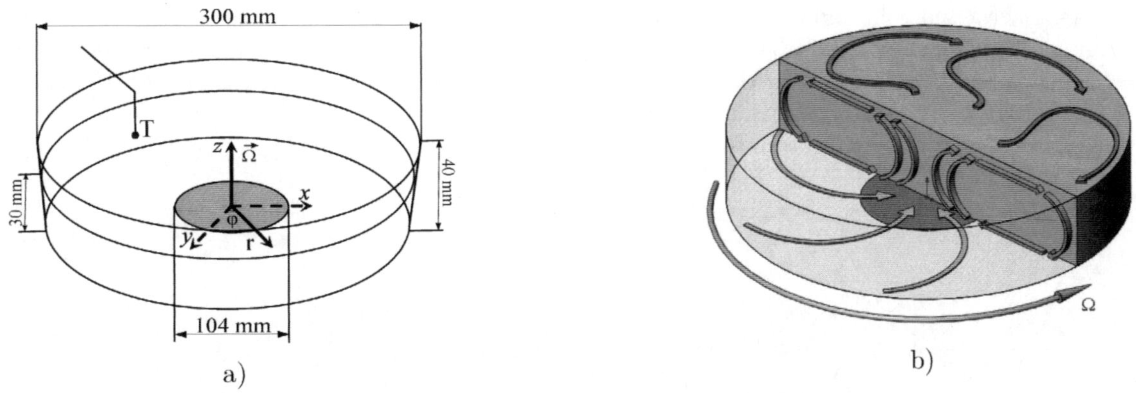

Figure 1. a - experimental model, dimensions and location of the coordinate system. T - thermocouple for control of the mean temperature; b - scheme of large-scale circulation.

Table 1. The values of the main parameters.

Experiment	P (W)	Ω (rad s^{-1})	ν (cSt)	Gr_f	Re	Pr
1	19	0.048	5.1	6×10^6	9.3	57
2	19	0.081	5.1	6×10^6	14.7	57
3	19	0.11	5.1	6×10^6	19.9	57
4	19	0.17	5.1	6×10^6	30.6	57
5	13.5	0.13	2.8	1.4×10^7	46	34

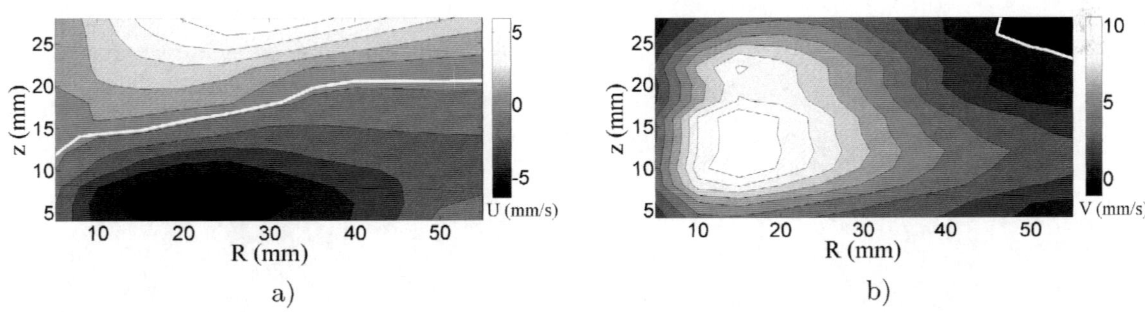

Figure 2. Mean radial (a) and azimuthal (b) velocity fields in a vertical cross-section over heating area, $Gr_f = 1.4 \cdot 10^7$, $Re = 15.6$. Thick solid white line shows the border between positive and negative values of velocity.

the free surface and moves toward the periphery where the cooled fluid moves downward along the side wall. Large-scale advective flow occupies the whole vessel (figure 1(b)).

The cyclonic vortex formation in the laboratory system can be described by following scenario. Large-scale radial circulation leads to the angular momentum transport and the angular momentum exchange on the solid boundaries. Convergent flow in the lower layer brings the fluid parcels with large values of angular momentum from the periphery to the center and produces cyclonic motion (figure 1(b)), lower horizontal cross-section). In the upper layer situation is opposite - divergent flow takes the fluid with low values of angular momentum to the periphery resulting in anticyclonic motion (figure 1(b), upper horizontal cross-section).

Friction in the viscous boundary layers leads to the sink of angular momentum in the part of the bottom occupied by cyclonic flow and produces source of angular momentum on the sidewalls when anticyclonic flow comes to the periphery. Zero net angular momentum flux on the solid boundaries is the necessary condition for the steady-state regime [7]. The general structure of laboratory vortex (figure 2) is similar to the observational data [9] excluding the side wall area.

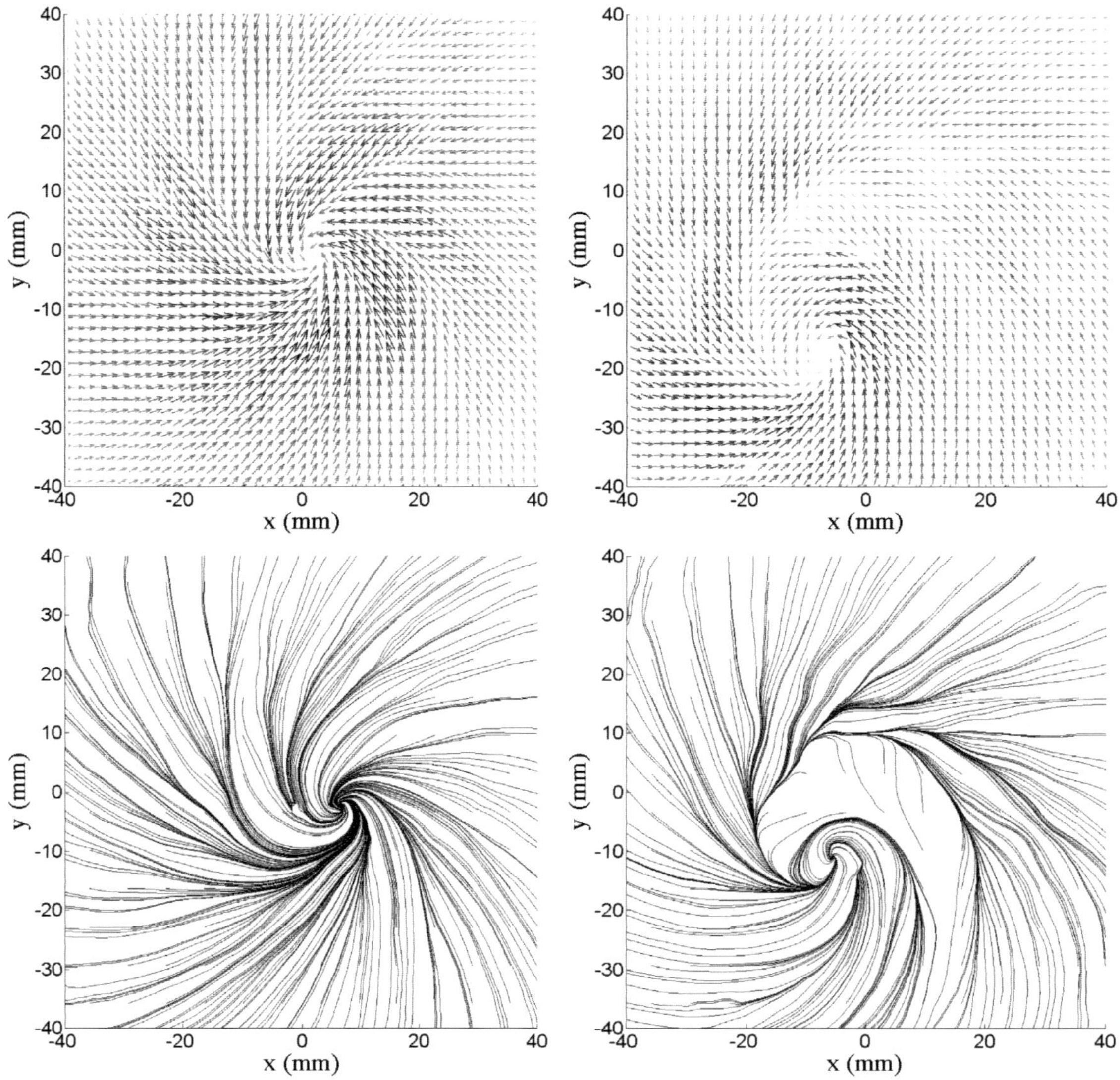

Figure 3. Instantaneous vector velocity fields (upper panel) and streamline flow fields (lower panel), left - experiment 2, right - experiment 4, $z = 3$ mm.

Series of experiments [6,8] showed that stable localized cyclonic vortex exist in a short range of governing parameters. Vortex becomes unstable at low value of viscosity and fast rotation of the model. Examples of instantaneous vector velocity field and streamline flow fields for the stable and unstable cases are shown in figure 3. In the case of developed cyclonic vortex there are several branches of radial inflow that comes very near to the center of the model figure 3 (left). Radial inflow transports angular momentum from the large radii to the small radii and its symmetry or asymmetry strongly influences the vortex structure. This result is

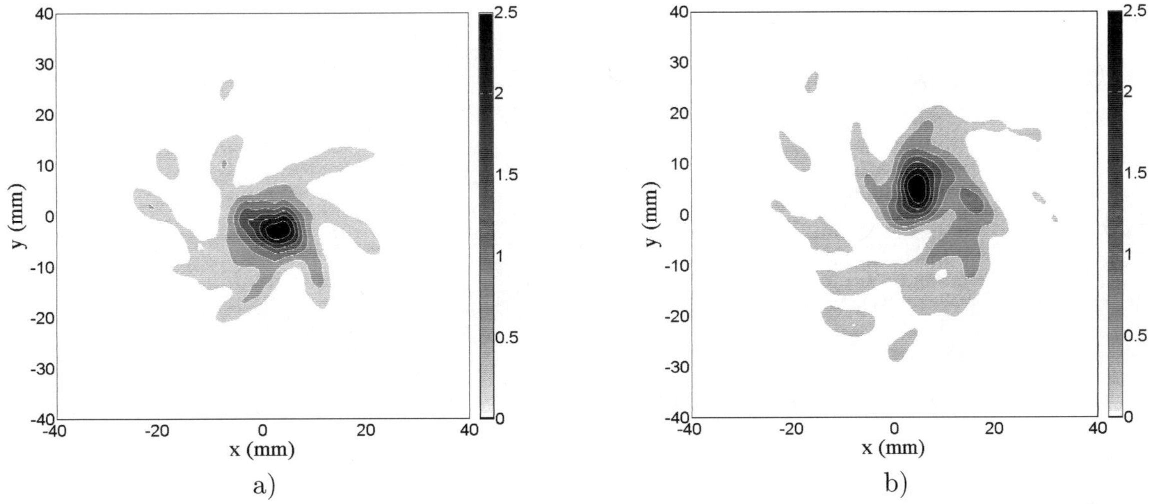

Figure 4. Instantaneous vorticity fields (s^{-1}), a - experiment 2, b - experiment 3, $z = 3$ mm.

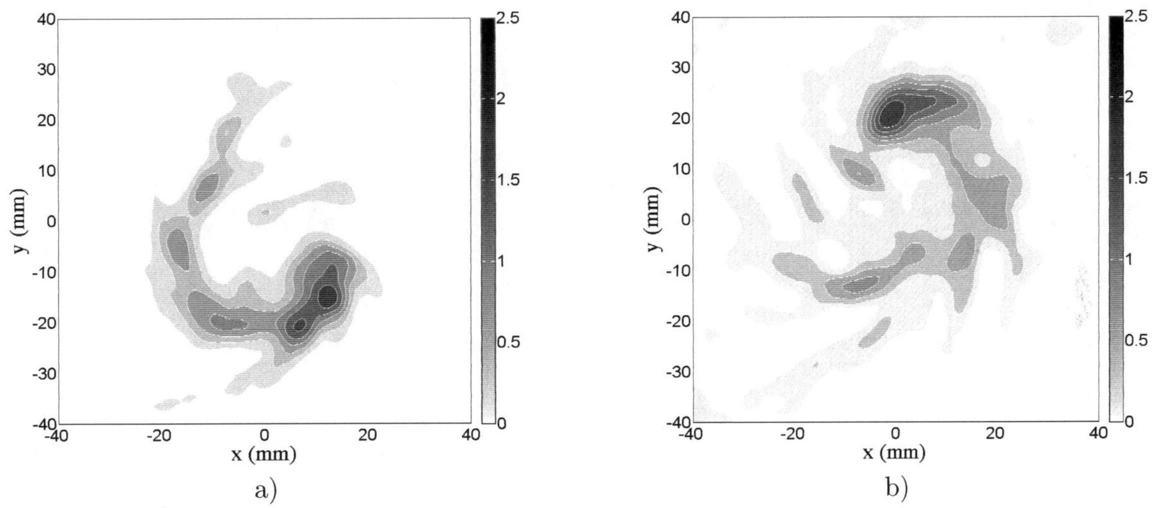

Figure 5. Two instantaneous vorticity fields (s^{-1}) for experiment 4, $z = 3$ mm.

in a good agreement with numerical simulation [10]. Increasing of rotational Reynolds number (2) by decreasing of viscosity or increasing of rotation rate leads to the instability of radial inflow. It changes its structure as shown in figure 3 (right). Instead of several branches which provide inflowing from different directions it is transformed into one big spiral branch that does not reach the center. As a result of this asymmetrical flux of angular momentum a localized cyclonic vortex is formed at some distance from the center. Radial inflow and cyclonic vortex slowly precessed around axis of rotation and in the mean vector field it appears as a vortex ring. The instantaneous vorticity fields for different values of Reynolds number (Gr_f and Pr are fixed) are shown in figure 4 and in figure 5. We see that transformation of the vortex structure is a gradual process, when we increase Re from 15.6 to 21.1 the radial inflow has already the dominant direction but still comes close to the center. After further increasing of Re up to 32.3 we achieve a state with localized at some radius vortex (figure 5). The radial inflow changes its direction (not periodically) so location of the vortex shown in figure 5 also changes. Figure 6

IOP Conf. Series: Materials Science and Engineering **208** (2017) 012040 doi:10.1088/1757-899X/208/1/012040

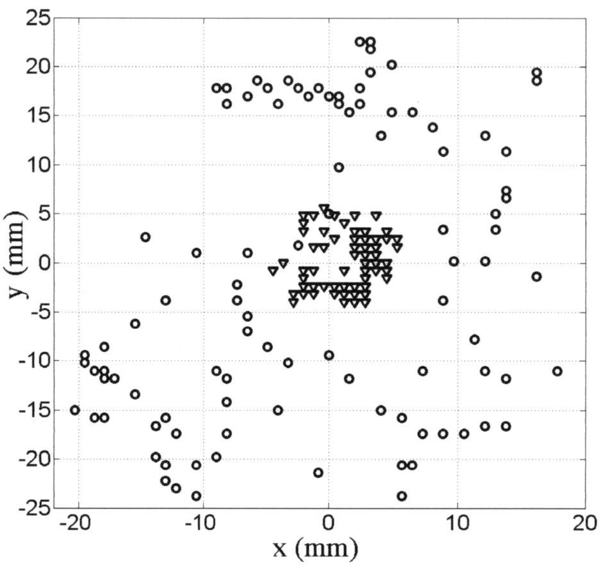

Figure 6. Locations of the cyclonic vortex center, triangles - experiment 2, circles - experiment 4, $z = 3$ mm.

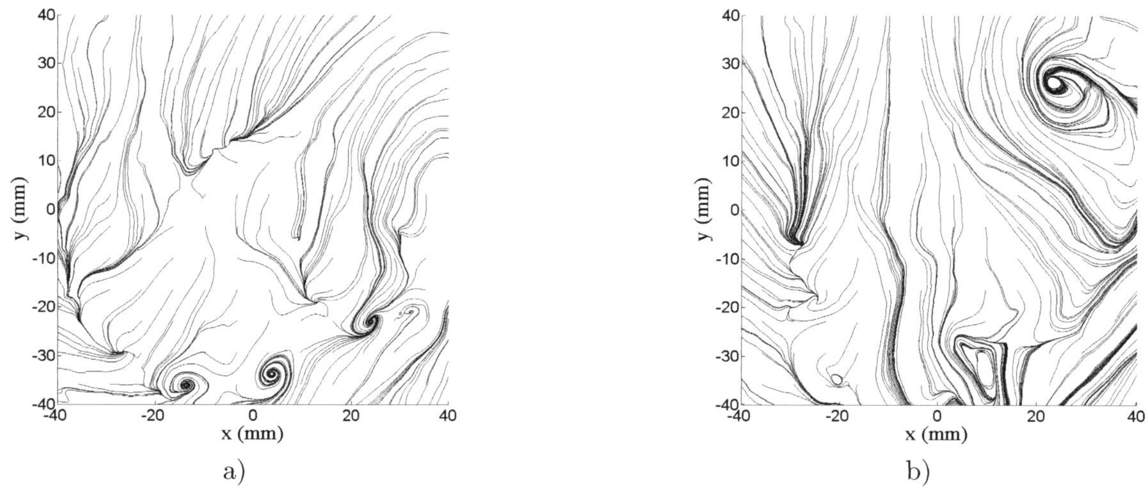

Figure 7. Instantaneous streamline flow fields (a,b) for experiment 5, $z = 3$ mm.

shows that locations of the center of the vortex for the stable case are concentrated near the axis of rotation but for unstable case they are dispersed at larger radii. For illustration of the flow structure at higher values of Re and Gr_f we used working fluid with lower viscosity (2.8 cSt instead of 5.1 cSt as in most of described experiments). For the fixed rotation rate and heating power it leads to increasing of Re and Gr_f and further destabilization of the flow. The flow becomes more chaotic with one or several vortices of different size (figure 7). Concluding the description of the vortex instability we want to emphasize two important features of the vortex evolution. The first one is that the instability of the vortex is tightly connected with changing structure of the radial inflow so the study of this process requires 3D approach. Another essential

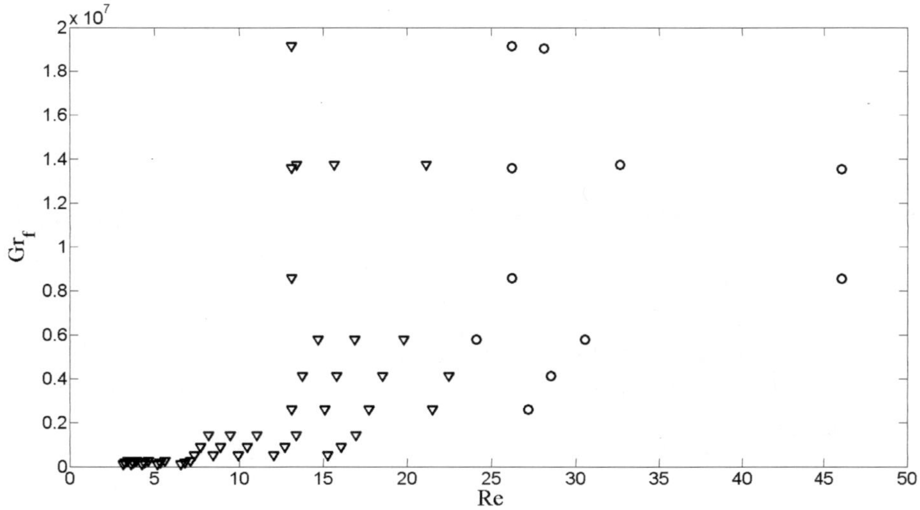

Figure 8. Map of regimes. Experiments with stable cyclonic vortex which is localized in the center are shown by triangles, experiments with unstable vortical flows are shown by circles.

feature of the described evolution of the cyclonic vortex is its strong dependence on Re. We present map of regimes in figure 8 which clearly shows that there is critical value of rotational Reynolds number $Re \approx 23$ for the cyclonic vortex instability.

4. Conclusions

Present study showed that instability of the vortex is tightly connected with a structure of the radial inflow. Up to moderate values of rotational Reynolds number the radial flows consist of several branches which transport angular momentum to the center. When Re exceeds the critical value radial inflow changes its structure and appears as one wide branch which does not reach the center. As a result the vortex which slowly moves around the center is formed instead of the vortex localized in the center. Further increase of Re leads to the chaotic state with several vortices which appears at different locations near the periphery of the heating area. The short interval of existence of localized intensive cyclonic vortex explains small number of laboratory studies of large-scale atmospheric vortices such as tropical cyclones. Also we assume that if described instability takes place in real tropical cyclones than it can be a reason why from a multiple cases of large-scale vortical disturbances in a favourable environment only a few can become a tropical cyclone.

The financial support of the grant RFBR 16-31-00150 is gratefully acknowledged.

References

[1] Hadlock RK and Hess SL 1968 *J. Atmos. Sci.* **2** 161-177
[2] Brickman D and Kelley DE 1993 *J. Dyn. Atmos. Oceans* **19** 389-405
[3] Brickman D 1995 *J. Phys. Oceanogr.* **1** 2609-2623
[4] Boubnov BM and Van Heijst GJF 1994 *J. Exp.Fluids* **3-4** 155-164
[5] Bogatyrev GP 1990 *Letters of JETP* **51** 557-559
[6] Bogatyrev GP 2009 *Monograph Laboratory model of a tropical cyclone (in Russian).*
[7] Batalov et al. 2010 *J. Geophys.Astrophys.Fluid Dynam.* **4** 349-368
[8] Sukhanovskii et al. 2016 *Quart J.R.Met.Soc.* **316** 23-33
[9] Zhang JA et al. 2011 *J. Mon.Wea.Rev.* **8** 2523-2535
[10] Evgrafova et al. 2016 *J. Computational continuum mechanics* **4** 498-508

IOP Conf. Series: Materials Science and Engineering **208** (2017) 012041 doi:10.1088/1757-899X/208/1/012041

Thermal convection of liquid metal in the titanium reduction reactor

A Teimurazov[1], **P Frick**[1] **and F Stefani**[2]
[1]Institute of Continuous Media Mechanics UB RAS, 1, Acad. Korolev St., Perm, 614013, Russia
[2]Helmholtz-Zentrum Dresden – Rossendorf, Bautzner Landstraße 400, 01328 Dresden, Germany

E-mail: tas@icmm.ru

Abstract. The structure of the convective flow of molten magnesium in a metallothermic titanium reduction reactor has been studied numerically in a three-dimensional non-stationary formulation with conjugated heat transfer between liquid magnesium and solids (steel walls of the cavity and titanium block). A nonuniform computational mesh with a total of 3.7 million grid points was used. The Large Eddy Simulation technique was applied to take into account the turbulence in the liquid phase. The instantaneous and average characteristics of the process and the velocity and temperature pulsation fields are analyzed. The simulations have been performed for three specific heating regimes: with furnace heaters operating at full power, with furnace heaters switched on at the bottom of the vessel only, and with switched-off furnace heaters. It is shown that the localization of the cooling zone can completely reorganize the structure of the large-scale flow. Therefore, by changing heating regimes, it is possible to influence the flow structure for the purpose of creating the most favorable conditions for the reaction. It is also shown that the presence of the titanium block strongly affects the flow structure.

1. Introduction

Liquid metals are characterized by low Prandtl numbers, that is by low ratios of viscosity to thermal diffusivity. This implies, in general, a weak contribution of thermo-gravitational convection to heat transfer. However, recent studies have shown that convection can strongly affect turbulent heat transport in turbulent natural convective [1] or even forced flows [2] of liquid metals. Moreover, convective flows in metals can affect significantly technological processes in metallurgy. For example, the buoyancy force can influence the metallothermic reduction of titanium, as was shown in a simplified axisymmetric computation in [3]. The retort for titanium reduction is a cylindrical vessel with a diameter up to 2 m and a height up to 4 m containing liquid magnesium at a temperature of about 850°C. Titanium tetrachloride is supplied from the top to the magnesium surface. The chemical reaction of titanium reduction is highly exothermic and runs predominantly on the surface. It produces spongy titanium and magnesium dichloride that sink to the bottom of the retort. The fact that the large mass and dimensions of the vessel and the very high temperatures hamper any direct experimental measurements inside the reactor is one of the motivations for the numerical study of the process. A first fully three-dimensional numerical study of the structure of convective flows of molten magnesium at realistic Rayleigh numbers has revealed a strong turbulent flow [4].

IOP Conf. Series: Materials Science and Engineering **208** (2017) 012041 doi:10.1088/1757-899X/208/1/012041

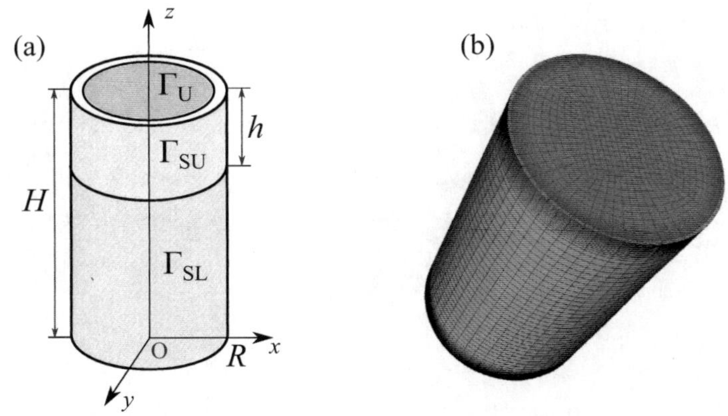

Figure 1. Scheme of the computational domain: (a) – general view with the coordinate system with the designated boundaries; (b) – computational mesh structure with solid (grey color) and fluid (blue color) regions.

The titanium reduction process takes more than two days during which the titanium sponge gradually occupies the retort. A stable reaction should be facilitated by controlling the supply of titanium tetrachloride, by draining the magnesium chloride, and by cooling and heating the furnace. At present, up to 5% of the production cycles are rejected due to contingencies associated with the disruption of the magnesium chloride settling process, the emergence of titanium sponge on the magnesium surface, and local retort overheating. Up to present, attempts to enhance the sponge production by local reactor cooling have ignored the flows inside the reactor [5].

In this paper we continue the numerical study of convective flows in the reactor for metallothermic reduction of titanium for different configurations of reactor cooling and heating zones and different levels of titanium sponge. Our goal is to reveal how strongly the large-scale magnesium circulation in the reactor can be affected by the change of heating and by the growth of the titanium mass.

2. Formulation of the problem and the mathematical model

We consider a flow of liquid magnesium inside the cylindrical vessel made of steel with a wall thickness of 25 mm. During the process the combined effect of the furnace heaters in the lower part of the vessel, the exothermic chemical reaction on the top surface of the magnesium layer, and the cooling on the side walls in the reaction zone by airflow leads to substantial temperature gradients which generate a convective flow inside the reactor. The structure of this convective flow may have significant influence on the reaction process. The height of the vessel is $H = 2.5$ m, the radius $R = 0.8$ m. The height of the upper part of the side surface in the reaction zone, which is cooled by air, is varied from $h = 0.15$ m to 0.7 m. A scheme of the computational domain is shown in figure 1a. The simulations were done with and without a titanium block inside the vessel in its lower part. This titanium block is assumed to be a solid body (not a porous medium). The contribution of the second liquid phase, magnesium dichloride, is disregarded.

We used the software package OpenFoam Extend 3.2 [6], namely *conjugateHeatFoam* solver. It allows to simulate conjugate heat transfer between the incompressible fluid (liquid magnesium), the steel walls of the retort (heat conductivity $k = 26$ Wm^{-1}K^{-1}), and the solid titanium block ($k = 15$ Wm^{-1}K^{-1}). The convective parameters of the fluid, corresponding to molten magnesium at a temperature 850°C, are: Prandtl number $Pr = 0.0095$, [7], kinematic viscosity $4.17 \cdot 10^{-7}$ m^{2}s^{-1}, and thermal expansion coefficient $\beta = 1.69 \cdot 10^{-4}$ K^{-1}.

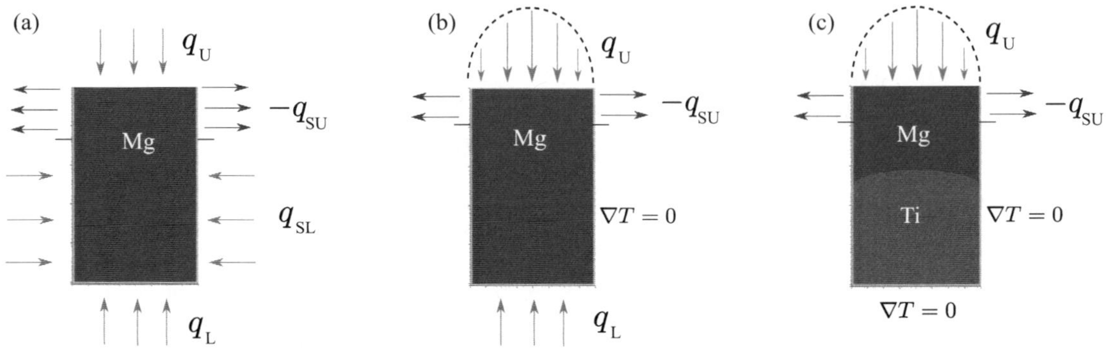

Figure 2. Schemes of the heat flux distribution at the boundaries of the computational domain for various heating configurations: TBC-1 (a), TBC-2 (b), and TBC-3 (c).

Our numerical code solves the 3D equations of thermo-gravitational convection in the Boussinesq approximation for the fluid phase. We use the LES (Large Eddy Simulation) approach to take into account the turbulence, specifically the Smagorinsky-Lilly model model [8, 9] with the Smagorinsky constant $C_s = 0.14$ and turbulent Prandtl number $Pr_t = 0.9$. The mesh has a regular block structure with a total of 3.7 million nodes with additional mesh refinement near the boundaries (see figure 1b). The terms with time derivatives are discretised using an implicit Euler scheme. The convective terms are calculated by the TVD (Total Variation Diminishing) scheme [10]. The numerical simulations were performed on the Triton supercomputer of ICMM UB RAS.

Figure 1a designates the boundary faces: Γ_U is the upper surface with area S_U, Γ_{SU} is the part of the cooled side surface with area S_{SU}, Γ_{SL} is the lower part of the side surface with area S_{SL}, and Γ_L is the base with area S_L.

Three different variants of thermal boundary conditions (TBCs), which correspond to three possible heating configurations inside the reactor during the titanium production process, were considered. The first configuration, TBC-1 (figure 2a), combines the heating from above at the boundary Γ_U with a constant heat flux q_U resulting from the chemical reaction with a thermal power $Q_R = 205$ kW, the heating of the lower part of the side surface Γ_{SL} with a heat flux Γ_{SL} coming from the side heaters with the total power $Q_{SL} = 329$ kW, and the heating from below at the boundary Γ_L with a heat flux q_L from the lower heater with the power $Q_L = 94$ kW. All of this heat is removed from the retort by cooling the boundary Γ_{SU} with height $h = 0.7$ m. Thus, the heat flux densities on the boundaries for TBC-1 are:

$$q_U = \frac{Q_R}{S_U}; \qquad q_{SL} = \frac{Q_{SL}}{S_{SL}}; \qquad q_L = \frac{Q_L}{S_L}; \qquad q_{SU} = -\frac{(Q_R + Q_{SL} + Q_L)}{S_{SU}}. \tag{1}$$

This heating regime corresponds to the situation with furnace heaters operating at full capacity and maximum air cooling efficiency. This case is most revealing for qualitative estimates of the maximum flow velocities inside the reactor, although it should be noted that during the real process also some specific sections of the furnace heater can be switched on simultaneously.

The second heating configuration, TBC-2 (figure 2b), is closer to the typical reactor operation and combines heating on Γ_U by the chemical reaction with only the heaters on the bottom of the vessel being switched on, whereas heaters on the side surface Γ_{SL} are switched off. Also note that the heat flux at the upper boundary Γ_U is distributed not uniformly but with a parabolic law with a maximum at the center and a minimum on the periphery. Such a temperature non-uniformity can emerge in the reactor during the actual titanium reduction process due to nonuniform titanium tetrachloride sputtering on the magnesium surface. The cooling zone has a height of $h = 0.3$ m in this case. Thus, the heat flux densities for TBC-2 are:

$$q_U = \frac{2Q_R}{S_U}\left(1 - \frac{x^2+y^2}{R^2}\right); \qquad q_{SL} = 0; \qquad q_L = \frac{Q_L}{S_L}; \qquad q_{SU} = -\frac{(Q_R + Q_L)}{S_{SU}}. \qquad (2)$$

The third heating configuration, TBC-3 (figure 2c), implies a nonuniform heat flux q_U with parabolic law specified at the upper boundary Γ_U of the cylinder . A constant heat flux q_{SU} is removed from the side surface Γ_{SL} with height $h = 0.15$ m. The lower part of the side surface Γ_{SL} and the base Γ_L are assumed to be thermally insulated. In this case the solid titanium block occupies the lower half part of the vessel, which corresponds to later stages of the process. Consequently, the heat flux densities on the outer boundaries for TBC-3 are:

$$q_U = \frac{2Q_R}{S_U}\left(1 - \frac{x^2+y^2}{R^2}\right); \qquad q_{SL} = q_L = 0; \qquad q_{SU} = -\frac{Q_R}{S_{SU}}. \qquad (3)$$

Free-slip conditions are specified for the velocity at the upper boundary Γ_U. This is in contrast to the no-slip boundary condition which was used in [4]. It is known that films emerge on a free surface in liquid metals, and it is often assumed in simulations that there is some sticking on that boundary. However, in the case of a chemical reaction on the surface the use of free-slip velocity boundary condition seems to be more reasonable. The no-slip conditions are specified for the velocity at all other boundaries which are "fluid to steel walls" and "fluid to titanium block" boundaries.

3. Results

All simulations were performed in dimensional form, with the governing parameters values corresponding to the actual titanium reduction process. The only parameter which is not exactly known is the intensity (heat transfer coefficient) of air cooling in the reaction zone. It is also known that there are several types of reactors with different heights h of the cooling zone and varying intensities of the cooling. In the real reactor the separate zone heaters in the furnace are switched on and off dynamically, depending on the temperature values measured at certain control points on the outer sidewall of the retort. In our simulations we assume that the integral thermal balance in the reactor is guaranteed, i.e. for every considered TBCs the amount of heat which goes into the vessel (by chemical reaction and furnace heaters) is equal to the amount of heat which is removed from the reactor (by air cooling).

Figure 3 presents the results of the simulations with the TBC-1 heating configuration (figure 2a). The snapshot of the instantaneous velocity field (figure 3a) shows that the turbulent flow occupies the entire volume of the liquid metal. The maximum instantaneous velocity V_{max} near the boundary reaches 18 cm s^{-1}. The time-averaged velocity field (figure 3b) demonstrates a two-vortex structure of the convective flow, with one ring vortex in the upper part of the retort and one in its lower part. A maximum mean velocity $< V >_{max} \approx$ 12 cm s^{-1} is observed near the boundaries. The temperature field (figure 3d) reflects the unsteady pattern of the motion caused by the near-wall ascending from the heated lower part of the reactor and by the descending flow arising in the cooled part of the retort. The intensity of turbulence can be characterised by the distribution of the energy of turbulent pulsations or by the standard deviation (SD). Figures 3c,d show the velocity and temperature SD fields. The most pronounced pulsations are observed in the region near the interface between heated and cooled parts of the sidewall. Strong temperature pulsations also exist in the upper part of the retort in the reaction zone.

It should be noted that in our previous paper [4] a similar heating configuration was studied. In this work we made the following changes: (a) – more accurate physical properties of liquid

IOP Conf. Series: Materials Science and Engineering **208** (2017) 012041 doi:10.1088/1757-899X/208/1/012041

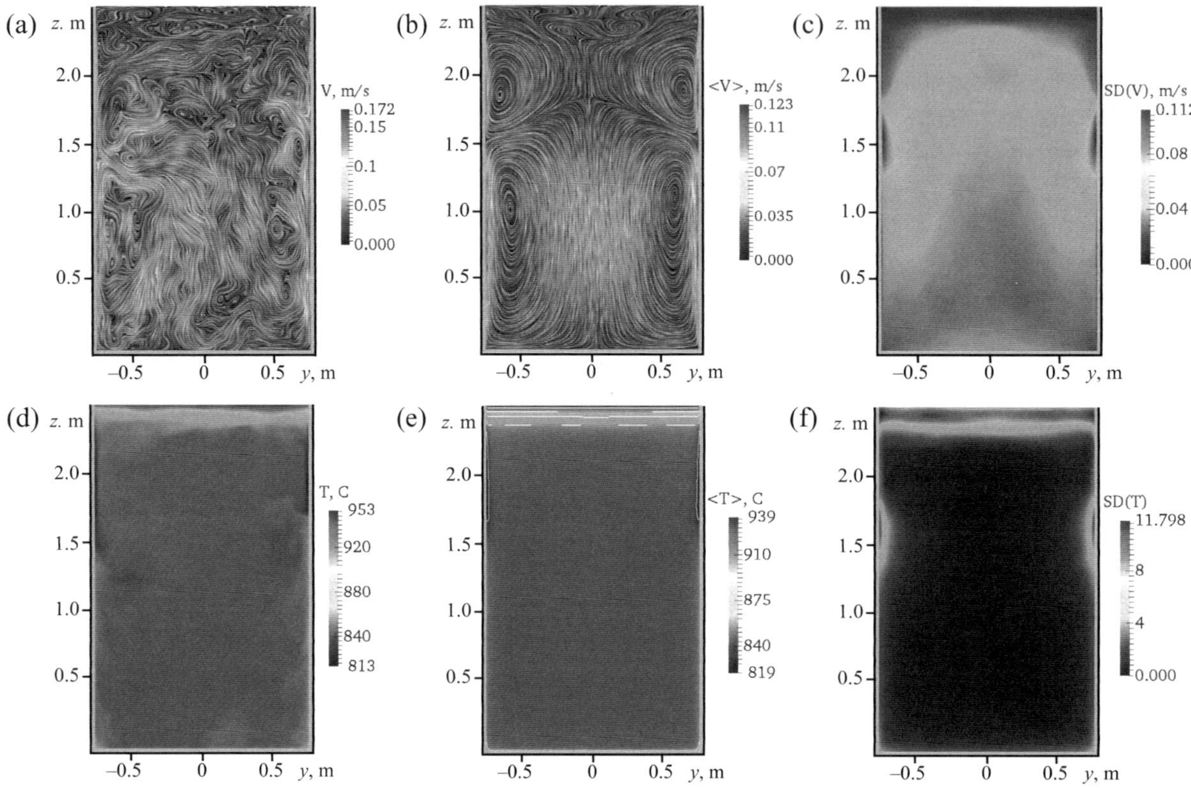

Figure 3. Results for the TBC-1 heating configuration in the yOz plane: (a) – instant velocity field, (b) – time-averaged velocity field, (c) – standard deviation of velocity field, (d) – instant temperature field, (e) – time-averaged temperature field, (f) – standard deviation of temperature field.

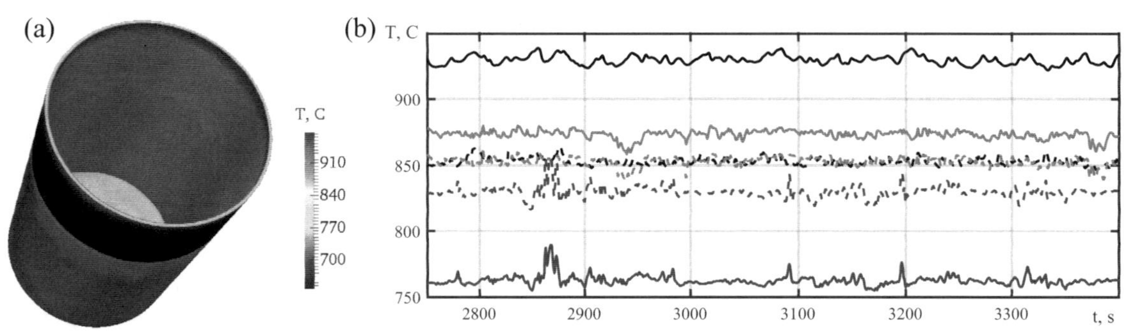

Figure 4. Results for the TBC-1 heating configuration: (a) – instant temperature field on the walls of the steel cavity, (b) – time series for variations of the temperature for the points at a distance of 10 mm from side wall (dashed lines) and for the points on the steel sidewall. Black line – z = 0.3 m, red line – z = 1.8 m, blue line – z = 2.3 m.

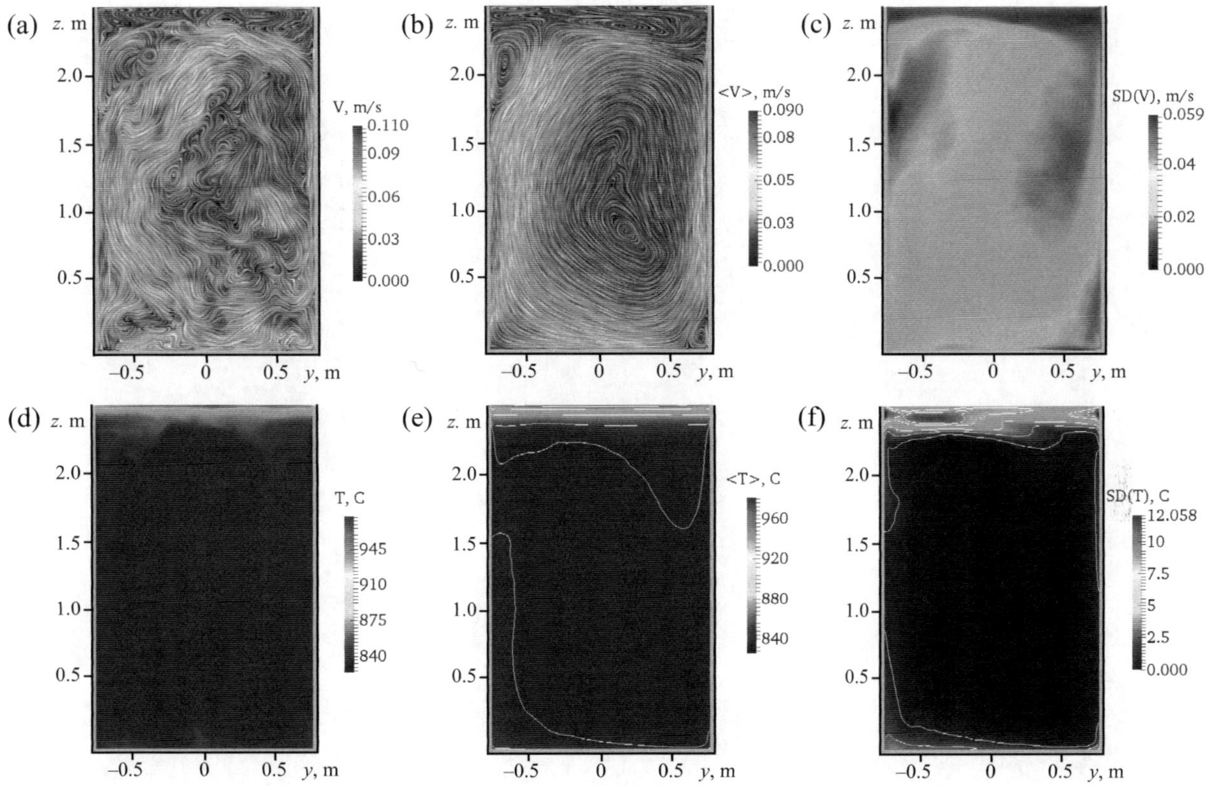

Figure 5. Same as figure 3 for the TBC-2 heating configuration.

magnesium according to [7]; (b) – use of free-slip boundary condition for the velocity on the top surface instead of no-slip boundary condition; (c) – taking into account the influence of steel walls of the vessel.

Although the steel walls of the vessel with thickness 25 mm are not expected to strongly influence the heat transfer and flow structure, they should be taken into account for two reasons. First, applying heat fluxes to steel walls instead of applying it directly to the liquid metal leads to a more accurate imposition of temperature BCs. Figure 4a demonstrates the instant temperature field on the walls of the steel cavity. It can be seen that the temperature drop near the interface between the cooled and heated parts of the lateral surface of the vessel and becomes essentially less pronounced on the inner surface. Second, the temperature evolution at several points on the outer sidewall surface of the retort are the only temperature measurements which are available during the real process of titanium production. The conjugate heat transfer simulations allow us to obtain temperature distributions on the outer sidewall of the retort, which can be compared with real data. Figure 4b shows the time series for variations of the temperature for the points on the the steel sidewall (solid lines) and inside the vessel at a distance of 10 mm from the wall (dashed lines).

Next, we consider the heat flux configuration TBC-2 (figure 2b), which is representative for most of the process. It combines the nonuniform heating from the top due to the reaction with the heating from below due to bottom heaters. Figure 5 shows that the flow structure in this case is very different from the TBC-1 regime. A large-scale circulation (LSC) with two small

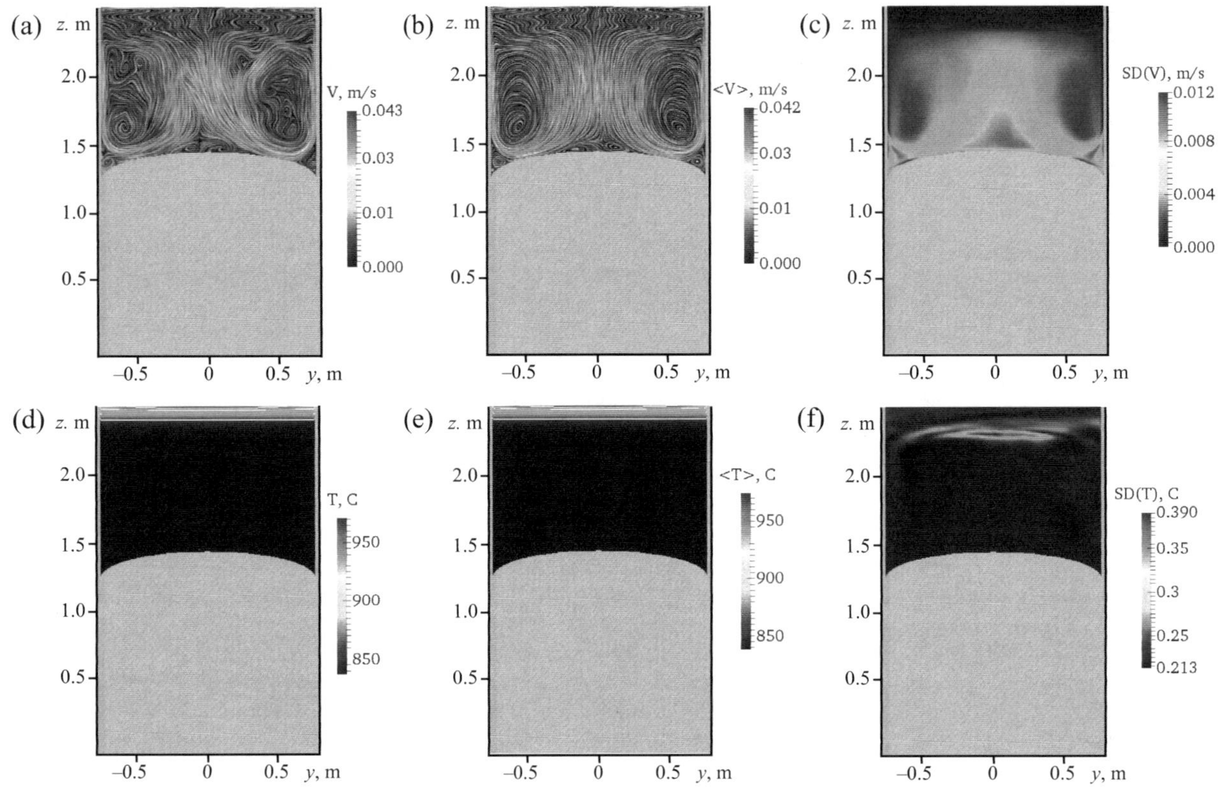

Figure 6. Same as figure 3 for the TBC-3 heating configuration. The grey domain in the lower part of the cylinder corresponds to the titanium block.

counter-rotating vortexes in the topmost and bottommost cylinder parts can be clearly seen on the instant and time-averaged velocity fields (figures 5a,b). It is well known that in the classical Rayleigh-Bénard problem (convection in a vessel heated from bottom) a large-scale circulation appears on the background of fully developed turbulence [11]. This LSC can reveal a variety of dynamical regimes (see, i.e. [12] for a cylinder with unit aspect ratio, or [13] for cubic geometry). The intensity of the velocity field is lower than in the case TBC-1, the maximum instant velocity $V_{max} \approx 14$ cm s^{-1} and the average velocity is $< V >_{max} \approx 9$ cm s^{-1}. The most intense velocity pulsations are observed in the regions with interaction of the LSC with the counter-rotating rolls (figure 5c), the strongest temperature pulsations are located near the top of the cavity in the "reaction zone" (figure 5f).

Note that in previous simulations [4] we had argued that if the cooling intensity on the sidewall were not strong enough a uniform vertical temperature gradient would establish in the upper part of the retort (see figure 6 in [4]). In a real reactor this would lead to overheating, given that the temperature on the vessel walls should not exceed 950°C. Since the performance of the cooling system is not exactly known we varied the height of the cooling zone h (and thus the heat flux density on the boundary Γ_{SU}) in order to maintain thermal balance and to prevent overheating. For TBC-2 the height of the cooling zone is $h = 0.3$ m, which guarantees that there is no uniform vertical temperature gradient in the upper part of the retort (figure 5d).

The configuration TBC-3 (figure 2c) models the flow structure in the presence of a titanium block in the lower part of the vessel. This corresponds to the middle stages of the process when the titanium occupies approximately half of the vessel volume. The furnace heaters are switched off and we have only the nonuniform heating from above. The velocity fields demonstrate one ring vortex structure flow in the liquid magnesium (figures 6a,b). The vortex intensity reaches its maximum (4.5 cm s^{-1}) near the boundaries, which is much lower than in the cases TBC-1,2. The distribution of velocity pulsations against a background of the mean flow (figure 6c) shows that the strongest pulsations appear in the regions where the descending flow along the boundaries collides with the surface of the solid block.

Note that in reality the titanium block is a porous medium, and magnesium trickles through the titanium sponge to the bottom of the retort. Yet, as was shown in [3], the flow intensity inside the titanium block is four orders of magnitude lower than the magnesium flow in the vessel. Therefore, in the present work the titanium block is assumed to be solid.

The temperature field in this heating configuration is almost stationary (figures 6d,e), the temperature pulsations are very weak and concentrated in the layer close to the upper part of the retort (figure 6f).

4. Conclusions

The structure of the convective flow of molten magnesium in a titanium reduction reactor was numerically studied in a three-dimensional non-stationary formulation with conjugated heat transfer between liquid magnesium and solids (steel walls of the cavity and titanium block), for different heating and cooling regimes. A detailed analysis was performed for three configurations of the heating rate control: with furnace heaters operating at full power, with furnace heaters switched on at the bottom of the vessel only, and with switched-off furnace heaters.

The most important finding of our study concerns the fact that the localization of the cooling zone can completely reorganize the structure of the large-scale flow. Therefore, by changing heating regimes, it is possible to influence the flow structure for the purpose of creating the most favorable conditions for the reaction. It was also shown that the presence of the titanium block strongly affects the flow structure.

Acknowledgments

This study was partially supported by the Government of Perm Krai, project No C-26/060 11.03.2016.

References

[1] Frick P, Khalilov R, Kolesnichenko I, Mamykin A, Pakholkov V, Pavlinov A and Rogozhkin S 2015 *Europhys. Lett.* **109** 14002
[2] Zikanov O, Listratov Y I and Sviridov V G 2013 *J. Fluid Mech.* **720** 486–516
[3] Tsaplin A I and Nechaev V N 2013 *Computational continuum mechanics* **6** 483–490
[4] Teimurazov A S and Frick P G 2016 *J. Appl. Mech. Tech. Phys.* **57** 1264–1275
[5] Wang W, Wu F and Jin H 2017 *Heat Mass Transfer* **53** 465–473
[6] Weller H G, Tabor G, Jasak H and Fureby C 1998 *Comput. Phys.* **12** 620–631
[7] Smithells C J, Gale W F and Totemeier T C 2004 *Smithells Metals Reference Book* 8th ed (Elsevier Butterworth-Heinemann)
[8] Smagorinsky J 1963 *Mon. Weather Rev.* **91** 99
[9] Deardorff J W 1970 *J. Fluid Mech.* **41** 453–480
[10] Versteeg H K and Malalasekera W 2007 *An Introduction to Computational Fluid Dynamics: The Finite Volume Method* (Pearson Education, Harlow)
[11] Ahlers G, Grossmann S and Lohse D 2009 *Rev. Mod. Phys.* **81** 503–537
[12] Brown E and Ahlers G 2009 *J. Fluid Mech.* **638** 383–400
[13] Vasiliev A, Sukhanovskii A, Frick P, Budnikov A, Fomichev V, Bolshukhin M and Romanov R 2016 *Int. J. Heat Mass Transfer* **102** 201 – 212

IOP Conf. Series: Materials Science and Engineering **208** (2017) 012042 doi:10.1088/1757-899X/208/1/012042

A comparison of the two approaches of the theory of critical distances based on linear-elastic and elasto-plastic analyses

A I Terekhina[1], O A Plekhov[1], A A Kostina[1] and L Susmel[2]

[1]Institute of Continuous Media Mechanics UB RAS, 1, Acad. Korolev St., Perm, 614013, Russia

[2] Department of Civil and Structural Engineering, The University of Sheffield, Sheffield S1 3JD, United Kingdom

E-mail: terekhina.a@icmm.ru

Abstract. The problem of determining the strength of engineering structures, considering the effects of the non-local fracture in the area of stress concentrators is a great scientific and industrial interest. This work is aimed on modification of the classical theory of critical distance that is known as a method of failure prediction based on linear-elastic analysis in case of elasto-plastic material behaviour to improve the accuracy of estimation of lifetime of notched components. Accounting plasticity has been implemented with the use of the Simplified Johnson-Cook model. Mechanical tests were carried out using a 300 kN electromechanical testing machine Shimadzu AG-X Plus. The cylindrical un-notched specimens and specimens with stress concentrators of titanium alloy Grade2 were tested under tensile loading with different grippers travel speed, which ensured several orders of strain rate. The results of elasto-plastic analyses of stress distributions near a wide variety of notches are presented. The results showed that the use of the modification of the TCD based on elasto-plastic analysis gives us estimates falling within an error interval of ±5-10%, that more accurate predictions than the linear elastic TCD solution. The use of an improved description of the stress-strain state at the notch tip allows introducing the critical distances as a material parameter.

1. Introduction

Nowadays the problem of determining the strength of engineering structures, considering the effects of the non-local fracture in the area of stress concentrators becomes especially acute due to the fact that this problem of fracture of materials under different loading conditions deals with a large number of applications in different engineering fields. This problem is significant in all the cases where geometrical features in structures give rise to localized stress concentration which may generate a crack leading to failure or to a shortening of the assessed lifetime of structures.

Today the method of failure prediction in cases where stress concentrations are present, known as the Theory of Critical Distances (TCD) [1], is one of the most popular methods. The fundamental idea on which the Theory of Critical Distances is based was first proposed in 1958 by Neuber [2] assumed that fatigue failure occurs when the mean stress with some distance from the notch root is equal to the fatigue strength of smooth specimen. One year later (1959), Peterson assumed that the fatigue failure occurs when the stress at one point which has a critical distance from the notch root is equal to the fatigue strength of smooth specimen [3]. In 1969 Novozhilov introduced a necessary and sufficient criterion for estimating the strength of an elastic body weakened by a crack in the form of the average stress limitation in the cohesive zone length d ahead of the crack tip [4]. The above-mentioned ideas were the basis of four formalized method of the TCD such as the Point, the Line, the Area, and the Volume Method [5]. These four methods use one characteristic material length parameter, so-called critical distance L, to predict both brittle fracture and fatigue strength. The TCD takes a starting point the assumption that that the strength of notched components can be estimated by directly post-processing the entire linear-elastic stress field acting on the material in the vicinity of the stress concentrator [6,7]. It has also been proved that the TCD can be used to predict static fracture in ductile metallic materials containing various stress raisers and subjected to both uniaxial and multiaxial loading [8]. Moreover, in papers [9, 10] was shown that Theory of Critical Distances is suitable for predicting the strength of notched metallic materials subjected to dynamic loading.

In this study considered the modification of the classical theory of critical distance in cases of elasto-plastic material behavior for improve the accuracy of estimation of lifetime of notched components, because the material behavior being, by nature, highly non-linear and cannot be described in the framework of the linear theory of elasticity.

The paper is organized as follows. Section 2 is devoted to Theory of Critical Distances based on the linear-elastic analysis. There is considered the fundamentals of the theory of critical distances. The materials and experimental conditions are described. The results of applying this theory to strength prediction are presented. Section 3 describes of modification of the classical theory of critical distance in cases accounting of elasto-plastic material behavior. The parameters of the Simplified Johnson Cook model are identified. The results of using elasto-plastic analysis for strength evaluation are presented.

2. Theory of Critical Distances based on linear-elastic analysis

2.1. The fundamentals of the classical theory of critical distances

According to the theory of critical distances the failure of notched components can be predicted using linear-elastic stress information in a critical region close to the notch tip. Failure initiates when applied stress provides an existence of area with stress higher ultimate stress of the material. In this work the spatial size of this area is determined according to Point method (Figure 1) of the Theory of Critical Distances [4]. The empirical relationship in terms of the Point Method to predict the fracture of laboratory samples with a complex geometry is represented by equation (1). The Point Method postulates that a notched component is failure when the effective stress σ_{eff} at a given distance from the stress raiser apex equals the material plain inherent strength.

$$\sigma_{eff} = \sigma_1 (r = L/2, \theta = 0) = \sigma_0 \tag{1}$$

where r, θ - polar coordinates, σ_1 - is the range of the maximum principal stress, L - material characteristic length, which is determined according to the relationship:

$$L = \frac{1}{\pi}\left(\frac{K_{Ic}}{\sigma_0}\right)^2 \tag{2}$$

where K_{Ic} is the plane strain fracture toughness and σ_0 is the material inherent strength.

An experimental way to determine L is by testing samples containing notches of different sharpness is proposed [6]. This procedure is summarized in Figure 2. As postulated by the PM, the coordinates of the point at which the two linear-elastic stress-distance curves, plotted in the incipient failure condition, intersect each other allow us estimated directly the material length parameter L and inherent strength σ_0.

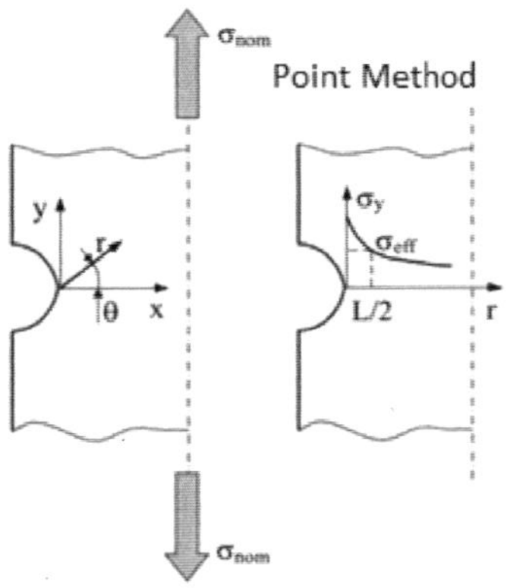

Figure 1. Local systems of coordinates and effective stress according to the PM

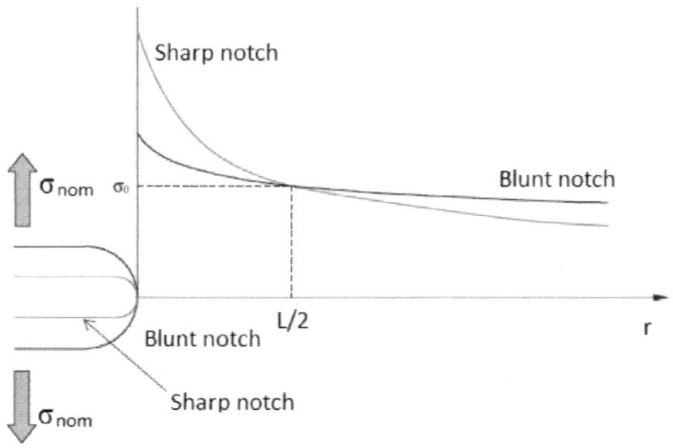

Figure 2. Determination of length scale parameter L through experimental results.

2.2. *Materials and experimental conditions*

Thirty-three cylindrical samples of titanium alloy Grade2 were tested under tensile loading with different grippers travel speed, which ensured several orders of strain rate. Tensile tests were carried out using a 300 kN electromechanical testing machine Shimadzu AG-X Plus. Measurement of strain during materials testing was carried out using video extensometer TRViewX240S f12.5. For this study three types of cylindrical specimens were used with different stress concentrators such as semi-circular edge notches with radius 1 mm and 2 mm, V-shaped notches and un-notched (plain) specimens. For all specimens, the notches depth was kept constant. The geometry of specimens is shown in Figure 3. Loading tensile specimens occurred with range of strain rate 10^{-4} - 10 s^{-1}. During each test the failure force and time of failure were determined. The failure force was taken equal to the maximum force recorded during each test. The stress fields requiring to calculation of the effective stress according to definitions TCD were determined by solving Finite-Element models done by the finite-element package Abaqus SE.

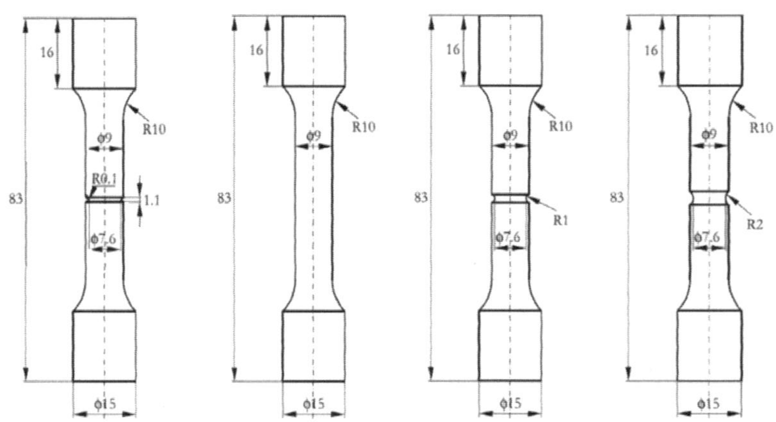

Figure 3. Geometry of specimens.

The results generated by testing un-notched and notched specimens titanium alloy Grade2 are summarized in Tables 1 in terms of failure force, F_f, time to failure, T_f, nominal loading rate, $\dot{F}$, and nominal strain rate $\dot{\varepsilon}_{nom}$.

Table 1. Summary of the experimental results generated by testing cylindrical samples of Grade2.

	r_n (mm)	Grip speed (mm/s)	F (kN)	T_f (s)	$\dot{F}$ (kN/s)	Displacement (mm)	$\dot{\varepsilon}_{nom}$ (1/s)
S 1	Plain	0.01	28.4	813.8	0.035	3.3939	0.0002
S 2	Plain	0.01	28.7	812.5	0.035	3.2740	0.0002
S 5	Plain	0.1	29.9	72.1	0.415	2.8747	0.002
S 6	Plain	0.1	29.8	70.9	0.420	2.7509	0.002
S 7	Plain	5	32.3	1.6	20.19	0.8956	0.051
S 8	Plain	5	32.5	1.6	20.31	0.7801	0.049
S 10	2	0.01	25.7	553	0.046	1.9571	0.003
S 11	2	0.01	26.2	550.3	0.048	1.9766	0.003
S 12	2	0.01	26	529.2	0.049	1.7608	0.003
S 13	2	0.1	27.4	50.3	0.545	1.5788	0.027
S 14	2	0.1	27.6	51.7	0.534	1.705	0.027
S 15	2	0.1	27.3	49.3	0.554	1.5962	0.027
S 16	2	5	29.9	1.1	27.182	0.2663	0.38
S 17	2	5	29.8	1.2	24.833	0.2991	0.427
S 18	2	5	29.9	1.1	27.182	0.2938	0.399
S 19	1	0.01	27.2	618.6	0.044	2.6024	0.0046
S 20	1	0.01	27	610	0.044	2.5759	0.0047
S 23	1	0.1	28.3	60.4	0.469	2.3252	0.0448
S 24	1	0.1	27.9	60	0.465	2.3752	0.0443
S 25	1	5	30.6	1.3	23.538	0.5421	0.884
S 26	1	5	30.7	1.4	21.929	0.6001	0.925
S 27	1	5	30.4	1.4	21.714	0.6041	1.08
S 28	0.1	0.01	27.13	641.6	0.043	2.7095	0.0078
S 29	0.1	0.01	27.2	585.8	0.046	2.2454	0.0078
S 30	0.1	0.1	28.3	59.9	0.472	2.4256	0.076
S 31	0.1	0.1	28.4	60.3	0.471	2.3841	0.077
S 32	0.1	5	31.1	1.3	23.923	0.4565	1.43
S 33	0.1	5	30.8	1.4	22	0.6893	1.8

2.3. *Results*

To use the Theory of Critical Distances to re-analyse the results generated by testing the notched cylindrical samples of titanium alloy Grade2, the initial assumption was made that the inherent strength could be taken equal to the corresponding plain material strength, that is:

$$\sigma_0\left(\dot{\varepsilon}_{nom}\right)=\sigma_f\left(\dot{\varepsilon}_{nom}\right) \quad (3)$$

According to the plain results reported in Table 1, $\sigma_f\left(\dot{\varepsilon}_{nom}\right)$ was expressed as follows:

$$\sigma_f\left(\dot{\varepsilon}_{nom}\right)=539.22\cdot\dot{\varepsilon}_{nom}^{0.0121}\ \text{(MPa)} \quad (4)$$

The chart of Figure 4a shows the linear-elastic stress–distance curves plotted, under quasi-static loading ($\dot{\varepsilon}_{nom}=0.0078\,\text{s}^{-1}$), in the incipient failure condition. As shown in Figure 4a, the use of the material ultimate tensile strength (σ_{UTS} = 506.46 MPa) resulted in a value for the critical distance equal to 1.702 mm. The same strategy (Figure 4b) was followed also to estimate the critical distance value under $\dot{\varepsilon}_{nom}=1.43\,\text{s}^{-1}$ and resulted in a critical distance value of 2.734 mm. Therefore, the two critical distance values allowed us to obtain the following relationship:

$$L\left(\dot{\varepsilon}_{nom}\right)=2.646\cdot\dot{\varepsilon}_{nom}^{0.09095}\ \text{(mm)} \quad (5)$$

By making use of power laws (5) the effective stress was then calculated, in the incipient failure condition, according to Point Method Theory of Critical Distances (equation 1). The results of this final re-analysis are summarized in Figure 4c, where the error is calculated as:

$$Error = \frac{\sigma_{eff}(\dot{\varepsilon}_{nom}) - \sigma_0(\dot{\varepsilon}_{nom})}{\sigma_0(\dot{\varepsilon}_{nom})} \quad (6)$$

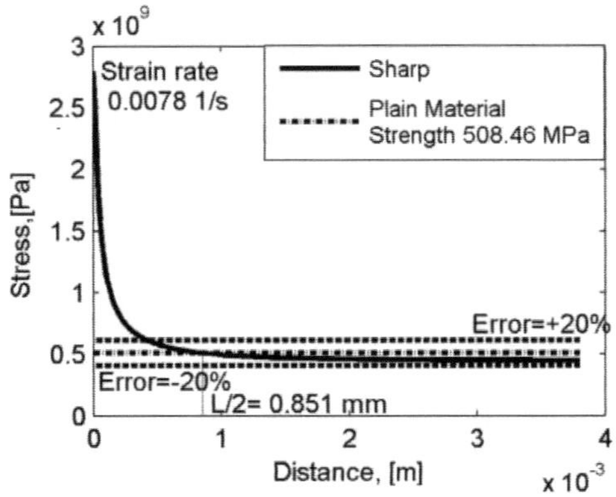

(a)

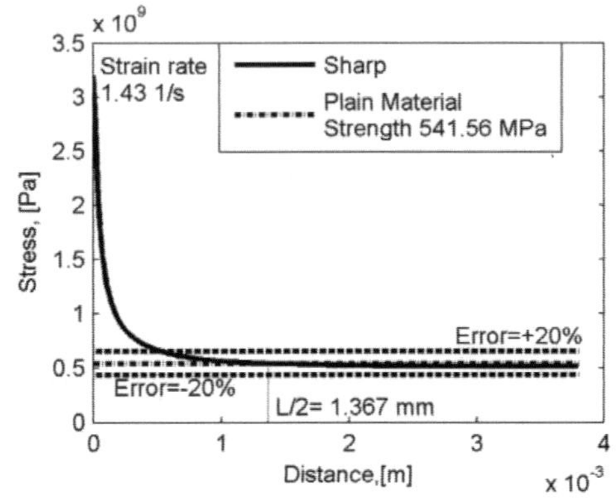

(b)

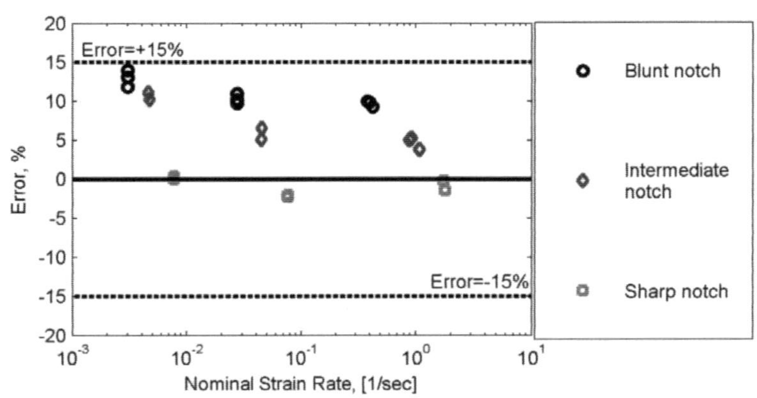

(c)

Figure 4. Local linear-elastic stress fields under $\dot{\varepsilon}_{nom} = 0.0078\,\mathrm{s}^{-1}$ and $\dot{\varepsilon}_{nom} = 1.43\,\mathrm{s}^{-1}$ (b); accuracy of the Point Method of TCD in predicting the strength (c and d).

The error diagrams reported in Figure 4c prove that the TCD was capable of accurately estimating also the strength of the notched specimens of titanium alloy Grade2, with the estimates falling within an error interval of ±15%.

3. Theory of Critical Distances based on elasto-plastic analysis

In this study considered the modification of the classical theory of critical distance in cases of elasto-plastic material behavior for improve the accuracy of estimation of lifetime of notched components, because the material behavior being, by nature, highly non-linear and cannot be described in the framework of the linear theory of elasticity. Accounting plasticity has been implemented with the use of the Simplified Johnson-Cook model.

3.1. The Simplified Johnson-Cook model

Johnson-Cook equation (7) consists of three brackets which defines the effect of the strain, strain rate and temperature on the flow curve of the tested material. The adiabatic heating effect is considered negligible for the tension tests as the material necks down at relatively low strains before any significant adiabatic heating. The first bracket includes the parameters related with the flow curve at the reference strain rate. The first bracket ($A+B\varepsilon^n$) is determined from the quasi-static tensile test at the reference strain rate of 0.0002 s^{-1}. The plastic strain is calculated by subtracting the elastic strain from the total strain as depicted in Figure 5.

$$\sigma = \left(A + B\varepsilon^n\right)\left(1 + C\ln\frac{\dot{\varepsilon}}{\dot{\varepsilon}_0}\right)\left(1 - \frac{T_m - T}{T_m - T_R}\right) \tag{7}$$

For the reference strain rate stress values are fitted with ($A+B\varepsilon^n$) as shown in Figure 6. The fitting is performed until about 0.08 strain since above this strain level necking starts. The fitting parameters for the tensile, A, B and n are shown in the same graph: A=363.1 MPa, B=389.89 MPa and n=0.435. For the second parentheses which show the strain rate effect on the deformation, the yield stress values are drawn as function strain rate as shown in Figure 7. At last to check the reliability of the model parameters we compared with the real tests result with the model results as in Figure 8 model and experimental stress values are very much similar.

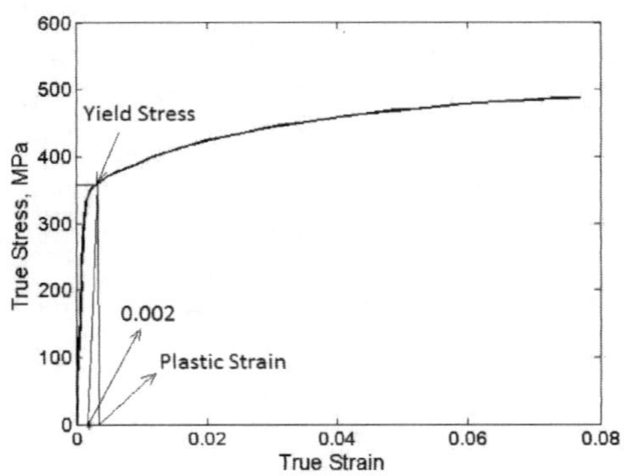

Figure 5. True stress-strain curve of tensile test at reference strain rate.

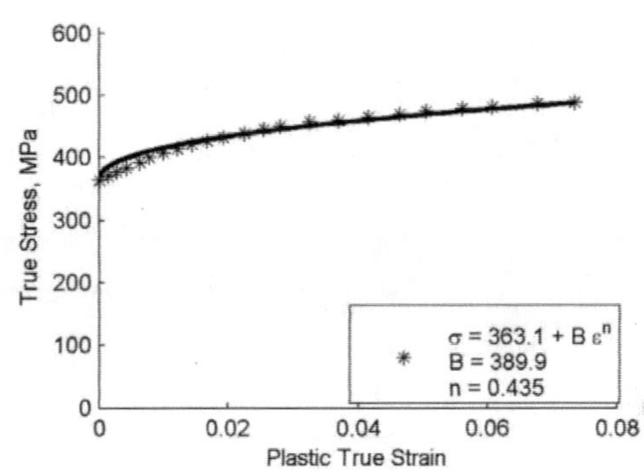

Figure 6. True plastic stress versus true plastic strain.

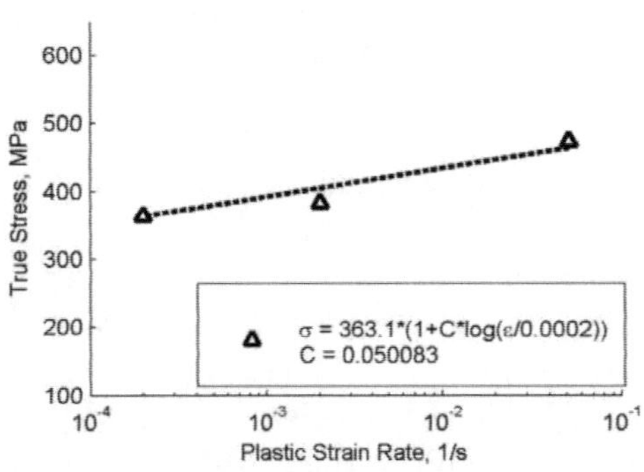

Figure 7. Yield strength vs. logarithmic strain rate curve.

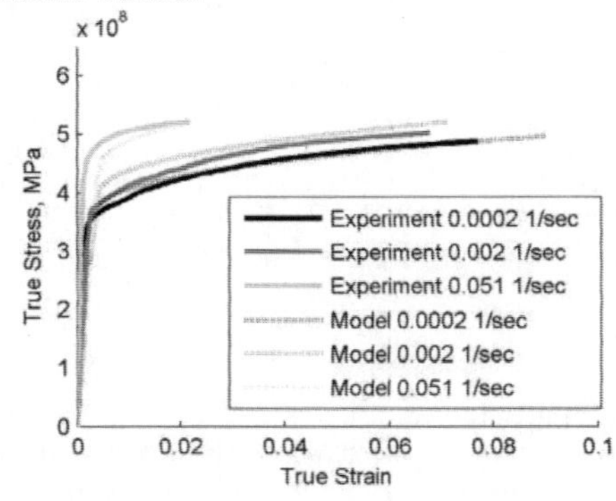

Figure 8. Johnson–Cook curve fitting result vs. experimental result.

3.2. The *Theory of Critical Distances based on elasto-plastic analysis*

The same analysis was carried out as in the previous section. However, for determine the value of the critical distance elasto-plastic stress fields were used. The chart of Figure 9 shows the elasto-plastic stress–distance curves plotted, under different strain rates. Determining the value of the critical distance for different strain rates, we can see that, in contrast to linear-elastic analysis, the value of the critical distance is a constant equal 0.5 mm.

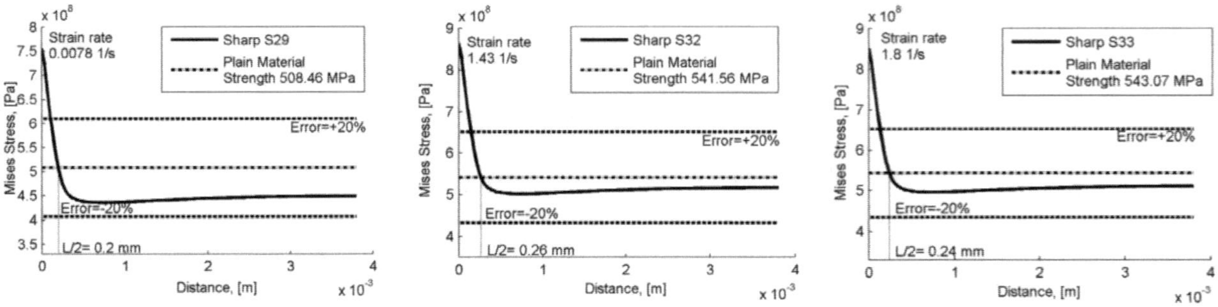

(a) (b) (c)

Figure 9. Local elasto-plastic stress fields under $\dot{\varepsilon}_{nom} = 0.0078\,\mathrm{s}^{-1}$ (a); $\dot{\varepsilon}_{nom} = 1.43\,\mathrm{s}^{-1}$ (b); $\dot{\varepsilon}_{nom} = 1.8\,\mathrm{s}^{-1}$ (c).

By making use value of critical distances the effective stress was then calculated, in the incipient failure condition, according to Point Method Theory of Critical Distances (equation 1). The results of this final re-analysis are summarized in Figure 10, where the error is calculated according equation (6):

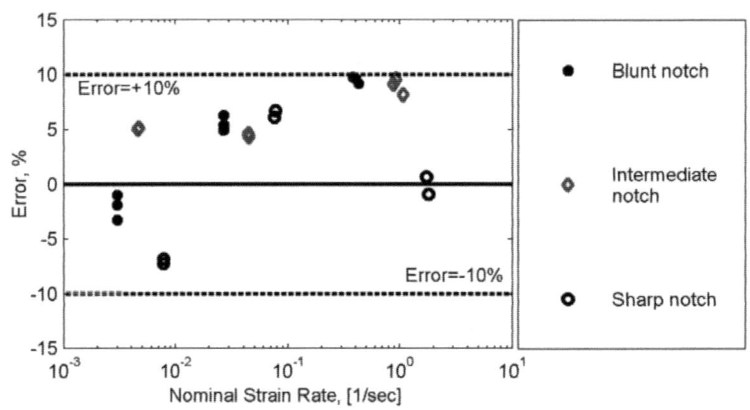

Figure 10. Accuracy of the TCD based on elasto-plastic analysis in predicting the strength.

4. Conclusions

In this study modification of the classical theory of critical distance in cases of elasto-plastic material behavior was proposed. Accounting plasticity has been implemented with the use of the Simplified Johnson-Cook model. The parameters for this model were identified from experimentally determined stress-strain curves by curve fitting techniques. The results of elasto-plastic analyses of stress distributions near a wide variety of notches are presented. The results showed that the use of the modification of the TCD based on elasto-plastic analysis gives us estimates falling within an error band of ±5-10%, that more accurate predictions than the linear elastic TCD solution. The use of an improved description of the stress-strain state at the notch tip allows introducing the critical distances as a material parameter.

Acknowledgments
The reported study was funded by the Russian Foundation for Basic Research in the framework of research projects No. 16-31-00156, No. 16-51-48003 and No. 16-48-590148

References

[1] Taylor D 2007 The Theory of Critical Distances: A New Perspecive in Fracture Mechanics (Amsterdam: Elsevier Science Limited)
[2] Neuber H 1958 *Kerbspannungslehre*, 2nd ed (Berlin: Springer-Verlag)
[3] Peterson R E 1959 Notch-sensitivity *Metal Fatigue*, edited by Sines G and Waisman J L (New York: MacGraw Hill) pp 293-306
[4] Novozhilov V V 1969 *Journal of Applied Mathematics and Mechanics* **33** 201–210
[5] Taylor D 2008 Engineering Fracture Mechanics **75** 1696-1705
[6] Taylor D 1999 *International Journal of Fatigue* **21** 413-420
[7] Taylor D 2006 *Structural Integrity and Durability* **1** 145
[8] Susmel L and Taylor D 2010 *Engineering Fracture Mechanics* **77** 470-478
[9] Yin T, Tyas A, Plekhov O, Terekhina A and Susmel L 2015 *Materials and Design* **69** 197–212
[10] Yin T, Tyas A, Plekhov O, Terekhina A and Susmel L 2014 *Fratturaed Integrita Strutturale* **30** 220-225

IOP Conf. Series: Materials Science and Engineering **208** (2017) 012043 doi:10.1088/1757-899X/208/1/012043

A functional-type a posteriori error estimate of approximate solutions for Reissner-Mindlin plates and its implementation

Maxim Frolov, Olga Chistiakova

Department of Applied Mathematics, Institute of Applied Mathematics and Mechanics, Peter the Great St. Petersburg Polytechnic University, 195251, Polytechnicheskaya st. 29, St. Petersburg, Russia

E-mail: `frolov_me@spbstu.ru`

Abstract. Paper is devoted to a numerical justification of the recent a posteriori error estimate for Reissner-Mindlin plates. This majorant provides a reliable control of accuracy of any conforming approximate solution of the problem including solutions obtained with commercial software for mechanical engineering. The estimate is developed on the basis of the functional approach and is applicable to several types of boundary conditions. To verify the approach, numerical examples with mesh refinements are provided.

1. Introduction

Different approaches to adaptive solution of boundary-value problems based on a posteriori error estimation are rapidly developing and form one of the modern research areas. Although this field is of interest for more than 30 years and the corresponding literature is vast, there is a very restricted set of papers on error estimation in the theory of plates. Most likely, the first paper on this subject appeared in 1999 [1] – authors used the constitutive relation method to propose an error estimate.

Nowadays, classical approaches (such as residual-based or postprocessing-based) dominate in the literature [2]–[18], but all these methods are essentially based on the assumption that Galerkin approximation is considered.

The functional approach of Repin and co-workers [19]–[23] is fully reliable even for commercial tools, which behave as a "black box", due to its ability to control accuracy of non-Galerkin approximations both with and without locking. The first reliable error estimate for Kirchhoff-Love plates based on this approach [24] was implemented numerically in [25]. In comparison to all functional-type error majorants obtained earlier for Kirchhoff-Love and Reissner-Mindlin plates (see [26]–[29]), the main advantage of the new result [30] consists in less restrictive sets of admissible fields.

2. Reissner-Mindlin plates

In case of linear problems the respective mathematical model describes the bending of linearly elastic plates of small to moderate thickness in terms of pair of variables in $\Omega \subset \mathbb{R}^2$: u — scalar-valued function (displacement) and θ — vector-valued function (rotations). The equilibrium equations in this case are as follows:

$$\begin{cases} -\mathbf{Div}\,(\mathrm{C}\varepsilon(\theta)) = \gamma & \text{in } \Omega, \\ -\mathbf{div}\gamma = g & \text{in } \Omega, \\ \gamma = \lambda t^{-2}(\nabla u - \theta) & \text{in } \Omega, \end{cases} \tag{1}$$

where t — thickness of a plate; $\lambda = \frac{Ek}{2(1+\nu)}$; $\varepsilon(\theta) = \frac{1}{2}(\nabla\theta + (\nabla\theta)^T)$; gt^3 represents the transverse loading; C — tensor of bending moduli; E and ν — material constants; k — correction factor.

Boundary conditions of two types are considered. Let Γ_{D} and Γ_{S} be two non-intersecting parts of the boundary Γ. Γ_{D} is a hard clamped part ($u = 0$, $\theta \cdot n = 0$, $\theta \cdot s = 0$ on Γ_{D}), Γ_{S} is a free part ($\partial u/\partial n = \theta \cdot n$, $n \cdot \mathrm{C}\varepsilon(\theta)n = 0$, $s \cdot \mathrm{C}\varepsilon(\theta)n = 0$ on Γ_{S}), where n is the outward unit normal vector to the boundary, s is the counterclockwise unit tangent vector. If Γ_{S} is a soft simply supported part, one has $u = 0$ on Γ_{S} instead of $\partial u/\partial n = \theta \cdot n$ on Γ_{S}.

Weak formulation of this problem has the following form: find a triple $(u, \theta, \gamma) \in U \times \Theta \times Q$

$$\begin{cases} \int\limits_\Omega \mathrm{C}\varepsilon(\theta) : \varepsilon(\varphi)\, d\Omega - \int\limits_\Omega \gamma \cdot \varphi\, d\Omega = 0, & \forall \varphi \in \Theta := \left\{\varphi \in \mathbb{W}_2^1(\Omega, \mathbb{R}^2) \mid \varphi = 0 \text{ on } \Gamma_{\mathrm{D}}\right\}, \\ \int\limits_\Omega \gamma \cdot \nabla w\, d\Omega = \int\limits_\Omega g w\, d\Omega, & \forall w \in U := \left\{w \in \mathbb{W}_2^1(\Omega) \mid w = 0 \text{ on } \Gamma_{\mathrm{D}}\right\}, \\ \int\limits_\Omega \left(\lambda^{-1}t^2\gamma - (\nabla u - \theta)\right) \cdot \tau\, d\Omega = 0, & \forall \tau \in Q := \mathbb{L}_2(\Omega, \mathbb{R}^2), \end{cases} \tag{2}$$

where $\mathbb{W}_2^1$ and $\mathbb{L}_2$ are standard denotations for Sobolev and Lebesgue spaces respectively. The domain Ω is assumed to be bounded and connected with Lipschitz-continuous boundary, $g \in \mathbb{L}_2(\Omega)$. C is symmetric and there exist real constants ξ_1 and ξ_2 such that $\xi_1|\varkappa|^2 \leq \mathrm{C}\varkappa : \varkappa \leq \xi_2|\varkappa|^2 \quad \forall \varkappa \in \mathbb{M}^{2\times2}_{sym}$, $|\varkappa|^2 = \varkappa : \varkappa$, where $\mathbb{M}^{2\times2}_{sym}$ — space of symmetric tensors of 2nd rank. For numerical tests the plate material is assumed to be isotropic, which implies $\mathrm{C}\varkappa = \frac{E}{12(1-\nu^2)}\left((1-\nu)\varkappa + \nu\,\mathrm{tr}\varkappa\,\mathbb{I}\right)$, where $\mathbb{I}$ — identity tensor of 2nd rank.

3. A reliable *a posteriori* error estimate

To obtain an error estimate which is robust to some hidden details of commercial software and even possible bugs of a program code, it is necessary to consider an arbitrary pair of conforming approximations $(\tilde{u}, \tilde{\theta})$ without any additional assumptions or restrictions. Therefore, the following natural error measures are introduced: $e_{\tilde{u}} = u - \tilde{u}$, $e_{\tilde{\theta}} = \theta - \tilde{\theta}$, $e_{\tilde{\gamma}} = \gamma - \tilde{\gamma}$.

Functional approach is based on the introduction of additional variables with appropriate mechanical (physical) meaning. In case of Reissner-Mindlin plates they could be a vector $\tilde{\tilde{y}}$ and a symmetric tensor $\tilde{\tilde{\varkappa}}$. The estimate proposed in [30] is exact for $\tilde{\tilde{\varkappa}} = \mathrm{C}\varepsilon(\theta)$ and $\tilde{\tilde{y}} = \gamma$. Recent numerical results of the first author obtained for classical and Cosserat elasticity show that explicit symmetry condition for tensor variable with standard finite elements yields unsatisfactory results, but implicit symmetry condition with $\tilde{\tilde{\varkappa}}$ split into parts $[\tilde{\tilde{\varkappa}}^1,\ \tilde{\tilde{\varkappa}}^2]$ and finite element approximations for mixed methods provides reliable estimates with efficient implementation. If so, one has the same functional space for computations:

$$\tilde{\tilde{y}},\ \tilde{\tilde{\varkappa}}^1,\ \tilde{\tilde{\varkappa}}^2\ \in\ \mathbb{H}(\Omega, \mathbf{div}) := \left\{y \in \mathbb{L}_2(\Omega, \mathbb{R}^2) \mid \mathbf{div} y \in \mathbb{L}_2(\Omega)\right\}. \tag{3}$$

Main result of [30] is the following inequality:

$$|\!|\!| e_{\tilde{\theta}} |\!|\!|^2 + \lambda^{-1}t^2\|e_{\tilde{\gamma}}\|^2_\Omega \leq \hat{a}^2 + \lambda^{-1}t^2\hat{b}^2, \tag{4}$$

$$\begin{aligned} \hat{a} = {} & |\!|\!|\mathrm{C}^{-1}\mathbf{sym}(\tilde{\tilde{\varkappa}}) - \varepsilon(\tilde{\theta})|\!|\!| + \mathfrak{c}_I\|\mathbf{skew}(\tilde{\tilde{\varkappa}})\|_\Omega + \\ & + \mathfrak{c}_{II}\mathfrak{c}_{III}\sqrt{|\Omega|\ \|g + \mathbf{div}\tilde{\tilde{y}}\|^2_\Omega\ +\ |\Gamma_{\mathrm{S}}|\ \|\tilde{\tilde{y}} \cdot n\|^2_{\Gamma_{\mathrm{S}}}} + \\ & + \mathfrak{c}_{IV}\sqrt{|\Omega|\ \|\tilde{\tilde{y}} + [\mathbf{div}\tilde{\tilde{\varkappa}}^1, \mathbf{div}\tilde{\tilde{\varkappa}}^2]\|^2_\Omega\ +\ |\Gamma_{\mathrm{S}}|\ \|[\tilde{\tilde{\varkappa}}^1 \cdot n, \tilde{\tilde{\varkappa}}^2 \cdot n]\|^2_{\Gamma_{\mathrm{S}}}}, \end{aligned} \tag{5}$$

$$\hat{b} = \|\tilde{\tilde{y}} - \tilde{\gamma}\|_\Omega + \mathfrak{c}_{III}\sqrt{|\Omega|\ \|g + \mathbf{div}\tilde{\tilde{y}}\|_\Omega^2\ +\ |\Gamma_S|\ \|\tilde{\tilde{y}} \cdot n\|_{\Gamma_S}^2}, \tag{6}$$

where

$$\|e_{\tilde{\gamma}}\|_\Omega^2 := \int_\Omega |e_{\tilde{\gamma}}|^2\, dx, \qquad \|\!|e_{\tilde{\theta}}\|\!|^2 := \int_\Omega \mathbb{C}\varepsilon(e_{\tilde{\theta}}) : \varepsilon(e_{\tilde{\theta}})\, dx \tag{7}$$

and the auxiliary constants are mesh-independent and come from the following inequalities:

$$\begin{gathered} \|\nabla\varphi\|_\Omega^2 \le \mathfrak{c}_I^2 \|\!|\varphi\|\!|^2, \quad \|\varphi\|_\Omega^2 \le \mathfrak{c}_{II}^2 \|\!|\varphi\|\!|^2, \\ \tfrac{1}{|\Omega|}\|w\|_\Omega^2 + \tfrac{1}{|\Gamma_S|}\|w\|_{\Gamma_S}^2 \le \mathfrak{c}_{III}^2 \|\nabla w\|_\Omega^2, \quad \tfrac{1}{|\Omega|}\|\varphi\|_\Omega^2 + \tfrac{1}{|\Gamma_S|}\|\varphi\|_{\Gamma_S}^2 \le \mathfrak{c}_{IV}^2 \|\!|\varphi\|\!|^2. \end{gathered} \tag{8}$$

4. Implementation and numerical results

Well-known Raviart–Thomas approximation was introduced in [31] on a basis of mapping of the reference square $\hat{\mathcal{K}} = (-1,1) \times (-1,1)$ to arbitrary quadrilateral (degrees of freedom are normal components of a vector-field in the midpoints): $\mathcal{RT}_0(\hat{\mathcal{K}}) = \mathcal{P}_{1,0}(\hat{\mathcal{K}}) \times \mathcal{P}_{0,1}(\hat{\mathcal{K}})$, where $\mathcal{P}_{i,j}(\hat{\mathcal{K}})$ – space of polynomials over $\hat{\mathcal{K}}$ power of i or less on $\hat{x}_1$ and j – on $\hat{x}_2$. Here $\hat{x}_1$ and $\hat{x}_2$ are coordinates in a local element coordinate system. In 2005 a new type of approximation was proposed by Arnold, Boffi and Falk [32]. It is based on $\mathcal{ABF}_0(\hat{\mathcal{K}}) = \mathcal{P}_{2,0}(\hat{\mathcal{K}}) \times \mathcal{P}_{0,2}(\hat{\mathcal{K}})$ – a wider space that includes the original degrees of freedom on edges and two additional internal degrees of freedom on every element.

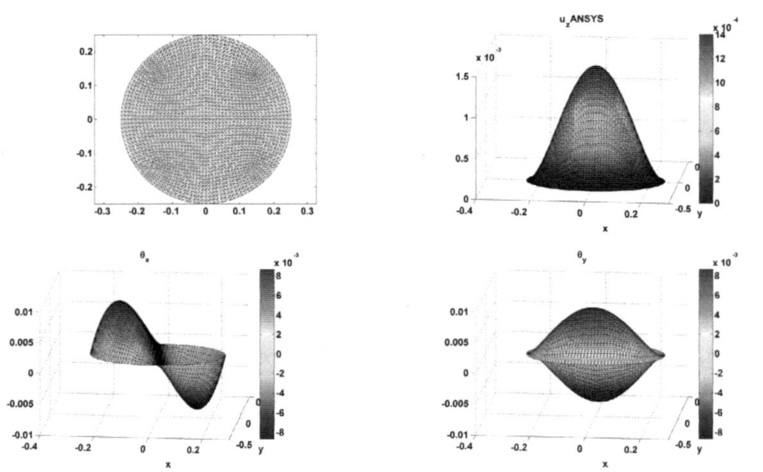

Figure 1. Example 1: ANSYS solution (auxiliary mesh, displacement and rotations).

In engineering practice the problem is usually solved using commercial software, which is convenient and easy to use, but keeps the implementation hidden. Moreover, algorithms may vary and behave different from version to version. Next, the variety of the elements may be confusing as well as a long list of relevant papers in documentation. For example, ANSYS Mechanical provides SHELL, SOLID and other finite elements for plate modeling and papers such as [33] and [34] are cited in ANSYS manual. Thus, given a numerical solution, it is hard or even impossible to tell if some incorrect settings or a locking effect appeared during computations. So, it is an obvious demand for a tool that would provide reliable and efficient error control even on coarse meshes and would be applicable to non-Galerkin approximations as

IOP Conf. Series: Materials Science and Engineering **208** (2017) 012043 doi:10.1088/1757-899X/208/1/012043

well as for Galerkin ones. We suggest functional-type error estimates as such a tool and provide some numerical experiments to verify this.

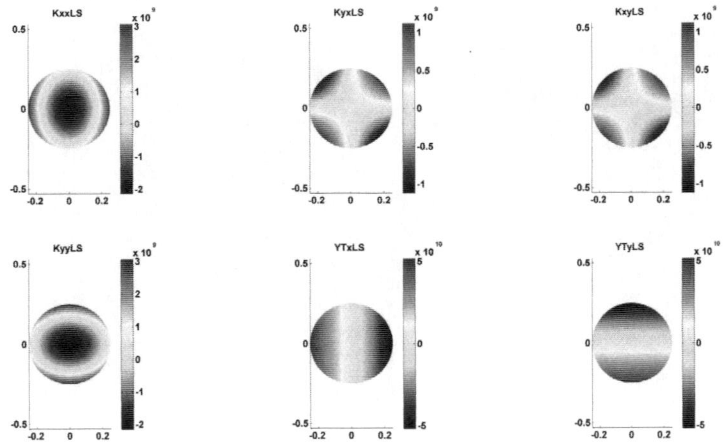

Figure 2. Example 1: components of $\tilde{\tilde{\varkappa}}$ marked by KxxLS, KyxLS, etc. and components of $\tilde{y}$ marked by YTxLS and YTyLS, where "LS" means that fields come from solving of the corresponding linear system of algebraic equations.

4.1. Example 1. Circular plate.
As a first numerical example a circular plate is considered. For material and shape parameters ANSYS Verification Manual (VM138, Timoshenko S.) was used. Radius of the plate is 0.25 m, thickness — 0.0025 m, $E = 2e11$ N/m^2, $\nu = 0.3$. The plate is uniformly loaded (6585.175 N/m^2). ANSYS solution of this problem (see figure 1, where x, y, z are spatial coordinates) was obtained using SHELL181-element, and the results of error estimation are set out in table 1. In figure 1 auxiliary splitting of a quadrilateral mesh into triangles is used for output. From the results collected in table 1 one can conclude that overestimation of the true error (square root of the left-hand side of (4)) by the majorant (square root of the right-hand side of (4)) is a stable non-increasing constant. Corresponding free variables are presented in figure 2. If Galerkin approximations are used for this problem, locking effect takes place – the displacement is underestimated over than 5 times

Table 1. Example 1: error estimation for subsequent refinements.

DOFs	123	435	1635	6339
Error	0.689e5	0.359e5	0.181e5	0.842e4
Estimate	0.101e6	0.527e5	0.266e5	0.133e5
Overestimation	1.5	1.5	1.5	1.5

As a posteriori error estimates are usually represented as a sum of local errors, they may be used for the construction of adaptive algorithms of mesh refinement. The main idea is to reduce

IOP Conf. Series: Materials Science and Engineering **208** (2017) 012043 doi:10.1088/1757-899X/208/1/012043

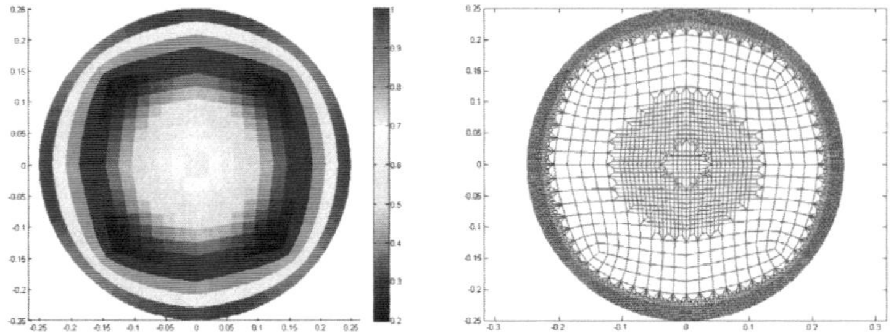

Figure 3. Example 1: local error indicator for a coarse mesh and an adaptive refinement.

Table 2. Example 2: error estimation for subsequent refinements.

DOFs	195	675	2499	9603
Error	0.175e3	0.953e2	0.487e2	0.242e2
Estimate	0.399e3	0.211e3	0.108e3	0.559e2
Overestimation	2.3	2.2	2.2	2.3

computational costs by refining only areas with a high estimated error instead of the whole mesh. Thus, a reliable error indicator is necessary for any adaptation algorithm. Indicator based on the majorant from (4) and an adapted mesh are shown in figure 3. Method proposed in [35] was used in mesh refinements for quadrilaterals.

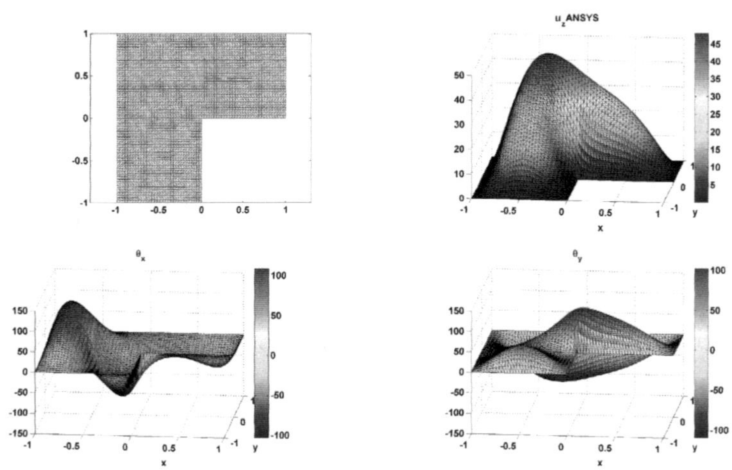

Figure 4. Example 2: ANSYS solution (auxiliary mesh, displacement and rotations).

4.2. Example 2. L-shaped plate.

Next example is L-shaped plate with moderate thickness and artificial material constants selected to provide a moderate locking. The side of the plate is 2 m, thickness — 0.1 m, $E = 1$ N/m^2, $\nu = 0.3$. The plate was uniformly loaded (1 N/m^2). ANSYS solution is depicted in figure 4, the results of error estimation by (4) are presented in table 2, while corresponding free variables are shown in figure 5. All previous conclusions are also valid for this case: error overestimation does not increase with the number of DOFs. Adaptive algorithm was also implied for this example: see figure 6 for initial functional-type error indicator and obtained refined mesh.

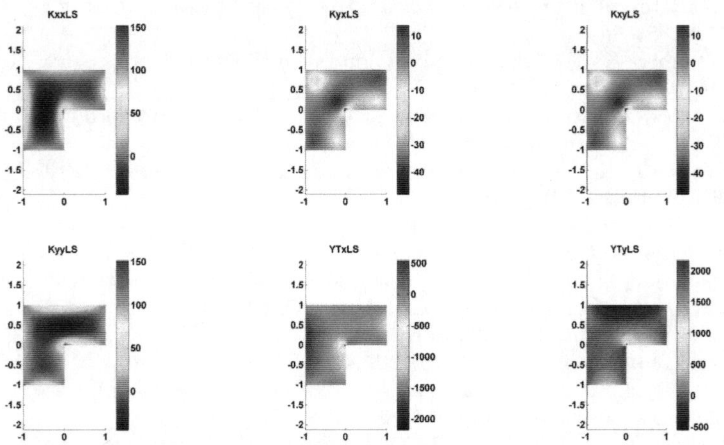

Figure 5. Example 2: components of $\tilde{\tilde{\varkappa}}$ marked by KxxLS, KyxLS, etc. and components of $\tilde{y}$ marked by YTxLS and YTyLS, where "LS" means that fields come from solving of the corresponding linear system of algebraic equations.

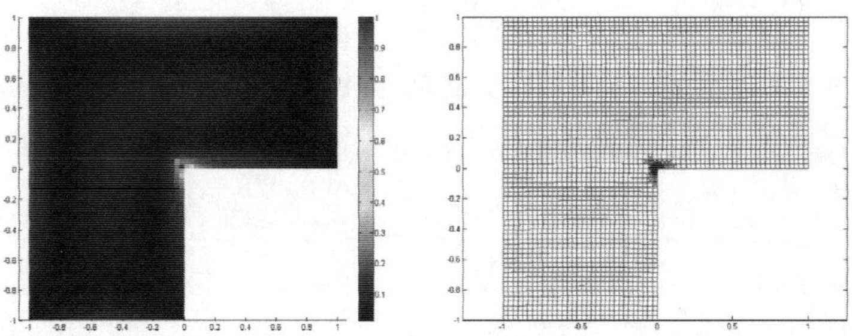

Figure 6. Example 2: local error indicator for a coarse mesh and an adaptive refinement.

5. Conclusions

From the presented results we conclude that the functional approach to error estimation of approximate solutions for Reissner-Mindlin plates is suitable even for coarse meshes. Arnold-Boffi-Falk approximation provides an efficient implementation of obtained majorant. Therefore, the approach forms a good independent error control tool, which is applicable to commercial software.

Acknowledgments
Work is supported by the Grant of the President of the Russian Federation MD-1071.2017.1.

References
[1] Boisse P, Perrin S, Coffignal G and Hadjeb K 1999 Error estimation through the constitutive relation for Reissner-Mindlin plate bending finite elements *Comput. Struct.* **73** 615–27
[2] Liberman E 2001 A posteriori error estimator for a mixed finite element method for Reissner-Mindlin plate *Math. Comput.* **70** 1383–96
[3] Carstensen C and Weinberg K 2001 Adaptive mixed finite element method for Reissner-Mindlin plate *Comput. Methods Appl. Mech. Eng.* **190** 51–52
[4] Carstensen C 2002 Residual-based a posteriori error estimate for a nonconforming Reissner-Mindlin plate finite element *SIAM J. Numer. Anal.* **39** 2034–44
[5] Carstensen C and Weinberg K 2003 An adaptive non-conforming finite-element method for Reissner-Mindlin plates *Int. J. Numer. Methods Eng.* **56** 2313–30
[6] Ludwig M, Shaw S and Whiteman J 2003 Adaptive numerical solution of thick plates using first-order shear deformation theory. Part I: Error estimates *Numer Methods Partial Differ Equ* **19** 44-66
[7] Boroomand B, Ghaffarian M and Zienkiewicz O 2004 On application of two superconvergent recovery procedures to plate problems *Int. J. Numer. Methods Eng.* **61** 1644–73
[8] Lovadina C and Stenberg R 2005 A posteriori error analysis of the linked interpolation technique for plate bending problems *SIAM J. Numer. Anal.* **43** 2227–49
[9] Carstensen C and Schöberl J 2006 Residual-based a posteriori error estimate for a mixed Reissner-Mindlin plate finite element method *Numer. Math.* **103** 225-50
[10] Beirão da Veiga L, Niiranen J and Stenberg R 2007 A family of C^0 finite elements for Kirchhoff plates. I: Error analysis *SIAM J. Numer. Anal.* **45** 2047–71
[11] Beirão da Veiga L, Chinosi C, Lovadina C and Stenberg R 2008 A priori and a posteriori error analysis for a family of Reissner-Mindlin plate elements *BIT* **48** 189–213
[12] Carstensen C and Hu J 2008 A posteriori error analysis for conforming MITC elements for Reissner-Mindlin plates *Math. Comput.* **77** 611-32
[13] Hu J and Huang Y 2010 A posteriori error analysis of finite element methods for Reissner-Mindlin plates *SIAM J. Numer. Anal.* **47** 4446-72
[14] Beirão da Veiga L, Niiranen J and Stenberg R 2010 A posteriori error analysis for the Morley plate element with general boundary conditions *Int. J. Numer. Methods Eng.* **83** 1–26
[15] Carstensen C, Xie X, Yu G and Zhou T 2011 A priori and a posteriori analysis for a locking-free low order quadrilateral hybrid finite element for Reissner-Mindlin plates *Comput. Methods Appl. Mech. Eng.* **200** 1161–75
[16] Creuse E, Nicaise S and Verhille E 2011 Robust equilibrated a posteriori error estimators for the Reissner-Mindlin system *Calcolo* **48** 307–35
[17] Castellazzi G, de Miranda S and Ubertini F 2011 Patch based stress recovery for plate structures *Comput. Mech.* **47** 379–94
[18] Beirão da Veiga L, Niiranen J and Stenberg R 2013 A posteriori error analysis for the postprocessed MITC plate elements *SIAM J. Numer. Anal.* **51** 1–23
[19] Repin S and Xanthis L S 1996 A posteriori error estimation for elastoplastic problems based on duality theory *Comput. Methods Appl. Mech. Engrg.* **138** 317–39
[20] Repin S 2000 A posteriori error estimation for variational problems with uniformly convex functionals *Math. Comput.* **69** 481–500
[21] Neittaanmäki P and Repin S 2004 *Reliable Methods for Computer Simulation – Error Control and A Posteriori Estimates* (Amsterdam: Elsevier)
[22] Repin S 2008 *A Posteriori Estimates for Partial Differential Equations* vol 4 (Berlin: Radon Series on Computational and Applied Mathematics, de Gruyter)
[23] Mali O, Neittaanmäki P and Repin S 2014 *Accuracy verification methods: Theory and algorithms* vol 32 Computational Methods in Applied Sciences (Berlin: Springer).
[24] Neittaanmäki P and Repin S 2001 A posteriori error estimates for boundary-value problems related to the biharmonic operator *East-West J. Numer. Math.* **9** 157–78
[25] Frolov M, Neittaanmäki P and Repin S 2004 *Proc. of the European congress on computational methods in applied sciences and engineering* (Jyväskylä, Finland)

[26] Frolov M 2004 On efficiency of the dual majorant method for the quality estimation of approximate solutions of fourth-order elliptic boundary value problems *Russ. J. Numer. Anal. Math. Model.* **19** 407–18.

[27] Repin S and Frolov M 2006 Estimation of deviations from the exact solution for the Reissner-Mindlin plate problem *J. of Math. Sci.* **132** 331–8

[28] Frolov M, Neittaanmäki P and Repin S 2006 Guaranteed functional error estimates for the Reissner-Mindlin plate problem *J. of Math. Sci.* **132** 553–61

[29] Frolov M 2010 Functional a posteriori error estimates for certain models of plates and beams *Russ. J. Numer. Anal. Math. Model.* **25** 117–29

[30] Frolov M 2014 *Proc. of 10th Int. Conf. Mesh methods for boundary-value problems and applications* (Kazan, Russia) pp 610–5

[31] Raviart P and Thomas J 1977 *Lecture Notes in Mathematics* (Berlin: Springer)

[32] Arnold D, Boffi D and Falk R 2005 Quadrilateral $H(\mathrm{div})$ finite elements *SIAM J. Numer. Anal.* **42** 2429–51

[33] Bathe K and Dvorkin E 1986 A formulation of general shell elements – the use of mixed interpolation of tensoral components *Int. J. Numer. Methods Eng.* **22** 697–722

[34] Simo J and Armero F 1992 Geometrically nonlinear enhanced strain mixed methods and the nethod of incompatible modes *Int. J. Numer. Methods Eng.* **33** 1413-49

[35] Karavaev A and Kopysov S 2013 A refinement of unstructured quadrilateral and mixed meshes *Vestn. Udmurtsk. Univ. Mat. Mekh. Komp. Nauki* **4** 62–78

IOP Conf. Series: Materials Science and Engineering **208** (2017) 012044 doi:10.1088/1757-899X/208/1/012044

Natural convection in a liquid metal locally heated from above

R Khalilov, I Kolesnichenko, A Teimurazov, A Mamykin and P Frick

Institute of Continuous Media Mechanics UB RAS, 1, Acad. Korolev St., Perm, 614013, Russia

E-mail: khalilov@icmm.ru

Abstract. A convective flow of liquid sodium generated nearby a hot round in the upper solid end face of a vertical cylinder has been studied experimentally and numerically. A developed turbulent flow is observed in the upper part of the cylinder. Strong velocity pulsations penetrate in the bulk of the metal up to a distance of about the diameter of the cylinder. Mean velocity fields reveal a toroidal vortex, which is localized in a narrow upper zone. Numerical simulations were done for two types of thermal boundary conditions (BCs): fixed temperature and fixed homogeneous heat flux on both heat exchangers. Experimental values of time-averaged velocity and temperature in the vortex are in good agreement with numerical data. The size and the intensity of the vortex weakly depend on BCs. The whole bulk of the metal is not involved in the motion. The temperature field depends much more on the BCs. Under fixed heat fluxes the temperature pulsations become much stronger and penetrate essentially deeper in the liquid metal, though the flow is slightly stronger under fixed boundary temperature. The considered flow is supposed to be a simplified model of the liquid magnesium flow in a reactor of metallothermic titanium reduction.

1. Introduction

Homogeneous heating of fluid from above provides a stable stratification and does not cause convective flows. Local heating on the free surface of fluid activates nearby the surface vigorous motion due to both surface-tension-driven (thermocapillary) and buoyancy-driven (thermo-gravitational) convection. To isolate the latter mechanism one consider a covered surface (solid upper boundary). Some exact solutions are known for simplest statement, e.g. for laminar flow near a linear heat source disposed on the surface of liquid [1]. In some technological apparatus, intense turbulent convective flows are produced by strong local heating from above. This is for example the case of flows generated in the reactor of metallothermic titanium reduction [2].

The reactor for titanium reduction is a cylindrical vessel with a diameter up to 2 m and a height up to 4 m filled by liquid magnesium at a temperature of about 850°C. The chemical reaction of titanium reduction starts when titanium tetrachloride $TiCl_4$ is supplied to the magnesium surface. The reaction is highly heat-producing and runs predominantly on the surface producing spongy titanium and magnesium dichloride $MgCl_2$ that sink to the retort bottom. The strong release of heat (1700 kJ per 1 kg of titanium tetrachloride), compensated by cooling at the sidewall of the reactor produces a radial temperature gradient which can reach hundreds of degrees per meter. However, it is unknown how strong can be the generated convective flow and how deep penetrate the velocity and temperature pulsations. The large mass

IOP Conf. Series: Materials Science and Engineering **208** (2017) 012044 doi:10.1088/1757-899X/208/1/012044

and dimensions of the vessel and very high temperatures complicate the direct experimental measurements inside the reactor, which is one of the motivations for numerical study of the process [2, 3]. Adequate simulations of turbulent convective flows of liquid metals under high governing parameters, typical for industrial systems, become possible not long ago [4, 5] and numerical codes require verification using reliable experimental data.

Liquid metals are characterized by high thermal diffusivity (low Prandtl number), which reduces the contribution of convection in heat transfer. It means that intense convective flow can be provided by a strong heat injection only. The experimental studies of liquid metal convection is also made difficult by complexity of thermal and dynamical measurements in liquid metals. There are few experimental works on convective flows, made on mercury or sodium (e.g., [6, 7, 8]). A convective flow in $GaInZn$ alloys heated locally from above and affected by magnetic field has been studied in context of the problem of electron beam evaporation of liquid metals [9].

In this work we consider a flow of liquid metal inside a vertical cylinder. The central part of the upper end face is heated, while the circumference of the same face is cooled. In this statement, the problem can be considered as a simple hydrodynamic model of a real reactor and allows both experimental and numerical study of the convective flow. Experimentally we study the flow of liquid sodium in a cylinder of diameter $D = 200$ mm and height $H = 700$ mm, which can be considered as infinite deep because the flow is concentrated in the very top of the cylinder. Numerically we study the convective flow in a three-dimensional non-stationary formulation, which allows us to analyze the instantaneous and average characteristics of the flow and the velocity and temperature pulsation fields. We aim to validate our numeric code using experimental data and to find out the structure of the convective vortex flow, analyzing how deep the velocity and temperature pulsations penetrate in the liquid metal bulk.

2. Methods

2.1. Experiment

The experimental setup designed to study the convection in liquid sodium with local heating from above consists of a cylindrical channel 1 (figure 1a), a heat exchanger 2, a pouring system 3, an expansion vessel 4 and a frame on which the installation is fixed. The channel 1 is made from stainless steel. The wall thickness is 8 mm, the channel length is 700 mm, the inner diameter is 200 mm. The channel is installed on the frame vertically. The bottom of the channel is connected to the pouring system. On the side wall of the channel there are ten locks 5, forming two vertical rows, opposite each other. Through these locks, the thermocouples of diameter 1 mm are inserted inside the channel. Five thermocouples are inserted to a depth of 100 mm (D1–D5), and the other five to a depth of 10 mm (N1–N5). The distance from the upper thermocouple to the end face is 40 mm. The distance between two adjacent thermocouples is 30 mm. To determine the temperature distribution along the entire cylinder, two sets of additional thermocouples (External 1 and 2 on figure 1a) with a step of 75 mm in each set are installed on channel's wall from the outside. The angle between the sets is 90°. One more additional thermocouple (Bottom) measured the temperature at the bottom of the cylinder. At the upper boundary of the cylindrical channel, a flange is mounted to connect the heat exchanger. From the outside the channel is covered by mineral wool insulation with a thickness of about 50 mm, and additionally covered with aluminum foil. The internal volume of the unit can be connected with vacuum and argon systems.

The heat exchanger at the upper boundary of the channel provides heating of sodium in the central region and cooling in the peripheral region (figure 1b). Heating is carried out by spiral wire 6, which is wrapped around the copper cylinder 7 of diameter 40 mm. The lower end face of this cylinder contacts the liquid sodium, and the upper end face is connected to the expansion vessel 4. There is a small hole through the cylinder axis for passing the sodium to the expansion vessel. Additionally three thermocouples (H1–H3) are placed in the hole. These thermocouples

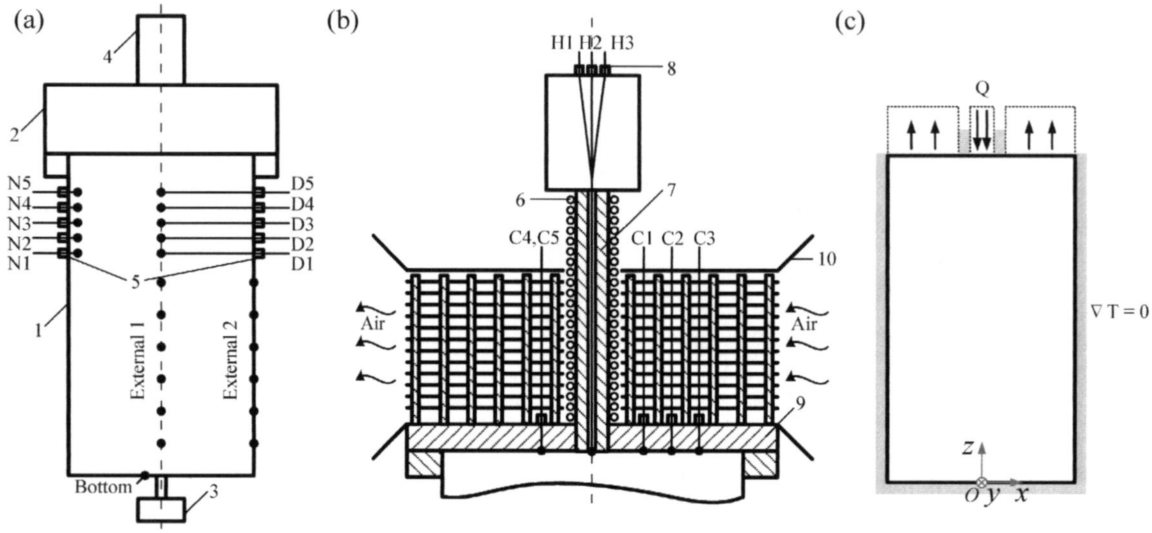

Figure 1. Experimental setup (a), heat exchanger (b) computational domain with boundary conditions (c).

measure the temperature at the interface between the copper cylinder and liquid sodium. The thermocouples are going out through the locks 8 installed on the cover of the expansion vessel.

The main element of the cooling system is a 20 mm thickness copper plate 9. In the copper plate, five thermocouples are installed through special locks. Three locks are placed on the line along the channel radius with 18 mm pitch (C1–C3). Two other thermocouples (C4, C5) are placed in azimuth with an angle of 120° at the radius of the middle thermocouple C2. These thermocouples measure the sodium temperature at the boundary of the cooling region. The cooling system is placed into the box-type casing of the external air-flux system 10.

Neighboring thermocouples in the rows were used to estimate the mean velocity by cross-correlation analysis. The correlation method is useful to estimate the velocity in turbulent nonisothermal flows in which the temperature perturbations move with the fluid [10, 8]. In this method, the average fluid velocity in the volume between two thermocouples are determined using the position of the maximum of the cross-correlation function, which gives the mean transfer time, and the distance between the thermocouples.

2.2. Mathematical model

The computational domain is a cylinder with the height $H = 700$ mm and the diameter $D = 200$ mm (figure 1c). The convective parameters of the fluid, corresponding to liquid sodium at a temperature 125°C, are: Prandtl number $Pr = 1.01 \cdot 10^{-2}$, kinematic viscosity $\nu = 6.72 \cdot 10^{-7}$ m^2/s, and thermal expansion coefficient $\beta = 2.195 \cdot 10^{-4}$ K^{-1}. The numerical code solves the Oberbeck-Boussinesq equations of thermogravitational convection. The problem was solved in a three-dimensional nonstationary formulation. Simulations were done using a numerical mesh with a total of 2.4 million grid points. The mesh was non-uniform with a higher density of points near the boundaries in order to resolve the boundary layers. The Large Eddy Simulation (LES) approach is used for modelling turbulence, namely the Smagorinsky-Lilly model [11] with the Smagorinsky constant $C_s = 0.14$ and turbulent Prandtl number $Pr_t = 0.9$. All the simulations were run using the free and open source finite volume code OpenFOAM 4.0. The PISO (Pressure Implicit with Splitting of Operators) algorithm based on the pressure correction procedure is used to solve the derived system of equations. The terms with time derivatives are discretised using an implicit Euler scheme. The convective terms are calculated by the TVD (Total Variation Diminishing) scheme [12]. The Courant number in our computations does not exceed 0.5. The numerical simulations were performed using Uran supercomputer of IMM UB RAS.

The simulations were done with two types of boundary conditions (BC) for temperature on the top surface. BC I – first-type (Dirichlet) thermal BC, fixed temperature on the heater equal to 350°C and fixed temperature on the cooler equal to 125°C. BC II – second-type (Neumann) BC, fixed homogeneous heat flux on the heater and on the cooler. The total flux $Q = 1$ kW, thereby the heat flux density on the heater is $q_h = 795.8$ kW/m^2, the heat flux density on the cooler is $q_c = -37.9$ kW/m^2. The ring with inner diameter 40 mm and with outer diameter 80 mm, which is placed between the heater and cooler, is heat insulated. The sidewalls are heat insulated. The no-slip velocity conditions are applied at all boundaries.

3. Results

Experimental setup allowed us to set the power of the heater and the air flow rate in the cooler. Setting these parameters we waited for a saturated regime, which established in some hours after the beginning of heating. Then the record of signals from all thermocouples was performed during few hours more. Figure 2 shows time-averaged temperature distribution along the cylinder (a) and the power spectrum of temperature pulsations for the uppermost thermocouple, set on the channel axis, D5 (b). The form of the power spectrum of temperature pulsations for this thermocouple indicates a developed turbulent flow in this area. The spectra of other thermocouples show a transient flow regime, and the thermocouple N1 (figure 3b) displays very weak pulsations of temperature (if a flow exists in its vicinity, than it is a laminar flow). Thus, the turbulent convective flow is recorded only by the upper probe under the hot spot of the heat exchanger. The slope of "$-5/3$" is shown on the graphs as the reference point and not for discussing the existence of the inertial interval in the spectrum, which can be expected at higher frequencies. In the range of distances, covered by thermocouples, the mean temperature gradually decreases to the cylinder bottom, following a similar law inside the cylinder and on the outer side of the cylinder wall (figure 2a).

To estimate the mean vertical velocity along both lines of thermocouples, we have calculated the cross-correlation function for the signals of neighbouring thermocouples. The location of the maximum of cross-correlation and the distance between corresponding thermocouples allow to estimate the vertical velocity averaged over the region in the between of thermocouples and under the time. The accuracy of this estimation in our case is low because the pulsations are stronger than the mean values. The estimations done gave values of average velocity in the range $5 - 15$ mm/s.

In numerical simulations we tried to reproduce the conditions of the experiment as close as possible, but the temperature distribution on the faces in the experiments is non-uniform and it is difficult to measure the actual temperature field on the heater and cooler. In the experimental setup the total power of the heater was fixed, but neither the distribution of heat flux density, nor the distribution of the temperature on the upper boundary could be fixed or controlled, because the thermal diffusivity of the copper is only a little higher than the thermal diffusivity of sodium. Moreover, the position of the heater, which was installed inside the cooler, made impossible to avoid some direct heat flux from the heater to cooler. Therefore, the numerical study was performed for two idealized boundary conditions, fixed temperature on the surface of the heater and cooler (BC I), and fixed homogeneous heat flux on both heat exchangers (BC II).

Figure 4 shows the temperature fields obtained in numerical simulations with BC II for horizontal cross-sections on the distance 20 mm and 40 mm from the top surface. The temperature distribution near the top surface in both instant and time-averaged fields (figure 4a,b) has a region with pronounced temperature maximum right below the heater. The

IOP Conf. Series: Materials Science and Engineering **208** (2017) 012044 doi:10.1088/1757-899X/208/1/012044

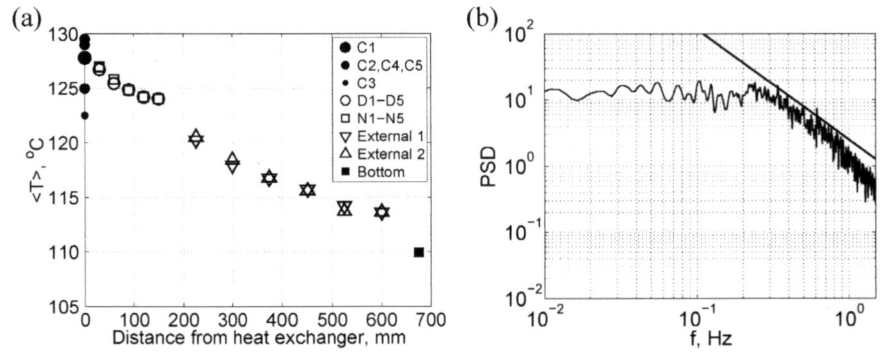

Figure 2. Time-averaged temperature distribution along the cylinder (a) and power spectrum of temperature pulsations for the thermocouple D5 (b).

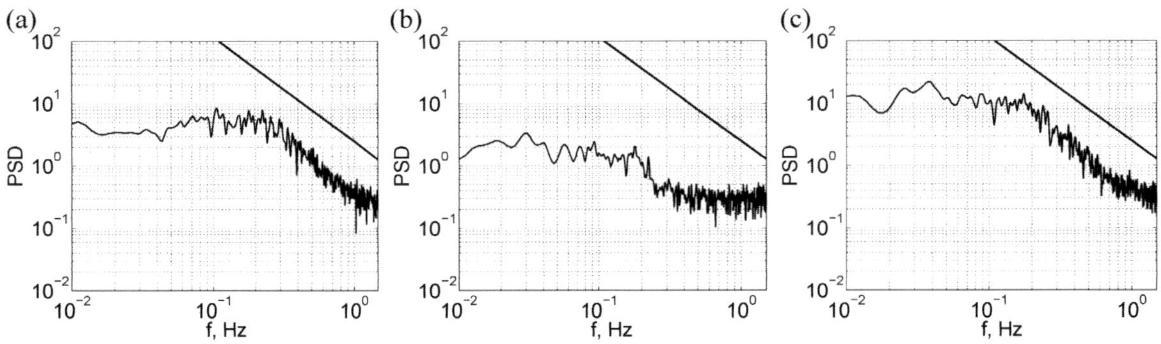

Figure 3. Power spectra of temperature pulsations for thermocouples D1 (a), N1 (b) and N5 (c).

field of temperature standard deviation (figure 4c) shows that the most intense pulsations occur in the ring region which separates the heater and cooler, and in the region close to the sidewalls. On the distance 40 mm from the top, where upper thermocouples were placed in the experiment, the temperature distributions (figure 4d-f) are very different from the previous case: there is no temperature maximum in the center and the amplitude of temperature variations is much lower.

Figure 5 presents the velocity fields in the vertical cross-section. The instant velocity field simulated with BC II (figure 5a) reveals a developed turbulent flow in the upper part of the cylinder. The intensity of nonstationary vortices decreases with depth and vanishes at a depth of the order of the cylinder diameter. The time-averaged velocity field (figure 5b) demonstrates a ring vortex near the upper face and a very weak counter-rotating vortex under it. The standard deviation of velocity fields (figure 5c) shows that the strongest pulsations are concentrated in the area of the upper vortex and that strong pulsations are observed on the distance up to 200 mm from the top face.

The difference in the flow structure between simulations with BC II and BC I can be revealed by comparing the top panel (a)-(c) and the bottom panel (d)-(f) in figure 5. In case of fixed temperature on the top face (BC I) the height of the upper vortex is lower but its intensity is higher (figure 5e). The figure 5f shows that the vortex itself becomes more stable, however the most pronounced pulsations occur in the central part of the vessel, in the region below the heater.

Figure 6a demonstrates time-averaged profiles of vertical velocity component $< V_z >$ along

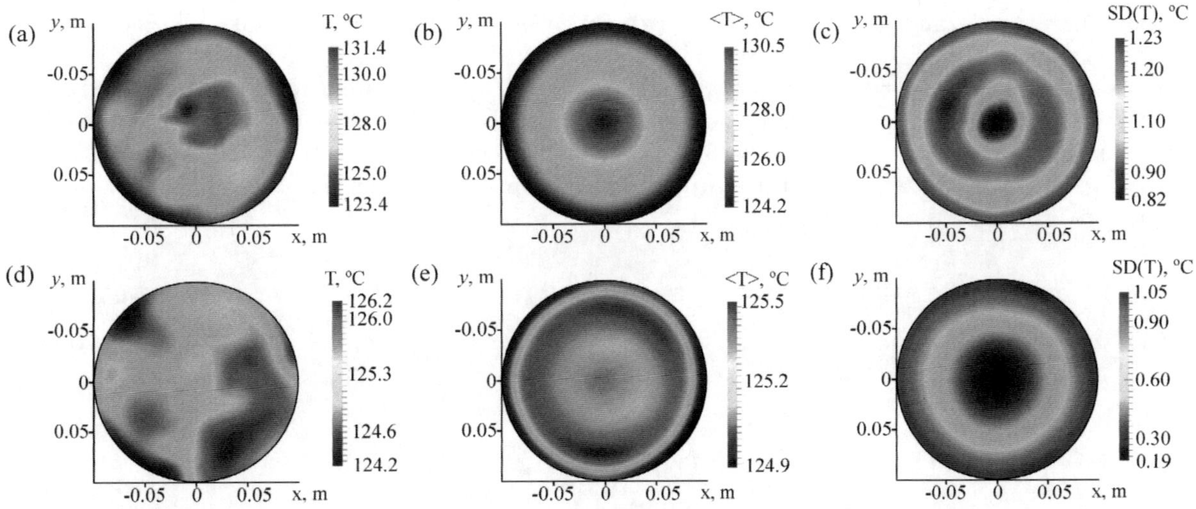

Figure 4. Results of simulations with BC II in the xOy plane: $z = 0.68$ m (a) – (c), $z = 0.66$ m (d) – (f). Instant temperature field (a), (d); time-averaged temperature field (b), (e); standard deviation of temperature field (c), (f).

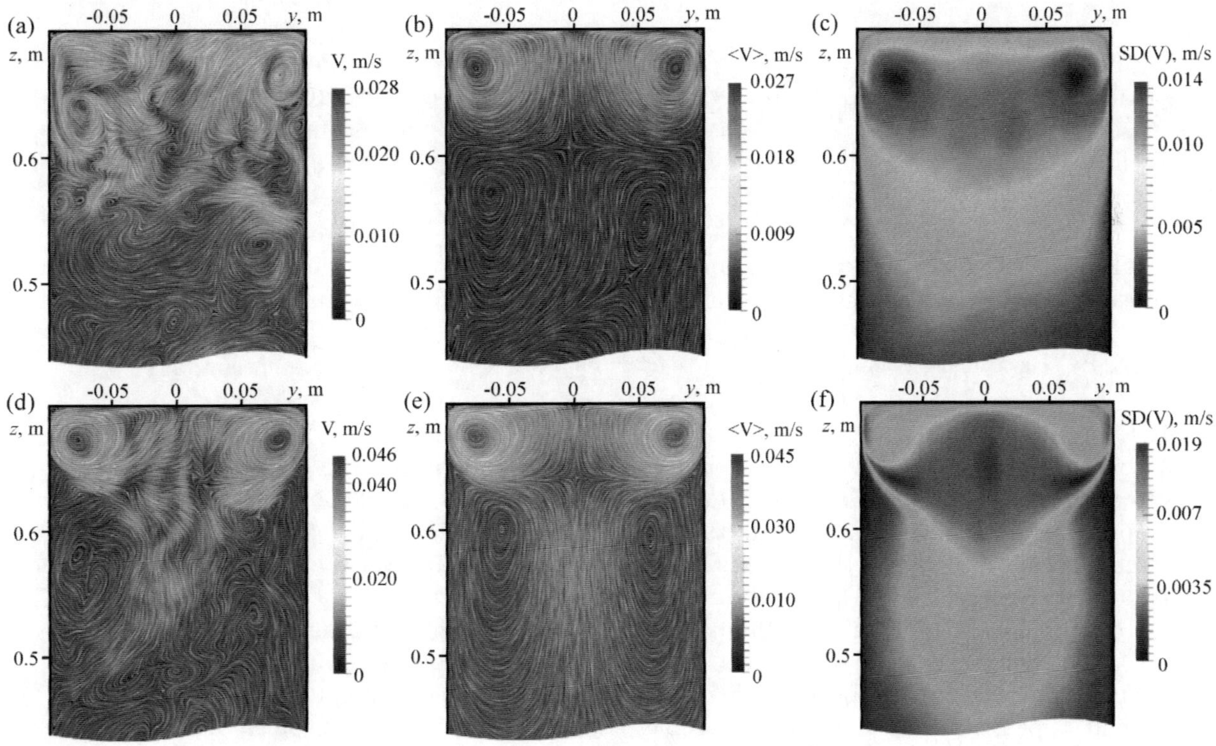

Figure 5. Results of simulations with BC II (a) – (c) and with BC I (d) – (f) in the yOz plane, only upper part of the fields is shown. Instant velocity field (a),(d); time-averaged velocity field (b),(e); standard deviation of velocity field (c), (f).

IOP Conf. Series: Materials Science and Engineering **208** (2017) 012044 doi:10.1088/1757-899X/208/1/012044

the cylinder axis and along the line at a distance of 10 mm from the wall. One sees again that the changing of BC II to BC I leads to change in localisation of the upper vortex: it pins more tight to the upper heat exchanger and the vertical velocity rises. Note that the maximum velocity on the axis is observed at the distance 15 mm from the top, while on the side line the maximum is at 40 mm from the top face. This is the distance where the upper thermocouples were installed in the experimental setup. In the region where lower thermocouples were placed the velocity values are much lower.

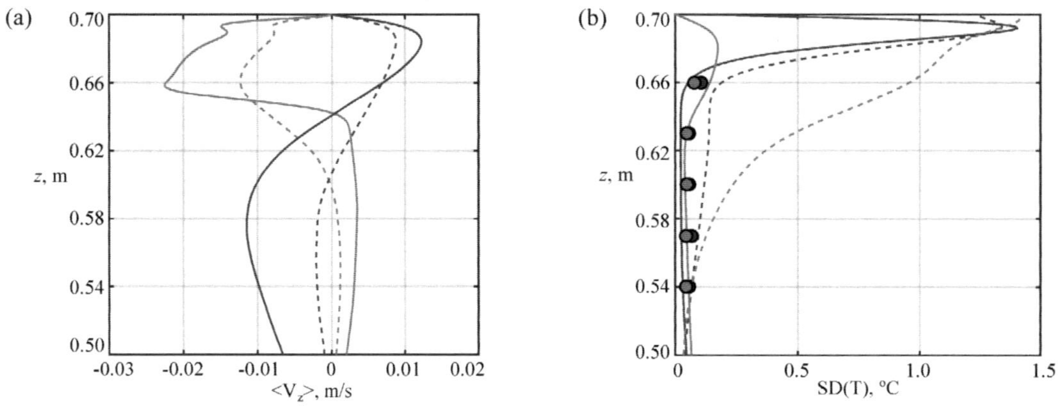

Figure 6. Time-averaged profiles of vertical velocity component $< V_z >$ (a) and profiles of standard deviation of temperature (b). Results of simulations with BC I are shown by solid lines, simulations with BC II – by dashed lines. Blue lines – profiles along the cylinder axis, red lines – profiles along the line at a distance of 10 mm from the wall. Experimental data are shown by circles.

The profiles of temperature pulsations obtained in simulations and in experiment are shown in figure 6b. Results of simulations show that temperature pulsations under BC II are stronger than under BC I, especially along the line near the sidewall. The level of temperature pulsations under BC I is much closer to the experimental results.

4. Conclusions

A convective flow of liquid metal generated nearby a hot round in the upper solid end face of a vertical cylinder has been studied experimentally and numerically. The considered flow is supposed to be a simplified model of the liquid magnesium flow in a reactor of metallothermic titanium reduction though the typical Grashof number in the reactor is about three order higher as in the experiment (10^{12} versus 10^9).

Ten thermocouples installed in the experimental setup along two vertical lines (on the cylinder axis and near the sidewall) allowed one to control the temperature in range of depth from 40 to 160 mm. Two upper thermocouples demonstrated strong temperature pulsations with developed turbulent spectra, while lower thermocouples shows strong reduction of pulsations with depth. The mean temperature distribution reveal in this range of depths a weak gradient, about 20 K/m, corresponding to stable stratification. The cross-correlation analysis of temperature pulsations allowed us to estimate the mean vertical velocity along both lines of thermocouples. The highest velocity is found around the upper thermocouples and is estimated as about 0.01 m/s.

The numerical study was performed for two idealized boundary conditions, fixed temperature on the surface of the heater and cooler (BC I), and fixed homogeneous heat flux on both heat exchangers (BC II). Obtained velocity fields show that a developed turbulent flow exists in the upper part of the cylinder. Strong velocity pulsations penetrate up to distance of about 200 mm from the upper boundary. Mean velocity fields reveals one toroidal vortex, which is localized in a narrow upper zone, mainly above the upper thermocouples (40 mm from the upper boundary). The size and the intensity of this vortex weakly depends on the boundary conditions. The experimental estimations of the value of the mean velocity nearby the upper thermocouples agree with numerical data. The whole bulk of the metal is not involved in the motion. The temperature field depends much more on the boundary conditions, what is common for the liquids with small Prandtl number. Under fixed heat fluxes (BC II) the temperature pulsations become much stronger and penetrate essentially deeper in the liquid metal, though the flow is slightly stronger under BC I.

Acknowledgement

This study was partially supported by the Government of Perm Krai, project No C-26/060 11.03.2016. We thank Andrey Vasiliev, Alexander Pavlinov and Alexander Shestakov for help in experiment preparation and performance.

References

[1] Evdokimova O A and Zimin V D 1970 *Izvestija AN SSSR, Mekhanika jidkosti i gaza* 139–142
[2] Teimurazov A and Frick P 2016 *J. Appl. Mech. Tech. Phys.* **57** 1264–1275
[3] Tsaplin A I and Nechaev V N 2013 *Computational continuum mechanics* **6** 483–490
[4] Scheel J and Schumacher J 2016 *J. Fluid Mech.* **802** 147–173
[5] Shishkina O and Horn S 2016 *J. Fluid Mech.* **790** 790 R3–1–12
[6] Grossmann S and Lohse D 2002 *Phys. Rev. E* **66** 016305
[7] Ahlers G, Grossmann S and Lohse D 2009 *Rev. Mod. Phys.* **81** 503–537
[8] Frick P, Khalilov R, Kolesnichenko I, Mamykin A, Pakholkov V, Pavlinov A and Rogozhkin S 2015 *Europhys. Lett.* **109** 14002
[9] Karcher C, Kolesnikov Y, Andreev O and Thess A 2002 *Eur. J. Mech. B/Fluids* **21** 75–90
[10] Belyaev I A, Genin L G, Listratov Y I, Melnikov I A, Sviridov V G, Sviridov E V, Ivochkin Y P, Razuvanov N G and Shpansky Y S 2013 *Magnetohydrodynamics* **49** 177–190
[11] Smagorinsky J 1963 *Mon. Weather Rev.* **91** 99
[12] Versteeg H K and Malalasekera W 2007 *An Introduction to Computational Fluid Dynamics: The Finite Volume Method* (Pearson Education, Harlow)

IOP Conf. Series: Materials Science and Engineering **208** (2017) 012045 doi:10.1088/1757-899X/208/1/012045

Investigation of vortex ring formation with account of generator piston motion

I V Khramtsov, VV Palchikovskiy, A A Siner, Yu V Bersenev

Perm National Research Polytechnic University, Perm, 614990, Komsomolsky prospect, 29, Russia

E-mail: igorhrs92@mail.ru

Abstract. The formation and dynamics of such an elementary object of fluid dynamics as a vortex ring was investigated. Intense turbulent vortex rings created by a piston generator were considered. For qualitative determination of vortex ring properties, the law of motion for the generator's piston was experimentally established. This law was used in the numerical simulation, which was performed for the axisymmetric problem formulation with the use of ANSYS Fluent software. The obtained results correspond to the self-similar law of vortex ring motion and experimental results.

1. Introduction

The existing problem of aviation noise is related above all to its deleterious effect on humans. Community noise of civil aircrafts is regulated by ICAO requirements and is an important issue for aviation industry despite significant achievements in its mitigation. The possibilities of further aviation noise reduction with traditional approaches are limited by difficulties in their implementation; therefore, acoustics of modern aircrafts has a significant effect on their competitiveness. Further work in this direction requires the development of new approaches and ideas based on a deeper understanding of the physical processes responsible for the generation of noise by turbulent flows.

In subsonic turbulent jets, which are one of the main noise sources of aircraft engines, noise is generated by turbulent vortices. The central problem in a jet noise study is related to the presence in turbulent jets of vortices with different scales interacting with each other, which leads to distortion of the radiation pattern. Thus, it is extremely important to investigate the fundamental problems of sound radiation by turbulent flows for the case of an isolated vortex.

A good example of such an isolated vortex is a vortex ring. To study aeroacoustic characteristics of the given vortex, a vortex ring generator was developed in the Laboratory of Noise Generation Mechanisms and Modal Analysis at Perm National Research Polytechnic University (PNRPU). Preliminary investigations of the noise of intense vortex rings with a Reynolds number of ~ 10^5 in the anechoic chamber of PNRPU confirmed the main results observed in previous experiments: the acoustic radiation of the vortex ring is concentrated in a rather narrow frequency band, the carrier frequency of the peak in the spectrum shifts over time into the low-frequency region, etc. [1].

The major observed features of vortex ring dynamics and acoustic radiation can be explained by the developed theoretical models of vortex rings [2-4]. A model of noise radiation by an intense vortex ring[2-4]allows relating the noise with the average parameters of the vortex ring flow (dimensions, velocity and vorticity). A deeper understanding of the noise radiation mechanisms could be obtained from the knowledge of nonstationary processes in the vortex ring cores. However, experimental data on the structure of the vortex ring and, particularly, nonstationary processes in the core of the vortex, are essentially limited.

One way to study this flow is numerical simulation. The solution of the problem of motion and oscillations of a vortex ring by direct numerical simulation in a three-dimensional formulation is limited by the power of modern supercomputers and represents a formidable computational problem. One of the subproblems that must be solved at the preliminary stage is the problem of the formation of the vortex ring and its motion at the initial section of the trajectory.

This problem can be solved in an axisymmetric formulation, assuming the axisymmetric formation and motion of the vortex at the initial section of the trajectory [5]. However, the properties of this vortex essentially depend on the conditions of its formation. In this regard, for the qualitative verification of this problem, a technique was developed for the experimental determination of the conditions of the creation of a vortex and its further motion [4]. Subsequently, the results obtained with the help of this technique were used in the numerical simulation for setting the initial and boundary conditions. Validation was performed by comparing the trajectories of the vortex motion obtained numerically and during the experiment.

2. Experimental study

In the experimental study of the vortex ring noise, the most important requirement is identification of the noise of vortex ring proper against the background noise, including the structural noise of the vortex ring generator, the reflection of sound from hard surfaces, etc. To do this, the vortex ring must be sufficiently intense so that the sound radiated by it can be distinguished against a background noise, and the way to create vortex ring, on the contrary, be as quiet as possible. It can be achieved by using a special piston generator and the anechoic chamber [6].

The vortex ring generator is a polished hollow steel cylinder with a diameter of 0.16 m, to which a conical nozzle with an exit diameter of 0.04 m is fixed. Inside the cylinder, the lightweight fluoroplastic piston moves freely. Before starting the vortex, the piston is extended from generator for a fixed distance of 0.06 m. Figure1 shows a part of the vortex ring generator and the recording system (view from the anechoic chamber).

Figure 1. Photo of the recording system. 1 – nozzle of the vortex ring generator;
2 – microphones recording the motion of the vortex ring (A, B).

The vortex ring is created by pulsed ejection of a portion of gas through the nozzle by the piston. Then, the formed vortex structure separates from the nozzle edge and moves in space along the axis coinciding with the piston axis. The piston is driven by the impact of a heavy hammer on the piston rod. To avoid the reverse motion of the piston, the vortex ring generator is equipped with a special brake assembly, which fixes the rod when the piston reaches its extreme position.

In this work, a joint registration of the generator piston motion and the vortex ring motion was carried out. The piston motion was realized with an accelerometer mounted on the outer wall of the piston. To determine the vortex ring motion, the method was utilized [4] based on microphones installed at a small distance from the generator axis, fixing the peak of the pressure at the time of passage of the vortex near the corresponding

IOP Conf. Series: Materials Science and Engineering **208** (2017) 012045 doi:10.1088/1757-899X/208/1/012045

microphone. These microphones were located only at the initial section of the trajectory, where the motion of the vortex ring can be assumed axisymmetric. The scheme of the experiment is shown in figure2.

To record the deviation of the vortex flight trajectory from the generator axis, a rectangular screen from a thin wire with silk streamers with dimensions of 2×1 m was used. This screen was set at a distance of 6.5 m from the nozzle exit section so that the center of the screen was on the axis of the generator. During the passage of the vortex ring through the screen, the silk streamers deflected, making the position of the ring visible. It made it possible to select for the further investigation only the correctly flying vortex rings passing through the center of the screen.

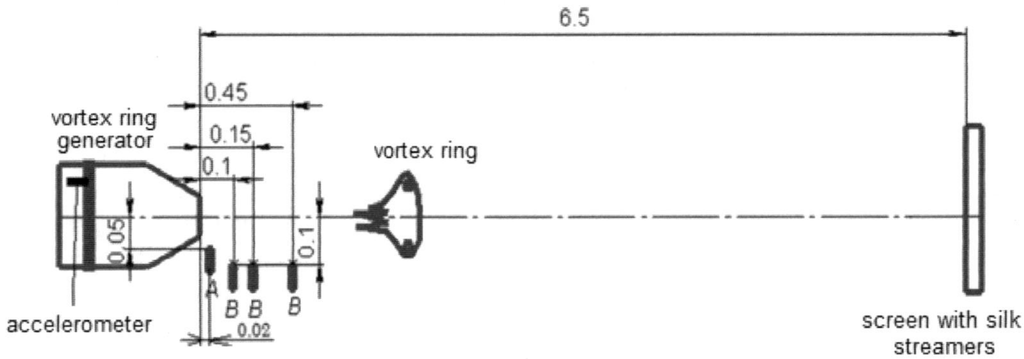

Figure 2. Scheme of the experiment.

Since each created vortex ring has individual properties (velocity, deviation from the expected trajectory, dimensions) due to the percussive method of starting the piston, a series of experiments was conducted. Then, an analysis of the time realizations of the signals of the accelerometer and microphones A and B (figure 2) was carried out. The signals from the accelerometer were integrated to obtain the velocity and displacement of the piston. This allowed us to use in the numerical simulation the real law of motion of the piston, whereas in some studies, theoretical laws are used, which could be poorly consistent with the actual movement of the piston [7]. The obtained dependences of acceleration, velocity and displacement of the generator piston are shown in figure 3.

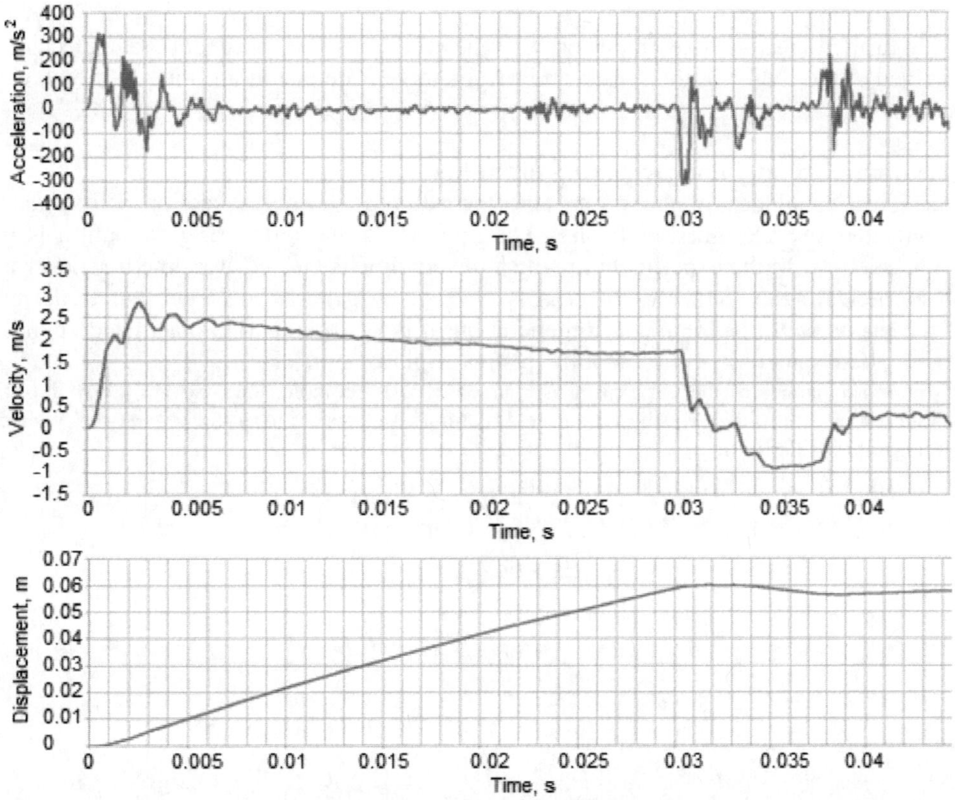

Figure 3. Dependences of acceleration, velocity and displacement of the generator piston for one realization.

Analysis of the signals of microphones A and B determined the time of flight of the vortex ring near the corresponding microphone. The signals received from these microphones are shown in figure 4.

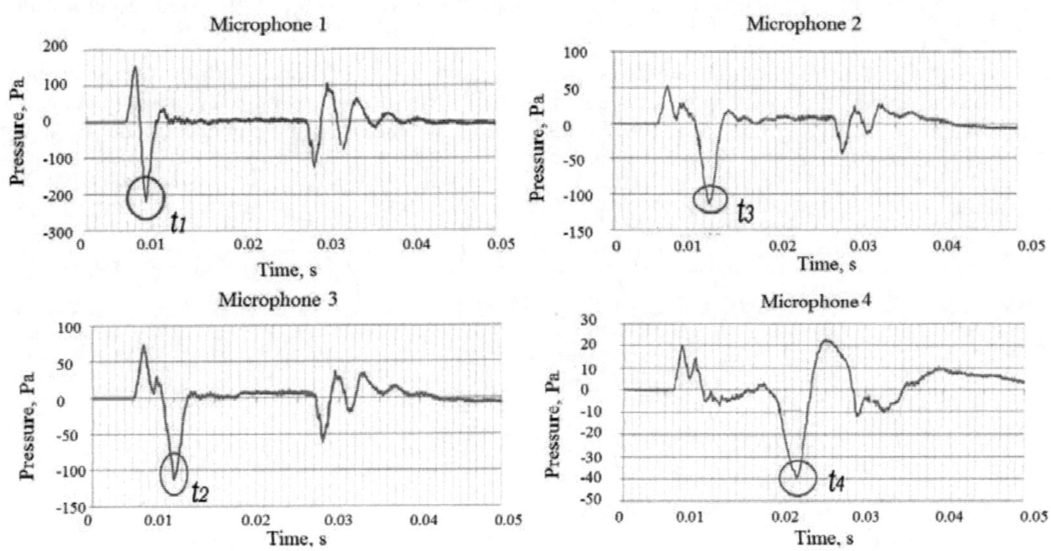

Figure 4. Dependence of pressure on microphones A and B on time for one realization of the vortex ring flight.

Thus, for further analysis, one characteristic realization of the experiment was selected, in which the ring arrived at the center of the screen and parameters of the piston and ring motion were close to the average ones.

3. Numerical simulation of the vortex ring formation

The numerical simulation of the vortex ring formation at the initial section of its trajectory was carried out in the ANSYS Fluent software. The process of the vortex ring formation is performed directly, by analogy with the experiment in which the air flow is ejected from the nozzle by a movable piston. In numerical simulation, this process is realized by generation of computational grid in the region of the piston wall displacement. Regeneration of the grid is carried out in accordance with the experimentally determined dependence of the piston displacement.

Formation and dynamics of the vortex ring is characterized by complex nonstationary turbulent flows. The numerical solution of the problem was based on the direct numerical solution of the nonstationary Navier-Stokes equations for compressible fluid:

$$\begin{gathered}\frac{\partial\rho}{\partial t}+\nabla\cdot(\rho\vec{v})=0\\ \frac{\partial}{\partial t}(\rho\vec{v})+\nabla\cdot(\rho\vec{v}\,\vec{v})=-\nabla p+\nabla\cdot\tau\\ \frac{\partial}{\partial t}(\rho E)+\nabla\cdot(\vec{v}(\rho E+p))=\nabla\cdot\left(\chi\nabla T+\tau\cdot\vec{v}\right)\end{gathered}\qquad(1)$$

The viscous stress tensor τ is defined as

$$\tau=\eta\left[(\nabla\vec{v}+\nabla\vec{v}^{T})-\frac{2}{3}\nabla\cdot\vec{v}I\right]\qquad(2)$$

and total energy E is

$$E=h-\frac{p}{\rho}+\frac{v^2}{2}\qquad(3)$$

Here ρ is density; $\vec{v}$ is velocity vector; p is pressure; T is temperature; h is enthalpy; η is molecular viscosity; χ is conductivity; I is unit tensor.To complete the system (1) - (3), the state equation of the perfect gas is used.

The geometric model used in computations is internal region of the generator and the region behind the nozzle exit section wherein the vortex ring forms and moves. The size of this region is 21 nozzle-exitdiameters, since in this section of trajectory the flow could be assumed axisymmetric. The geometric model is shown in figure 5.

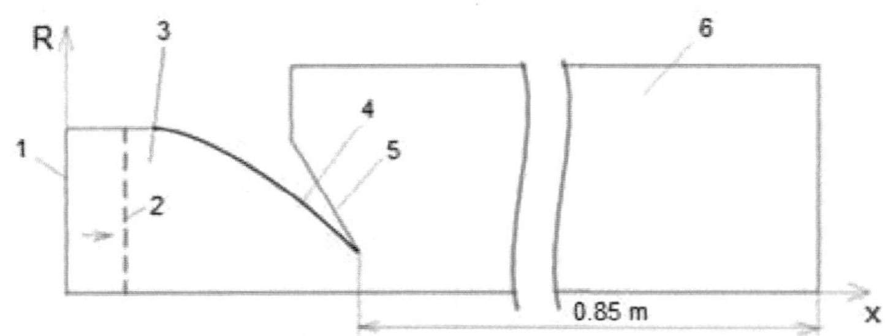

Figure 5. Geometric model of the computational domain. 1 – movable piston wall; 2 – piston stopping; 3 – internal region of the generator; 4 – internal nozzle wall; 5 – external nozzle wall; 6 – region of vortex ring motion.

The numerical solution of the problems of vortex gas dynamics requires a sufficiently fine computational grid. At the same time, typical size of the vortex for the normal resolution requires 40-50 computational cells. The radius of the vortex ring core is taken as the typical size, which according to experimental data for similar vortex rings is 8-9 mm. For this reason, the average size of the computational cell in the region of vortex motion is accepted equal to 0.2 mm. At the wall of the nozzle, the computational grid is finer so that the dimensionless parameter y^+ satisfies the condition $y^+ < 1$. The grid contains 540 000 computational cells and it is shown in figure 6.

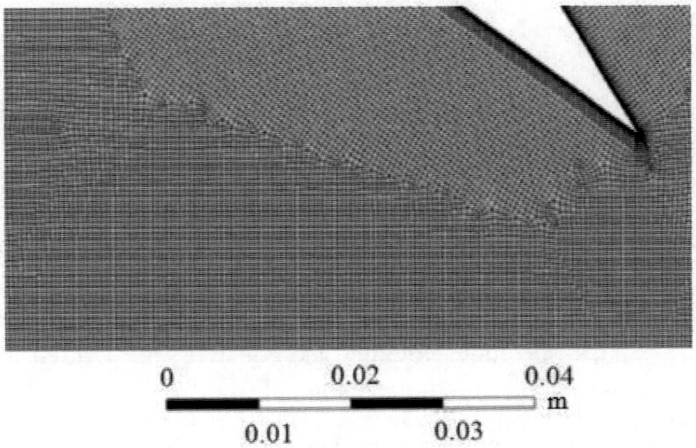

Figure 6. Computational grid in the region of the nozzle exit section.

The computations are carried out under normal environmental conditions (pressure 101325 Pa and temperature 295K). The flow is considered single-phase, viscous and compressible. The working medium is air, whose properties are described by the law of ideal gas.

At the boundary of the outer region of the vortex motion the "Outlet" boundary condition with zero overpressure is used. Internal and external walls of the generator and movable wall of the generator piston are simulated by boundary conditions "Wall". It is assumed that the walls are perfectly smooth. The boundary conditions are shown in figure 7.

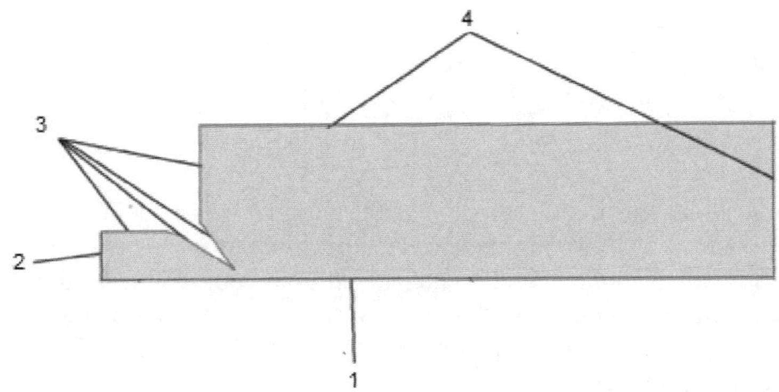

Figure 7. Boundary conditions. 1 – axis; 2 – movable wall; 3 – wall; 4 – outlet.

The mathematical formulation of boundary conditions based on the following equations:

- on axis

$$v_n\big|_b = 0; \quad \frac{\partial v_\tau}{\partial n}\Big|_b = 0; \quad \frac{\partial T}{\partial n}\Big|_b = 0; v\big|_b = v_\tau\big|_b \tag{4}$$

- on movable wall

$$v_n|_b = \{u_p(t)\},\ [\text{m/s}];\quad \frac{\partial v_\tau}{\partial n}|_b = \frac{v_\tau}{\kappa}\ln\left(\varepsilon\frac{\rho v_\tau y}{\mu}\right);\quad \frac{\partial T}{\partial n}|_b = 0 \tag{5}$$

- on wall

$$v_n|_b = 0;\quad \frac{\partial v_\tau}{\partial n}|_b = \frac{v_\tau}{\kappa}\ln\left(\varepsilon\frac{\rho v_\tau y}{\mu}\right);\ \frac{\partial T}{\partial n}|_b = 0 \tag{6}$$

- on outlet

$$p|_b = p|_{t=0};\quad \frac{\partial T}{\partial n}|_b = 0 \tag{7}$$

$$\text{if } v|_b \le 0,\ \text{then } \frac{\partial v}{\partial n}|_b = 0;\quad \text{if } v|_b > 0,\ \text{then } v|_b = v_\tau|_b$$

Here u is piston velocity; ε is rate of turbulent energy dissipation; n is a normal. Lower indexes n, τ are a normal and a tangential component respectively; b is a boundary component; $t=0$ is the initial value.

Solution of the problem (1) - (7)in Fluent is executed by the finite volume method. The computational domain is divided into finite volumes and equations are written in integrated form for each cell. For time integration the second-order implicit scheme is used. Time step is equal to 1/65536 s. The computations are conducted during 1900 time steps. The details on the scheme of solution of nonstationary Navier-Stokes equations in Fluent are presented in [8].

4. Results

The evaluation of the results was carried out using a comparison of the trajectory of the vortex ring motion obtained in experiment and numerical simulation, as well as with the self-similar law of the vortex ring displacement [9]. The self-similar law of motion is a semi-empirical theory of vortex ring motion and is well correlated with empirical models proposed by other authors [10, 11]. The displacement of the vortex ring and its radius obtained in numerical simulation were evaluated in the point at which the pressure, temperature and density were minimal, and the vorticity maximum was observed in the same region. Equations (8) represent the law of vortex ring motion

$$L(t) = \frac{R_0}{\alpha}\left[\left(1 + \frac{4\alpha U_0}{R_0}t\right)^{\frac{1}{4}} - 1\right]\qquad R(t) = R_0 + \alpha L(t) \tag{8}$$

Here R_0,U_0 are initial radius and initial velocity of the vortex ring, respectively; α is a factor of the vortex ring expansion rate. To determine the displacement of the vortex ring in accordance with the equation (8) the initial radius and the velocity determined in the numerical simulation were substituted into this system. The initial velocity of the vortex ring was determined from equation (9) as the difference in the positions of the ring in sections with coordinates$x_1 = 0.1$ m and $x_2 = 0.15$ m (where microphones have been set) divided by time

$$U_0 = \frac{\Delta x}{\Delta t} \tag{9}$$

The initial radius was determined at the time of the passage of the ring through the section at $x_1 = 0.1$ m. The factor α was determined graphically according to figure 8, so that the dependence of the radius on time would well describe the results of numerical simulation. It leads to the following values of parameters, used in the self-similar motion law: $R_0 = 0.033$ m, $U_0 = 24.8$ m/s, $\alpha = 0.005$.

Figure 8 shows a comparison of the results of numerical simulation, experiment, and self-similar law. For convenient presentation of results, the origin is shifted to the point corresponding to the position of the first microphone A (figure 2), and the time is measured from this point. As can be seen, the results of numerical simulation, experiment, and the self-similar theory of vortex ring dynamics are well matched.

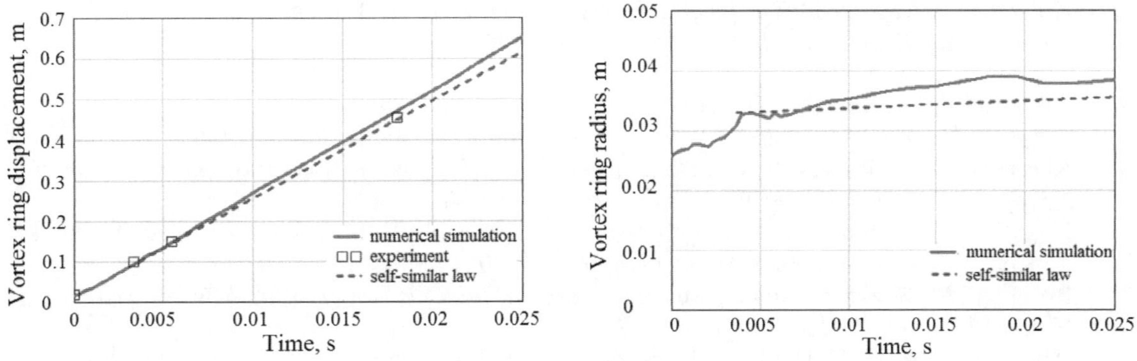

Figure 8. Comparison of results of numerical simulation, experiment and self-similar law.

Figure 9 shows the distribution of the density gradient module at time point of 0.015 s. It can be seen that in addition to the main vortex, secondary vortices of lower intensity are also observed. The process of disintegration of the secondary vortices differs significantly from the axisymmetric one, which leads to an increase in the discrepancy between the computational results and the experiment on the study of the vortex ring motion.

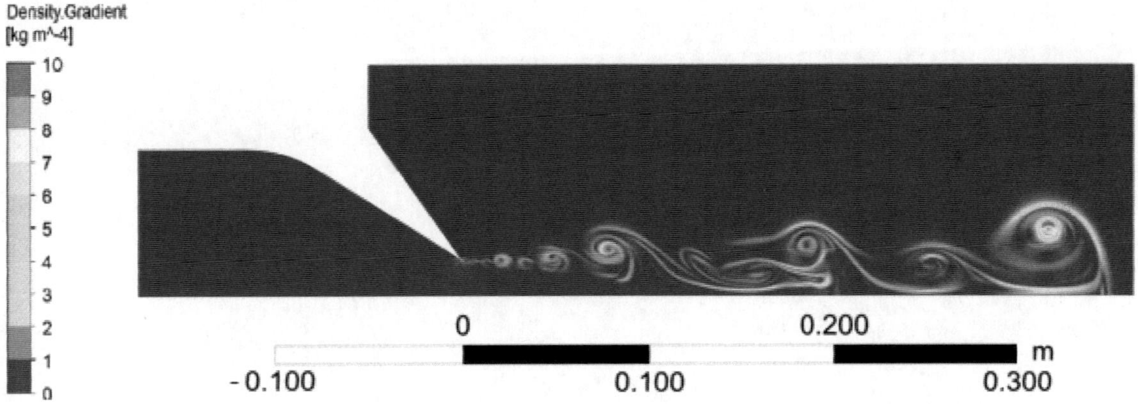

Figure 9. The distribution of the density gradient at time point of 0.015 s.

5. Conclusion

1. An experimental technique for simultaneous determination of the motion of a vortex ring and a piston of a vortex ring generator is developed.
2. The law of the piston motion obtained in experiment was used for numerical simulation of the vortex ring dynamics. The carried out validation of the numerical model showed that the results of simulation are in good agreement with the results of the experiment and the self-similar theory of motion of the vortex ring.

Acknowledgments

The authors are grateful to Prof. V.F. Kopiev for the problem statements and further fruitful discussions.

The results have been obtained in the framework of the government task "Researchers working in scientific laboratories created by the government program "Megagrants", contract No.9.7942.2017/P220.

Experiments were carried out on unique scientific installation "Acoustic anechoic chamber with

IOP Conf. Series: Materials Science and Engineering **208** (2017) 012045 doi:10.1088/1757-899X/208/1/012045

aerodynamic noise sources".

References

[1] Kopiev V F, Zaytsev M Yu, Palchikovskiy V V, Khramtsov I V and Bersenev Yu V*PNRPU Aerospace Engineering Bulletin*. 2016 **45** 133-151
[2] Zaytsev M Yu and Kopiev V F *Akusticheskiy zhurnal*. 1993 **39** 1068-1075
[3] Zaitsev M Yu and Kopiev V F *TsAGI Science Journal*. 1998 **29** 83-91
[4] Zaitsev M Yu, Kopiev V F and Kotova A N Acoustical Physics. 2000 **47** 699-706
[5] Khramtsov I V, Pisarev P V, Palchikovskiy V V, Bulbovich R V and Pavlogradskiy V V *Applied Mechanics and Materials*. 2015 **770** 483-48
[6] Kopiev V F, Palchikovskiy V V, Belyaev I V, Bersenev Yu V, Makashov S Yu, Khramtsov I V, Korin I A and Sorokin E V *Acoustical Physics*. 2017 **63** 114-126
[7] Rosenfeld M, Rambod E and Gharib M *Journal of Fluid Mechanics*. 1998 **376** 297-318
[8] Fluent 12.0 User's Guide, February 2008
[9] Lavrentev M A and Shabat B V The problems of hydrodynamics and their mathematical models (Moscow, Nauka) 1973 p 416
[10] Johnson GM *AIAA Journal*. 1971**9** 763-764
[11] Maxworthy T. *Journal of Fluid Mechanics*. 1974 **64** 227-239

IOP Conf. Series: Materials Science and Engineering **208** (2017) 012046 doi:10.1088/1757-899X/208/1/012046

Numerical estimation of various influence factors on a multipoint hydrostatic leveling system

R V Tsvetkov[a], V V Yepin and A P Shestakov

Institute of Continuous Media Mechanics UB RAS, 1, Acad. Korolev St., Perm, 614013, Russian Federation

E-mail: [a] flower@icmm.ru

Abstract. Among various methods that allow controlling quasi-static vertical displacements of structures, the hydrostatic leveling method remains relevant. A multi-point hydrostatic leveling systems allows controlling the vertical displacement field with the required spatial resolution. It is assumed that the liquid level in the each measuring vessel has the same absolute elevation. However, it is influenced by various external factors, which are difficult to eliminate when implementing the method in real constructions. Consequently, it is necessary to assess the influence of these factors and develop methods aimed at their reduction or compensation. A mathematical model, describing the viscous incompressible fluid motion located in a multi-point level with absolutely rigid walls, is presented in the paper. The experiments performed with two point levels of different lengths, as well as analytical estimates of other authors, made it possible to estimate the degree of confidence of the model and the boundaries of applicability. The influence of non-uniform density of a liquid on the liquid level in a 4-point hydrostatic level of different topologies is numerically estimated using the model. An estimation of transient processes in the level, caused by an air pressure surge over one of the measuring vessels is carried out.

1. Introduction

Among various methods that allow controlling quasi-static vertical displacements of different objects, a hydrostatic leveling system (HLS) has several advantages due to which it remains relevant and is used to monitor various structures: colliders [1], dams [2], tunnels, buildings [3,4], etc.

The hydrostatic level is a system of communicating measuring vessels filled with a liquid. The liquid level in these vessels determines the plane relative to which quasi-static vertical displacements can be controlled. The use of the multipoint HLS allows one to control vertical displacement fields with the required spatial resolution. In HLS measurements, it is assumed that the liquid level in the measuring vessels is the same. However, its position is influenced by various factors described by Pellissier [5] in 1965: non-uniform liquid density or temperature, air pressure, external mechanical influences and etc. Methods for reducing the influence of these factors are known, however, the structural features of real structures generally do not allow one to implement all of them (minimum level difference, minimum temperature gradient). While under laboratory conditions and in special cases it is possible to obtain an error of the order of 1 μm, for example using the half-filled pipe method, which is less sensitive to temperature variations [6]. However, the half-filled pipe method is difficult to implement in most real designs, that is why in this case the "fully filled pipe system" is used, for which it is necessary to assess the influence of negative factors, as well as proposals to reduce their influence or compensation. In the literature, the estimates of the influence of these factors for 2-point longbase levels (consisting of 2 sensors) are described [7]. For the multipoint level in [8], it is suggested to evaluate transient processes based on the electronic-hydraulic analogy, by replacing the measuring and connecting vessels (segments) with an electrical circuit with RLC elements. However, different variants of connection topologies of the multipoint level are not considered, and are not analyzed. Therefore, modeling of liquid flow processes in multipoint levels of various topologies is relevant and significant for assessing the influence of negative factors on level readings and its reduction.

2. Modeling of a Multipoint hydrostatic leveling system

The application of modern three-dimensional finite-element approaches for solving transient problems of liquid motion in hydrostatic level tubes requires considerable computational resources. Since this requires sufficient spatial and temporal discretization, determined by dimensions of real levels, the length of their connecting tubes is much greater than its diameter. Consequently, in order to evaluate liquid flow processes in the levels of different configurations, less resource-intensive approaches should be used.

2.1. Hydraulic Model review

Let's consider an incompressible liquid with density ρ and viscosity μ, which is in a multipoint level with absolutely rigid walls. Figure 1 shows a fragment of the hydrostatic level, consisting of measuring (vertical) and connecting (horizontal) segments. The measuring and connecting segments of the level have a cylindrical shape. Along the segment, with a constant cross-section, the liquid velocity υ and density ρ does not change. Let's consider the liquid pressure *P* at the segment connection point (node).

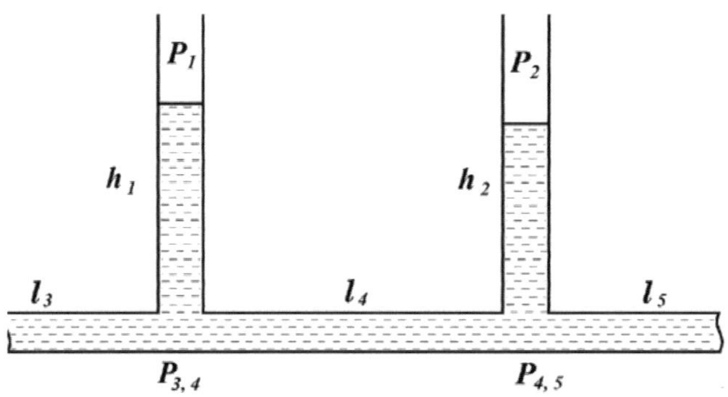

Figure 1. A fragment of the hydrostatic level.

The parameters of real levels and the velocity of fluid motion in them correspond to the numbers Re << 2100 (for cylindrical tubes), i.e. laminar flow. In the hydraulic model approximation, we write the system of equations for a level, consisting of several segments. The Poiseuille law with some correction coefficient *k*, depending on frequency is used. For the Poiseuille law, $k = 1$. In the measuring vessel, the height of the liquid level h_i (the filled part of the vessel) is variable, and in the connecting segment the length l_j is constant. The system of equations consists of the balance of forces (pressures) equations for the measuring (1) and connecting (2) segments:

$$\rho_1 \cdot h_1 \cdot \frac{\mathrm{d}\upsilon_1}{\mathrm{d}t} + P_{3,4} - P_1 - \rho_1 \cdot g \cdot h_1 - \frac{8\mu \cdot h_1 \cdot \upsilon_1}{R_1^2} k_1 = 0 \tag{1}$$

$$\rho_4 \cdot l_4 \cdot \frac{\mathrm{d}\upsilon_4}{\mathrm{d}t} + P_{3,4} - P_{4,5} - \rho_4 \cdot l_4 \cdot g \cdot \delta_4 - \frac{8\mu \cdot l_4 \cdot \upsilon_4}{R_4^2} k_4 = 0 \tag{2}$$

where $P_{4,5}$ is the fluid pressure at the node between the connecting segments *4* and *5*;
P_1, P_2 are the air pressures in the measuring vessel 1 and 2 above the liquid;
υ_i is the liquid velocity in the segment *i*;
R_i is the radius of the segment *i*;
δ_4 is the coefficient of the liquid weight, depending on the direction of the connecting segment 4;
k_i is the frequency-depended friction coefficient in the segment *i*.

The equation of the relationship between the height of the liquid level h and the velocity υ of its motion in the measuring vessel 1 is:

$$\frac{\mathrm{d}h_1}{\mathrm{d}t} = \upsilon_1 \tag{3}$$

The continuity equations, corresponding to the equality of volumetric flow rates at the node between the connecting segments 3 and 4, have the form:

$$-\upsilon_3 R_3^2 - \upsilon_1 R_1^2 + \upsilon_4 R_4^2 = 0 \tag{4}$$

The system of equations (1) - (4), with respect to the variables h, υ with the following initial conditions:

$$h_i = h_i^0, \quad \upsilon_i = \upsilon_i^0 \tag{5}$$

for every specific topology of the level is solved numerically using the Matlab® software.

2.2. Model verification

In order to verify the presented model, the problem of free damped oscillations of a 2-point level with a constant cross section of the measuring and connecting parts, i.e. a U-tube, was numerically solved. The total length of the liquid column (L) for the U-tube consists of two measuring segments heights h and one connecting segment l. The resulting damped oscillations can be characterized by the values of the damping ratio λ and the period of oscillations T. According to [9] for the Poiseuille flow, they are equal:

$$\lambda = \frac{4 \cdot \mu}{\rho \cdot R^2} \tag{6}$$

$$T = 2 \cdot \pi \sqrt{\frac{L}{2g}} \cdot \left(\frac{1}{\sqrt{1 - \lambda^2 L / 2g}} \right) \tag{7}$$

There are other approaches to describe dissipation processes, arising during a laminar non-stationary flow in tubes [10-11]. Comparisons of the calculated values of λ and T with analytical results for U-shaped tubes were carried out: with dissipation in the Poiseuille, Ogawa [10], Popov [11] law approximation, with analytical results of Manukin [7] for the longbase level, as well as with own experiments.

For the experiments, a 2-layer PVC fiber-reinforced tube was used as the connecting and measuring vessel. The inner diameter of the tube is 12.5 mm, the outer one is 18.5 mm. The height of each measuring segment in the experiments was 0.3 m, and the length of the connecting section changed. Tubes which total length L are 50 m and 20 m, were studied and that corresponded to the real segments size. A short tube, where L equaled 1.1 m, was also investigated to estimate the boundaries of applicability of the models. The temperature of the liquid (water) during the experiments was 21÷23 °C, which corresponded to its density of 998 kg/m^3, viscosity - 0.001 Pa·s. Liquid level oscillations, with an initial amplitude of about 20 mm, was recorded by a video camera.

Further, for each frame, the position of the liquid level was determined using the image recognition algorithm described in [12]. As a result, evolutions of the liquid level height were obtained. It should be noted that during each experiment the liquid had time to make several complete oscillations, aperiodic process was not observed. Based on the processing of the experimental data, T and λ were evaluated. For a better understanding, the results are shown in Table 1.

Table 1. Comparison of the results for a 2-point HLS (U-tube) on different models.

	L = 1.1m		L = 20m		L = 50m	
Type of Model	T (s)	λ (s^{-1})	T (s)	λ (s^{-1})	T (s)	λ (s^{-1})
Classic (Poiseuille)	1.49	0.103	6.38	0.103	10.165	0.103
Popov	1.49	0.273	6.38	0.154	10.26	0.131
Ogawa	1.49	0.397	6.37	0.369	10.17	0.351
Manukin	1.16	0.077	7.21	0.077	11.51	0.077
Numerical (k = 1)	1.49	0.103	6.37	0.103	10.165	0.103
Experiment	1.46±0.05	0.49±0.01	7.0±0.12	0.142±0.005	11.45±0.15	0.120±0.005

The presented results show that:

- for k = 1, the numerical results coincide with the analytical ones (Poiseuille law), which indicates the correct numerical implementation of the model;
- the considered models, except Ogawa, do not accurately describe dissipative processes in the short levels (for the conditions under consideration);
- for the long levels (about hundreds of meters), the oscillation period should be estimated by the Manukin model;
- for the average range levels (about tens meters) with an error of 10%, it is possible to estimate the oscillation period according to the proposed model (Numerical), and to estimate damping ratio, the correction coefficient should be used, depending on the oscillation frequency ω in the Poiseuille law according to the Popov model:

$$k = 0.4 + \frac{R}{4}\sqrt{\frac{\omega \cdot \rho}{2 \cdot \mu}} \tag{8}$$

The coefficient can be used to describe more accurately damping processes in equations (1) – (2).

3. The influence of an air pressure surge on the liquid level in HLS

Let's consider a level consisting of 4 measuring vessels (with the heights of the liquid levels h_1, h_2, h_3 and h_4), connected together by the connecting segments. The diameter of the connecting segment is 12.5 mm, the measuring one is 50 mm. The lengths of the segments between the sensors No. 1 and No. 2; No. 2 and No. 3; No. 3 and No. 4; No. 4 and No. 1 are 10, 20, 15, 25 m, respectively.

Figure 2. The location of the level on the foundation with a sequential connection circuit.

Figure 3. The location of the level on the foundation with a closed circuit.

For instance, let us consider 2 options: serial connection from sensor No. 2 through sensors No. 3, No. 4 and up to No. 1 (without any connection from No. 1 to No. 2), shown in figure 2, as well as the connection diagram where all 4 sensors are closed in a circuit shown in figure 3. The initial conditions of the problem, as well as the characteristics of the level are presented in Table 2.

Table 2. Task parameters.

υ^0 (m/s)	$h^0_{1\text{-}4}$ (m)	R_l (m)	R_h (m)	ρ (kg/m^3)	μ (Pa·s)
0	0.1	0.006125	0.025	999	0.001

Let the excess air pressure $P_{h,2}$ over sensor No. 2 change abruptly at $t = 5$ s by +100 Pa relative to the air pressure over the other sensors. Consequently, a transient process occurs, after which the values of h are set on new levels. The results of the change in the liquid levels in the sensors are shown in Figures 4 and 5.

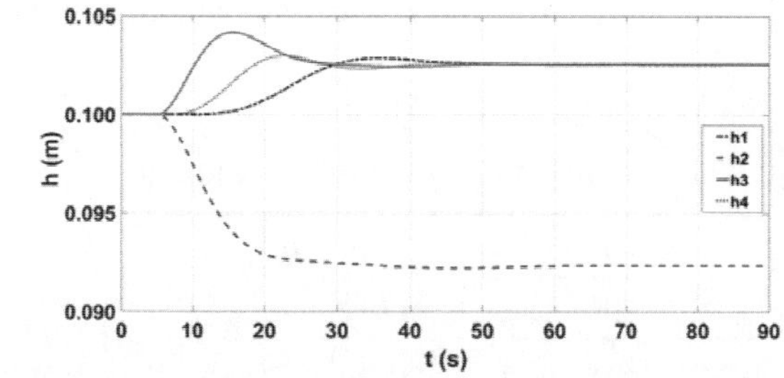

Figure 4. The change in the liquid level heights on the sensors with a sequential circuit.

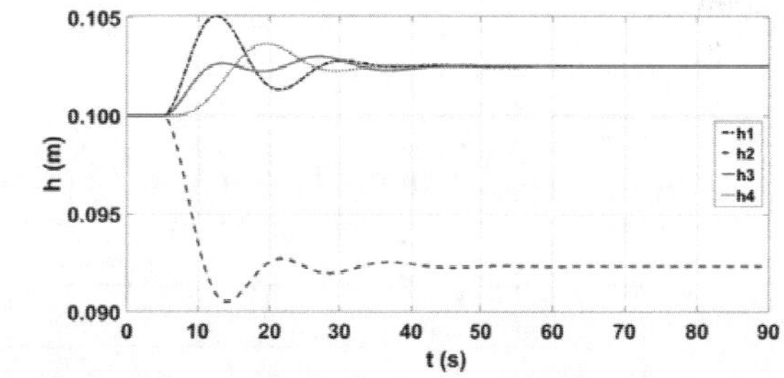

Figure 5. The change in the liquid level heights on the sensors with a closed circuit.

As expected, the change in air pressure over one of the sensors causes a measurement error of 10 mm. The presented graphs demonstrate different nature and frequencies of the transient processes that occur when the gas pressure changes over one of the sensors with different connection circuits. In the case of the sequential circuit, after an air pressure change, the liquid levels are set to new values with an error of ± 0.1 mm in 44 s from the time of the onset of the transient process, and in the presence of a closed circuit in a somewhat shorter time, which is equal to 36 s.

4. The influence of non-uniform density of a liquid on the liquid level in HLS

One of the main factors, affecting the position of the liquid level in the level is non-uniformity of liquid density of its different parts. It can be caused by a non-uniform temperature of the environment. Within the framework of the suggested model, the levels, for which liquid density is constant for each segment, and can be different in various segments, are considered. For example, firstly let us consider a simple case - a 2-point-level, and after that a multi-point one with different topologies.

4.1. 2-point HLS case

On the monitored objects due to structural features, situations, where the connecting tube between the sensors is not located on the same level, but goes on two levels - with periodic ascents and descents, are possible. This case corresponds to a 2-point level of the following configuration (Figure 6).

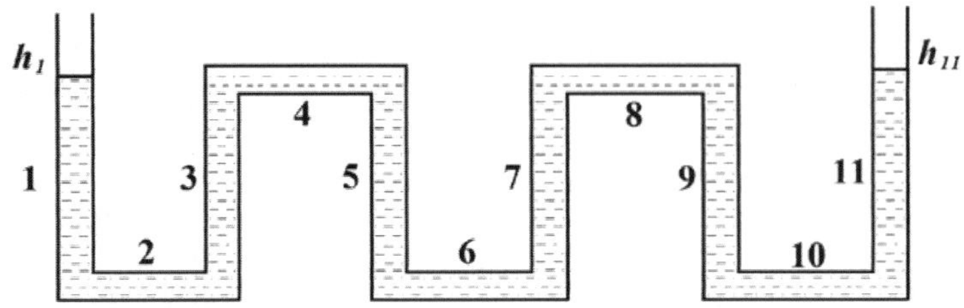

Figure 6. The scheme of a 2-point level of 11 segments.

The level, presented in figure 6, consists of segments of the same length of 1 m. The diameter of all the vessels is 0.0125 m, the initial liquid velocities in all the segments are zero. As a liquid, we choose water, whose viscosity is 0.001 Pa s and the Poiseuille law of friction (k = 1). Let us investigate the influence of the change in the density of each segment on the difference of the liquid level heights Δh. Let ρ_1 = 999 kg/m^3, and ρ_2 = 998 kg/m^3, which corresponds to temperatures of 16 °C and 21 °C. For each density change, the problem (1) - (5) is solved and on the basis of the obtained results the difference of the steady-state values of the liquid levels in the measuring vessels is calculated. Results for a few cases are presented in Table 3.

Table 3. The difference between liquid level heights depending on density in the segments.

The value of liquid density in the segment											Δh (m)
1	2	3	4	5	6	7	8	9	10	11	
ρ_1	ρ_1	ρ_1	ρ_1	ρ_1	ρ_1	ρ_1	ρ_1	ρ_1	ρ_1	ρ_1	0
ρ_2	ρ_1	ρ_1	ρ_1	ρ_1	ρ_1	ρ_1	ρ_1	ρ_1	ρ_1	ρ_1	0.001
ρ_2	ρ_1	ρ_2	ρ_1	ρ_1	ρ_1	ρ_1	ρ_1	ρ_1	ρ_1	ρ_1	0
ρ_1	ρ_1	ρ_2	ρ_1	ρ_1	ρ_1	ρ_1	ρ_1	ρ_1	ρ_1	ρ_1	0.001
ρ_2	ρ_2	ρ_2	ρ_1	ρ_2	ρ_1	ρ_2	ρ_1	ρ_1	ρ_1	ρ_1	0
ρ_2	ρ_2	ρ_2	ρ_1	ρ_2	ρ_1	ρ_1	ρ_1	ρ_2	ρ_1	ρ_1	0.002
ρ_2	ρ_1	ρ_1	ρ_1	ρ_2	ρ_1	ρ_1	ρ_1	ρ_2	ρ_1	ρ_1	0.003
ρ_2	ρ_1	ρ_2	ρ_1	ρ_2	ρ_1	ρ_2	ρ_1	ρ_2	ρ_1	ρ_2	0

The presented results show that the change in liquid density in the horizontal segments does not influence on the liquid level, as distinct from liquid density in the vertical ones. However, the effect of different temperature (density) of the vertical segments can be summed up or compensated, depending on their mutual position. This should be taken into account when mounting hydrostatic leveling systems and multiple ascents and descents of the connecting tube in the presence of a temperature gradient along it should be avoided. Or measuring the temperature of each vertical section and implementing software compensation should be performed.

4.2. 4-point HLS case

Let us consider the levels of Section 3, the schemes of which are shown in figure 2 and figure 3. They consist of 4 measuring vessels (sensors): connected between each other in series and closed in a circuit. The segment between sensors 2 and 3 is supplemented by a single level difference of 1 m, similar to that shown in figure 6. Liquid density in all the segments is equal to ρ_1. At time t = 0, liquid density on the left side of the level difference changes from ρ_1 to ρ_2. The parameters of this task are given in Table 4.

Table 4. Task parameters.

υ^0 (m/s)	$h^0_{1\text{-}4}$ (m)	R_l (m)	R_h (m)	ρ_1 (kg/m^3)	ρ_2 (kg/m^3)	μ (Pa·s)
0	0.1	0.006125	0.025	998	999	0.001

The results of the change in the liquid level heights in the measuring segments, caused by the change in density, are presented in Figures 7 and 8.

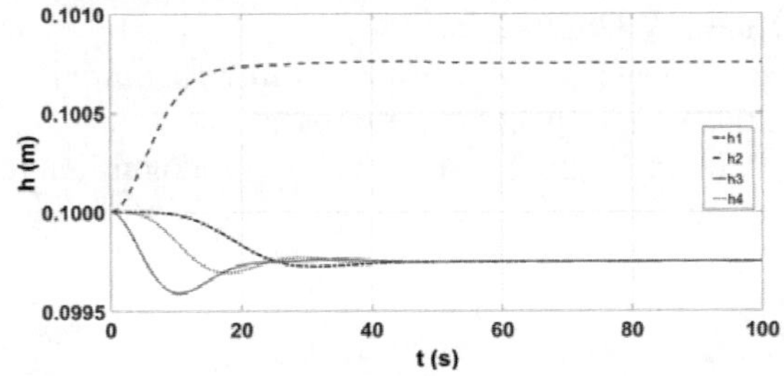

Figure 7. The change in the liquid level heights on the sensors with a sequential circuit.

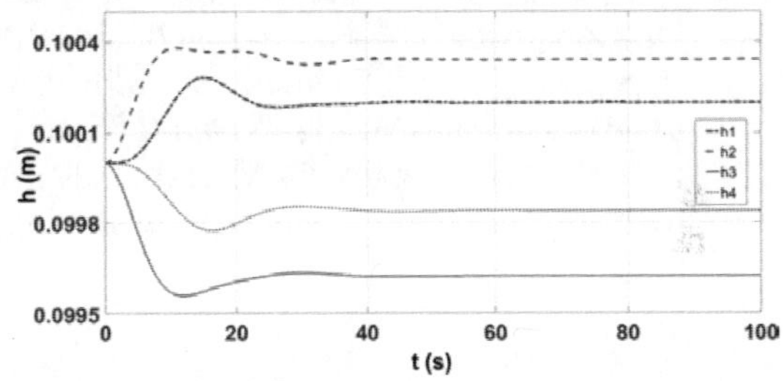

Figure 8. The change in the liquid level heights on the sensors of the level with a closed circuit.

Figure 7 and figure 8 demonstrate that the considered topologies of the level affect both nature of the transient processes and the values of the liquid levels of in the measuring vessels. In the first case (figure 7), the height of the liquid level on sensor No. 2 changed by 1 mm from the values of the other sensors, as in the problem, considered in § 4.1 (line 4 in Table 3). In the second case, the liquid level heights on all the sensors took different values, but the maximum level difference was only 0.68 mm (between sensors No. 2 and No. 3). This should be taken into account when designing HLS and, if possible, to close the hydraulic circuit.

5. Conclusions

The article presents a hydraulic model for a multipoint hydrostatic level. The conducted experiments with two point levels allowed estimating the degree of confidence of the model and the boundaries of its applicability. With the help of this model, it is possible to perform numerical estimations of transient processes in multipoint levels, with a number of sensors and topologies and various sizes. The proposed model also makes it possible to evaluate the influence of non-uniform density of liquid and air pressure on the level readings.

Acknowledgments

This research was done at the expense of the Russian Foundation for Basic Research (grant № 16-48-590025).

References

[1] Meier E, Geiger A, Ingensand H, Licht H, Limpach P, Steiger A and Zwyssig R 2010 Hydrostatic Levelling System: measuring at the system limits *J. of App. Geodesy* **4** 91

[2] Meier E, Huggenberger P and Ingensand H 1998 Precision monitoring of displacement over large areas *Hydropower & Dams* **5** 77

[3] Shardakov I N, Shestakov A P, Tsvetkov R V and Yepin V V 2016 The hydrostatic level method for continuous monitoring of building foundations *Solid State Phenomena* **243** 105

[4] Choi H J, Seo K W, Gil K H, Kim S H and Kang H S 2016 *Journal of the Korean Physical Society* **69** 998
[5] Pellissier P F Hydrostatic Leveling Systems 1965 *IEEE Transactions* **12** 19
[6] Volk J et al. Hydrostatic level sensors as high precision ground motion instrumentation for Tevatron and other energy frontier accelerators 2012 *J. Instrumentation* **7** P01004
[7] Manukin A B, Kazantseva O S, Bekhterev S V, Matyunin V P and Kalinnikov I I Long base hydrostatic level 2014 *Seismic Instruments* **50** 238
[8] Singatulin S, Volk J and Shiltsev V A 2006 *Proc. of 9-th IWAA* (SLAC California, United States) FR005
[9] Jayasinghe D A P, Letelier M and Leutheusser H J 1974 Frequency-dependent friction in oscillatory laminar pipe flow *International Journal of Mechanical Sciences* **16** 819
[10] Ogawa A, Tokiwa S, Mutou M. et al. Damped oscillation of liquid column in vertical U-tube for Newtonian and non-Newtonian liquids 2007 *J. Therm. Sci.* **16** 289
[11] Popov D N 1976 *Dinamika i regulirovanie gidro- i pnevmosistem* [*Dynamics and regulation of hydro - and pneumatic systems*] (Moscow: Mashinostroenie) p 424 (in Russian).
[12] Epin V V, Tsvetkov R V and Shestakov A P 2016 Application of feature recognition to hydrostatic leveling systems *Measurement Techniques* **59** 405

IOP Conf. Series: Materials Science and Engineering **208** (2017) 012047 doi:10.1088/1757-899X/208/1/012047

On location of piezoelectric element in a smart-structure: numerical investigation and experiment

D Oshmarin[a], M Iurlov

Department of complex problems of deformable solids mechanics, Institute of Continuous Media Mechanics UB of RAS, Acad. Korolev str., 1, Perm, 614013, Russian Federation

E-mail: a oshmarin@icmm.ru

Abstract. In this paper, based on some example problems it was demonstrated that in examining the possibilities of smart structure applications, the matter of considerable researchers' concern is the problem of location of piezoelectric elements in the structure to allow effective realization of its smart functions in the framework of the specified strategy of structure control and target purposes (vibration damping, defectoscopy, etc.) The numerical and experimental investigations have shown that for structures with the elements made of piezoelectric materials, it is more convenient to use as a parameter, specifying the best location of the piezoelectric element for damping the vibrations at the prescribed frequency, the coefficient of electromechanical coupling, which is evaluated by the values of eigenfrequencies of the structure in the short-circuit and open-circuit regimes. The values of eigenfrequencies of vibrations are evaluated by solving the problem of natural vibrations of electromechanical systems by the finite element method using the applied ANSYS package. The investigation were conducted for a thin-walled aluminum shell in the form of half-cylinder.

1. Introduction

In designing of smart structures, which involve elements made of piezoelectric materials, one of the most important question is how the piezoelectric elements can be located in the structure to ensure the most effective realization of the function of the smart structure in compliance with the target goal (control of dynamic behavior, monitoring of the structure state, collection and accumulation of energy, etc.).

In one of the first works devoted to determining the location of a piezoelectric element, which would allow the maximum damping of the oscillations at a given frequency [1], it was shown that the zones associated with the highest values of average strains are the most appropriate for this purpose. In [2] the main currently available approaches to optimization of piezoelectric element arrangement are systematized and analyzed, allowing using it most effectively to achieve the goal. It should be emphasized that for the most part, the works cited in the review paper [2] investigate smart systems that realize the strategy of active control, in which case the requirements on the operating efficiency of piezoelectric elements are very high. It should be noted that improper location of piezoelements in a structure can upset the regime of its steady operation and eventually leads to its failure. Although at present, there are hundreds of works devoted to this problem [2], the question of the optimal piezoelement arrangement in structures is of immediate interest due to an abundance of possible application fields and purposes they are intended to serve.

At the same time, for systems designed to fulfill the strategy of passive control, the problem of determining the optimal location of piezoelectric elements in the system in most cases is reduced to determining the place where the structure deformation leads to the maximum value of the electromechanical coupling coefficient [3-5], which is defined by the eigenfrequencies of the structure under short-circuit and open-circuit regimes in the form proposed in [6]. Despite the fact that different authors apart from the above coefficient offer a number of other parameters for passive systems (deformation energy [3, 7], the level of the electric charge [3], the level of the electric potential [7], etc.), priority is given to the coefficient of electromechanical coupling [3-5, 6].

In this paper, we determine an optimal location of the piezoelectric element in the structure, which provides damping of its vibrations at a certain frequency, based on the coefficient of electromechanical coupling. The eigenfrequencies of structure vibrations are evaluated by solving the problem of natural vibrations of electroelastic bodies by the finite element method using the applied ANSYS package (license agreement ANSYS Academic Research Mechanical and CFD № 1064623). The optimal arrangement of the piezoelement in the structure, determined numerically, is confirmed experimentally.

2. The technique for determining the optimal location of piezoelectric element

The object of investigation is the structure with a peiezoelectric element, connected to the external electric circuit. The task is to select a location for the piezoelectric element such, that provides the most intensive damping of structure vibrations at a certain frequency.

As it follows from the analysis of the literature, the deformation of the piezoelectric element in the structure specified by the vibration mode of the system at the prescribed frequencies depends on its location, which directly affects its possibilities for transforming the mechanical energy to the electrical one. Recent works [7, 8] showed that, as a characteristic parameter for determining the optimal location of the piezoelectric element in the structure, it is best to use a relative quantity of the electric potential generated on the ungrounded surface of the piezoelectric element under deformation. However, further investigations have shown that the selection of the best location for the piezoelectric element for spatial structures cannot always rely on the relative values of the electric potential, which have been obtained by solving the natural vibration problem. This is because of the fact that the relative value has meaning only within the confines of one particular computation performed for estimation of the operating efficiency of the piezoelectric element at different vibration modes. A comparison between the values of the electric potential obtained by solving the natural vibration problem for different locations, dimensions, physico-mechanical characteristics of the element, etc, is not always correct.

Therefore, to estimate the efficiency of operation of the piezoelectric element placed on the main structure, in [6] it has been suggested that the generalized coefficient of the electromechnical coupling *K*, determining the amount of the mechanical vibration energy transformed into the electric energy, can be used as a criterion of the optimal element location:

$$K = \sqrt{\omega_{o/c}^2 - \omega_{s/c}^2 \Big/ \omega_{s/c}^2} \qquad (1)$$

where $\omega_{o/c}$, $\omega_{s/c}$ are the eigenfrequencies of the structure with piezoelectric elements in the open circuit regime and short-circuit regimes. The open-circuit regimes are realized in the case when one of the electrode-covered surfaces of the piezoelectric element is grounded (its potential is equal to zero), and the other surface is at no load. In the short-circuit regime, zero potentials are set at both surfaces.

In works [7-8], the correlation between the value of the electric potential and electromechanical coupling coefficient is demonstrated by finding the best variant of piezoelement arrangement allowing the most effective damping of several vibration modes. In this case, from the viewpoint of energy transformation, the coefficient *K* has proved to be the most informative. With all this in mind, it has been suggested that the evaluation of the optimal location of the piezoelectric element be made based on the maximum value of the electromechanical coupling coefficient *K*.

3. The obtained results

To demonstrate the applicability of the proposed condition, we conducted a series of numerical and experimental investigations with the aim of finding a piezoelement location such that could provide its most efficient work (i.e. provide maximum value of the received signal) when connected to an external electric circuit to provide damping of structure vibrations at a certain frequency. The investigations were made for a thin-walled shell in the form of a half cylinder (fig.1). The generatrices of the shell were clamped along their entire length with the aid of flange plates.

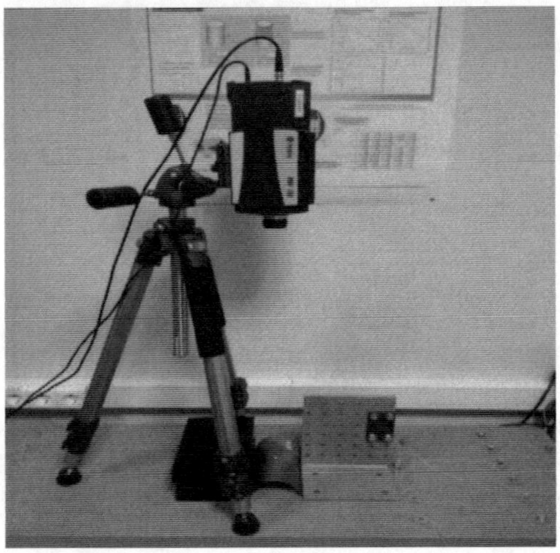

a

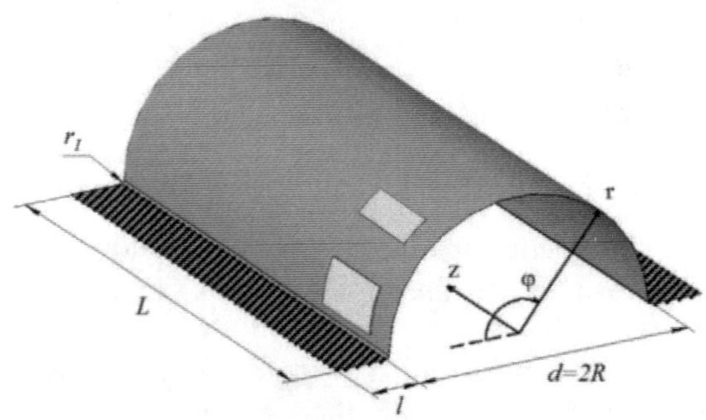

b

Figure 1. Experimental scheme (a). Computational scheme for modeling (b).

The shell with geometrical dimensions: radius $R = 0.044$ m, length $L = 0.17$ m, thickness $h = 15 \cdot 10^{-5}$ m (fig.1) was made of 3104 aluminum. In computations, it was assumed that the material of the shell is elastic and isotropic and has the following physicomechanical characteristics: $E = 6.5 \cdot 10^{10}$ Pa, $\nu = 0.33$, $\rho = 2800$ kg/m^3. For experiments we used piezoelectric elements of rectangular shape, which were made of LZT-19 (Lead zirconate titanate) piezoceramics and had the following standard dimensions: 70x15x0.72 mm, 50x20x0.36 mm. The upper and lower surfaces of piezoelectric elements were covered with electrodes. The polarization axis coincided with the direction of the normal to the electrode-covered surface.

In computation, the material of piezoelectric elements was assumed to be elastic. The physicomechanical characteristics claimed by the manufacturer, have the following values: elastic modulus $C_{11} = C_{22} = 13.9 \cdot 10^{10}$ Pa, $C_{12} = 7.78 \cdot 10^{10}$ Pa, $C_{13} = C_{23} = 7.43 \cdot 10^{10}$ Pa, $C_{33} = 11.5 \cdot 10^{10}$ Pa, $C_{44} = 3.06 \cdot 10^{10}$ Pa, $C_{55} = C_{66} = 2.56 \cdot 10^{10}$ Pa, piezoelectric coefficients $\beta_{31} = \beta_{32} = -5.2$ C/m^2, $\beta_{33} = 15.1$ C/m^2, $\beta_{52} = \beta_{61} = 12.7$ C/m^2, dielectric coefficients $e_{11} = e_{22} = 6.45 \cdot 10^{-9}$ F/m, $e_{33} = 5.62 \cdot 10^{-9}$ F/m, density $\rho = 7700$ kg/m^3, the vacuum permittivity $e_0 = 8.85 \cdot 10^{-12}$ F/m.

3.1. Evaluation of eigenfrequencies of the shell

At the first stage we performed simulation to evaluate eigenfrequencies of the structure vibrations in the absence of the piezoelectric element. The oscillations were excited by an impact applied to the shell and were fixed by a laser vibrometer PDV-100 in a single- measurement regime during 6 seconds at the point ($z = L/2$, $\varphi = 15^0$) (fig.1,a).

Table1 presents the values of eigenfrequencies determined numerically by solving the natural vibration problem and experimentally in the case of impact-excited vibrations.

Table 1. Eigenfrequencies of the shell without piezoelectric elements

Eigenfrequencies, Hz		Difference, %
Numerical computation	Experiment	
77	79	2.44
172 173	191	11.18
288	-	-
320	305	4.65
492 495	458	7.52

The results obtained numerically and experimentally for a shell without a piezoelectric element, demonstrate that the computational model describes quite adequately the actual experiment. It should be noted that a particular type of external action can not excite all vibration modes, in particular, in the example considered, the 4th mode of vibration is not fixed. In the spectrum of eigenfrequencies determined numerically there are three pairs of frequencies, the values of which are very close but not multiple (the vibration modes at these frequencies are different). In the realized experiment, these frequencies failed to separate from each other. Therefore, in the comparison of numerical and experimental results a pair of close frequencies aligned to a single frequency determined experimentally (for example, for frequencies 172.31 Hz and 173.55 Hz – 191.56 Hz, and for frequencies 492.90 Hz and 495.21 Hz – 457.96 Hz).

3.2. Determination of placement of the actuator

To ensure stability and repeatability of the experiment the excitation of vibrations is executed by a piezoelectric element, acting as an actuator, the possibility of which is demonstrated in [9]. As an actuator, we used a rectangular 30x15x0.72 mm piezoelectric element.

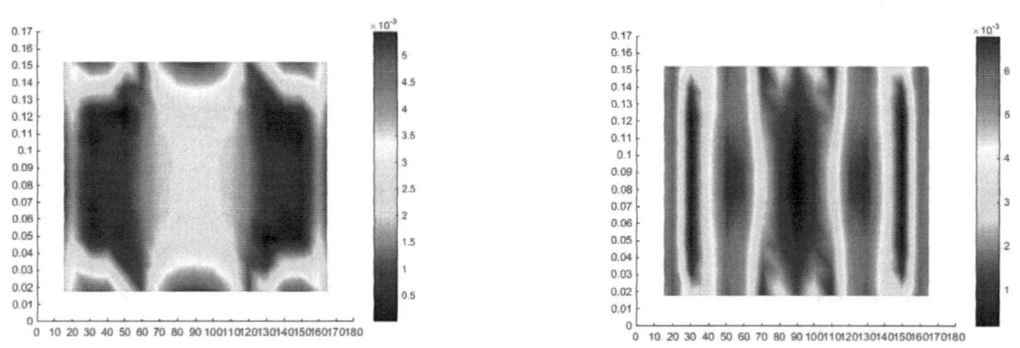

a) the frequency 172 Hz b) the frequency 492.90 Hz

Figure 2. Patterns of distribution of the coefficients K in relation to the location of the actuator for the vibration modes corresponding to the frequencies of 172 Hz and 493 Hz.

Based on the solution of the natural vibration problem were determined eigenfrequencies of the structure in the open- and short- circuit regimes, the coefficients of electromechanical coupling *K* and to obtain the distribution of its values depending on the location of the actuator for the first six eigenfrequencies.

As an example, fig.2 shows the pattern of distribution of the coefficient *K* for vibration modes corresponding to the frequencies of 172 Hz and 493 Hz in relation to the location of the actuator. For illustration purpose, the cylindrical surface of the shell is unfolded in the plane; the angular coordinate φ, is plotted along the x-axis and the longitudinal coordinate z is plotted along the y-axis. In the distribution pattern, each point corresponds to the location of the center of actuator mass.

Numerical simulation was used to select the location of the actuator, which provides the excitation of the largest number of vibration modes in the given frequency range. Based on the analysis of the obtained *K* distribution, it was decided to place the coordinates of the center of the actuator mass are at points of 15°, 27 mm.

A series of experiments conducted by determining the frequency spectrum when the oscillations are excited by an actuator located in accordance with the results of a numerical study. A signal was set to actuator from the frequency generator through the E-413.D2 voltage amplifier for piezoelectric elements. The vibrations were registered by a laser vibrometer PDV-100.

Table 2. Eigenfrequencies of shells fitted with an actuator

№ of frequency	Frequency, Hz		Difference, %
	Numerical computation	Experiment	
1	83	80	4.13
2	171	155	9.49
3	183	178	3.14
4	288	312	8.23
5	310	338	8.78
6	469	459	2.14

Table 3. Resonance frequencies and vibration amplitudes registered in the experiment for a shell with the actuator

№	Experiment			Electromechanical coupling coefficient *K*1000* (calculated)
	Frequency (Hz)	Vibration velocity (μm/s)	Displacement(nm)	
1	80	15	30	0.32
2	155	43	14	2.02
3	178	654	600	1.53
4	312	1500	550	3.48
5	338	45	17	1.41
6	459	1400	590	5.46

IOP Conf. Series: Materials Science and Engineering **208** (2017) 012047 doi:10.1088/1757-899X/208/1/012047

Table 2 gives the values of eigenfrequencies of the shell with an actuator, which were calculated in the framework of the natural vibration problem in the short-circuit regime and determined experimentally in the case of vibration excitation by the piezoelectric element.

From the results presented in Table 2 it follows that connection of the piezoelectric element to the shell causes a frequency shift in the original spectrum. It is experimentally possible to obtain all frequencies in the considered range. The maximum difference between the frequency values obtained numerically and experimentally is not more than 11%.

Table 3 displays the experimentally obtained velocity of vibrations and displacements at corresponding resonances. Also Table 3 shows the calculated values of the electromechanical coupling coefficient K for frequencies obtained numerically, with the chosen position of the piezoelectric element-actuator. In this case, the largest value of K takes on the 6th frequency (in the experiment 459 Hz).

3.3. Determination of the placement of the damping piezoelectric element

In addition to being used as an actuator, the piezoelectric element can be used for passive damping of oscillations when to its electroded surfaces attached the external electrical circuit consisting of resistance, inductance and capacitance. In this embodiment, in order to provide the maximum damping at the corresponding frequency, it is necessary to find the location of the piezoelectric element, at which the maximum value of the electromechanical coupling coefficient K.

As an example, let us consider the problem of choosing the location of the piezoelectric element, which provides the maximum damping of the 6th frequency.

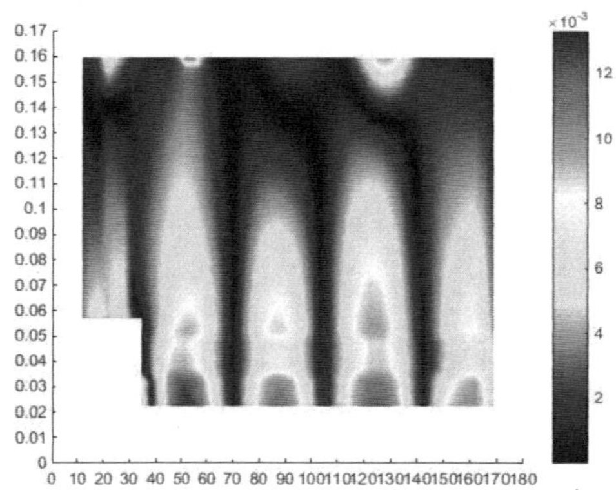

Figure 3. Patterns of distribution of the electromechanical coupling coefficient for the 6th frequency, depending on the location of the damping piezoelectric element.

To implement passive damping using an external electrical circuit, a rectangular piezoceramic element measuring 30x10x0.36 mm was chosen.

Figure 3 shows the results of numerical calculations of the electromechanical coupling coefficient K for the sixth frequency for various arrangements of the damping piezoelectric element when the actuator corresponds to the results given in Table 3. The cut rectangular region in Fig. 3 corresponds to the position occupied by the actuator.

On the basis of an analysis of the distribution patterns of the coefficient K, a damping piezoelectric element was placed for experimental studies at a point with coordinates (φ = 52 °, z = 27 mm), where K takes on the largest value.

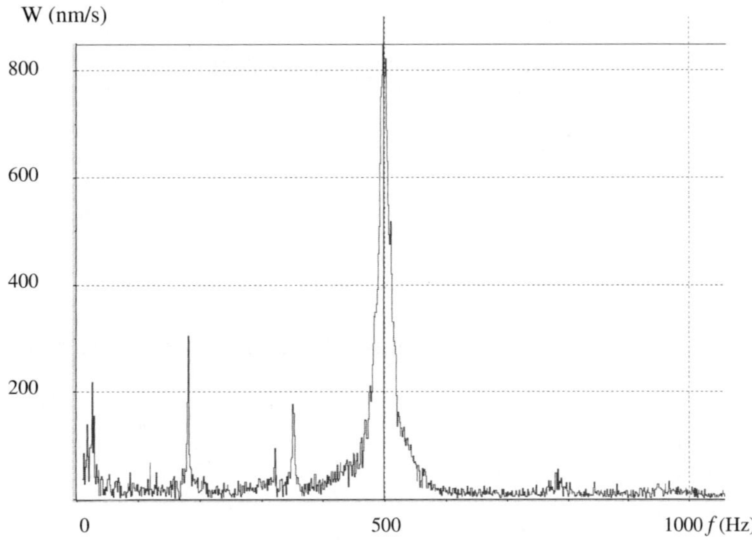

Figure 4. Frequency response of the displacement velocity (W), taken by a laser vibrometer at the center of the shell (from peak to peak).

Table 4. The amplitudes of the electrical voltage on the second piezoelectric element and the coefficient of electromechanical coupling at different frequencies.

Frequency (Hz)	Amplitudes of voltage (V)	Electromechanical coupling coefficient *K*1000* (calculated)
80	0.3	5.01
178	0.4	8.90
320	0.7	10.84
352	0.7	6.64
498	1.2	13.30

It should be noted that changing the spectrum of natural frequencies as a result of adding the second element does not change the qualitative conclusions obtained on the basis of the results presented in Table 3. This is confirmed by the frequency response of the displacement velocity (W) shown in Fig. 4, taken by a laser vibrometer, where the maximum values of W are attained at the sixth frequency.

Table 4 shows the experimental values of the electric potential at the second piezoelectric element and the values of the electromechanical coupling coefficient, obtained as a result of numerical calculations.

These results demonstrate the qualitative agreement of numerical and experimental conclusions about the optimum location of the piezoelectric element for damping a given (sixth) eigenfrequency.

Conclusions

The experimental and numerical investigations performed for smart-structures based on the application of piezoelectric elements have shown that it is essential to conduct a study on the arrangement of piezoelectric elements, providing optimal damping of vibrations, energy harvesting, etc.

The numerical and experimental confirmed, that as a parameter, determining the best location of the piezoelectric element for damping the structure vibrations at the prescribed frequency, can be used the coefficient of electromechanical coupling, which is evaluated by the values of eigenfrequencies of the structure in the short-circuit and open-circuit regimes.

Acknowledgments

The work is supported by the RFBR (projects No 15-01-03976-a, No 17-01-590152-r-ural-a).

References

[1] Crawley E F, de Luis J. 1987 Use of piezoelectric actuators as elements of intelligent structures *AIAA Journal*, **25:10** 1373–1385

[2] Gupta V, Sharma M, Thakur N. 2010 Optimization Criteria for Optimal Placement of Piezoelectric Sensors and Actuators on a Smart Structure: A Technical Review *J. Int. Mat. Syst. and Struct.* **21:8** 1227-1243

[3] Bachmann F, Bergamini A, Ermanni P. 2012 Optimal piezoelectric positioning a strain-energy based finite element approach, *J. Int. Mat. Syst. and Struct.* **23:14** 1575–1591

[4] Ducarne J, Thomas O, Deu J.-F. 2013 Placement and dimension optimization of shunted piezoelectric patches for vibration reduction *Journal Sound and Vibration* **331** 3286–3303

[5] Belloli A, Ermanni P. 2007 Optimum placement of piezoelectric ceramic modules for vibration suppression of highly constrained structures *Smart Mater. Struct.* **16** 1662–1671

[6] Hagood N W, von Flotow A. 1991 Damping of structural vibrations with piezoelectric materials and passive electrical networks *Journal of Sound and Vibration* **146:2** 243-268

[7] Sevodina N V, Yurlova N A, Oshmarin D A. (2015) The optimal placement of the piezoelectric element in a structure based on the solution of the problem of natural vibrations *Solid State Phenomena* **243** 67-74

[8] Iurlova N A, Matveenko V P, Oshmarin D A, Sevodina N V, Yurlov M A. (2016) Layout optimization of piezoelectric elements with external electric circuits in smart constructions based on solution of the natural vibrations problem *ECCOMAS Congress 2016 VII European Congress on Computational Methods in Applied Sciences and Engineering Proceedings, 2016* **1** 1920-1929

[9] Yurlov M A (2012) Experimental determination of dissipative properties of electro viscoelastic systems with external electric circuits *PNRPU Mechanics Bulletin* **1** 179-194.

Institute of Physics
Dirac House, Temple Back
Bristol BS1 6BE UK

ISSN: 1757-8981
ISBN 978-1-5108-4223-6

9 781510 842236

CURRAN ASSOCIATES INC.
proceedings.com